# AUTOMOTIVE TECHNOLOGY
## Fuels and Emissions



Barry Hollembeak

Technical Training Inc.
Rochester Hills, Michigan

CENGAGE Learning™

Australia • Canada • Mexico • Singapore • Spain • United Kingdom • United States

**Automotive Technology: Fuels & Emissions**
**Barry Hollembeak**



ISBN-10: 81-315-1418-8
ISBN-13: 978-81-315-1418-4

Cengage Learning India Private Limited
418 F.I.E. Patparganj
New Delhi-110092

Tel: 91-11-43641111
Fax: 91-11-43641100

Cengage Learning products are represented in Canada by Nelson Education, Ltd.

For product information, visit www.cengage.co.in

Printed in India
First Indian Reprint 2011

**NOTICE TO THE READER**

Publisher does not warrant or guarantee any of the products described herein or perform any independent analysis in connection with any of the product information contained herein. Publisher does not assume, and expressly disclaims, any obligation to obtain and include information other than that provided to it by the manufacturer.

The reader expressly warned to consider and adopt all safety precaution that might be indicated by the activities herein and to avoid all potential hazards. By following the instructions contained herein, the reader willingly assumes all risks in connection with such instructions.

The publisher make no representation or warranties of any kind, including but not limited to, the warranties of fitness for particular purpose or merchantability, nor are any such representation implied with respect to the material set forth herein, and the publisher takes no responsibility with respect to such material. The publisher shall not liable for any special, consequential, or exemplary damages resulting in whole or part, from the readers' use of, or reliance upon this material.

# Contents

# Preface to India Edition

## The Series

Welcome to Cengage Learning's, a state-of-the-art series designed to respond to today's automotive instructor's and students' needs. This series offers current, concise information on ASE and other specific subject areas, combining classroom theory, diagnosis, and repair into one easy-to-use volume.

You'll notice several differences from a traditional textbook. First, a large number of short chapters divide complex material into small chunks. Instructors can give tight, detailed reading assignments that students will find easy to digest. These short chapters can be taught in almost any order, allowing instructors to pick and choose the material that best reflects the depth, direction, and pace of their individual classes.

Books in this series also feature an art-intensive approach to suit today's visual learners—images drive the chapters. From drawings to photos, you will find more art to better understand the systems, parts, and procedures under discussion. Look also for helpful graphics that draw attention to key points in features like You Should Know and Interesting Fact.

Just as important, each book starts off with a section on Safety and Communication, which stresses safe work practices, tool competence, and familiarity with workplace "soft skills," such as customer communication and the roles necessary to succeed as an automotive technician.

## A Note to the Student

There are now more computers on a car than aboard the first spacecraft and even gifted backyard mechanics long ago turned their cars over to automotive professionals for diagnosis and repair. That is a statement about the nation's need for the knowledge and skills you will develop as you continue your studies. Whether you eventually choose a career as a certified or licensed technician, service writer, manager, or automotive engineer—or even decide to open your own shop—hard work will give you the opportunity to become one of the 840,000 automotive professionals providing and maintaining safe and efficient automobiles on our roads. As a member of a technically proficient, cutting-edge workforce, you will fill a need and—even better—you will have a career to feel proud of.

### Features of the Text

- **Section Openers** provide the student with a **Section Table of Contents** and **Objectives** to focus the learner on the section's goals.
- **Interesting Facts** appear in the section openers to spark student attention with industry trivia or history.
- An **Introduction** orients the reader at the beginning of each chapter. **Technical terms** are bolded in the text upon first reference and are defined.
- **You Should Know** informs the reader whenever special safety cautions,warnings, or other important points deserve emphasis.
- **Review Questions** are structured in a variety of formats, including ASE style, challenging students to prove they have mastered the material.
- A comprehensive **Index** helps instructors and students pinpoint information in the text.


## Acknowledgments

Many individuals were involved in the production of this text. First, I would like to thank my wife Judy for her countless hours of helping me with art manuscript preparation and for putting up with me during this period. I would also like to thank David Boelio, Matthew Thouin, Barbara Diaz, Cheri Plasse, and Jack Erjavec from Cengage Learning for their support and guidance. Finally, I would like to thank the following individuals for reviewing the manuscripts. Their technical expertise was invaluable in the creation of this text.

Steve Bertram
Palomar College
San Marcos, CA

Dana Brewer
Wyoming Technical Institute
Laramie, WY

Christopher Van Stavoren
Pennsylvania College of Technology
Williamsport, PA

Paul Witek
Western Technical Institute
El Paso, TX

# About the Author

Barry Hollembeak has been involved with the automotive repair industry for more than thirty years. He entered the automotive technology field in 1974 as a service technician, worked in both fleet and independent repair shops for six years, and then opened his own repair facility and operated it for eight years. After selling the business, he retired from being a technician and concentrated his efforts on fulfilling his desire to become a teacher and author. After teaching postsecondary automotive technology for five years, he became Manager of Education at Denver Automotive and Diesel College in Denver, Colorado, a position he held for two years. After leaving the postsecondary education field, Barry entered into manufacturer training. For the past twelve years he has been a technical trainer providing instruction to dealership technicians. A Master and Advanced Engine Performance Certified Technician with ASE, he is also a member of the Colorado Automotive Teachers Society (CATS) and is involved with the College Automotive Program, AYES, and Curriculum Enhancement with Industry Application at Colorado State University. He has also been a guest speaker for NATEF, CATS, and several other organizations. In his free time he enjoys working at his ranch in eastern Colorado with his family, Judy, Kristina, Daniel, and Kimberly. He also enjoys horseback riding with his two grandsons, Brandon and Jacob.

SECTION

# 1 Safety and Communication

## Section Objectives

After you have read, studied, and practiced the contents of this section, you should be able to:

- Explain how safety is a part of professionalism.
- List and describe personal safety responsibilities.
- List the different types of eye-protection devices, and explain the proper application of each.
- Lift and carry heavy objects properly.
- Classify fires and fire extinguishers.
- Explain the proper storage of fuels.
- Explain the proper use of the fire extinguisher.
- Demonstrate proper procedures for raising a vehicle using a floor jack and safety stands.
- Demonstrate proper procedures for raising a vehicle using a hoist.
- Demonstrate the ability to properly run the engine in the shop.
- Define hazardous materials.
- Explain the right-to-know law or workplace hazardous materials information systems (WHMIS).
- Determine what constitutes hazardous waste and how to properly dispose of it.
- Describe the basic safety rules and precautions associated with servicing fuel and emission systems.
- Describe some of the special tools used to service fuel and emission systems.

### Interesting Fact

*In 2002, the automotive service industry recorded 27,700 injuries. Of this number, 12,600 resulted in days away from work or restrictions of work duties.*

# CHAPTER 1 Safe Work Practices

## Introduction

Being a professional technician is more than being knowledgeable about automotive systems; it is also an attitude. Being a professional technician includes having an understanding of all the hazards that may exist in the workplace. One of the most obvious traits of a professional is the ability to work productively and safely. This is where knowledge becomes very important. You need it to be productive, and you need it to ensure your own safety and the safety of others.

Safety is everyone's business. However, never assume the person working next to you is as conscientious as you are. You must be aware of what is going on around you at all times. As a professional technician, you must perform your work in a manner that protects not only yourself but others in the workplace. This chapter discusses the safety concerns associated with working in an automotive repair shop and the safety concerns associated with the vehicle's fuel and emission systems. In addition to basic shop safety, working on the vehicle's fuel and emission systems presents many special concerns.

## Personal Safety

Personal safety encompasses all aspects of preventing injury, including awareness, attitudes, and dress. All three of these are manifested through neat work habits. Cleaning up spills, keeping tools clean, and organizing the tools and materials in the shop all help to prevent accidents. Rushing to complete a job may result in a lack of consideration for personal safety and may ultimately cause an accident. Taking time to be neat and safe is rewarded by fewer accidents, higher customer satisfaction, and better pay.

### Dress and Appearance

Nothing displays professional pride and a positive attitude more than the way you dress **(Figure 1)**. Customers are demanding a professional atmosphere in the service shop. Your appearance instills customer confidence as well as expresses your attitude toward safety. The wearing of proper and neat clothing can prevent injuries.

Long hair is a serious safety concern. Very serious injury can result if hair becomes caught in rotating machinery, fan belts, or fans. If your hair is long enough to touch the bottom of your shirt collar, it should be tied back and tucked under a hat.

Jewelry has no place in the automotive shop. Rings, watches, bracelets, necklaces, earrings, and so forth can cause serious injury. The gold, silver, and other metals used in jewelry are excellent **conductors** of electricity. Conductors are materials that can support the flow of electricity. Your body is also a good conductor. When electrical current flows

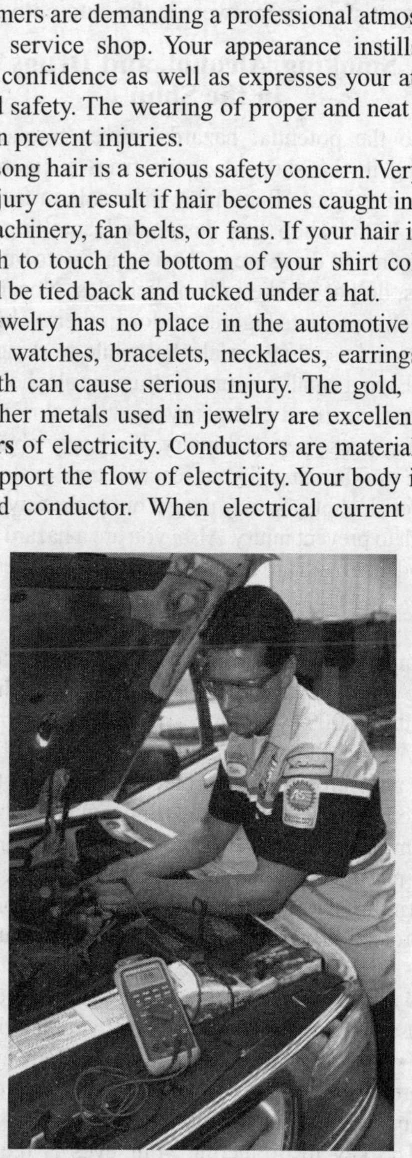

**Figure 1** *A professional automotive technician takes pride in his or her appearance.*

through a conductor, it generates heat. The heat can be great enough to cause severe burns. Jewelry can also get caught in moving parts, causing serious cuts. Necklaces can cause serious injury or even death if they become caught in moving equipment.

You should wear shoes or boots that will protect your feet in the event that something falls or you stumble into something. It is a good idea to wear safety shoes or boots with steel toes and shanks. Most safety shoes also haveslip-resistant soles. Tennis and jogging shoes provide little protection and are not satisfactory footwear in the automotive shop.

## Smoking, Alcohol, and Drugs in the Shop

Due to the potential hazards, never smoke when working in the shop. A spark from a cigarette or lighter may ignite flammable materials in the workplace. If the shop has designated smoking areas, smoke only in these areas. As a courtesy to your customers, do not smoke in their vehicles. Nonsmokers may not appreciate cigarette odor in their vehicles.

Do not use drugs or alcohol while working in the shop. Even a small amount of drugs or alcohol affects reaction time. In an emergency situation, slow reaction time may cause personal injury. If a heavy object falls off the workbench and your reaction time is slowed by drugs or alcohol, you may not get out of the way quickly enough to prevent injury. Also, you are a hazard to your co-workers if you are not performing at your best.

## Eye Protection

The importance of wearing proper eye protection cannot be overemphasized. Every working day there are more than 1,000 eye injuries, many resulting in blindness. Almost all of these injuries are preventable. The safest and surest way to protect your eyes is to wear the proper eye protection any time you enter the shop. At a minimum, wear eye protection when grinding, using power tools, hammering, cutting, chiseling, or performing service under the car. In addition, wear eye protection when doing any work that can cause sparks, dirt, or rust to enter your eyes, and when you are working around chemicals. Remember, just because you are not doing the work yourself does not mean you cannot suffer an eye injury. Many eye injuries are caused by a co-worker. Wear eye protection any time you are near an eye hazard.

The key to protecting your eyes is the use of *proper* eye protection. Prescription glasses do not provide adequate protection. Regular glasses are designed to impact standards that are far below those that are required in the workplace. The lens may stop a flying object, but the frame may allow the lens to pop out and hit your face, causing injury. In addition, regular glasses do not provide side protection.

There are many types of eye protection **(Figure 2)**. One of the best ways to protect your eyes is to wear **occupational safety glasses**. Occupational safety glasses are designed with special high impact lenses and frames and provide for side protection. These glasses are light and comfortable. They are constructed of tempered glass or safety plastic lenses and have frames that prevent the lenses from being pushed out upon impact. They have side shields to prevent the entry of objects from the side. Occupational safety glasses are available in prescription lenses so they can be worn instead of regular corrective-lens glasses.

**Safety goggles** fit snugly around the area of your eyes to prevent the entry of objects and to provide protection from liquid splashes. Safety goggles provide eye protection from all sides because they fit against the face and forehead to seal off the eyes from outside elements. The force of impact on the lens is distributed throughout the entire area where the safety goggles are in contact with your face and forehead. Safety goggles are designed to fit over regular glasses.

**Face shields** are used when there is potential for sparks, flying objects, or splashed liquids, which can cause neck, facial, and eye injuries. Face shields are made with a clear plastic shield that protects the wearer's entire face. The plastic is not as strong or as impact resistant as that of occupational safety glasses

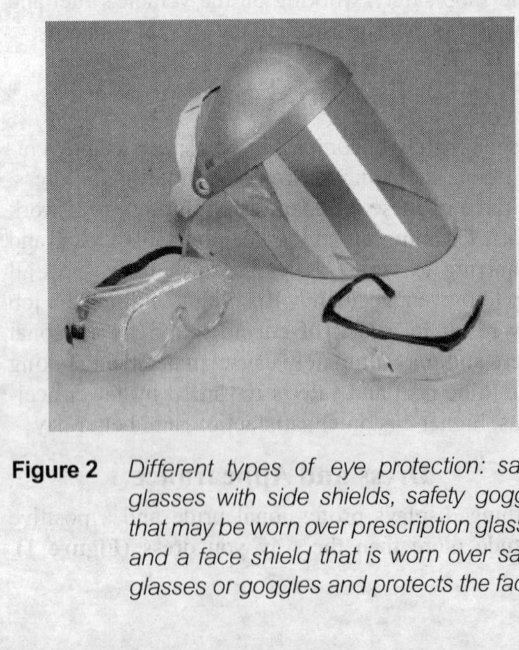

**Figure 2** *Different types of eye protection: safety glasses with side shields, safety goggles that may be worn over prescription glasses, and a face shield that is worn over safety glasses or goggles and protects the face.*

or safety goggles. If there is a danger of high-impact objects hitting the face shield, then wear safety glasses under the face shield.

Safety glasses provide little or no protection against chemicals. When working with chemicals, such as battery acid, refrigerants, cleaning solutions, and so forth, safety goggles should be worn. Full-face shields are not intended to provide primary protection for your eyes. They are designed to provide primary protection for your face and neck and should be worn in addition to eye protection.

Before removing your eye protection, close your eyes. Pieces of metal, dirt, or other foreign material may have accumulated on the outside. These particles could fall into your eyes when you remove your glasses or shield.

## Eyewash Fountains

Eye injuries may occur in various ways in an automotive shop. The following are some common types of eye injuries:

- Thermal burns from excessive heat
- Irradiation burns from excessive light such as from an arc welder
- Chemical burns from strong liquids such as gasoline or battery electrolyte
- Foreign material in the eye
- Penetration of the eye by a sharp object
- A blow from a blunt object

Wearing safety glasses and observing shop safety rules will prevent most eye accidents. If a chemical gets in your eyes, it must be washed out immediately to prevent a chemical burn. An eyewash fountain is the most effective way to wash the eyes, and every shop should be equipped with some eyewash facility **(Figure 3)**. Be sure you know the location of the eyewash fountain in the shop.

**Figure 3** *An eyewash fountain is the most effective way to wash the eyes.*

## First Aid Kits

First aid kits should be clearly identified and conveniently located **(Figure 4)**. These kits contain such items as bandages and ointment required for minor cuts. All shop personnel must be familiar with the location of first aid kits. At least one of the shop personnel should have basic first aid training, and this person should be in charge of administering first aid and keeping first aid kits filled.

## Hand Protection

Good hand protection is often overlooked. A scrape, cut, or burn can seriously impair your ability to work for many days. A well-fitting pair of heavy work gloves should be worn while grinding, welding, or handling chemicals or high-temperature components. Special rubber gloves are recommended for handling **caustic** chemicals that have the ability to destroy or eat through regular glove material and are considered extremely corrosive.

Many technicians wear latex, vinyl, or nitrile gloves to help protect their hands and to keep them clean. These are similar to the type of gloves doctors and dentists wear during examinations. Latex gloves are cheap to buy and provide good hand protection. However, some people are allergic to latex. If you wear latex gloves and develop a rash or redness on your hands, discontinue use. Vinyl gloves are also available and provide good resistance to tears and many non aggressive liquids. They are also latex free so those who are allergic to latex can wear them. At a higher cost, nitrile gloves are latex-free synthetic rubber gloves that are superior to latex or vinyl in puncture resistance. In

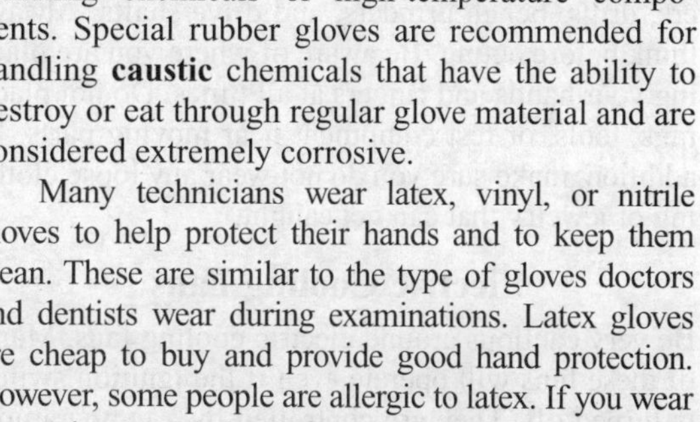



**Figure 4** *First aid kit.*

addition, nitrile gloves resist a wide range of chemicals that are harmful to either latex or vinyl.

Latex, vinyl, or nitrile gloves should be worn if you have an open cut or other injury on your hand to prevent infection and the spread of diseases. In addition, you should wear these gloves if you are required to render first aid or medical assistance to someone who is injured. Because of the serious nature of blood-borne pathogens (disease and infection-causing microorganisms carried by blood and other potentially infectious materials), it is important that you take every precaution to protect yourself regardless of the perceived status of the individual you are assisting. In other words, whether or not you think the blood/body fluid is infected with blood-borne pathogens, treat it as if it were.

### Rotating Belts and Pulleys

Many times the technician must work around rotating parts such as generators, power steering pumps, air pumps, water pumps, and air conditioner (A/C) compressors. Other rotating equipment or components of concern incvlude tire changers, spin balancers, drills, bench grinders, and drive shafts. Always think before acting. Be aware of where you are placing your hands and fingers at all times. Do not place rags, tools, or test equipment near moving parts. In addition, make sure you do not wear any loose clothing or jewelry that can get caught.

### Electric Cooling Fans

Be very cautious around electric cooling fans. Many of these fans will operate even if the ignition switch is turned off. They are controlled by a temperature-sensing unit in the engine block or radiator and will turn on any time the coolant temperature reaches a certain point. Before working on or around an electric cooling fan, disconnect the electrical connector to the fan motor or the negative battery cable.

### Lifting and Carrying

Back injuries are one of the most crippling injuries in the industry, yet most of them are preventable. Improper lifting practices cause most occupational back injuries. You can avoid these injuries by following a few simple lifting guidelines:

1. Do not lift a heavy object by yourself. Get help from someone else.
2. Do not lift more than you can handle. If the object is too heavy, use the proper equipment to lift it.
3. Do not attempt to lift an object if there is not a good way to hold onto it. Study the object to determine the best balance and grip points.
4. Do not lift with your back. Your legs have some of the strongest muscles in your body. Usc them to lift with.
5. Place your body close to the object. Keep your back and elbows straight.
6. Make sure you have a good grip on the object. Do not attempt to readjust the load once you have lifted it. If you are not comfortable with your balance and grip, lower the object and reposition yourself.
7. When lifting, keep the object as close to your body as possible. Keep your back straight and lift with your legs.
8. While carrying the object, do not twist your body to change directions. Use your feet to turn your whole body in the new direction.
9. To set the load down, keep the object close to your body. Bend at the knees and keep your back straight. Do not bend forward or twist.
10. If you need to place the object onto a shelf or benchtop, place an edge of the object on the surface and slide it into place. Do not lean forward.

## Fire Hazards and Prevention

Fires are classified by the types of materials that are involved **(Figure 5)**. Technicians should be able to locate the correct **fire extinguisher** to control the fire types that they are likely to experience **(Figure 6)**. Fire extinguishers are portable apparatuses containing chemicals, water, foam, or special gas that can be discharged to extinguish a small fire. Technicians must also be able to fight a fire in an emergency. Labels on the fire extinguisher indicate the types of fires that it will put out. Become familiar with the use of a fire extinguisher.

> **You Should Know**
>
> *The following procedure for fire extinguisher use is not intended to be used as a lab exercise unless expressly directed by your instructor. Use the following discussion as a guide to become familiar with fire extinguisher use. You must be willing to fight a fire in the shop, if the occasion arises.*

The proper use of a fire extinguisher is very important **(Figure 7)**. It is possible to deplete an extinguisher and still not put out even the smallest of fires if the extinguisher is used improperly. Procedures vary depending on the agents used. Technicians must become familiar with all the extinguishers

| | Class of Fire | Typical Fuel Involved | Type of Extinguisher |
|---|---|---|---|
| Class A Fires (green) | **For Ordinary Combustibles** Put out a Class A fire by lowering its temperature or by coating the burning combustibles. | Wood<br>Paper<br>Cloth<br>Rubber<br>Plastics<br>Rubbish<br>Upholstery | Water*[1]<br>Foam*<br>Multipurpose dry chemical[4] |
| Class B Fires (red) | **For Flammable Liquids** Put out a Class B fire by smothering it. Use an extinguisher that gives a blanketing flame-interupting effect; cover whole flaming liquid surface. | Gasoline<br>Oil<br>Grease<br>Paint<br>Lighter fluid | Foam*<br>Carbon dioxide[5]<br>Halogenated agent[6]<br>Standard dry chemical[2]<br>Purple K dry chemical[3]<br>Multipurpose dry chemical[4] |
| Class C Fires (blue) | **For Electrical Equipment** Put out a Class C fire by shutting off power as quickly as possible and by always using a nonconducting extinguishing agent to prevent electric shock. | Motors<br>Appliances<br>Wiring<br>Fuse boxes<br>Switchboards | Carbon dioxide[5]<br>Halogenated agent[6]<br>Standard dry chemical[2]<br>Purple K dry chemical[3]<br>Multipurpose dry chemical[4] |
| Class D Fires (yellow) | **For Combustible Metals** Put out a Class D fire of metal chips, turnings, or shaving by smothering or coating with a specially designed extinguishing agent. | Aluminum<br>Magnesium<br>Potassium<br>Sodium<br>Titanium<br>Zirconium | Dry powder extinguishers and agents only |

*Cartridge-operated water, foam, and soda-acid types of extinguishers are no longer manufactured. These extinguishers should be removed from service when they become due for their next hydrostatic pressure test.

Notes:

(1) Freeze in low temperatures unless treated with antifreeze solution, usually weighs over 20 pounds, and is heavier than any other extinguisher mentioned.

(2) Also called ordinary or regular dry chemical (solution bicarbonate).

(3) Has the greatest initial fire-stopping power of the extinguishers mentioned for class B fires. Be sure to clean residue immediately after using the extinguisher so sprayed surfaces will not be damaged (potassium bicarbonate).

(4) The only extinguishers that fight A, B, and C class fires. However, they should not be used on fires in liquified fat or oil of appreciable depth. Be sure to clean residue immediately after using the extinguisher so sprayed surfaces will not be damaged (ammonium phosphates).

(5) Use with caution in unventilated, confined spaces.

(6) May cause injury to the operator if the extinguishing agent (a gas) or the gases produced when the agent is applied to a fire is inhaled.

**Figure 5** *A guide to fire classification and fire extinguisher types.*

equipped in the shop. To use a multipurpose dry chemical extinguisher, follow these steps:

1. Hold the fire extinguisher in an upright position.
2. Pull the safety pin from the handle.
3. Free the hose from its retainer and aim it at the base of the fire.

### You Should Know

*Do not risk your life fighting a fire. If it is evident that the fire is out of control, get out. Always be aware of where you are and the location of the nearest exit. Do not open garage doors in the event of a fire because the extra oxygen will intensify the flames.*

**Figure 6** *Know the location and types of fire extinguishers that are available in the shop.*

**Figure 7** *Know how to use a fire extinguisher properly.*

4. Stand 8 feet from the fire. Do not approach any closer to the fire than this distance.
5. Squeeze the lever while sweeping the hose from side to side. Keep the hose aimed at the *base* of the fire.

## Fuel Storage

Gasoline is so commonly found in automotive repair shops that its dangers are often forgotten or overlooked. A slight spark or an increase in heat can cause a fire or explosion. Gasoline is a very explosive and powerful liquid. One exploding gallon of gasoline has a force equal to fourteen sticks of dynamite. It is the expanding vapors from gasoline that are extremely dangerous. These vapors are present even in cold temperatures. Gasoline vapors are heavier than air. Therefore, when an open container of gasoline is sitting about, the vapors spill out over the sides of the container onto the floor. These fumes are more **flammable** than liquid gasoline and can easily support combustions and explode.

Never smoke around gasoline since even hot ashes can ignite the gasoline. If an engine has a gasoline leak, or if you have caused a leak by disconnecting a fuel line, stop the leak and clean up the spilled gasoline immediately. While stopping the leak, be extra careful not to cause sparks. If any rags are used in the cleanup of the gasoline they must be placed in an approved container **(Figure 8)**. Due to extreme fire hazards, it is important to immediately clean up any gasoline spilled on the floor. Also, many of the compounds in petroleum are toxic, especially if they are in high concentrations.

**Figure 8** *Dirty rags and towels must be stored in approved containers.*

The chemicals in petroleum that do not evaporate quickly are biodegradable. Optimum degradation occurs if the gasoline is diluted and there is enough air, water, and nutrients for the microbes. These properties of gasoline are an advantage in the cleanup and disposal of small spills. Spreading absorbent material such as kitty litter, sand, ground corn cobs, straw, sawdust, wood chips, peat, synthetic absorbent pads, or even dirt can stop the flow and soak up the gasoline. Keep in mind that the absorbent does not make gasoline nonflammable.

Brooms can be used to sweep up the absorbent material and put it into buckets, garbage cans, or barrels. Remember to control ignition sources. Be aware of local laws concerning gasoline spills. Some states or municipalities require notification of any gasoline spill larger than 5 gallons.

Gasoline should always be stored in approved containers **(Figure 9)** and never in glass containers. If the glass container is knocked over or dropped, a terrible explosion can occur. Approved gasoline storage cans have a flash-arresting screen at the outlet. These screens prevent external ignition sources from igniting the gasoline within the can while the gasoline is being poured.

Follow these safety precautions regarding gasoline containers:

1. Always use approved gasoline containers that are painted red for proper identification.
2. Do not fill gasoline containers completely full. Always leave the level of gasoline at least 1 inch (25 mm) from the top of the container. This action allows expansion of the gasoline at higher temperatures. If gasoline containers are completely full, the gasoline will expand when the temperature increases. This expansion forces gasoline from the can and creates a dangerous spill.

**Figure 9** *Store any type of combustible materials in an approved safety cabinet.*

3. If gasoline containers must be stored, place them in a well-ventilated area such as a storage shed. Do not store gasoline containers in your home or in the trunk of a vehicle.
4. When a gasoline container must be transported, be sure it is secured against upsets. Do not transport or fill gasoline containers on plastic bed liners. Static electricity can be generated and ignite the vapors.
5. Do not store a partially filled gasoline container for long periods of time, because it may give off vapors and produce a potential danger.
6. Never leave gasoline containers open, except while filling or pouring gasoline from the container.
7. Do not prime an engine with gasoline while cranking the engine.
8. Never use gasoline as a cleaning agent.

Diesel fuel is not as **volatile** (vaporizes or explodes) as gasoline, but it should still be stored and handled in the same way as gasoline. It is also not as refined as gasoline and tends to be a dirty fuel. It normally contains impurities, including active microscopic organisms that can be highly infectious. If diesel fuel happens to get on an open cut or sore, thoroughly wash it immediately. If it gets into your eyes, flush them immediately and get medical help.

## Solvents and Rags

Cleaning solvents are not as volatile as gasoline, but they are still flammable. They should be treated and stored in the same way as gasoline. Whenever using solvents, wear eye protection.

Oily and greasy rags can also cause fires. Used rags should be stored in an approved container and should never be thrown out with normal trash. Like gasoline, oil is a hydrocarbon and can ignite with or without a spark or flame.

## Lifting the Vehicle

There will be times when it will be necessary to raise the vehicle off the floor to perform a service or repair. The vehicle can be lifted by use of either floor jacks or a hoist. Each method has unique safety precautions and rules that must be followed.

> **You Should Know**
>
> *Not only is safety a primary issue when lifting a vehicle, but it is important to protect your customer's vehicle from damage by taking the time to confirm the proper lifting procedure. Locating the hoist pads, the floor jack, or the jack stands in an improper location on the vehicle may damage the frame or body of the vehicle.*

## Floor Jack and Safety Stand Use

Floor jacks are portable hydraulic tools used to lift the vehicle a short distance off the floor, or when only a portion of the vehicle needs to be raised **(Figure 10)**. Before using the floor jack, make sure it has a sufficient rating to lift and sustain the weight of the vehicle. Next, inspect it for proper lubrication, proper hydraulic fluid level, and for damage that would compromise its safe use. If the jack does not pass any one of these inspections, tag it and notify your supervisor immediately.

To lift the entire vehicle, begin by placing the vehicle's transmission into Park. Place it in first gear if the vehicle has a manual transmission. Set the parking brake and place wheel blocks around the rear wheels. Position the floor jack under the front of the vehicle at a location strong enough to support the weight. Many jack manufacturers and service manuals provide illustrations for the **vehicle lift points** of the type of vehicle being lifted. Vehicle lift points are the areas structurally strong enough to sustain the stress of lifting. If this information is not available, always place the floor jack on major strength parts. These areas include the frame, cross member, or differential. If you are in doubt about the proper liftpoint, ask your instructor/supervisor. Never lift on sheet metal or plastic parts. The jack should be centered between the front tires and positioned so that the lift will be straight up and down.

> **You Should Know**
>
> *If you are lifting only one wheel of the vehicle, be careful not to lift it so high that it can slip off the jack saddle.*

Operate the jack until the jack saddle contacts the vehicle lift point. Check for good contact. If

**Figure 10** *Floor jacks are used to raise the vehicle a short distance off the floor.*

things look good, lift the front of the vehicle a couple of inches off the floor. Recheck the position of the jack. Continue to check the jack position throughout the lifting procedure. If the vehicle or jack begins to lean, lower the jack and reset it. Lift the vehicle to the required height. Do not lift higher than is necessary.

### You Should Know

*Never place blocks of wood, and so on, between the vehicle frame and floor jack to obtain additional lift. If additional lift is required, place the blocks under the floor jack.*

The floor jack is to be used only to lift the vehicle off the floor. It is not intended to support the vehicle while someone is under it. Use jack stands (floor stands) to support the vehicle. Use one jack stand for each quarter of the vehicle that is lifted **(Figure 11)**. Place the jack stand under the frame or a major support component of the vehicle. Slowly lower the vehicle onto the stands while making sure that they do not tilt.

Place the floor jack under the rear of the vehicle. Follow the same procedure to raise the rear of the vehicle. Use two safety stands to support the rear of the vehicle. When the vehicle is properly lifted and supported by safety stands, it is safe to work under the vehicle.

Use the same lift points to lower the vehicle. Once one end of the vehicle is on the floor, place wheel blocks around those wheels. Then lower the other end.

### You Should Know

*Some vehicles with fiberglass or composite body structures may require special lifting considerations. Some of these may include opening the doors, deck lid, and hood before lifting. Others may not allow for opening any doors after the vehicle is lifted. Refer to the service manual and follow the procedures listed.*

**Figure 11** *Support the vehicle by jack stands located at each corner that is lifted. Do not rely on the floor jack to support the vehicle.*

## Using the Hoist

A hoist is used when the entire vehicle needs to be raised, usually high enough for the technician to stand underneath the vehicle **(Figure 12)**. When a vehicle is placed on the hoist, it must be centered. The balance of the vehicle must be taken into consideration, as well as the effects on balance if a heavy component is removed from the vehicle.

To lift the vehicle using a hoist, first locate the correct lift points in the service manual for the vehicle you are working on. Pay close attention to any warnings or special considerations listed there. Center the vehicle over the hoist; keep in mind the vehicle's center of gravity and balance point. Locate the hoist pads under the lift points. Adjust the pads so that the vehicle will be lifted level.

Lift the vehicle a few inches off the floor. Shake the vehicle while observing for signs of any movement. If the vehicle is not secure on the hoist, or unusual noises are heard while lifting, lower it to the floor and reset the pads.

### You Should Know

*If the vehicle is not level on the pads, or the pads are not in the proper position, lower the vehicle and readjust as needed. Never get under a vehicle that is not sitting properly on the hoist.*

Once the vehicle is at the desired height, lock the hoist. Do not get under the vehicle until the hoist locks have been set.

**Figure 12** *One type of hoist that may be used in the automotive shop. The hoist is used to raise the entire vehicle.*

To lower the vehicle, release the locks and put the control valve into the "lower" position. Once it is returned to the floor, push the contact pads out of the path of the tires.

## Running the Vehicle while in the Shop

Many times it will be necessary to run the engine while the vehicle is in the shop. This presents the possibility of **carbon monoxide (CO)** poisoning if it is not done safely. Carbon monoxide is an odorless, toxic gas that is the product of combustion; it is typically caused by excessive fuel or a lack of air. Most shops are equipped with a ventilation system that will remove the vehicle exhaust. If the shop is not equipped with a ventilation system, a hose must be routed from the vehicle's exhaust to outdoors.

Check the operation of the shop's ventilation system before connecting to the vehicle. Turn on the ventilation motor and place your hand over the hose. You should feel a strong, consistent vacuum. A weak vacuum indicates a restriction or a leak in the ventilation system. If you encounter a problem, notify your instructor.

### You Should Know

*Some of the early warning signs of carbon monoxide poisoning are headaches, dizziness, and blurred vision. If you experience any of these while in the shop, notify your instructor immediately, and get some fresh air. If symptoms persist, seek medical attention.*

Connect the ventilation hose to the vehicle's exhaust **(Figure 13)**. Place wheel blocks around a tire and place the transmission in Park or Neutral.

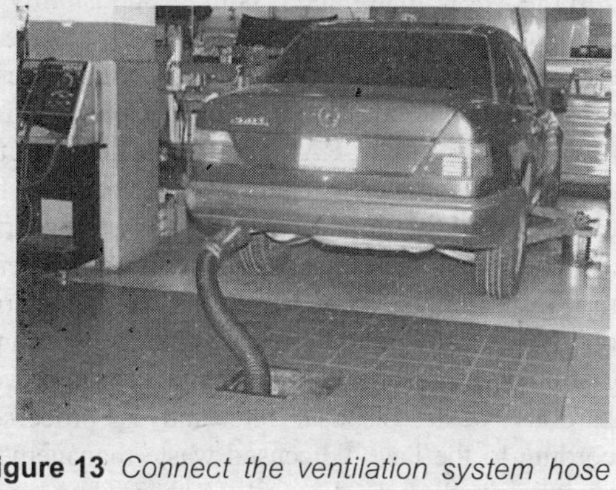

**Figure 13** *Connect the ventilation system hose to the vehicle's exhaust before starting the engine. Make sure the hose is secure and there are no kinks in the hose.*

Set the parking brake. The engine is now ready to be started. While running the engine, be aware of the warning signs of CO poisoning. Report any symptoms to your instructor immediately.

## Hazardous Materials

Many solvents and other chemicals used in an automotive shop have warning and caution labels that should be read and understood by everyone who uses them. These are typically considered **hazardous materials** since they can cause illness, injury, or death or pollute water, air, or land. Many service procedures generate what are known as hazardous wastes. Examples of hazardous waste are used or dirty cleaning solvents and other liquid cleaners.

### Right-to-Know Laws

In the United States, **"right-to-know" laws** concerning hazardous materials and wastes protect every employee in a workplace. The general intent of these laws is for employers to provide a safe workplace, as it relates to hazardous materials. The right-to-know law states that employees have a right to know when the materials they use at work are hazardous. These laws originated with the **Hazard Communication Standard** published by the Occupational Safety and Health Administration (OSHA) in 1983. This document was originally intended for chemical companies and manufacturers that required employees to handle hazardous materials in their work situations. At the present time, most states have established their own right-to-know laws. Meanwhile, the federal courts have decided to apply these laws to all companies, including automotive service shops.

Under the right-to-know laws, the employer has three responsibilities regarding the handling of hazardous materials by its employees. The first responsibility concerns employee training and the providing of information. All employees must be trained about the types of hazardous materials they will encounter in the workplace. All employees must be informed about their rights under legislation regarding the handling of hazardous materials. In addition, information about each hazardous material must be posted on **material safety data sheets (MSDS)** available from the manufacturer **(Figure 14)**. MSDS contain extensive information and facts about hazardous materials. The Canadian equivalent to the MSDS is the **workplace hazardous materials information systems (WHMIS)** used to inform workers of hazardous material facts.

The employer has a responsibility to place MSDS where they are easily accessible by all employees. The MSDS provide extensive information about the hazardous material such as:

- Chemical name
- Physical characteristics
- Protective equipment required for handling
- Explosion and fire hazards
- Other incompatible materials
- Health hazards such as signs and symptoms of exposure, medical conditions aggravated by exposure, and emergency and first aid procedures
- Safe handling precautions
- Spill and leak procedures

The second responsibility of the employer is to make sure that all hazardous materials are properly labeled. The label information must include health, fire, and reactivity hazards posed by the material, as well as the protective equipment necessary to handle the material. The manufacturer must supply all warning and precautionary information about hazardous materials, and the employee must read and understand this information before handling the material. Pay great attention to the information on the label. By doing so, you will use the substance in the proper and safe way, thereby preventing hazardous conditions.

The third responsibility of the employer is for maintaining permanent files regarding hazardous materials. These files must include information on hazardous materials in the shop, proof of employee training programs, and information about accidents such as spills or leaks of hazardous materials. The employer's files must also include proof that employees' requests for hazardous material information, such as MSDS, have been met. The employer must maintain a general right-to-know compliance procedure manual.

Employees have responsibilities as well. Employees must be familiar with the intended purpose of the substance, the recommended protective equipment, accident and spill procedures, and any other information regarding the safe handling of hazardous materials. This training must be given annually to employees and provided to new employees as part of their job orientation.

## Hazardous Waste

Waste is considered hazardous if it is on the Environmental Protection Agency (EPA) list of known and harmful materials, or if it has one or more of the following characteristics:

1. **Reactive.** Any material that reacts violently with water or other chemicals is considered hazardous. When exposed to low-pH acid solutions, if a material releases cyanide gas, hydrogen sulfide gas, or similar gases, it is hazardous.
2. **Corrosive.** If a material burns the skin or dissolves metals and other materials, it is considered hazardous.
3. **Toxic.** Materials are hazardous if they leach one or more of eight heavy metals in concentrations greater than 100 times primary drinking water standard.
4. **Ignitable.** A liquid is hazardous if it has a flash point below 140°F (60°C), and a solid is hazardous if it ignites spontaneously.

A complete list of EPA hazardous wastes can be found in the Code of Federal Regulations. It should be noted that no material is considered hazardous waste until the shop is finished using it and is ready to dispose of it. New oil is not a hazardous waste: however, used oil is. Once you drain oil from an engine, you have generated the waste and now become responsible for the proper disposal of this hazardous waste. There are many other wastes that need to be handled properly after you have removed them, such as batteries, brake fluid, transmission fluid, and engine coolant.

No fluids drained from a vehicle should be allowed to go into sewage drains. Some fluids, such as coolant, can be captured and recycled. Filters for fluids (transmission, fuel, and oil filters) also need to be handled in designated ways. Used filters need to be drained and then crushed or disposed of in a special shipping barrel. Most regulations demand that oil filters be drained for at least 24 hours before they are disposed of or crushed.

Federal and state laws control the disposal of hazardous waste materials. It is the responsibility of the employer and the employee to ensure everyone in the shop is familiar with these laws. Hazardous waste disposal laws include the **Resource Conservation and Recovery Act (RCRA)**. This law basically states that hazardous waste generators are responsible for the waste from the time it becomes a waste material until the proper waste disposal is completed. Therefore, the user must store hazardous waste material properly and safely and be responsible for the transportation of this material until it arrives at an approved hazardous waste disposal site and is processed according to the law. A licensed waste management firm normally does the disposal. The hazardous waste coordinator for the shop should have a written contract with the hazardous waste hauler.

```
HEXANE
====================================================
MSDS Safety Information
====================================================
Ingredients
====================================================
Name: HEXANE (N_HEXANE)
% Wt: >97
OSHA PEL: 500 PPM
ACGIH TLV: 50 PPM
EPA Rpt Qty: 1 LB
DOT Rpt Qty: 1 LB
====================================================
Health Hazards Data
====================================================
LD50 LC50 Mixture: LD50:(ORAL,RAT) 28.7 KG/MG
Route Of Entry Inds _ Inhalation: YES
Skin: YES
Ingestion: YES
Carcinogenicity Inds _ NTP: NO
IARC: NO
OSHA: NO
Effects of Exposure: ACUTE:INHALATION AND INGESTION ARE HARMFUL AND MAY BE FATAL.
INHALATION AND INGESTION MAY CAUSE HEADACHE, NAUSEA, VOMITING, DIZZINESS, IRRITATION
OF RESPIRATORY TRACT, GASTROINTESTINAL IRRITATION AND UNCONSCIOUSNESS. CONTACT
W/SKIN AND EYES  MAY CAUSE IRRITATION. PROLONGED SKIN MAY RESULT IN DERMATITIS (EFTS
OF OVEREXP)
Signs And Symptions Of Overexposure: HLTH HAZ:CHRONIC:MAY INCLUDE CENTRAL
NERVOUS SYSTEM DEPRESSION.
Medical Cond Aggravated By Exposure: NONE IDENTIFIED.
First Aid: CALL A PHYSICIAN. INGEST:DO NOT INDUCE VOMITING. INHAL:REMOVE TO FRESH AIR. IF
NOT BREATHING, GIVE ARTIFICIAL RESPIRATION. IF BREATHING IS DIFFICULT, GIVE OXYGEN.
EYES:IMMED FLUSH W/PLENTY OF WATER FOR AT LEAST 15 MINS. SKIN:IMMED FLUSH W/P LENTY
OF WATER FOR AT LEAST 15 MINS WHILE REMOVING CONTAMD CLTHG & SHOES. WASH CLOTHING
BEFORE REUSE.
====================================================
Handling and Disposal
====================================================
Spill Release Procedures: WEAR NIOSH/MSHA SCBA & FULL PROT CLTHG. SHUT OFF
IGNIT SOURCES:NO FLAMES, SMKNG/FLAMES IN AREA. STOP LEAK IF YOU CAN DO SO W/OUT
HARM. USE WATER SPRAY TO REDUCE VAPS. TAKE UP W/SAND OR OTHER NON_COMBUST MATL &
PLACE INTO CNTNR FOR LATER (SU PDAT)
Neutralizing Agent: NONE SPECIFIED BY MANUFACTURER.
Waste Disposal Methods: DISPOSE IN ACCORDANCE WITH ALL APPLICABLE FEDERAL, STATE AND
LOCAL ENVIRONMENTAL REGULATIONS. EPA HAZARDOUS WASTE NUMBER:D001 (IGNITABLE
WASTE).
Handling And Storage Precautions: BOND AND GROUND CONTAINERS WHEN TRANSFERRING LIQUID.
KEEP CONTAINER TIGHTLY CLOSED.
Other Precautions: USE GENERAL OR LOCAL EXHAUST VENTILATION TO MEET
TLVREQUIREMENTS. STORAGE COLOR CODE RED (FLAMMABLE).
====================================================
Fire and Explosion Hazard Information
====================================================
Flash Point Method: CC
Flash Point Text: _9F,_23C
Lower Limits: 1.2%
Upper Limits: 77.7%
Extinguishing Media: USE ALCOHOL FOAM, DRY CHEMICAL OR CARBON DIOXIDE. (WATER MAY BE
INEFFECTIVE).
Fire Fighting Procedures: USE NIOSH/MSHA APPROVED SCBA & FULL PROTECTIVE
EQUIPMENT (FP N).
Unusual Fire/Explosion Hazard: VAP MAY FORM ALONG SURFS TO DIST IGNIT SOURCES & FLASH
BACK. CONT W/STRONG OXIDIZERS MAY CAUSE FIRE. TOX GASES PRDCED MAY INCL:CARBON
MONOXIDE, CARBON DIOXIDE.
====================================================
```

**Figure 14** *An example of a Material Safety Data Sheet.*

The RCRA controls these types of automotive waste:

- Paint and body repair products waste
- Solvents for parts and equipment cleaning
- Batteries and battery acid
- Mild acids used for metal cleaning and preparation
- Waste oil, engine coolants, or antifreeze
- Air conditioning refrigerants
- Engine oil filters

Never, under any circumstances, use these methods to dispose of hazardous waste material:

- Pour hazardous wastes on weeds to kill them.
- Pour hazardous wastes on gravel streets to prevent dust.
- Throw hazardous wastes in a dumpster.
- Dispose of hazardous wastes anywhere but an approved disposal site.
- Pour hazardous wastes down sewers, toilets, sinks, or floor drains.
- Bury hazardous wastes in the ground.

## Fuel System Safety

Most modern fuel systems operate under pressure. Before attempting to perform any service requiring the disconnection of hoses or tubes, this pressure must be relieved. The service manual will provide the proper method of relieving the pressure. This may include removing the fuel pump relay or fuse and running the engine until it runs out of fuel, or using a hose connected to the fuel rail and catching the fuel into an approved container. Most gasoline fuel systems operate up to approximately 60 pounds per square inch (psi) (4 bar) of pressure. Diesel engine fuel systems can exceed 23,000 psi (1,586 bar). Compressed natural gas vehicles can have tank pressures up to 3,500 psi (241 bar) with line pressures of 140 psi (9.6 bar). One of the reasons the fuel system is pressurized is to provide better vaporization of the fuel. Opening lines or tubes without first relieving the pressure causes vaporized fuel to be expelled. This vaporized fuel is very easy to ignite.

These high pressures can cause serious personal injury. Due to the high pressures associated with fuel systems, never use your hands to locate a leak. The pressures may be high enough to cause severe cuts, severed fingers, or fuel injection into the bloodstream. Use a piece of cardboard to safely locate the source of the leak.

Throughout this textbook there will be many warnings and cautions given concerning fuel system service safety. In addition, the service manual will also have many similar warnings. It is imperative that you read and follow these instructions.

## Review Questions

1. Explain how safety is a part of professionalism.
2. Explain the four characteristics of a material that the EPA uses to classify a waste as being hazardous.
3. When working with chemicals, you should always wear ____________ ____________.
4. Hazardous waste generators are responsible for the waste from the time it becomes a waste material until the proper ____________ ____________ is completed.
5. Fire extinguisher use is being discussed. Technician A says electrical fires are extinguished with Class A fire extinguishers. Technician B says gasoline fires are extinguished with Class B fire extinguishers. Who is correct?
   A. Technician A only
   B. Technician B only
   C. Both Technician A and Technician B
   D. Neither Technician A nor Technician B
6. List the basic safety rules of proper lifting.
7. What are the three responsibilities of the employer outlined by the "right-to-know" laws?
8. One gallon of gasoline has a force equal to ____________ sticks of dynamite.
9. A vehicle is in the shop with the engine running. Technician A says it is safe to work around the electric cooling fan when the engine is off and the engine temperatures are low. Technician B says the electric cooling fans can only come on if the engine is running. Who is correct?
   A. Technician A only
   B. Technician B only
   C. Both Technician A and Technician B
   D. Neither Technician A nor Technician B
10. Which of the following statements concerning gasoline storage is true?
    A. Always use approved gasoline containers that are painted blue for proper identification.
    B. Prevent air from entering the container by filling gasoline containers completely.
    C. Always transport gasoline containers on plastic bed liners.
    D. Do not store a partially filled gasoline container for long periods of time.

CHAPTER

# 2 Special Tools and Equipment

## Introduction

In addition to normal hand tools, today's technician will need to use many special tools while diagnosing and servicing fuel and emission systems. Special tools are designed to perform a specified task. They make it easier and safer for the technician to perform certain tasks. Some of the special tools that may be used in diagnosing fuel and emission systems include: pressure gauge flow meters, evaporative system leak detectors (smoke machine), exhaust analyzers, and scan tools.

## Fuel Pressure Gauge

A pressure gauge **(Figure 1)** is used to measure the pressure in the fuel system. This tester is very important for diagnosing fuel-injection systems. These systems rely on very high fuel pressures, from 35 to 70 psi (2 to 4.8 bar). A drop in fuel pressure will reduce the amount of fuel delivered to the injectors and result in a lean air-fuel mixture.

A fuel pressure gauge is used to check the discharge pressure of fuel pumps, the regulated pressure of fuel-injection systems, and injector pressure drop. These tests can identify faulty pumps, regulators, or injectors. In addition, these tests can identify restrictions present in the fuel delivery system. Restrictions are typically caused by a dirty fuel filter, collapsed hoses, or damaged fuel lines.

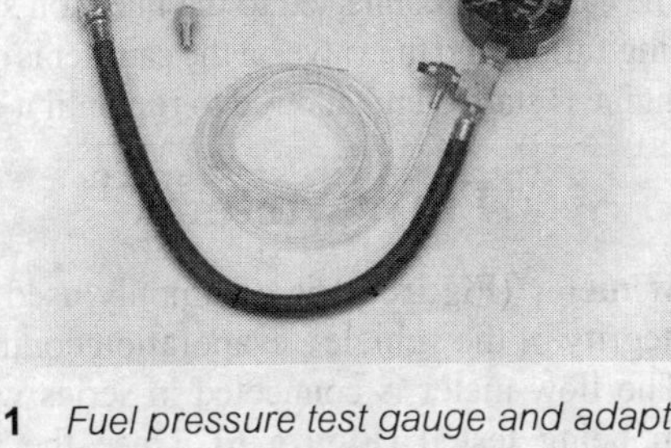

**Figure 1** *Fuel pressure test gauge and adapters.*

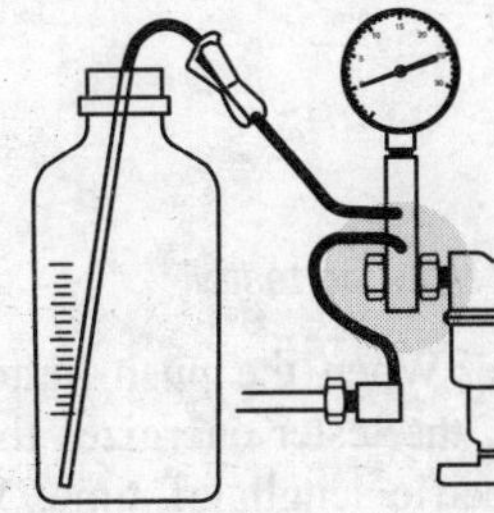

**Figure 2** *Checking fuel pump volume.*

Some fuel pressure gauges also have a valve and outlet hose for testing fuel pump discharge volume **(Figure 2)**. The manufacturer's specification for discharge volume will be given as a number of pints or liters of fuel that should be delivered in a certain number of seconds.

### You Should Know

*While testing fuel pressure, be careful not to spill gasoline: gasoline spills may cause explosions and fires, resulting in serious personal injury and property damage.*

## Injector Balance Tester

The injector balance tester is used to test the injectors in a port fuel-injected engine for proper operation. Each injector should flow the same amount of fuel through it and create the same amount of pressure drop in the fuel rail. The injector balance tester contains a timing circuit, and some injector balance testers have an off/on switch. A pair of leads on the tester must be connected to the battery with the correct polarity **(Figure 3)**. The injector terminals are disconnected, and a second double lead on the tester is attached to the injector terminals.

Before the injector test, the fuel pressure gauge is connected to the Schrader valve on the fuel rail, and the ignition switch should be cycled two or three times until the specified fuel pressure is indicated on

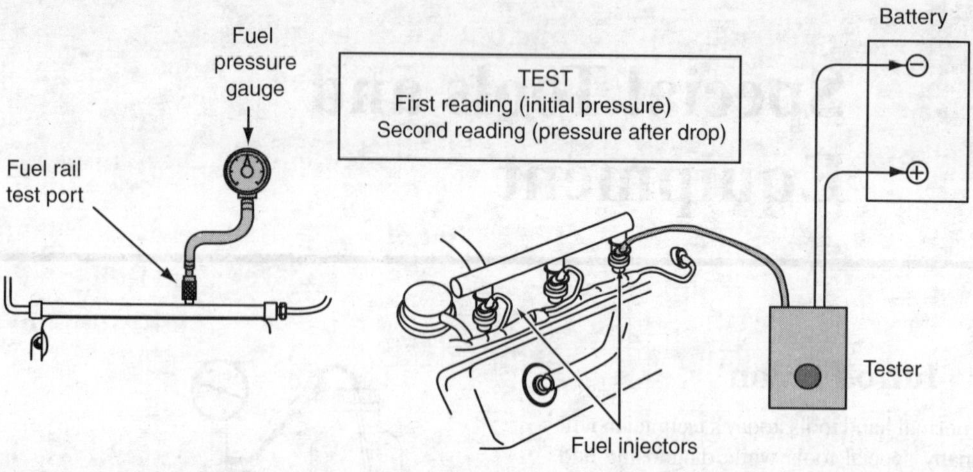


**Figure 3** *Injector balance tester.*

the pressure gauge. When the push button on the tester is depressed, the tester energizes the injector winding for a specific length of time. While the injector is energized, the technician records the pressure decrease on the fuel pressure gauge. This procedure is repeated on each injector.

Some vehicle manufacturers provide a specification of a maximum difference of 3 psi (20 kilopascals, or kPa) between the pressure readings after each injector is energized. If the injector orifice is restricted, there is not much pressure decrease when the injector is energized. Restricted injector orifices can cause acceleration stumbles, engine stalling, and erratic idle operation. The injector plunger is sticking open if excessive pressure drop occurs when the injector is energized. Sticking injector plungers may result in a rich air-fuel mixture.

**You Should Know**

*It is very difficult to gain access to the injector terminals of some engines to hook up the injector balance test. Some scan tools provide an activation test of the injectors that can be substituted for the injector balance tester.*

**You Should Know**

*Electronic fuel-injection systems are pressurized, and these systems require depressurizing prior to fuel pressure testing and other service procedures.*

## Fuel Injector Cleaner

Fuel injectors spray a certain amount of fuel into the intake system. If the fuel injector is dirty, not enough fuel will be sprayed. When these sensitive fuel injectors become partially clogged, fuel flow is restricted. Spray patterns are altered, causing poor performance and reduced fuel economy.

From the standpoint of cost savings, it is usually more desirable to clean sulfated and/or plugged fuel injectors instead of replacing them. There are different kinds of fuel injector cleaners available. One is a pressure tank that holds a mixture of solvent and unleaded gasoline. Follow the manufacturer's instructions for mixing, quantity, and safe handling of the mixture. The vehicle's fuel pump must be disabled, and on some vehicles the fuel line must be blocked between the pressure regulator and the return line. Then, the hose on the pressure tank is connected to the service port in the fuel system. The tank is pumped to build internal pressure. The in-line valve is then partially opened and the engine is started. It should run at approximately 2,000 rotations per minute (rpm) for about 10 minutes to clean the injectors thoroughly.

Another type of cleaner kit uses a pressurized canister **(Figure 4)**. The canister is filled with a premixed solvent solution. Use of the canister-type cleaner is similar to that of the pressure tank system described above but does not require mixing or pumping.

The canister is connected to the injection system's servicing fitting, and the valve on the canister is opened. The engine is started and allowed to run until it dies.

## Flow Meters

A **flow meter (Figure 5)** is commonly used to test the integrity of the vehicle's evaporation control system. The flow meter is connected in series with the system being tested **(Figure 6)**. Once the system

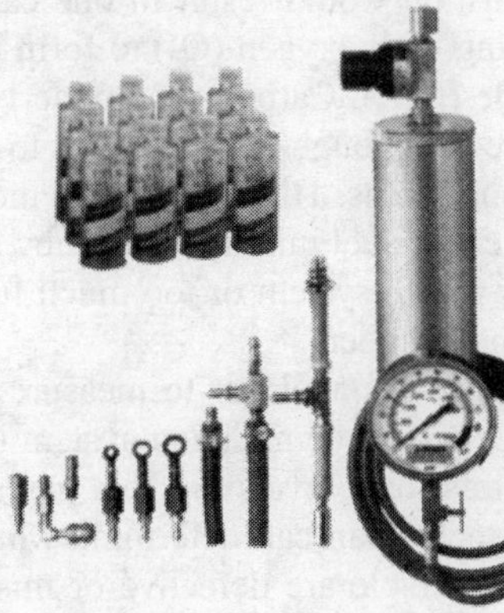

**Figure 4** *Pressurized canister used for injector cleaning.*

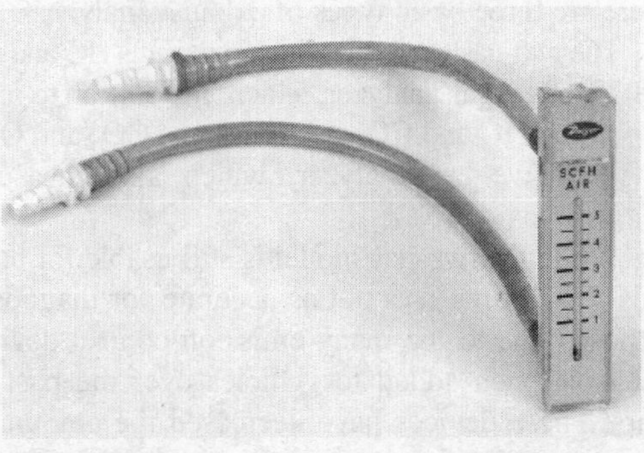

**Figure 5** *Flow meters are used to measure the flow rate of vapors.*

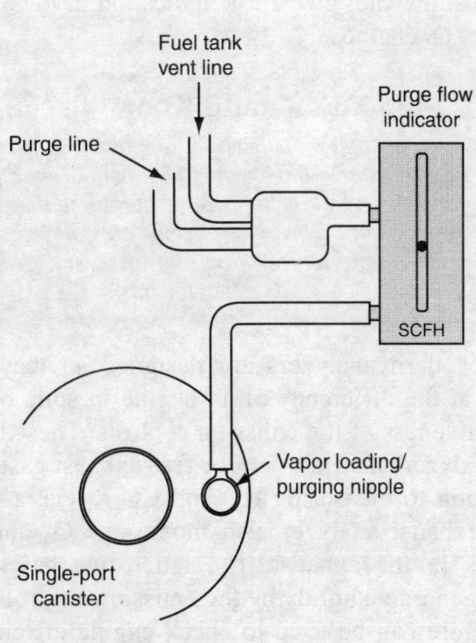


**Figure 6** *The flow meter is connected in series to the system being tested.*

becomes active, the flow meter will indicate the rate of flow. Flow rates above or below specifications indicate a problem with the system.

## Evaporative Emissions Leak Detector

The **evaporative emissions leak detector (EELD)** is a very useful tool for testing the integrity of the evaporative emission system and to locate any leaks **(Figure 7)**. The EELD can be used to pressurize the evaporative emission system and monitor the machine's flow meter to determine if there is a leak in the system. If a leak is suspected, the technician then can either listen for the leak or use soapy water to locate it. If these approaches fail to find the leak, the machine can be used to force smoke into the evaporative emission system. The technician then can use a very bright white light that the EELD is equipped with to locate the leak. If this still fails to locate the leak, the smoke has a dye tracer in it. The

**Figure 7** *The evaporative emissions leak detector makes the job of finding small leaks easier.*

technician will then use the special yellow glasses and the machine's ultraviolet lamp to find signs of the dye tracer and ultimately the source of the leak.

## Exhaust Analyzers

Testing the quality of the exhaust is both a procedure for testing emission levels and a diagnostic routine. One of the most valuable diagnostic aids is the **exhaust analyzer (Figure 8)**. The exhaust analyzer is a special tester that measures some of the emission gases produced by the engine. By viewing the quality of an engine's exhaust, a technician is able to look at the effects of the combustion process. Any defect can cause a change in exhaust quality. The amount and type of change serves as the basis of diagnostic troubleshooting.

The levels of **hydrocarbon (HC)** and carbon monoxide (CO) in the exhaust are a direct indication of engine performance. Hydrocarbon in the exhaust is raw, unburned fuel. HC emissions indicate that complete combustion is not occurring in the engine. Emission analyzers measure HC in parts per million (ppm) or grams per mile (Cg/mi). A high level of hydrocarbons indicates that a misfire or incomplete combustion is occurring and unburned fuel is leaving the cylinder. This could be caused by a fouled spark plug, a defective spark plug wire, a burned valve, an excessively lean air-fuel mixture, incorrect ignition timing, and so forth. All of these problems would decrease the effectiveness of combustion and allow raw fuel to leave the cylinder.

Carbon monoxide is an odorless, toxic gas that is the product of combustion and is typically caused by excessive fuel or a lack of air. CO is measured as a percent of the total exhaust. A high CO level results from incomplete combustion of the air-fuel mixture.

**Figure 8** *An exhaust gas analyzer is used to determine emission gases.*

Perfect combustion would result in one carbon (C) atom combining with oxygen ($O_2$) to form harmless carbon dioxide ($CO_2$). Carbon monoxide is formed when there are not enough oxygen atoms to combine with the carbon atoms. High CO levels indicate an excessively rich air-fuel mixture caused by a restriction in the air intake system or too much fuel being delivered to the cylinders.

By using the gas analyzer to measure the efficiency of the engine's combustion system (by measuring the emissions levels of the gases at the tailpipe), the technician can determine which components in the vehicle are defective or misadjusted by seeing which gas readings are too high or too low. Then the defective or misadjusted components can be repaired or brought back within specifications. There are three basic types of exhaust analyzers:

1. The two-gas analyzer, which measures HC and CO
2. The four-gas analyzer, which, in addition to measuring HC and CO, also measures $CO_2$ and $O_2$
3. The five-gas analyzer, which also measures nitric oxide (NO)

While the two-gas model is still usable for basic emission testing, it is not as accurate for diagnostic purposes due to the many emission control devices that have been added to vehicles over the past 30 years. These devices have decreased the amount of HC and CO in the exhaust. For example, catalytic converters alter the contents of the exhaust to the point that checking the HC and CO contents in the exhaust may not give a true indication of the operation of an engine.

### You Should Know

*If a two-gas analyzer is used for diagnostic purposes, it is best to sample the exhaust upstream of the catalytic converter to get a more accurate reading of engine efficiency. This can be done by removing the upstream oxygen sensor from the exhaust pipe and inserting the analyzer tube into the hole.*

Modern analyzers are designed so they can look at the efficiency of an engine in spite of the effectiveness of the emission controls. These types of analyzers measure four or five exhaust gases. In addition to measuring HC and CO levels, a four-gas exhaust analyzer also monitors $CO_2$ and $O_2$ levels in the exhaust. The latter two gases are changed only slightly by the emission controls and therefore can be used to check engine efficiency. Four-gas analyzers can be used to diagnose the following conditions:

- Rich or lean mixtures
- Air pump malfunctions
- Catalytic converter malfunction
- Blown head gaskets
- Faulty carburetors or injectors
- Intake manifold leaks
- Leaking exhaust gas recirculation (EGR) valves
- Excessive misfire
- Excessive spark advance
- Leaks or restrictions in the exhaust system

Presently, only a five-gas analyzer can detect oxides of nitrogen. These units do *not* measure nitrogen oxide ($NO_x$) directly. Instead, they measure NO, which is one of the components of $NO_x$. The NO reading is not an accurate measurement of the amount of $NO_x$, but a high NO reading does indicate that there is a $NO_x$ problem.

For the technician, $NO_x$ levels are determined by assumption and can only be measured with the vehicle under load. Make sure the vehicle does not have any driveability problems, particularly anything that might cause a lean condition (such as a vacuum leak). Also, check for anything that may cause the engine to run at higher-than-normal operating temperatures. This could include faulty thermostats, faulty water pumps, worn water pump drive belt, damaged fan, damaged or missing fan shroud, faulty electric fan operation, and so forth. Then, if all possible conditions are confirmed as being acceptable, if HC and CO levels are acceptable, and if the vehicle's emission control systems are in good working order (especially the EGR system and three-way catalytic converter), $NO_x$ levels can be assumed to be acceptable also.

## Engine Analyzers

When performing a complete engine performance analysis, an **engine analyzer** is used **(Figure 9)**. An engine analyzer houses all of the necessary test equipment. Although the term *engine analyzer* is often loosely applied to any multipurpose test meter, a complete engine analyzer incorporates most, if not all, of the test instruments used to diagnosis the engine's mechanical condition, ignition system, emission system, and fuel system. Most engine analyzers are based on a computer that guides a technician through the tests. Most will also do the work of the following tools:

- Compression gauge
- Voltmeter
- Pressure gauge
- Ohmmeter

**Figure 9** *An engine analyzer can perform many different functions.*

- Vacuum gauge
- Ammeter
- Vacuum pump
- Oscilloscope
- Tachometer
- Computer scan tool
- Timing light/probe
- Emissions analyzer

The microprocessors in some computerized engine analyzers are programmed with specifications for specific model vehicles. Diagnostic trouble codes have also been loaded into the analyzer's memory. Based on the input from the leads and connectors, the microprocessors can identify worn, misadjusted, or faulty components in all major engine systems. The analyzer will also list the probable causes of specific performance problems and will prompt, or guide, the technician step by step through a troubleshooting procedure designed to verify and correct the problem.

Commands and specifications can be entered into the analyzer on a keyboard. Specifications, commands, and test results are displayed on the cathode-ray tube (CRT) screen. Some analyzers will graphically display test results on their CRT screens. The analyzer's printer can also print out copies of the information that appears on the screen.

Many engine analyzers will perform a complete series of tests and record the results automatically. The analyzer compares all the test results to the vehicle manufacturer's specifications. When the test series is completed, the analyzer prints a report indicating those readings that were not within specifications. Many analyzers also provide diagnostic assistance for the problems indicated by the readings that were not within specifications.

However, the technician may select any test function, or functions, separately. Engine analyzer test capabilities vary depending on the equipment manufacturer, and technicians must familiarize themselves with the engine analyzer in their shop; for example, some analyzers have oscilloscope patterns and digital ignition readings, and others use digital ignition readings exclusively.

Some engine analyzers have vehicle specifications on diskette, and the technician enters the necessary information, such as model year and engine size, for the vehicle being tested. Specifications may be updated simply by obtaining a new diskette from the equipment manufacturer.

A phone modem is contained in some engine analyzers to provide networking capabilities. This phone modem allows the technician to download all technical reports and pattern reports of a specific problem vehicle to off-location technical support teams.

## Scan Tools

The introduction of power-train computer controls brought with it the need for tools capable of troubleshooting the systems. Several manufacturers provide a variety of computer **scan tools** that allow the technician to "talk" with the computer **(Figure 10)**. Scan tools are computerized test equipment used to read data streams of information from various vehicle onboard computers. Connected to the computer through diagnostic connectors, a scan tool can access diagnostic trouble codes (DTCs), run tests to check system operations, and monitor system activity. Trouble codes and test results are displayed on a light-emitting diode (LED) screen or are printed out on the scanner printer.

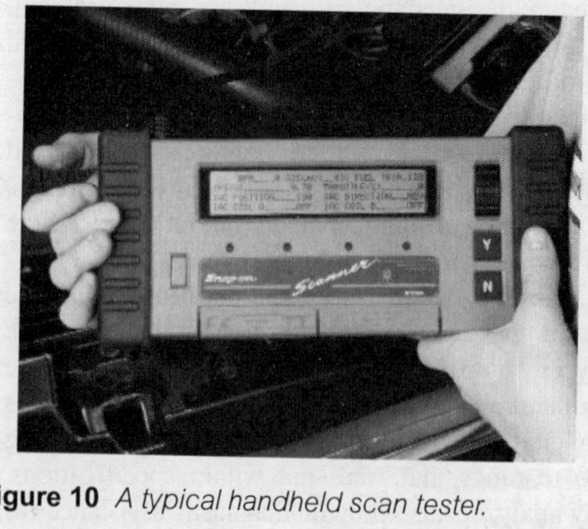


**Figure 10** *A typical handheld scan tester.*

A scan tool receives its testing information from several possible sources. Some scan tools have a programmable read-only memory (PROM) chip that contains all the information needed to diagnose specific model lines. This chip is normally contained in a cartridge, which is plugged into the tool. The type of vehicle being tested determines the appropriate cartridge that should be inserted. These cartridges contain the test information for that particular car. A cartridge is typically needed for each make and model vehicle. As new systems are introduced on new car models, a new cartridge is made available.

Scan testers have the capability to test many onboard computer systems such as engine computers, antilock brake computers, airbag computers, and suspension computers, depending on the year and make of vehicle and the type of scan tester. In many cases, the technician must select the computer system to be tested with the scan tester after the tester is connected to the vehicle.

The scan tester is connected to specific diagnostic connectors on various vehicles. Some manufacturers have one diagnostic connector and they connect the data wire from each onboard computer to a specific terminal in this connector. Other vehicle manufacturers have several different diagnostic connectors on each vehicle and each of these connectors may be connected to one or more onboard computers. A set of connectors is supplied with the scan tester to allow tester connection to various diagnostic connectors on different vehicles.

The scan tester must be programmed for the model year, make of vehicle, and type of engine. With some scan testers, this selection is made by pressing the appropriate buttons on the tester as directed by the digital tester display. On other scan testers, the appropriate memory card must be installed in the tester for the vehicle being tested. Some scan testers have a built-in printer to print test results, while other scan testers may be connected to an external printer.

As automotive computer systems become more complex, the diagnostic capabilities of scan testers continue to expand. Many scan testers now have the capability to store, or "freeze," data into the tester during a road test and then play back this data when the vehicle is returned to the shop.

Some scan testers now display diagnostic information based on the fault code in the computer memory. Service bulletins published by the scan tester manufacturer may be indexed by the tester after the vehicle information is entered in the tester. Other scan testers will display sensor specifications for the vehicle being tested.

## Special Hand Tools

Often a special tool that is designed to perform a specific function is needed to replace or repair a faulty component. Special sockets, wrenches, and small testing equipment are required for servicing many of today's vehicles.

### Oxygen Sensor Socket

This special socket helps remove and install oxygen sensors in the exhaust system. The tool looks like a deep socket that has a slot down one side to provide for the wires of the sensor **(Figure 11)**. Without this special feature the pigtail would have to be cut off the connector. Some exhaust systems mount the oxygen sensor in a location that is so easy to get at you can use a regular box end wrench to remove and install it. Other exhaust systems mount the oxygen sensor so inaccessibly you may still struggle to remove it with this special tool.

This special slotted socket will service all ⅞ in. (22 mm) hex oxygen sensors. Use with ½ in.-square drive tools or a ⅞ in. (22 mm) wrench. This socket can also be used on many vacuum switches.

### Sending Unit Sockets

Many sending unit sockets are similar to oxygen sensor sockets in that they provide a means of removing or installing the sensor without having to cut the unit's signal wire. Also, some sending units have a unique pattern and require a special shaped socket. One example of this is the oil pressure sending units used by some manufacturers. These require a special socket **(Figure 12)**.

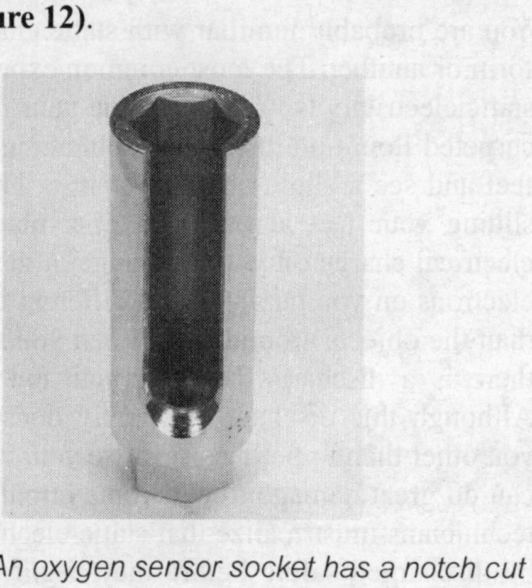

**Figure 11** *An oxygen sensor socket has a notch cut into it for the sensor's wires.*

**Figure 12** *Many sensors require the use of special tools. This tool is used to replace the oil pressure sending unit on many vehicles.*

### Pinch-Off Pliers

Pinch-off pliers are designed to pinch off hoses without damaging or cutting them **(Figure 13)**. This is useful when checking vacuum leaks, fuel leaks, and fuel pressure relief valves or having to prevent fluid flow while disconnecting a hose. The pliers squeeze the hose shut while the jaws swivel for parallel clamping. This applies an even force to the hose so it will not be damaged. Most pinch-off pliers include an automatic locking ratchet mechanism.

> **You Should Know**
>
> *Do not use pinch-off pliers on wire-reinforced hose or plastic lines.*

### Spark Plug Thread Tap

It may be necessary to clean or repair the spark plug threads in the cylinder head prior to installing new spark plugs. If the threads can be repaired, a spark plug thread tap can be used to "clean up" the threads.

**Figure 13** *Pinch-off pliers allows the technician to close off vacuum and fuel lines without damaging them.*

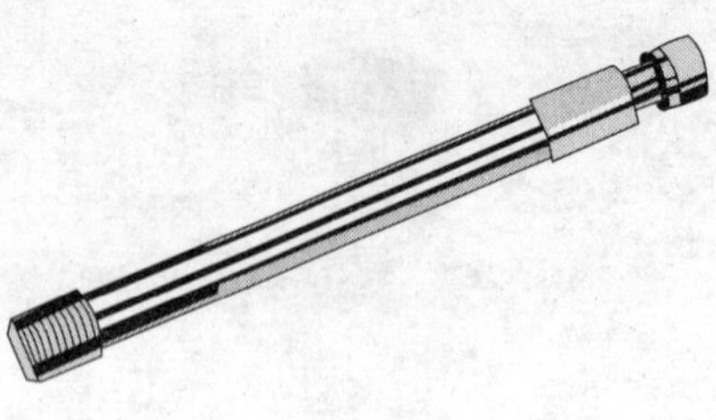

**Figure 14** *One style of spark plug thread tap. The tool is expanded out to the correct thread size.*

This special tap –allows repair to the spark plug threads without pulling the cylinder head. One design of this tool **(Figure 14)** uses an adjusting knob to expand the thread chaser to exactly 14 mm (most common size of spark plug threads). Debris and filings cling to the tool instead of falling into the cylinder head. This tool eliminates the chance of cross threading since it starts the rethreading at the "good" threads at the bottom of the spark plug port.

Another style has both 14 mm and 18 mm threads with a long neck to allow access to recessed spark plug cavities **(Figure 15)**. This tool is used with either a ⅜-in. drive tool or ⅝-in. socket.

## Spark Tester

With today's complex computer control systems used for the ignition system, a **spark tester** is an essential tool to prevent damage to the electronics while testing for secondary ignition voltage. The spark tester provides a quick way to check no-start cars for ignition problems. Two types of spark testers are commonly available. The neon bulb type **(Figure 16)** is connected between the spark plug wire and the spark plug. The tester flashes with each ignition pulse.

One style of tool **(Figure 17)** is used by simply unplugging the plug wire, connecting the tester to the spark plug cable, and clamping the tester to a

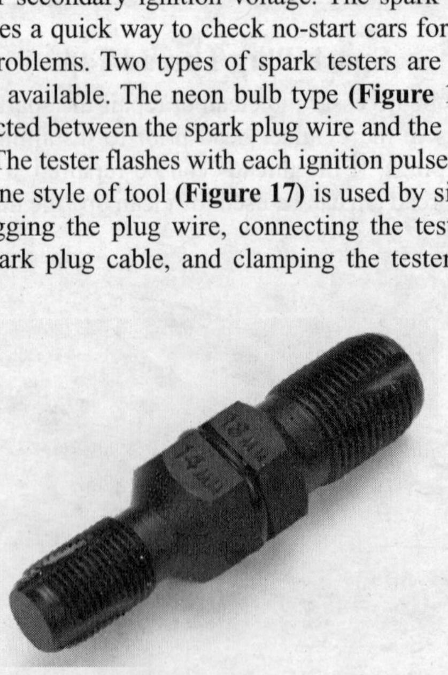

**Figure 15** *This thread chaser has both common-size spark plug threads on one tool.*

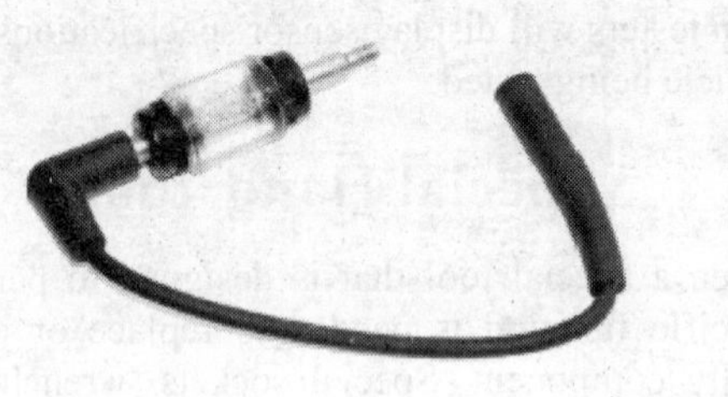

**Figure 16** *Spark tester with a neon light that will flash, indicating electrical current is flowing through the circuit.*

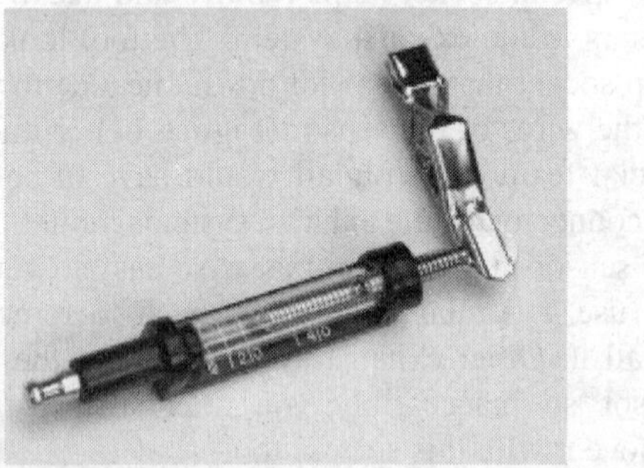

**Figure 17** *A simple spark tester.*

good engine ground. This type of tester requires that the air gap be adjusted to match the spark plug gap required for the engine. While cranking the engine, watch for spark. A properly working ignition system will be able to jump a spark across the tester's spark gap. Two testers are recommended for checking spark on a direct ignition system (DIS). A good DIS will fire two testers at once. By using two, you can check individual coil packs.

## Static Strap

You are probably familiar with static charges in one form or another. The most common experience with static electricity is when you slide your feet across a carpeted floor and then touch something. You might feel and see a slight static discharge. The action of sliding your feet across the carpet places a slight electrical charge on you. A change in the number of electrons on you puts you at a different charge level than the objects around you. When you touch them, there is a discharge between you and the object. Although this discharge generally does nothing to you other than to perhaps surprise you, it potentially can do great damage to electronic circuitry. Today's technicians must realize that static electricity has to be discharged safely before they begin working on an electronic component or processor.

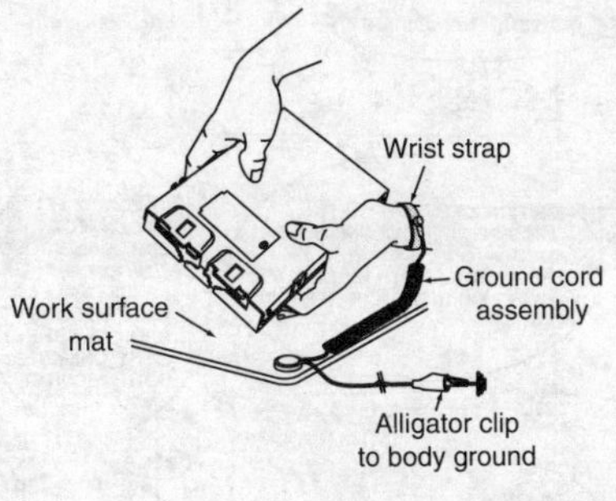


**Figure 18** *Static straps are needed to protect electronic devices.*

To effectively work on these circuits, some precautions are necessary. Generally these can be summarized by this statement. You must be at the same electrical potential as the component you are working on, and the vehicle you are working in. Many manufacturers suggest the use of static straps that connect the technician, the component, and the ground system of the vehicle together **(Figure 18)**. The theory behind this is to place all things that will touch at the same electrical potential so that a discharge will not take place. Even if you do not have all the special static straps, run jumper ground wires between the components and the vehicle and ground yourself to the vehicle before you begin working.

## Service Information

With today's complex fuel and emission systems, it would be impossible to repair every customer concern brought into the service bay without the proper service information. The **service manual** (either in paper form or electronic format) is one of the most valuable tools for today's technician. It provides information concerning engine identification, service procedures, specifications, and diagnostic data. In addition, the service manual provides information on wiring harness connections and routing, component location, and fluid capacities. Service manuals can be supplied by the vehicle manufacturer or through aftermarket suppliers.

It is also important to stay current with updated service information. This information is usually published as **Technical Service Bulletins (TSBs)**. These documents may provide information concerning fixes for a problem, new part numbers to replace a defective unit, corrections to service manual information, and general information about system operation **(Figure 19)**.

**ARTICLE BEGINNING**

TECHNICAL SERVICE INFORMATION

FALSE DTC P0121 (REPROGRAM PCM)

Model(s): 1997 Buick Century, Skylark
1997 Chevrolet Lumina Monte Carlo Malibu, Venture
1997 Oldsmobile Achieva, Cutlass, Cutlass Supreme, Silhouette
1997 Pontiac Grand Am, Grand Prix, Trans Sport with 3100/3400 V6 Engine (VINs M, E - RPOs L82, LA1)

Section: 6E - Engine Fuel & Emission
Bulletin No.: 77-65-14A
Date: May, 1998

NOTE: This bulletin is being revised to add additional models and calibration numbers. Please discard Corporate Bulletin 77-65-14 (Section 6E - Engine Fuel & Emission).

**CONDITION**

Some owners may experience a MIL (Malfunction Indicator Lamp) light illuminated on the vehicle's instrument panel. Additionally, the engine's normal controlled idle speed may be slightly elevated when the MIL is illuminated.

**CAUSE**

The current DTC (Diagnostic Trouble Code) P0121 is too sensitive. The rational check that the diagnostic calibration performs has been changed. Part of those changes involve eliminating the defaulted higher idle.

**CORRECTION**

Check the calibration identification number utilizing a scan tool device. Re-flash with the updated calibration if the current calibration is not one listed in this bulletin. If the vehicle already has the most recent calibration, then refer to the appropriate service repair manual to diagnose and repair for DTC P0121. Test drive the vehicle after repair to ensure that the condition has been corrected. The new calibrations are available from the GM Service Technology Group starting with CD number 6 for 1998.

IMPORTANT: Do not attempt to order the calibrations from GMSPO. The calibrations are programmed into the vehicles PCM via a Techline Tool device.

**Figure 19** *Technical Service Bulletins provide information concerning repairs of known problems.*

To obtain the correct information, you must be able to identify the engine you are working on. This may involve using the vehicle identification number (VIN). This number has a code for model year and engine. Which numbers are used varies among manufacturers, but the service manual will provide instructions for proper VIN usage. The service manual may also assist you in engine identification through the interpretation of casting numbers and marks on the block or cylinder head.

With the engine properly identified, you can retrieve the required information. In the past, each manufacturer and manual publisher used its own method of organizing its manuals. Recent guidelines now require manufacturer service manuals to have a standard organization **(Figure 20)**.

Procedural information provides the steps necessary to perform the task. Most service manuals include illustrations to guide the technician through the task **(Figure 21)**. To get the most out of the service manual, you must use the correct manual for the vehicle and system being worked on, and follow each step in order.

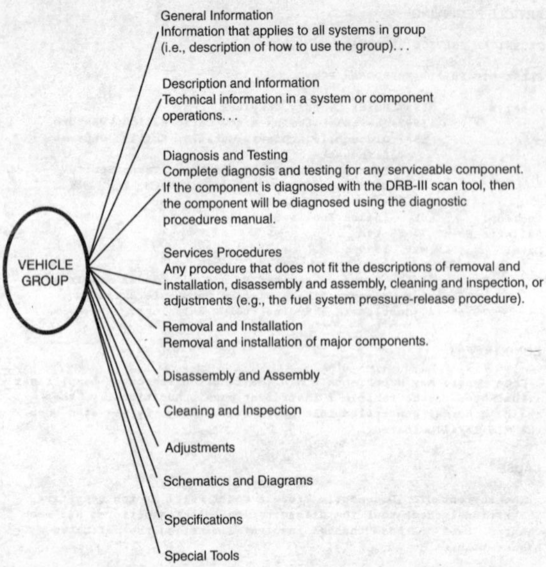


**Figure 20** *Service manuals use a uniform layout.*

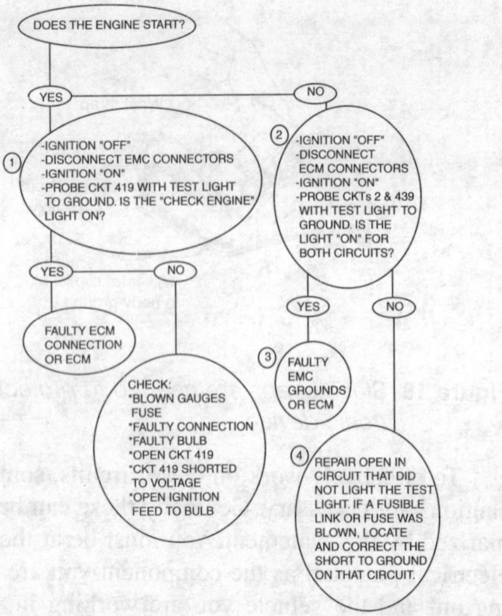


**Figure 22** *A diagnostic tree leads the technician to the cause of the fault.*

Some technicians lead themselves down the wrong trail by making assumptions and skipping steps.

Torque, end play, and clearance specifications may be located within the text of the procedural information. In addition, specifications may be provided in a series of tables. The heading above the table provides a quick reference to the type of specification information included.

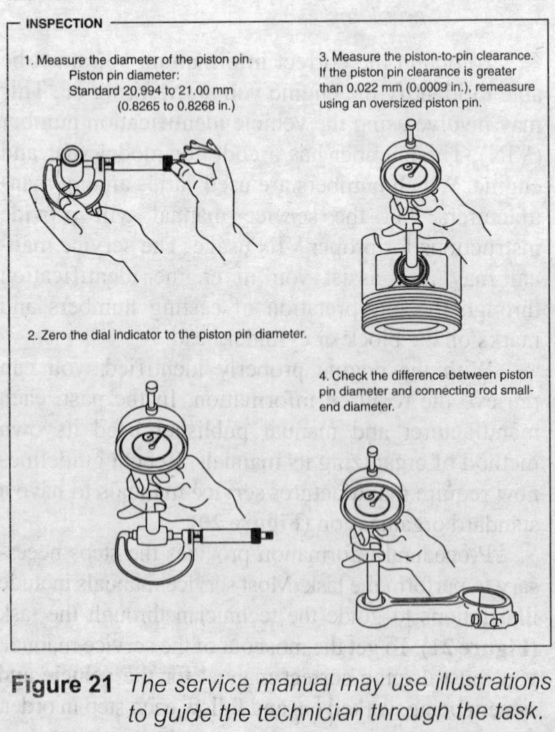


**Figure 21** *The service manual may use illustrations to guide the technician through the task.*

Diagnostic procedures are often presented in a chart form or a tree **(Figure 22)**. The tree guides you through the process as system tests are performed. The result of a test then directs you to a branch. Keep following the steps until the problem is isolated.

Since the service manual is divided into several major component areas, a table of contents is provided for easy access to the information. Each component area of the vehicle is covered under a section in the service manual. Using the table of contents identifies the section to turn to. Once in the appropriate section, a smaller, more specific table of contents will direct you to the page on which the information is located. Due to the extensive amount of information provided in service manuals and the number of component areas, the manual may be divided into several volumes.

Service and parts information is also provided through computer services **(Figure 23)**. This is a popular method since libraries of service manuals can take large areas of space. Computerized systems can have the information stored on disks, the computer can be connected to a central database, or the shop may subscribe to a Web-based supplier. The computer system helps the technician find the required information quicker and easier than with a book-type manual. Using the computer keyboard, light pen, touch-sensitive screen, or mouse, the technician makes choices from a series of menus on the monitor. If needed, the information can be printed to paper.

**Figure 23** *Computerized service information retrieval systems are replacing paper service manuals.*

Another tool that can be of assistance to the technician is hotline services. There are many companies and organizations that provide online assistance to technicians. The hotline assistants use database information along with factory service manuals, TSBs, and other sources to help the technician repair the vehicle.

Also, some companies and organizations provide the service of tracking the service history of vehicles. Manufacturers will generally do this for any of their vehicles that is serviced in one of their dealerships. This information lets the technician know if service that may be related to the present problem has been performed on the vehicle. Also, some Web sites will track if the vehicle has ever been "totaled" in an accident, stolen, or caught in a flood.

## Review Questions

1. Special fuel and emission tools are being discussed. Technician A says that an injector balance tester is used to measure the mass of air that is possible to enter the combustion chamber. Technician B says that a pressure gauge is used to measure the pressure in the intake manifold with the engine running. Who is correct?
   A. Technician A only
   B. Technician B only
   C. Both Technician A and Technician B
   D. Neither Technician A nor Technician B
2. Technician A says that an engine analyzer can be used to help determine the condition of the engine. Technician B says that an engine analyzer may be computer controlled and provide guided tests. Who is correct?
   A. Technician A only
   B. Technician B only
   C. Both Technician A and Technician B
   D. Neither Technician A nor Technician B
3. A ____________ ____________ is commonly used to test the integrity of the vehicle's evaporation control system.
4. Technician A says that by viewing the quality of an engine's exhaust, a technician is able to look at the effects of the combustion process. Technician B says that the levels of HC and CO in the exhaust are a direct indication of engine performance. Who is correct?
   A. Technician A only
   B. Technician B only
   C. Both Technician A and Technician B
   D. Neither Technician A nor Technician B
5. Technician A says that a spark tester is a spark plug with its center electrode removed. Technician B says that the spark tester provides a quick way to check no-start cars for ignition problems. Who is correct?
   A. Technician A only
   B. Technician B only
   C. Both Technician A and Technician B
   D. Neither Technician A nor Technician B
6. What is the purpose of the flow meter?
7. Describe the function of the injector balance tester.
8. Explain the use of the evaporative emissions leak detector (EELD).
9. What is the purpose of the scan tool?
10. Why is a static strap recommended?

CHAPTER

# 3 Electrical Test Equipment

## Introduction

Many failures associated with fuel and emission systems are within the electrical portion of the systems. In the course of diagnosing electrical failures, the technician will be required to use special equipment such as digital multimeters (DMMs) and lab scopes.

## Test Lights

A **test light** is used when the technician needs to "look" for electrical power in the circuit. The test light allows the technician to see if current is at a point in the circuit by lighting the light. There are two types of test lights commonly used in diagnosing electrical problems: nonpowered and self-powered.

A typical nonpowered test light has a transparent handle that contains a light bulb. A sharp probe extends from one end of the handle, while a ground wire with a clamp extends from the other end **(Figure 1)**. If the circuit is operating properly, clamping the lead of the test light to ground and probing the circuit at a point of voltage should light the lamp **(Figure 2)**.

A test light is limited because it does not display how much voltage is at the point of the circuit being tested. However, by understanding the effects of voltage drop, the technician will be able to interpret the brightness of the test light and relate the results to the expectations of a good circuit. If the lamp is connected after a voltage drop, the lamp will light dimly. Connecting the test lamp before the voltage drop should light the lamp brightly. The light should not illuminate at all if it is probing for voltage after the last resistance.

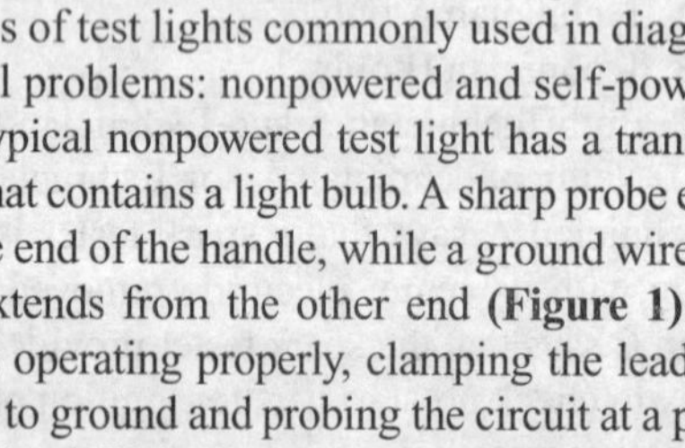

**Figure 1** *A typical test light used to probe for voltage in a circuit.*

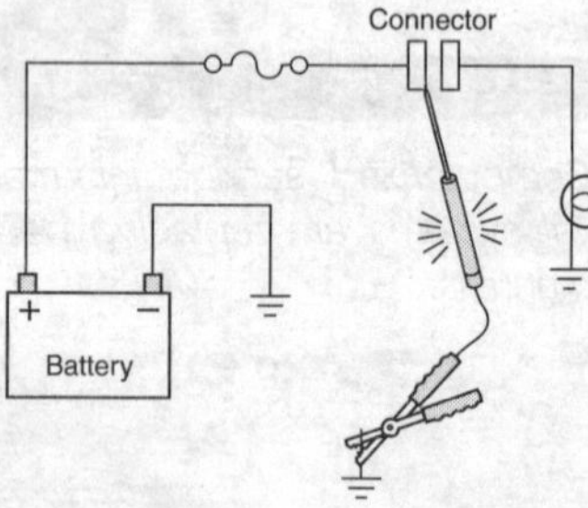


**Figure 2** *If voltage is present, the test light will illuminate.*

> **You Should Know**
>
> *Do not use a test light to probe for power in an electronic circuit. The increased current draw of the test light may damage the system components.*

Another type of circuit tester is the self-powered continuity tester **(Figure 3)**. The continuity tester has an internal battery that powers a light bulb. With the power in the circuit turned off or disconnected, the ground clip is connected to the ground terminal of the load component. By probing the feed wire, the light will illuminate if the circuit is complete (has continuity). If there is an opening in the circuit, the lamp will not illuminate.

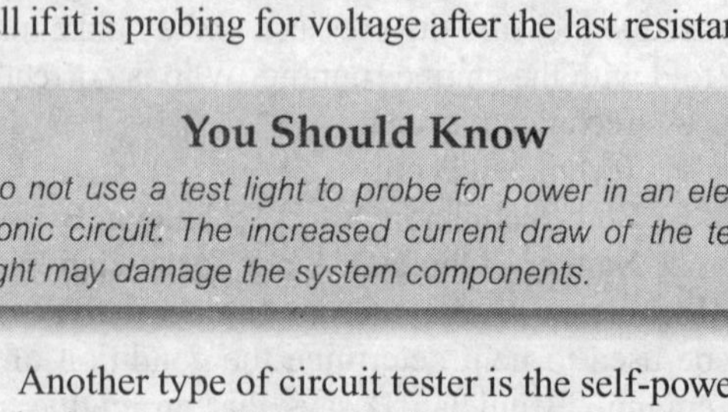

**Figure 3** *Self-powered continuity tester.*

### You Should Know

*Never use a self-powered test light to test the airbag system. The battery in the tester can cause the airbag to deploy.*

### You Should Know

*Do not connect a self-powered test light to a circuit that is powered. Doing so will damage the test light.*

## Injector Circuit Test Light

A special test light called a **noid light** can be used to determine if a fuel injector is receiving its proper voltage pulse from the computer. The wiring harness connector is disconnected from the injector and the noid light is plugged into the connector **(Figure 4)**. After disabling the ignition to prevent starting, the engine is turned over by the starter motor. The noid light will flash rapidly if the voltage signal is present. No flash usually indicates an opening in the power feed or ground circuit to the injector.

It is important to remember that just because the noid light flashes, it does not mean the circuit is at peak performance. The flashing light only indicates there is enough voltage present to light the bulb. This may not be enough to actually energize the injector.

## Logic Probes

Many computer controlled systems use a pulsed voltage to transmit messages or to operate a component. A standard or self-powered test light should not be used to test these circuits since it may damage the computer. However, a **logic probe (Figure 5)** can be used. A logic probe looks something like a test light

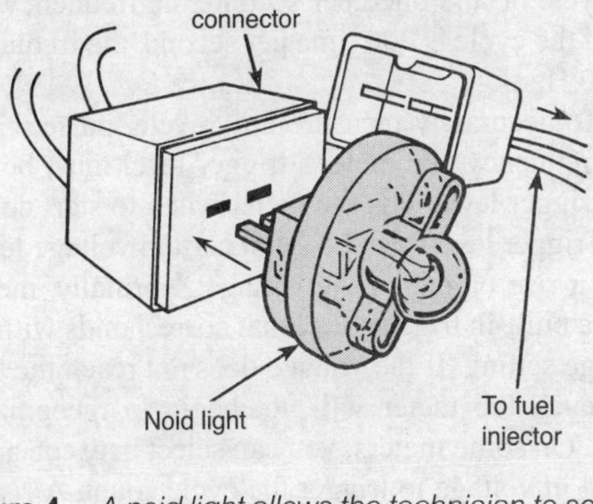


**Figure 4** *A noid light allows the technician to see the voltage pulses at the injector connector.*

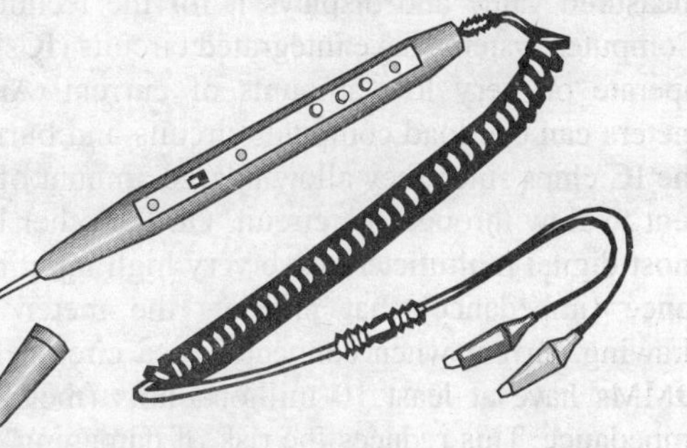

**Figure 5** *A typical logic probe.*

except it contains three different colored LEDs. The red LED will light if there is high voltage at the point in the circuit being tested. The green LED will light to indicate the presence of low voltage. The yellow LED indicates the presence of voltage pulses. If the voltage is a pulsed voltage from a high level to a low level, the yellow LED will be on and the red and green LEDs will cycle, indicating the change in voltage.

## Multimeters

The **multimeter** is one of the most versatile tools used to diagnose electrical systems. It can be used to measure voltage, resistance, and amperage. In addition, some types of multimeters are designed to test diodes and measure frequency, duty cycle, temperature, and rotation speed. Multimeters are available in analog (swing needle) and digital display.

### Digital Meters

With modern vehicles incorporating computer-controlled systems, **digital multimeters (DMMs)** are required **(Figure 6)**. Digital multimeters (also called DVOMs) display values using liquid crystal displays (LCDs) instead of a swinging needle. A DMM is basically a computer that determines the

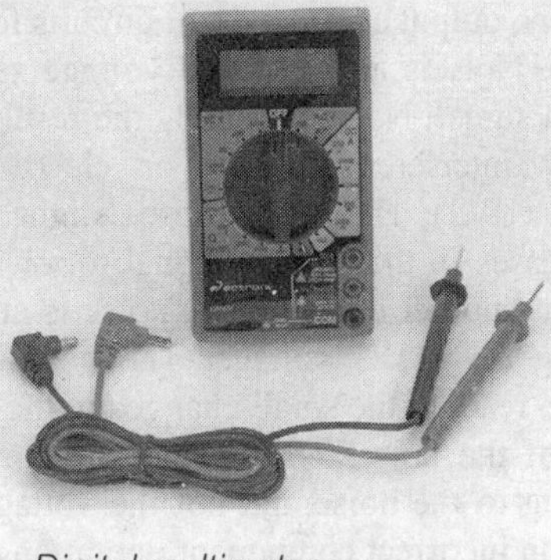

**Figure 6** *Digital multimeter.*

measured value and displays it for the technician. Computer systems have integrated circuits (ICs) that operate on very low amounts of current. Analog meters can overload computer circuits and burn out the IC chips since they allow a large amount of current to flow through the circuit. On the other hand, most digital multimeters have very high input resistance (impedance) that prevents the meter from drawing current when connected to a circuit. Most DMMs have at least 10 million ohms (megohms) impedance. This reduces the risk of damaging computer circuits and components.

**You Should Know**

*Not all DMMs are rated at 10 megohms of impedance. Be sure of the meter you are using to prevent electronic component damage.*

Digital meters rely on electronic circuitry to measure electrical values. The measurements are displayed with LEDs or on an LCD. Digital meters tend to give more accurate readings and are certainly much easier to read. Rather than reading a scale at the point where the needle lines up, digital meters simply display the measurement in a numerical value. This also eliminates the almost certain error caused by vviewing an analog meter at an angle.

## Additional DVOM Functions

Some multimeters feature additional functions besides measuring alternating current (AC) or direct current (DC) voltage, amperage, and ohms. Many multimeters can also measure engine rpm, ignition dwell, diode condition, distributor condition, frequency, and temperature.

Some meters have a min/max function that displays the minimum, maximum, and average values received by the meter during the time the test was being recorded. This feature is valuable when checking sensors, output commands, or circuits for electrical **noise**. Noiseis an unwanted voltage signal that rides on a signal. Noise is usually the result of radio frequency interference (RFI) or electromagnetic induction (EMI). The noise causes slight increases and decreases in the voltage signal to, or from, the computer. Another definition of noise is an AC signal riding on a DC voltage. The computer may attempt to react to the small changes in the signal as a result of the noise. This means the computer is responding to the noise and not the voltage signal, resulting in incorrect component operation.

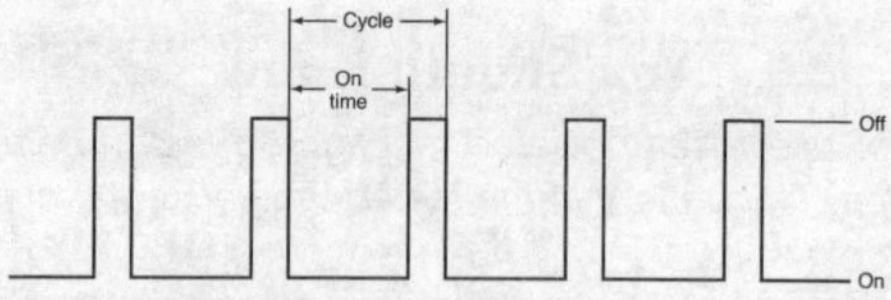


**Figure 7** *Duty cycle.*

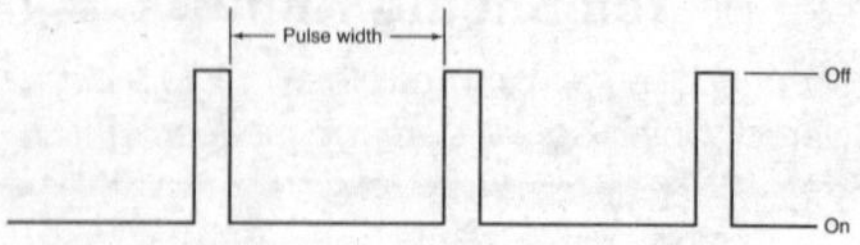


**Figure 8** *Pulse width.*

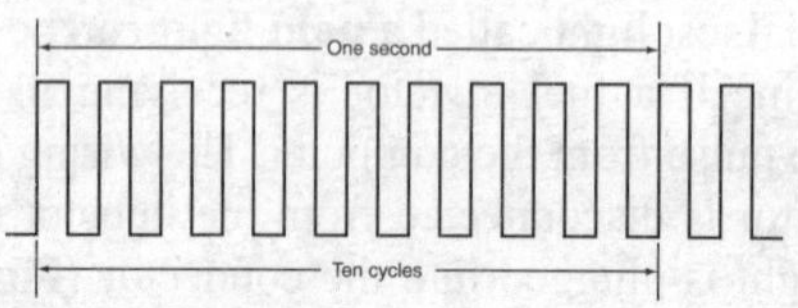


**Figure 9** *Frequency signal.*

Also, some multimeters may have the capabilities to measure **duty cycle**, pulse width, and frequency. Duty cycle is the measurement of the amount of "on" time as compared with the time of a **cycle** **(Figure 7)**. The duty cycle is displayed in a percentage. For example, a 60 percent duty cycle means that during one cycle the device is on 60 percent of the time and off 40 percent of the time.

**Pulse width** is similar to duty cycle except that it is a measurement of time the device is turned on within a cycle **(Figure 8)**. Pulse width is usually measured in milliseconds.

As previously mentioned, some multimeters can measure **frequency (Figure 9)**. Frequency is a measure of the number of cycles that occurs in 1 second. The higher the frequency, the more cycles that occur in a second. Frequency is measured in **hertz** (Hz). If the cycle occurs once per second, the frequency is 1 Hz. If the cycle is 300 time per second, the frequency is 300 Hz.

To accurately measure duty cycle, pulse width, and frequency, the meter's trigger level must be set. The trigger level tells the meter when to start counting. Trigger levels can be set at certain voltage levels or at a rise or fall in the voltage. Normally, meters have a built-in trigger level that corresponds with the voltage setting. If the voltage does not reach the trigger level, the meter will not begin to recognize a cycle. On some meters, you can select between a rise or fall in voltage to trigger the cycle count. A rise in voltage is a positive increase in voltage. This setting is used to monitor the activity of devices whose power feed is controlled by a computer. A fall in voltage is negative voltage. This setting is used to monitor ground-controlled devices.

## Lab Scopes

A **lab scope** is a visual voltmeter **(Figure 10)**. For many years, technicians have used scopes to diagnose ignition, fuel injection, and charging systems. These scopes, called "tune-up scopes," were normally part of a large diagnostic machine, although some were stand-alone units. In recent years, the lab scope has become the diagnostic tool of choice for many good technicians.

The lab scope is very useful in diagnosing many electrical problems quickly and accurately. Digital and analog voltmeters do not react fast enough to read systems that cycle quickly. The lab scope may be considered a very fast reacting voltmeter that reads and displays voltages. The scope allows the technician to view voltage over time **(Figure 11)**. These voltage readings appear as a voltage trace on the lab scope's screen.

An upward movement of the voltage trace on the lab scope screen indicates an increase in voltage, and a downward movement of this trace represents a decrease in voltage. As the voltage trace moves across the screen, it represents a specific length of time. These divisions set up a grid pattern on the screen. Time is represented by the horizontal movement of the waveform. Voltage is measured with the vertical position of the waveform. Since the scope displays voltage over time, the waveform moves from left (the beginning of measured time) to right (the end of measured time). The value of the divisions can be adjusted to improve the view of the voltage waveform. For example, the vertical scale can be adjusted so that each division represents 0.5 volt (V), and the horizontal scale can be adjusted so that each division equals 0.005 (5 milliseconds, or ms). This allows the technician to view small changes in voltage that occur in a very short period of time. The grid serves as a reference for measurements.

Since a scope displays actual voltage, it will display any electrical noise or disturbances that accompany the voltage signal **(Figure 12)**. This noise can cause intermittent problems with unpredictable results. When a computer receives a voltage signal with noise, it will try to react to the minute changes.

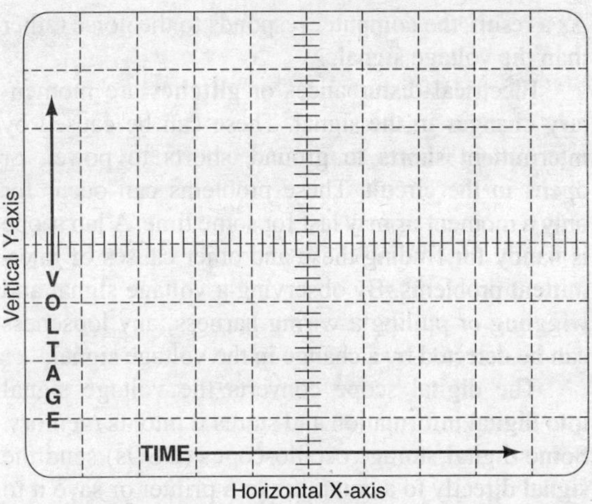


**Figure 11** *Grids on a scope screen serve as a time and voltage reference.*

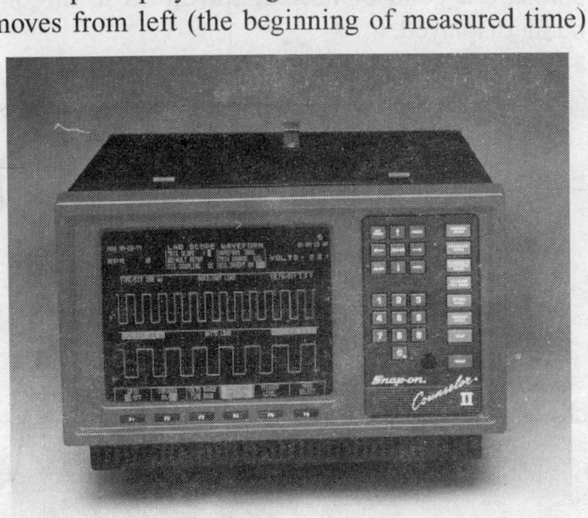


**Figure 10** *Lab scope.*

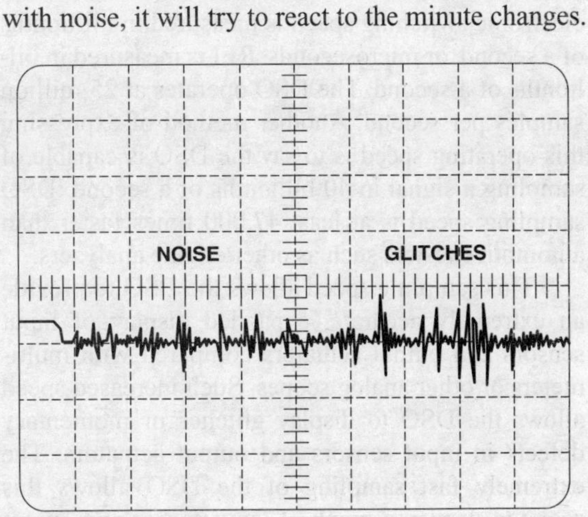


**Figure 12** *RFI noise and glitches showing as voltage signals.*

As a result, the computer responds to the noise rather than the voltage signal.

Electrical disturbances or **glitches** are momentary changes in the signal. These can be caused by intermittent shorts to ground, shorts to power, or opens in the circuit. These problems can occur for only a moment or may last for some time. A lab scope is handy for finding these and other causes of intermittent problems. By observing a voltage signal and wiggling or pulling a wiring harness, any looseness can be detected by a change in the voltage signal.

The digital scope converts the voltage signal into digital information and stores it into its memory. Some digital storage oscilloscopes (DSOs) send the signal directly to a computer or a printer or save it to a disk. To help with diagnosis, a technician can "freeze" the **captured signal** for closer analysis. A captured signal is a signal that is stored in a DSO's memory. DSOs also have the ability to capture low-frequency signals. Low-frequency signals tend to flicker when displayed on an analog screen. To display a clean waveform on an analog scope, the signal must be repetitive and occur in real time. The signal on a DSO is not quite real time. Rather, it displays the signal as it occurred a short time before.

This delay is actually very slight. Most DSOs have a sampling rate of 1 million samples per second. This is quick enough to serve as an excellent diagnostic tool. This fast sampling rate allows slight changes in voltage to be observed.

The greatest advantage of a DSO is the speed at which it samples electrical signals. Mechanical switching speed of switches and relays is measured in thousandths of a second, or milliseconds. Electrical/electronic switching speed is measured in millionths of a second, or microseconds. RFI is measured in billionths of a second. The DSO operates at 25 million samples per second. Another method of expressing this operating speed is to say the DSO is capable of sampling a signal in 40 billionths of a second. DSO sampling speed is at least 47,000 times faster than automotive testers such as other engine analyzers.

This sampling speed allows the DSO to provide an extremely accurate, expanded display of input sensors and output actuators compared with multimeters or other analog scopes. Such increased speed allows the DSO to display glitches or momentary defects in input sensors and output actuators. The extremely fast sampling of the DSO allows this scope to display a graph of input sensor and output actuator operation. Some DSOs have the capability to display two voltage traces across the screen

**Figure 13** *Dual-trace scope.*

**(Figure 13)**. Other DSOs, such as the SimuTech, display six voltage traces simultaneously.

A DSO uses an analog-to-digital (A/D) converter to digitize the input signal. Since digital signals are based on binary numbers, the trace appears slightly choppy when compared with an analog trace. But the voltage signal is sampled more often, which results in a more accurate waveform. The waveform is constantly being refreshed as the signal is pulled from the scope's memory. Remember that the sampling rate of a DSO can be as high as a million times per second.

The lab scope can be designed as dual trace scopes **(Figure 14)**. This means they both have the capability of displaying two traces at one time. By watching two traces simultaneously, you can watch the cause and effect of a sensor as well as compare a good or normal waveform to the one being displayed.

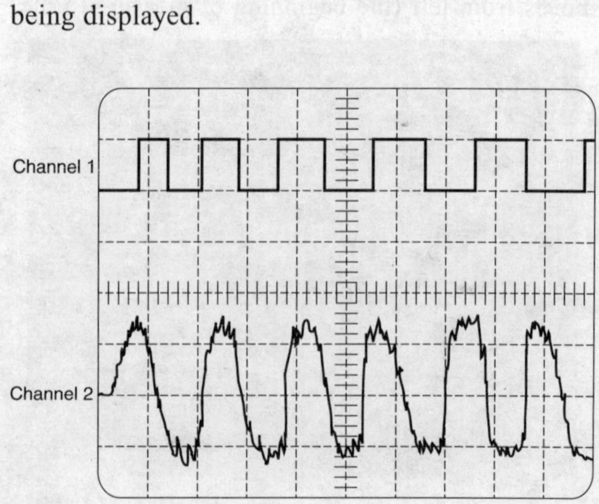


**Figure 14** *A dual-trace scope can show two patterns at the same time.*

## Review Questions

1. Technician A says that a 10 percent duty cycle indicates that the load device is turned on most of the time. Technician B says that the pulse width is measured in degrees. Who is correct?
   A. Technician A only
   B. Technician B only
   C. Both Technician A and Technician B
   D. Neither Technician A nor Technician B
2. Technician A says that test light is ideal for checking for voltage on a low-current, low-power circuit. Technician B says that to use an analog multimeter to test computer-controlled circuits. Who is correct?
   A. Technician A only
   B. Technician B only
   C. Both Technician A and Technician B
   D. Neither Technician A nor Technician B
3. Technician A says that most DMMs have low impedance. Technician B says that low impedance is required to prevent damage to low-voltage–low-current circuits. Who is correct?
   A. Technician A only
   B. Technician B only
   C. Both Technician A and Technician B
   D. Neither Technician A nor Technician B
4. What is the relationship between frequency and hertz?
5. Technician A says that the upward voltage traces on a lab scope screen indicate a specific length of time. Technician B says that a lab scope operates like a very fast reacting voltmeter. Who is correct?
   A. Technician A only
   B. Technician B only
   C. Both Technician A and Technician B
   D. Neither Technician A nor Technician B
6. Explain the purpose of a noid light.
7. Define electrical noise.
8. Define electrical glitches.

CHAPTER

# 4 Working as a Fuels and Emissions Technician

## Introduction

To be a successful automotive technician, you need to have good training, a desire to succeed, and be committed to becoming a good technician and a good employee. A good employee works well with others and strives to make the business successful. The required training is not just in the automotive field. Technicians need to have good reading, writing, and math skills. These skills will allow you to better understand and use the material found in service manuals and textbooks, as well as provide you with the basics for good communications with customers and others.

## Compensation

Technicians are typically paid according to their abilities. Most often, new or apprentice technicians are paid by the hour. While being paid they are learning the trade and the business. Time is usually spent working with a master technician or doing low-skilled jobs. As an apprentice learns more, he or she can earn more and take on more complex jobs. Once technicians have demonstrated a satisfactory level of skills, they can go on **flat rate**.

Flat rate is a pay system in which technicians are paid for the amount of work they do. Each job has a flat-rate time. Pay is based on that time, regardless of how long it takes to complete the job. To explain how this system works, consider a technician who is paid $15.00 per flat-rate hour. If a job has a flat-rate time of 3 hours, the technician will be paid $45.00 for the job, regardless of how long it takes to complete it. Experienced technicians beat the flat-rate time nearly all of the time. Their weekly pay is based on the time "turned in," not on the time spent. If the technician turns in 60 hours of work in a 40-hour workweek, he or she actually earned $22.50 for each hour worked. However, if he or she turned in only 30 hours of work in the 40-hour week, the hourly pay is $11.25.

The flat-rate system favors good technicians who work in a shop that has a large volume of work. The use of flat-rate times allows for more accurate repair estimates to the customers. It also rewards skilled and productive technicians.

## Workplace Relationships

When you begin a job, you enter into a business agreement with your employer. When you become an employee, you sell your time, skills, and efforts. In return, your employer pays you for these resources. As part of the employment agreement, your employer also has certain responsibilities:

- **Instruction and supervision.** You should be told what is expected of.you. A supervisor should observe your work and tell you if it is satisfactory and offer ways to improve your performance.
- **A clean, safe place to work.** An employer should provide a clean and safe work area as well as a place for personal cleanup.
- **Wages.** You should know how much you are to be paid, what your pay will be based on, and when you will be paid before accepting a job.
- **Fringe benefits.** When hired, you should be told what benefits to expect, such as paid vacations and employer contributions to health insurance and retirement plans.
- **Opportunity.** You should be given a chance to succeed and possibly advance within the company.
- **Fair treatment.** All employees should be treated equally, without prejudice or favoritism.

On the other side of this business transaction, employees have responsibilities to their employers. Your obligations as an employee to the employer include the following:

- **Regular Attendance.** A good employee is reliable.
- **Following Directions.** As an employee, you are part of a team, and doing things your way may not serve the best interests of the company.

- **Responsibility.** You must be willing to answer for your behavior and work. You need to also realize that you are legally responsible for the work you do.
- **Productivity.** Remember, you are paid for your time as well as your skills and effort.
- **Loyalty.** Loyalty is expected, and by being loyal you will act in the best interests of your employer, both on and off the job.

## Customer Relations

Another responsibility you have is good customer relations. Learn to listen and communicate clearly. Be polite and organized, particularly when dealing with customers. Always be as honest as you possibly can.

Look and present yourself as a professional, which is what automotive technicians are. Professionals are proud of what they do, and they show it. Always dress and act appropriately, and watch your language—even when you think no one is near.

Respect the vehicles you work on. They are important to the lives of your customers. Always return the vehicle to the owner in a clean, undamaged condition. Remember, a vehicle is the second largest expense a customer has. Treat it that way. It doesn't matter if you like the car. It belongs to the customer; treat it respectfully.

Explain the repair process to the customer in understandable terms. Whenever you are explaining something to a customer, make sure you do this in a simple way without making the customer feel stupid. Always show the customers respect and be courteous to them. Not only is this the right thing to do, but it also leads to loyal customers.

### Communicating With the Customer

Depending on the size of the service center that you are working in, you may or may not talk directly with the customer. A service advisor or manager might be in between the consumer and you, the technician. In either case, getting the correct information from the customer cannot be overemphasized. The customer is likely the person who was driving the vehicle when the problem occurred. The conversation that someone has with the customer can be extremely useful and save hours of fruitless work. A repair ticket that states "driveability problem" will require the technician to figure out the driving conditions that were present when the problem occurred. It is possible that the technician will not be able to duplicate the conditions and observe the problem. If the vehicle is returned to the consumer with the note of "no problem found," think about how frustrating this could be, especially if the customer experiences the problem the very next day. The conversation with the customer should reveal important information for the technician. When did the problem occur? What specifically did the vehicle do or not do? What was the outside temperature? Was the engine warm or cold? What driving conditions produced the problem? Can you duplicate the problem with the technician in the vehicle?

Think about how much more information the technician has in the second example. It is likely that less time will be necessary to fix the vehicle because the technician has a starting point. Once you get out in the field, try to develop your communication skills, especially your ability to listen to the customer. The customer's information, if you hear it, will save you countless hours of frustration and no doubt result in a better repair. Better repairs bring customers back to the service center the next time they need service.

## ASE Certification

An obvious sign of your knowledge and abilities, as well as your dedication to the trade, is Automotive Service Excellence (ASE) certification. The National Institute for Automotive Service Excellence has established a voluntary certification program for automotive, heavy-duty truck, auto body repair, and engine machine shop technicians. In addition to these programs, ASE offers individual testing in some specialty areas. This certification system combines voluntary testing with on-the-job experience to confirm that technicians have the skills needed to work on today's vehicles **(Figure 1)**. ASE recognizes two distinct levels of service capability—the automotive technician and the master automotive technician.

To become ASE certified, you must pass one or more tests that stress diagnostic and repair problems. There are several areas that cover the skills associated with fuel and emission. These areas include:

- **Engine Performance (A8).** This test covers the basics of engine performance, including diagnostic procedures involving compression testing, cylinder leakage testing, and general driveability problems.
- **ASE Advanced Engine Performance Specialist Test (L1).** The objective of this test is to identify those technicians who possess knowledge of the skills required to diagnose sophisticated driveability and emissions-related problems. The L1 test consists of questions that test technicians' knowledge of computer-controlled engine systems. In addition, the test measures technicians'

**Figure 1** *ASE's certification combines voluntary testing with on-the-job experience to confirm that technicians have the skills needed to work on today's vehicles.*

abilities at diagnosing the causes of high-emission failures. This is especially important in areas with basic inspection/maintenance (I/M) and enhanced I/M programs. Those registering for the L1 test must be currently certified in the regular Engine Performance (A8) category and satisfy the 2-year hands-on experience requirement.

- **ASE Alternate Fuels Series.** The Compressed Natural Gas Test (F1) consists of fifty multiple-choice questions. Additionally, the test will have up to ten pretest questions. The test was designed by industry "peer" groups with the needs and wants of alternate fuels technicians in mind. The questions are entirely job-related. They are not theoretical. The test covers:
  - Vehicle compatibility analysis
  - Parts fabrication
  - Equipment installation
  - Leak testing and repairs
  - Initial adjustments
  - Engine performance diagnosis
  - System diagnosis and repair
  - Driver orientation

After passing at least one exam and providing proof of 2 years of hands-on work experience, you become ASE certified. Retesting is necessary every 5 years to remain certified. A technician who passes one examination receives an automotive technician shoulder patch. The master automotive technician patch is awarded to technicians who pass all eight of the basic automotive certification exams. You may receive credit for 1 of the 2 years by substituting relevant formal training in one or a combination of the following:

- **High school training.** Three years of training may be substituted for 1 year of experience.
- **Post-high school training.** Two years of post–high school training in a trade school, technical institute, or community college may be counted as 1 year of work experience.
- **Short courses.** For shorter periods of post–high school training, you may substitute 2 months of training for 1 month of work experience.
- **Apprenticeship programs.** You may receive full credit for the experience requirement by satisfactorily completing a 3- or 4-year apprenticeship program.

Each certification test consists of forty to eighty multiple-choice questions. The questions are written by a panel of technical service experts, including domestic and import vehicle manufacturers, repair and test equipment and parts manufacturers, working automotive technicians, and automotive instructors. All questions are pretested and quality checked on a national sample of technicians before they are included in the actual test. Many test questions force the student to choose between two distinct repair or diagnostic methods. The test questions focus on basic technical knowledge, repair knowledge and skill, and testing and diagnostic knowledge and skill.

## Duties of a Fuels and Emissions Technician

As a fuel and emission technician, you will be responsible for servicing, troubleshooting, and repairing all of the components that make up the vehicle's fuel and emission systems. Most of these components require little maintenance other than periodic tune-ups and filter changes. Usually you will get involved with the customer's vehicle when it begins to experience driveability problems or poor fuel economy or when it fails an emission test.

## Diagnostics

The true measure of a good technician is an ability to find and correct the cause of a problem. Service manuals and other information sources will guide you through the diagnosis and repair of problems, but those guidelines will not always lead you to the exact cause of the problem. To do this you must use your knowledge and take a logical approach while troubleshooting. Diagnosis is not guessing, and it is more than following a series of interrelated steps in order to find the solution to a specific problem. Diagnosis is a way of looking at systems that are not functioning properly and finding out why. It is knowing how the system should work and deciding if it is working correctly. Through an understanding of the purpose and operation of the system, you can accurately diagnose problems.

Most good technicians use the same basic diagnostic approach. Simply because this is a logical approach, it can quickly lead to the cause of a problem. Logical diagnosis follows these steps:

1. Gather information about the problem.
2. Verify that the problem exists.
3. Thoroughly define what the problem is, and when it occurs.
4. Research all available information and knowledge to determine the possible causes of the problem.
5. Isolate the problem by testing.
6. Continue testing to pinpoint the cause of the problem.
7. Locate and repair the problem.
8. Verify the repair.

## Use and Interpretation of Vehicle Identification and Calibration Numbers

With the several different types and designs of engines used, today's technician must be informed and able to correctly identify the different engines and systems. Proper engine identification is required for the technician to obtain correct specifications, service procedures, and parts. Parts for the same engine size and same manufacture model year may have several different part numbers for one part, based on the engine emission package. For example, a 2003 5.7 liter engine made by the same manufacturer may have four different part numbers for the EGR valve depending on where the vehicle was originally marketed. Also, when determining emission limits and needed repairs, proper identification is critical.

### Engine Identification

The **vehicle identification number (VIN)** is one source for information concerning the engine. Usually, the VIN is located on a metal tag that is visible through the vehicle's windshield **(Figure 2)**. The VIN provides the technician with a variety of information concerning the vehicle **(Figure 3)**. The eighth digit represents the engine code. The service manual will provide information concerning interpretation of the engine code.

A **vehicle code plate** is also located on the vehicle body, usually inside the driver-side door or under the hood. This plate provides information concerning paint codes, transmission codes, and so forth **(Figure 4)**. The engine sales code is also stamped on the vehicle code plate.

The engine itself may have different forms of identification tags or stamped numbers **(Figure 5)**. In addition, color coding of engine labels may be used to indicate Transitional Low Emission Vehicles (TLEVs) and Flex Fuel Vehicles (FFVs). These numbers may provide build date codes required to order proper parts. Use the service manual to determine the location of any engine identification tags or stamps and for the methods to interpret them. For example: a Jeep engine with the build date code of *401HX23* would identify a 2.5 liter engine with a multipoint fuel injection system, 9:1 compression ratio built on January 23, 1994. This is determined by the first digit representing the year of manufacture (4 = 1994). The second and third digits represent the month of manufacture (01 = January). The fourth and fifth digits represent the engine type, fuel system, and compression ratio (HX = 2.5 liter engine with a multipoint fuel injection system, 9:1 compres-

**Figure 2** *The VIN can be seen through the windshield.*

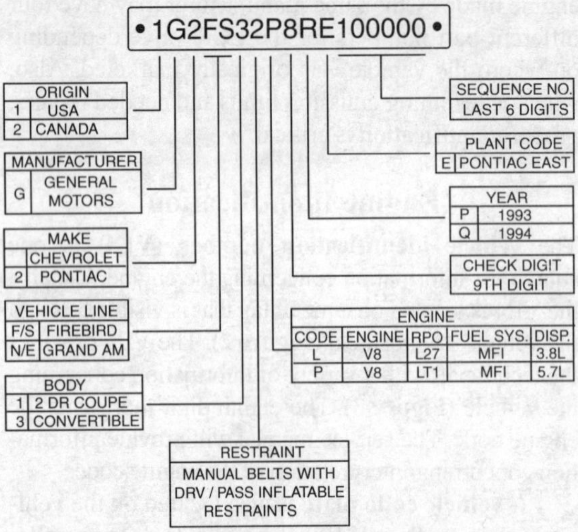


**Figure 3** *The VIN is used to provide required information about the vehicle and engine.*

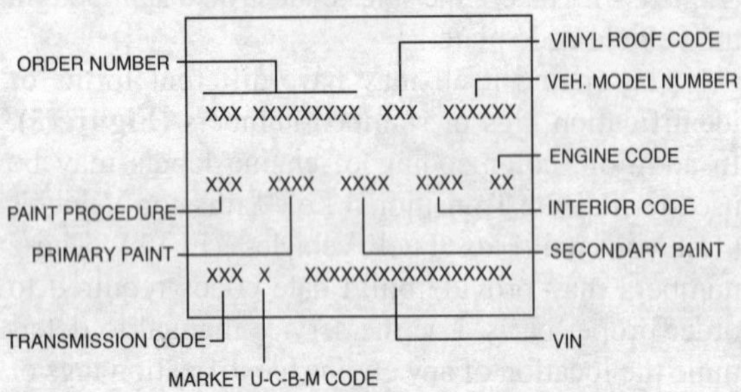


**Figure 4** *Vehicle code plates provide sales code information about the vehicle.*

sion ratio). The sixth and seventh digits represent the day of manufacture (23).

Some engines are manufactured with oversized or undersized components. This occurs due to mass production of the engines. Components that may be different sizes than normal include:

- Cylinder bores
- Camshaft-bearing bores
- Crankshaft main–bearing journals
- Connecting rod journals

Usually engines with oversized or undersized components (from the factory) are identified by a letter code stamped on the engine block **(Figure 6)**. This information is helpful when ordering parts and measuring for wear.

## Vehicle Emission Control Information Label

Every vehicle manufactured after 1979 is required to have a permanent under-the-hood Vehicle Emission

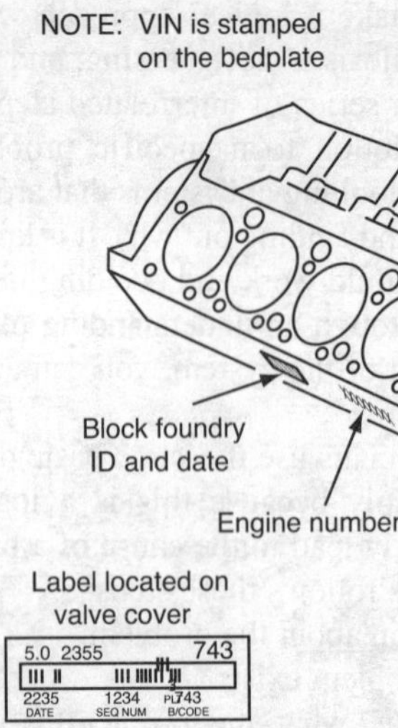


**Figure 5** *The engine may have identification numbers stamped onto it.*

Control Information label. This label contains information concerning engine size (displacement), engine family, and other emission information **(Figure 7)**.

## Emission-Tampering Laws

Emission tampering is a concern for all technicians who work with driveability issues. It is out of the scope of this book to cover every aspect of tampering laws since there are federal, state, city, and local laws pertaining to this issue. Just because your local area may not have strict emission laws does not mean there are no tampering laws. Federal laws regulate what constitutes emissions tampering. Basically, if it involves altering the emission system in any way it is tampering.

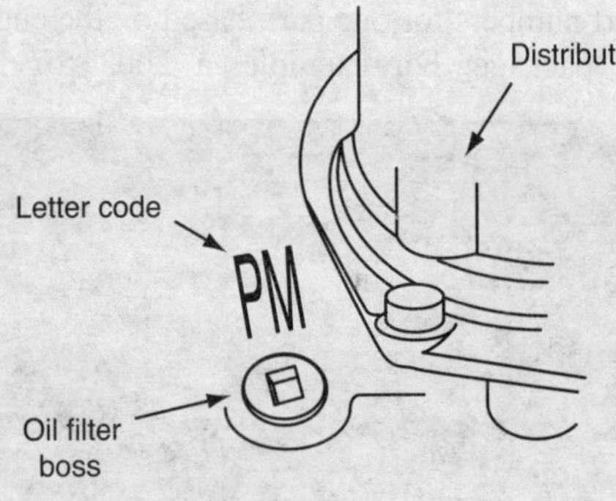


**Figure 6** *Additional markings indicate if components are undersized or oversized.*

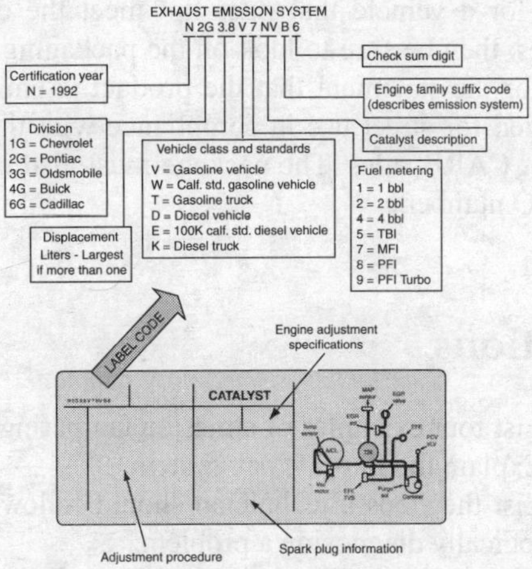


**Figure 7** *Common style Vehicle Emission Control Information label.*

### You Should Know

*Federal law makes it illegal for anyone to tamper with the vehicle's emission system. This includes disconnecting, removing, or disabling any emissions-related control device. The EPA (and most states) have aggressively prosecuted professional service facilities and technicians that have been found guilty of tampering. You may be liable for up to a $2,500 fine.*

Sometimes a service to the vehicle may not appear to be tampering, but in reality it is. For example, a common service is to perform an engine swap by installing a used motor. If the used motor is not installed with all emission-control devices that were installed on the year of the vehicle, then the technician and facility may be prosecuted for emissions tampering.

The Federal emission-tampering law does not include full time off-road type vehicles. This means race cars and show cars that are not also operated on the street. Also, vehicles that were not factory equipped with emission controls (most 1967 and earlier vehicles) are exempt. Examples of these would include antique cars, most classic cars, and many "muscle" cars.

Revisions to the Clean Air Act in 1990 broadened the definition of emissions tampering. Included in these revisions is the requirement that any replacement part that is installed on a nonexempt vehicle that is not a direct original equipment manufacture's (OEM) replacement must be EPA approved and emissions legal.

Although not an inclusive list, any of the following may be considered emissions tampering:

- Removing the catalytic converter
- Removing the EGR valve

### Interesting Fact

*Before the 1990 revision to the law it was not illegal for owners to tamper with the emission system of their own vehicles. The law only made it illegal for professional mechanics to remove or disconnect emission control devices.*

- Plugging the vacuum lines to the EGR valve
- Removing, disconnecting, or plugging the positive crankcase ventilation (PCV) valve
- Removing or disabling the air injection pump
- Modifying, removing, or replacing the OEM computer with a noncertified unit. This includes replacing or reprogramming the PROM chip with a noncertified programming
- Removing the filter restrictor in the fuel tank filler pipe
- Removing, disconnecting, or disabling the fuel vapor recovery system
- Altering the idle mixture adjustment
- Changing the OEM carburetor jet sizes
- Installing noncertified fuel injectors
- Altering the OEM ignition advance mechanism or timing curve
- Removing the OEM air cleaner and heat riser duct plumbing
- Installing an aftermarket intake manifold that does not have provisions for the OEM EGR valve
- Installing exhaust headers that lack provisions for OEM emission components, such as oxygen sensors
- Installing any air induction, fuel or ignition system component that is *not* emissions certified

Aftermarket parts manufacturers that make performance parts that are not direct OEM replacements must apply for special certification for any parts they want to sell as being emissions certified. The parts manufacturing company will submit their proposal to the California Air Resources Board (CARB). They must provide data from dynamometer testing that documents that a part does not adversely effect the exhaust emissions. After CARB reviews the supplied data, it may approve the product or ask for additional testing or data. If the product is approved, it is issued an executive order (EO) number. This number certifies the component as being in compliance with the applicable clean air regulations.

Once the EO number is obtained, the manufacture can market the part as emissions legal and the part can be used on a street-driven vehicle. Since the Federal Environment Protection Agency recognizes the CARB exemption program as meeting its "Memorandum A" requirements for certifying emissions-legal parts, the part is legal for street use in all 50 states.

All OEM replacement parts manufactured by reputable companies are emissions legal. However, be aware of "counterfeit" parts that are sold at a very low price. Also, if you are buying any performance parts for a vehicle that does not meet the exempt statues, then be sure to look on the packaging of the part for the statement that the product is emissions certified for street use in compliance with the EPA and/or CARB rules. The package must also include the EO number.

## Review Questions

1. What must a technician do in order to become certified as an ASE Master Automobile Technician?
2. What are your responsibilities as an employee to your employer?
3. Technician A says that after an individual passes a particular ASE certification exam, he or she is certified in that test area. Technician B says that all of the questions on an ASE certification exam are written as Technician A and Technician B questions. Who is correct?
   A. A only
   B. B only
   C. Both A and B
   D. Neither A nor B
4. Technician A says that emission-tampering laws are not intended to prevent vehicle owners from disabling their vehicles' emission systems. Technician B says that the installation of any performance part constitutes emission tampering. Who is correct?
   A. A only
   B. B only
   C. Both A and B
   D. Neither A nor B
5. List four examples of emission tampering.
6. Explain the flat-rate pay system.
7. List the steps a technician should follow while logically diagnosing a problem.
8. Technician A says that the VIN can be used to identify the engine. Technician B says that the vehicle code plate can be used to identify the engine. Who is correct?
   A. A only
   B. B only
   C. Both A and B
   D. Neither A nor B
9. What does an EO number mean?
10. Explain why it is necessary to get as much information as possible from the customer when diagnosing a problem.

SECTION

# 2 Computer Systems

## Section Objectives

After you have read, studied, and practiced the contents of this section, you should be able to:

- Describe the principle of analog and digital voltage signals.
- Explain the principle of computer communications.
- Describe the basics of logic gate operation.
- Describe the basic function of the central processing unit (CPU).
- Explain the basic method by which the CPU can make determinations.
- List and describe the differences in memory types.
- Describe the function of adaptive strategy.
- Describe the service precautions associated with servicing the computer system.
- Diagnose computer voltage supply and ground circuits.
- Distinguish between hard and intermittent codes.
- Perform flash code retrieval on various vehicles.
- Erase fault codes.
- Perform a complete visual inspection of the problem system.
- Enter diagnostics by use of a scan tool.
- Perform basic actuator tests.
- Perform basic potentiometer-type sensor tests.
- Perform basic thermistor-type sensor tests.
- Perform Hall-effect switch testing.
- Test diodes and properly interpret the readings.
- Replace computer PROM chips properly.
- List and describe the functions of the various inputs by the computer.
- List and describe the operation of output actuators.

### Interesting Fact

*Jean Joseph Lenoir created the first workable internal combustion engine in 1860. Dr. Nikolaus Otto designed the first successful four-stroke engine in 1866. All previous internal combustion engines did not compress the air-fuel mixture. They attempted to draw the mixture in during a downward movement of the piston, then ignite it. The expansion of the gases would force the piston down the remainder of its travel. This design was used in an attempt to make the pistons double acting (a power stroke each way). Otto's engine used the downward stroke to ingest the air-fuel mixture, then an upward stroke to compress it. Many laughed at his idea since only one stroke was used to produce power. However, comparing Otto's engine to other engines of that era, his was lighter, able to run almost twice as fast, and required only 7 percent of the cylinder displacement to produce the same amount of horsepower with the same amount of fuel. Both Lenoir's and Otto's engines were powered by cooking gas. In 1876 Otto successfully converted his engine to burn gasoline.*

CHAPTER

# 5 Introduction to the Computer

## Introduction

This chapter introduces the basic theory and operation of the digital computer used to control many of a vehicle's systems. The use of computers on automobiles has expanded to include control and operation of several functions such as fuel delivery, ignition systems, emission systems, climate control, lighting circuits, cruise control, antilock braking, electronic suspension systems, and electronic shift transmissions. Some of these functions are the responsibility of the power-train control module (PCM), while others are functions of what is known as a body computer module (BCM).

> **You Should Know**
>
> *When computer controls were first installed on the automobile, the aura of mystery surrounding these computers was so great that some technicians were afraid to work on them. Most technicians coming into the field today have grown up around computers and the fear is not as great. Regardless of your comfort level with computers, knowledge is the key to understanding their function. Although it is not necessary to understand all the concepts of computer operation in order to service the systems they control, knowledge of the digital computer will help you feel more comfortable when working on these systems.*

## Basic Computer Operation

A computer processes the physical conditions that represent information (data). The operation of the computer is divided into four basic functions:

1. **Input.** A voltage signal sent from an input device. This device can be a sensor or a switch the driver or technician activates.
2. **Processing.** The computer uses the input information and compares it with programmed instructions. The logic circuits process the input signals into output demands.
3. **Storage.** The program instructions are stored in an electronic memory. Some of the input signals are also stored for later processing.
4. **Output.** After the computer has processed the sensor input and checked its programmed instructions, it will put out control commands to various output devices. These output devices may be the instrument panel display or a system actuator. The output of one computer can also be used as an input to another computer.

Understanding these four functions will help today's technician organize the troubleshooting process. When a system is tested, the technician will attempt to isolate the problem to one of these functions.

## Analog and Digital Principles

Remembering the basics of electricity, voltage does not flow through a conductor; current flows, and voltage is the pressure that "pushes" the current. However, voltage can be used as a signal: For example, difference in voltage levels, frequency of change, or switching from positive to negative values can be used as a signal.

The computer can read only voltage signals. The computer follows a set of instructions, called a **program**, to achieve desired results. The program is "burned" into IC chips using a series of numbers. These numbers represent various combinations of voltages that the computer can understand. The voltage signals to the computer can be either analog or digital. Many of the inputs from the sensors are analog variables. An analog voltage signal is continuously variable within a defined range. For example, engine coolant temperature sensors do not change abruptly. The temperature varies in infinite steps from low to high. The same is true for several other inputs such as engine speed, vehicle speed, fuel flow, and so on.

Compared with an analog voltage representation, digital voltage patterns are square shaped because the transition from one voltage level to another is very

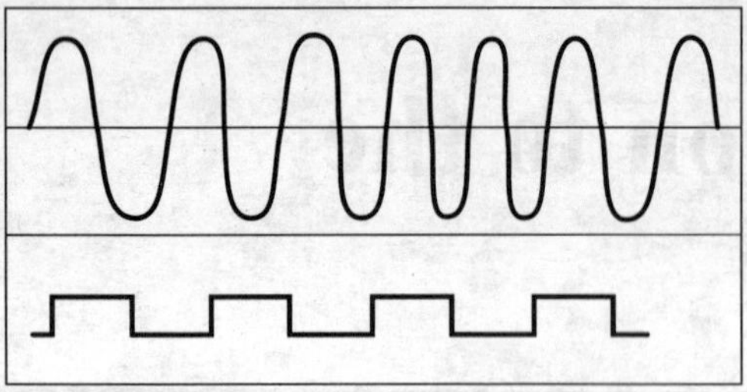


**Figure 1** *Analog voltage signals are constantly variable. Digital voltage patterns are either on or off. Digital signals are referred to as a square sine wave.*

abrupt **(Figure 1)**. A digital signal is produced by an on/off or high/low voltage. The simplest generator of a digital signal is a switch **(Figure 2)**. If 5 volts are applied to the circuit, the voltage sensor will read close to 5 volts (a high-voltage value) when the switch is open. Closing the switch will result in the voltage sensor reading close to 0 volt. This measuring of voltage drops sends a digital signal to the computer. The voltage values are represented by a series of digits, which create a **binary code**. Binary code is represented by the numbers 1 and 0. Any number and word can be translated into a combination of 1s and 0s.

## Binary Numbers

A transistor that operates as a relay is the basis of the digital computer. As the input signal switches from off to on, the transistor output switches from cutoff to saturation. The on and off output signals represent the binary digits 1 and 0.

The computer converts the digital signal into binary code by translating voltages above a given value to 1 and voltages below a given value to 0. As shown in **Figure 3**, when the switch is open and 5 volts is sensed, the voltage value is translated into a 1 (high voltage). When the switch is closed, lower voltage is sensed and the voltage value is translated into a 0. Each 1 or 0 represents one **bit** of information. Note, high voltage being translated to a 1 and low voltage to a 0 is given for explanation purposes. In

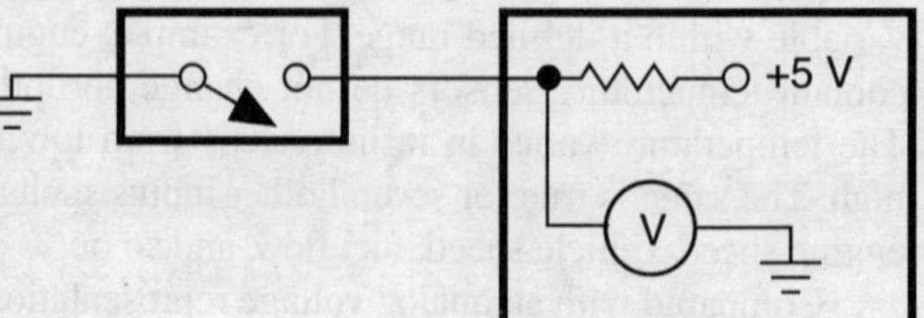


**Figure 2** *Simplified voltage sensing circuit that indicates if the switch is open or closed.*

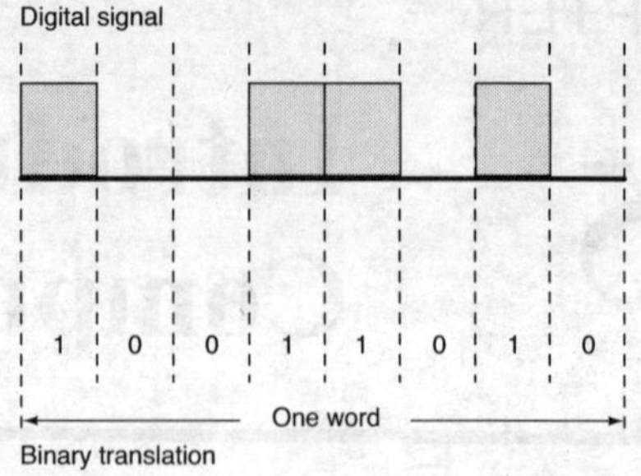


**Figure 3** *Each binary 1 and 0 is 1 bit of information; 8 bits equal 1 byte.*

some systems, a low voltage is represented by a binary 1 while a high voltage is equal to 0.

In the binary system, whole numbers are grouped from right to left. Because the system uses only two digits, the first portion must equal a 1 or a 0. To write the value of 2, the second position must be used. In binary, the value of 2 would be represented by 10 (one 2 and zero 1s). To continue, a 3 would be represented by 11 (one 2 and one 1). **Figure 4** provides a conversion of binary numbers to digital base ten numbers. If a thermistor is sensing 150ºF (65.5ºC), the binary code would be 10010110. If the temperature increases to 151ºF (66ºC), the binary code changes to 10010111.

The computer contains a crystal oscillator or **clock** that delivers a constant time pulse. The clock is a crystal that electrically vibrates when subjected to current at certain voltage levels. As a result, the chip produces a very regular series of voltage pulses. The clock maintains an orderly flow of information through the computer circuits by transmitting 1 bit of binary code for each pulse **(Figure 5)**. In this manner, the computer can distinguish between the binary codes such as 101 and 1001.

| Decimal number | Binary number code 8 4 2 1 | Binary-to-decimal conversion |
|---|---|---|
| 0 | 0000 | = 0 + 0 = 0 |
| 1 | 0001 | = 0 + 1 = 1 |
| 2 | 0010 | = 2 + 0 = 2 |
| 3 | 0011 | = 2 + 1 = 3 |
| 4 | 0100 | = 4 + 0 = 4 |
| 5 | 0101 | = 4 + 1 = 5 |
| 6 | 0110 | = 4 + 2 = 6 |
| 7 | 0111 | = 4 + 2 + 1 = 7 |
| 8 | 1000 | = 8 + 0 = 8 |
| 9 | 1001 | = 8 + 1 = 9 |

**Figure 4** *Binary number code conversion to base ten numbers.*

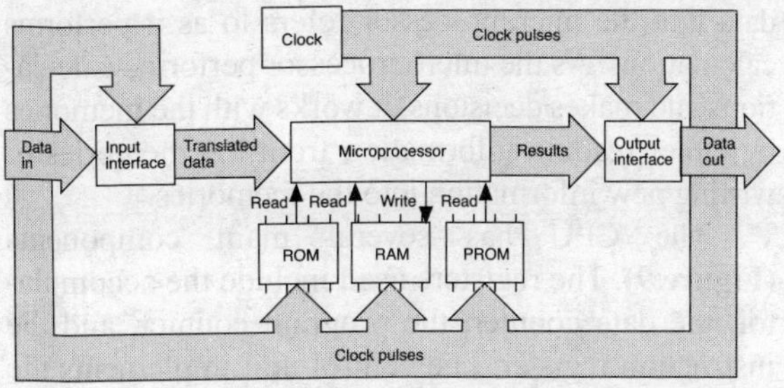


**Figure 5** *Interaction of the main components of the computer. All of the components monitor clock pulses.*

## Signal Conditioning and Conversion

The input and/or output signals may require conditioning in order to be used. This conditioning may include amplification and/or signal conversion.

Some input sensors produce a very low–voltage signal of less than 1 volt. These signals also have an extremely low current flow: therefore, this type of signal must be amplified, or increased, before it is sent to the microprocessor. This amplification is accomplished by the amplification circuit in the input conditioning chip inside the computer **(Figure 6)**.

For the computer to receive information from the sensor and to give commands to actuators, it requires an interface. The **interface** has two purposes: to protect the computer from excessive voltage levels and to translate input and output signals. The computer will have two interface circuits: input and output. The digital computer cannot accept analog signals from the sensors, and it requires an input interface to convert the analog signal to digital. The A/D converter continually scans the analog input signals at regular intervals. For example, if the A/D converter scans the throttle position sensor (TPS) signal and finds this signal at 5 volts, the A/D converter assigns a numeric value to this specific voltage. The A/D converter then changes this numeric value to a binary code **(Figure 7)**.

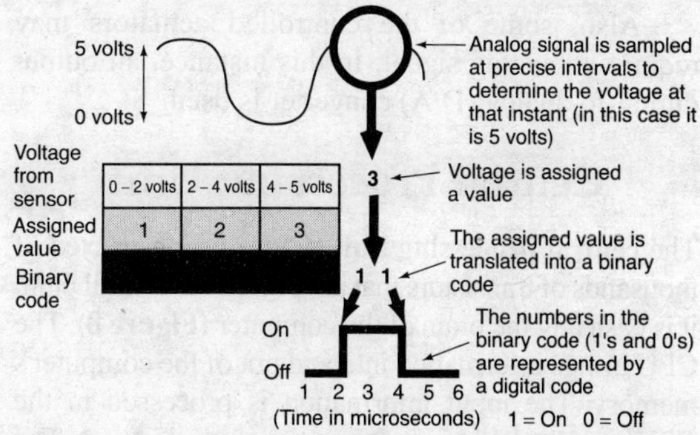


**Figure 7** *The A/D converter assigns a numeric value to input voltages and changes this numeric value to a binary code.*

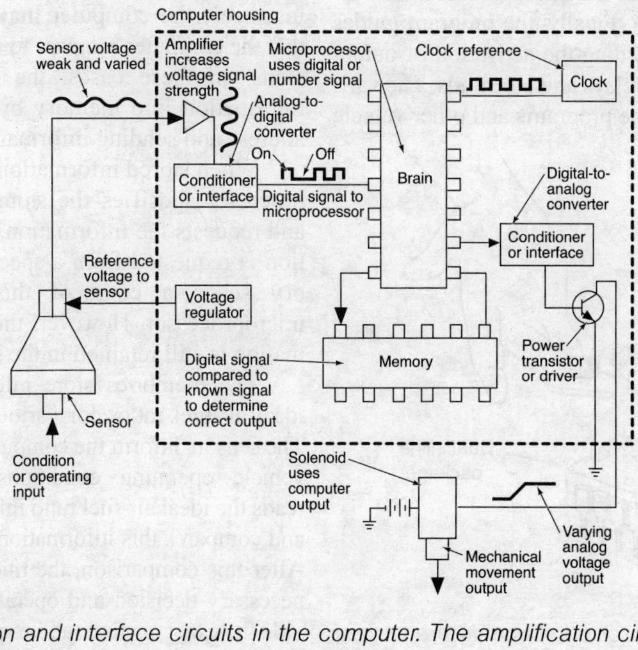


**Figure 6** *Amplification and interface circuits in the computer. The amplification circuit boosts the voltage and conditions it. The interface converts analog inputs into digital signals. The digital-to-analog converter changes the output from digital to analog.*

Also, some of the controlled actuators may require an analog signal. In this instance, an output digital-to-analog (D/A) converter is used.

## Central Processing Unit

The **central processing unit (CPU)** is constructed of thousands of transistors that are placed on a small chip; it is basically the brain of the computer **(Figure 8)**. The CPU brings information into and out of the computer's memory. The input information is processed in the CPU and checked against the program in memory. The CPU also checks memory for any other information regarding programmed parameters. The information obtained by the CPU can be altered according to the program instructions. The program may have the CPU apply logic decisions to the information. Once all calculations are made, the CPU will deliver commands to make the required corrections or adjustments to the operation of the controlled system.

> **You Should Know**
>
> *The terms microprocessor and central processing unit are basically interchangeable.*

The program guides the microprocessor in decision making. For example, the program may inform the microprocessor when sensor information should be retrieved and then tell the microprocessor how to process this information. Finally, the program guides the microprocessor regarding the activation of output control devices such as relays and solenoids. The various memories contain the programs and other vehicle

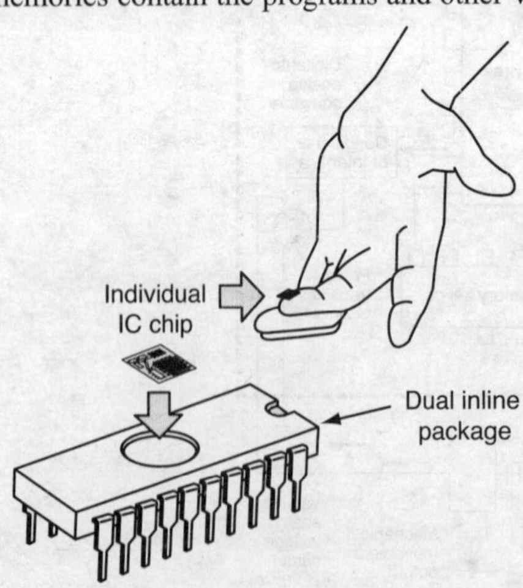


**Figure 8** *Microprocessor components are etched on an integrated circuit chip that is small enough to fit on a fingertip.*

data that the microprocessor refers to as it performs calculations. As the microprocessor performs calculations and makes decisions, it works with the memories by either reading information from the memories or writing new information into the memories.

The CPU has several main components **(Figure 9)**. The registers used include the accumulator, the data counter, the program counter, and the instruction register. The control unit implements the instructions located in the instruction register. The arithmetic logic unit (ALU) performs the arithmetic and logic functions.

## Computer Memory

The computer requires a means of storing both permanent and temporary memory. The memories contain many different locations. These locations may be compared to file folders in a filing cabinet, with each location containing one piece of information. An address is assigned to each memory location. This address may be compared to the lettering or numbering arrangement on file folders. Each address is written in a binary code, and these codes are numbered sequentially beginning with 0.

While the engine is running, the engine computer receives a large quantity of information from a number of sensors. The computer may not be able to process all this information immediately. In some instances, the computer may receive sensor inputs that the computer requires to make a number of decisions. In these cases, the microprocessor writes information into memory by specifying a memory address and sending information to this address.

When stored information is required, the microprocessor specifies the stored information address and requests the information. When stored information is requested from a specific address, the memory sends a copy of this information to the microprocessor. However, the original stored information is still retained in the memory address.

The memories store information regarding the ideal air-fuel ratios for various operating conditions. The sensors inform the computer about the engine and vehicle operating conditions. The microprocessor reads the ideal air-fuel ratio information from memory and compares this information with the sensor inputs. After this comparison, the microprocessor makes the necessary decision and operates the injectors to provide the exact air-fuel ratio required by the engine.

Several types of memory chips may be used in the computer:

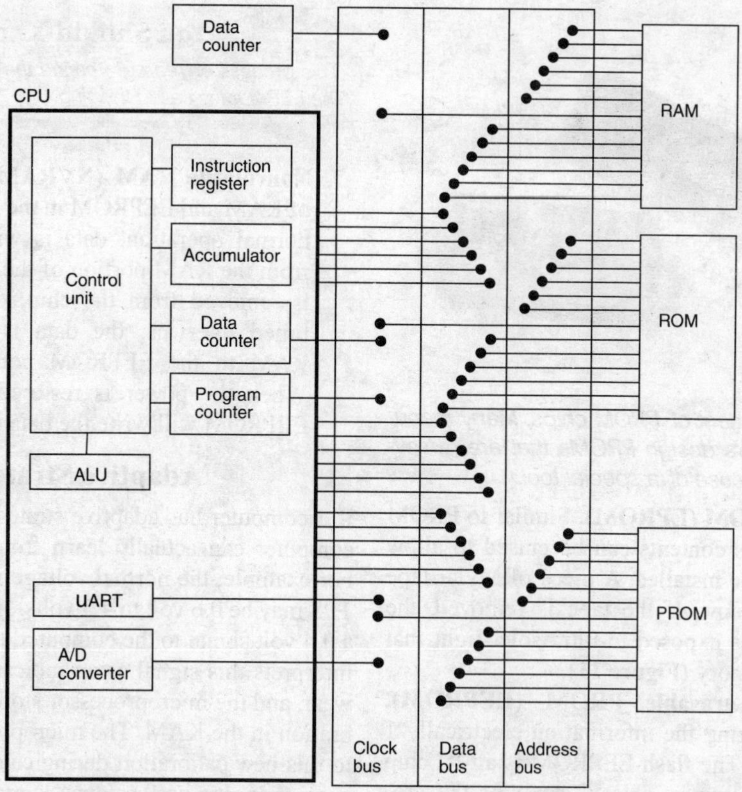


**Figure 9** *Main components of the computer and the CPU.*

1. **Read-only memory (ROM).** Contains a fixed pattern of 1s and 0s that represent permanent stored information. This information is used to instruct the computer on what to do in response to input data. The CPU reads the information contained in ROM, but it cannot write to it or change it. ROM is permanent memory that is programmed in. This memory is not lost when the computer loses power. ROM contains formulas, calibrations, ignition-timing tables, and so on. ROM contains the basic operating parameters for the vehicle.
2. **Random access memory (RAM).** Constructed from flip-flop circuits formed into the chip. The RAM stores temporary information that can be read from or written to by the CPU. RAM stores information that is waiting to be acted upon, and it stores output signals that are waiting to be sent to an output device. RAM can be designed as volatile or nonvolatile. In volatile RAM, the data is retained as long as current flows through the memory. RAM that is connected to the battery through the ignition switch loses its data when the switch is turned off (see item 7, nonvolatile RAM).
3. **Keep alive memory (KAM).** A version of RAM, KAM is connected directly to the battery through circuit-protection devices. For example, the microprocessor can read and write information to and from the KAM and erase KAM information. However, the KAM retains information when the ignition switch is turned off. KAM will be lost when the battery is disconnected, if the battery drains too low, or if the circuit opens. Nonvolatile RAM will retain its information if current is removed.
4. **Programmable read only memory (PROM).** Contains specific data that pertains to the exact vehicle in which the computer is installed. This information may be used to inform the CPU of the accessories that are equipped on the vehicle. The information stored in PROM is the basis for all computer logic. The information in PROM is used to define or adjust the *operating parameters* held in ROM. In many instances, the computer is interchangeable between models of the same manufacturers; however, the PROM is not. Consequently, the PROM may be replaceable and plug into the computer (**Figure 10**).

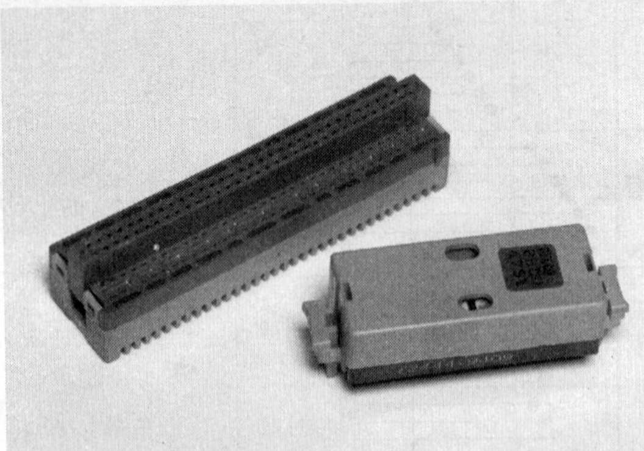

**Figure 10** *Assortment of PROM chips. Many manufacturers design PROMs that are removable by use of a special tool.*

5. **Erasable PROM (EPROM).** Similar to PROM except that its contents can be erased to allow new data to be installed. A piece of Mylar tape covers a window. If the tape is removed, the microcircuit is exposed to ultraviolet light that erases its memory **(Figure 11)**.
6. **Electrically erasable PROM (EEPROM).** Allows changing the information electrically 1 bit at a time. The flash EEPROM is an IC chip inside the computer. This IC contains the program the computer uses to provide control. It is possible to erase and reprogram the EEPROM without removing this chip from the computer. For example, when a modification to the PCM operating strategy is required, it is no longer necessary to replace the PCM. The flash EEPROM may be reprogrammed through the data link connector (DLC) using the manufacturer's specified diagnostic equipment.

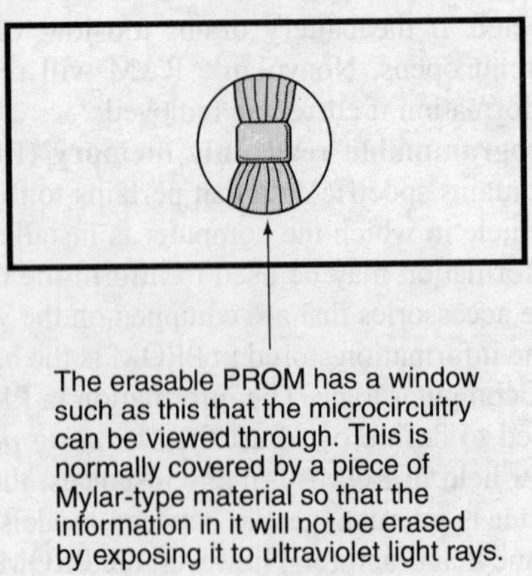


**Figure 11** *EPROM is erased when the ultraviolet ray contacts the microcircuitry.*

**You Should Know**

*This process of reprogramming the computer through the EEPROM is called flashing.*

7. **Nonvolatile RAM (NVRAM).** A combination of RAM and EEPROM in the same chip. During normal operation, data is written to and read from the RAM portion of the chip. If the power is removed from the chip, or at programmed timed intervals, the data is transferred from RAM to the EEPROM portion of the chip. When the power is restored to the chip, the EEPROM will write the data back to the RAM.

### Adaptive Strategy

If a computer has adaptive strategy capabilities, the computer can actually learn from past experience. For example, the normal voltage input range from a TPS may be 0.6 volt to 4.5 volts. If a worn TPS sends a 0.4 volt signal to the computer, the microprocessor interprets this signal as an indication of component wear, and the microprocessor stores this altered calibration in the KAM. The microprocessor now refers to this new calibration during calculations, and thus normal engine performance is maintained. If a sensor output is erratic or considerably out of range, the computer may ignore this input. When a computer has adaptive strategy, a short learning period is necessary under the following conditions:

1. After the battery has been disconnected
2. When a computer system component has been replaced or disconnected
3. On a new vehicle

## Information Processing

The air charge temperature (ACT) sensor input will be used as an example of how the computer processes information. If the air temperature is low, the air is denser and contains more oxygen per cubic foot. Warmer air is less dense and therefore contains less oxygen per cubic foot. The cold, dense air requires more fuel compared with the warmer air that is less dense. The microprocessor must supply the correct amount of fuel in relation to air temperature and density.

An ACT sensor is positioned in the intake manifold where it senses air temperature. This sensor contains a resistive element that has an increased resistance when the sensor is cold. Conversely, the ACT sensor resistance decreases as the sensor temperature increases. When the ACT sensor is cold, it sends a high analog voltage signal to the computer, and the A/D converter changes this signal to a digital signal.

When the microprocessor receives this ACT signal, it addresses the tables in the ROM. The lookup tables list air density for every air temperature. When the ACT sensor voltage signal is very high, the lookup table indicates very dense air. This dense air information is relayed to the microprocessor, and the microprocessor operates the output drivers and injectors to supply the exact amount of fuel required by the engine **(Figure 12)**.

## Logic Gates

**Logic gates** are the thousands of field-effect transistors (FETs) incorporated into the computer circuitry. These circuits are called logic gates because they act as gates to output voltage signals, depending on different combinations of input signals. The FETs use the incoming voltage patterns to determine the pattern of pulses leaving the gate. The following are some of the most common logic gates and their operation. The symbols represent functions and not electronic construction:

1. **NOT gate.** A NOT gate simply reverses binary 1s to 0s and vice versa **(Figure 13)**. A high input results in a low output, and a low input results in a high output.
2. **AND gate.** The AND gate has at least two inputs and one output. The operation of the AND gate is similar to two switches in series to a load **(Figure 14)**. The only way the light will turn on is if switches A *and* B are closed. The output of the gate will be high only if both inputs are high. Before current can be present at the output of the gate, current must be present at the base of both transistors **(Figure 15)**.
3. **OR gate.** The OR gate operates similarly to two switches that are wired in parallel to a light **(Figure 16)**. If switch A *or* B is closed, the light will turn on. A high signal to either input will result in a high output.
4. **NAND and NOR gates.** A NOT gate placed behind an AND or OR gate inverts the output signal **(Figure 17)**.

**You Should Know**

*The NOT gate is also called an inverter.*

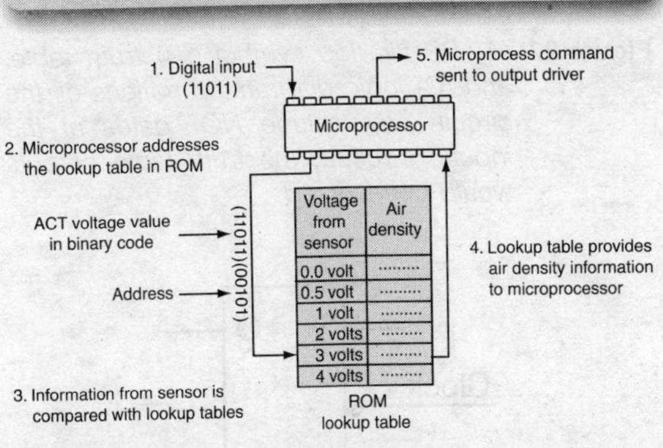


**Figure 12** *The microprocessor addresses the lookup tables in the ROM, retrieves air density information, and issues commands to the output devices.*

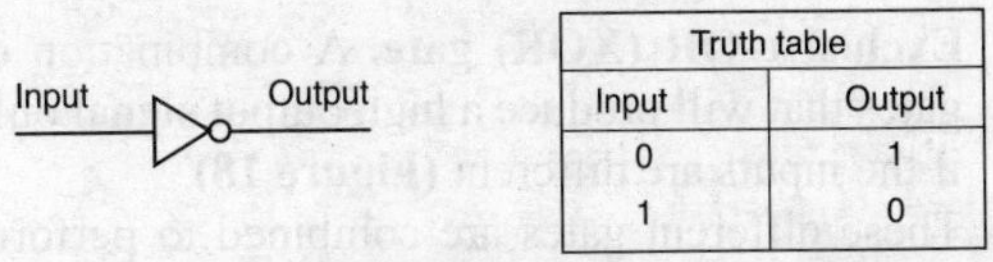


**Figure 13** *The NOT gate symbol and truth table. The NOT gate inverts the input signal.*

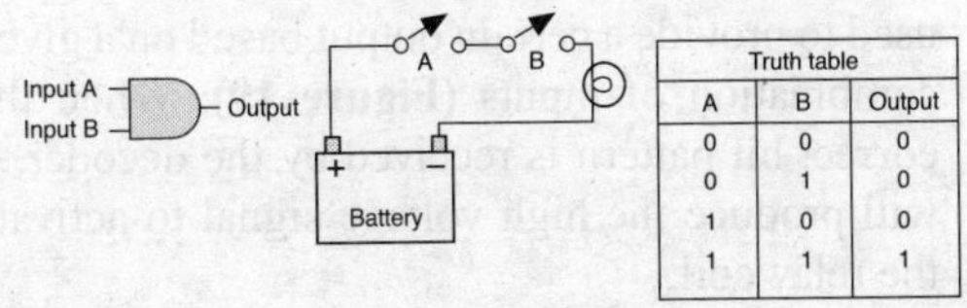


**Figure 14** *The AND gate symbol and truth table. The AND gate operates similarly to switches in series.*

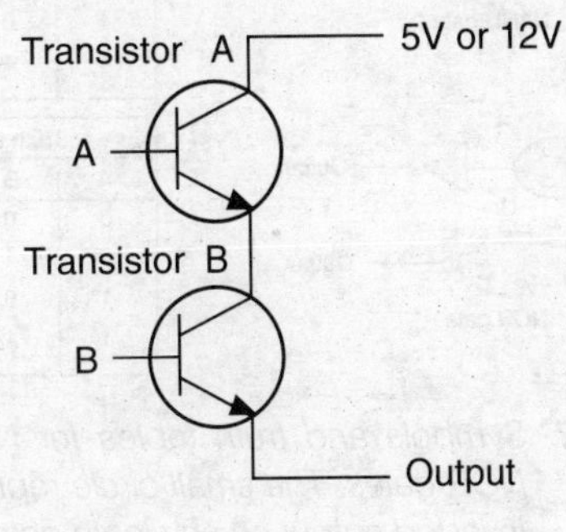


**Figure 15** *The AND gate circuit.*

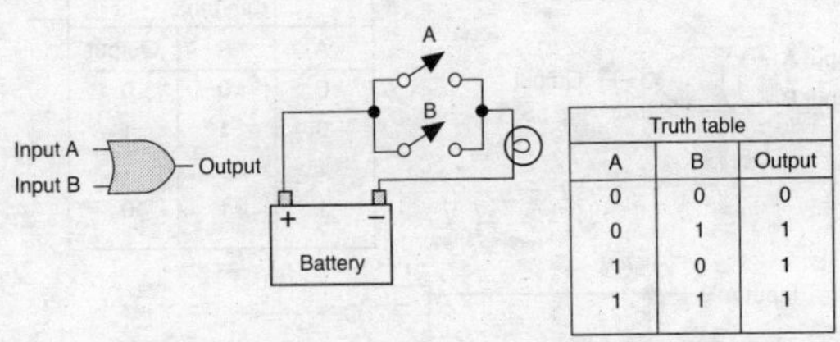


**Figure 16** *OR gate symbol and truth table. The OR gate is similar to parallel switches.*

5. **Exclusive-OR (XOR) gate.** A combination of gates that will produce a high output signal only if the inputs are different **(Figure 18)**.

These different gates are combined to perform the processing function. The following are some of the most common combinations:

1. **Decoder circuit.** A combination of AND gates used to provide a certain output based on a given combination of inputs **(Figure 19)**. When the correct bit pattern is received by the decoder, it will produce the high voltage signal to activate the relay coil.

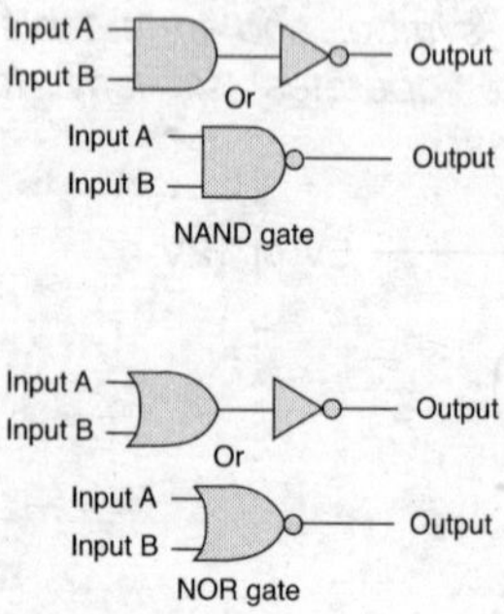


Truth table

| A | B | Output |
|---|---|---|
| 0 | 0 | 1 |
| 0 | 1 | 1 |
| 1 | 0 | 1 |
| 1 | 1 | 0 |

Truth table

| A | B | Output |
|---|---|---|
| 0 | 0 | 1 |
| 0 | 1 | 0 |
| 1 | 0 | 0 |
| 1 | 1 | 0 |

**Figure 17** *Symbols and truth tables for NAND and NOR gates. The small circle represents an inverted output on any logic gate symbol.*

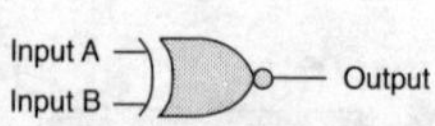


Truth table

| A | B | Output |
|---|---|---|
| 0 | 0 | 0 |
| 0 | 1 | 1 |
| 1 | 0 | 1 |
| 1 | 1 | 0 |

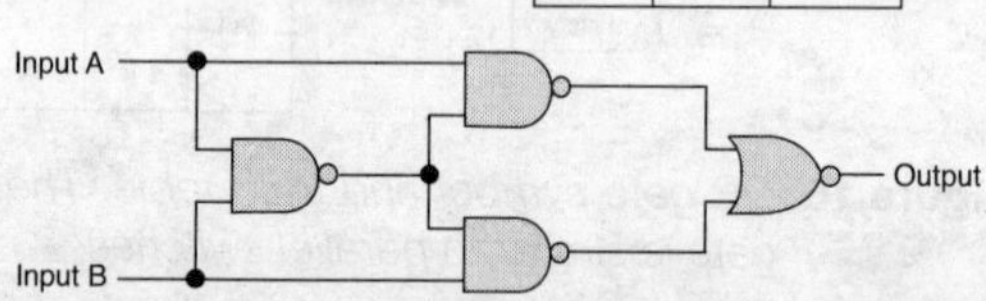


**Figure 18** *XOR gate symbol and truth table. An XOR gate is a combination of NAND and NOR gates.*

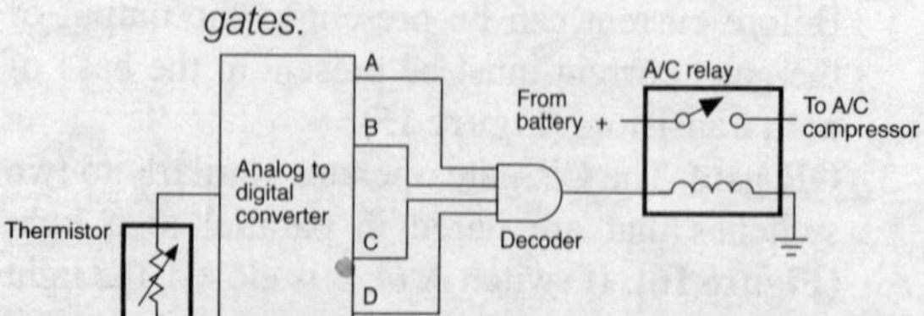


**Figure 19** *An example of a simplified converter circuit that uses a temperature-sensing circuit to turn on the air-conditioning compressor.*

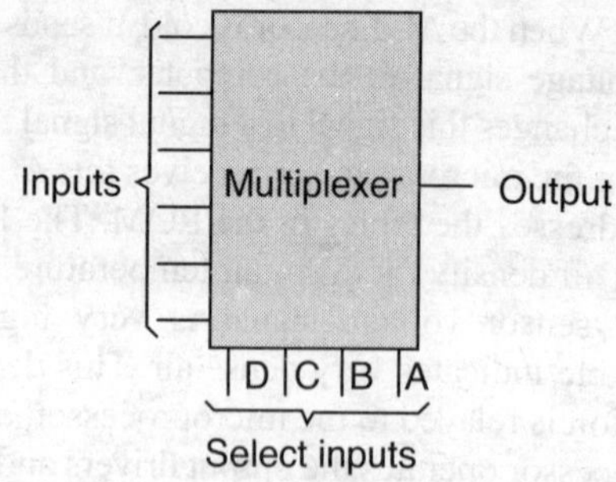


**Figure 20** *Selection at inputs D, C, B, A will determine which data input will be processed.*

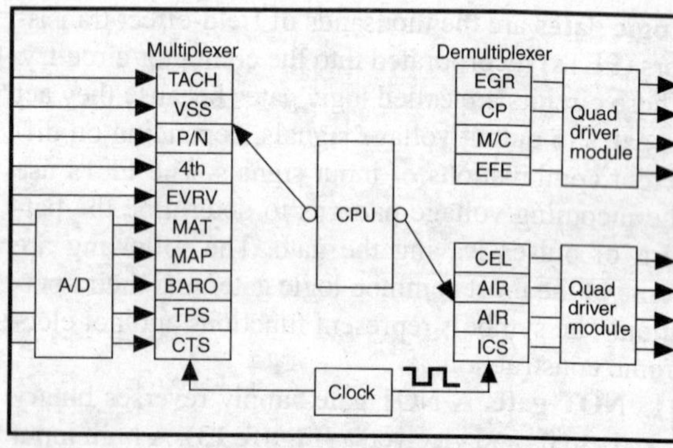


**Figure 21** *Block diagram representation of the MUX and DEMUX circuits.*

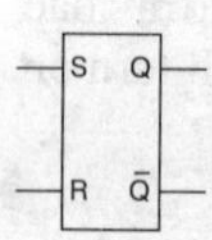


(A)

Truth table

| S | R | Q | Q̄ |
|---|---|---|---|
| 0 | 0 | Disallowed | |
| 0 | 1 | 1 | 0 |
| 1 | 0 | 0 | 1 |
| 1 | 1 | No change | |

(B)

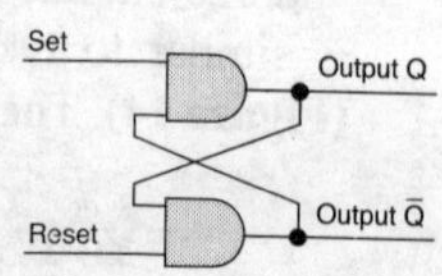


(C)

**Figure 22** *(A) RS flip-flop symbol, (B) truth table, and (C) logic diagram. Variations of the circuit may include NOT gates at the inputs; if used, the truth table outputs would be reversed.*

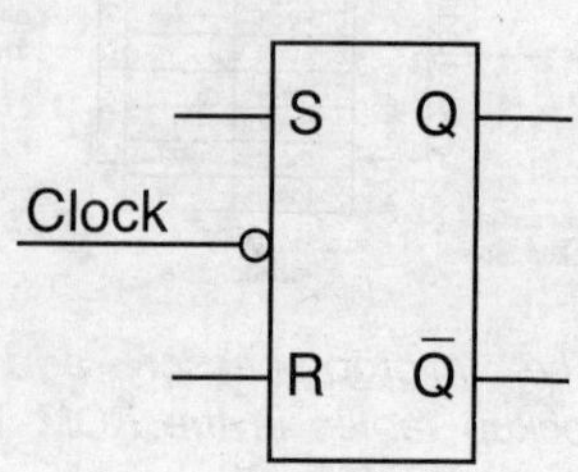


**Figure 23** *Clocked RS flip-flop symbol.*

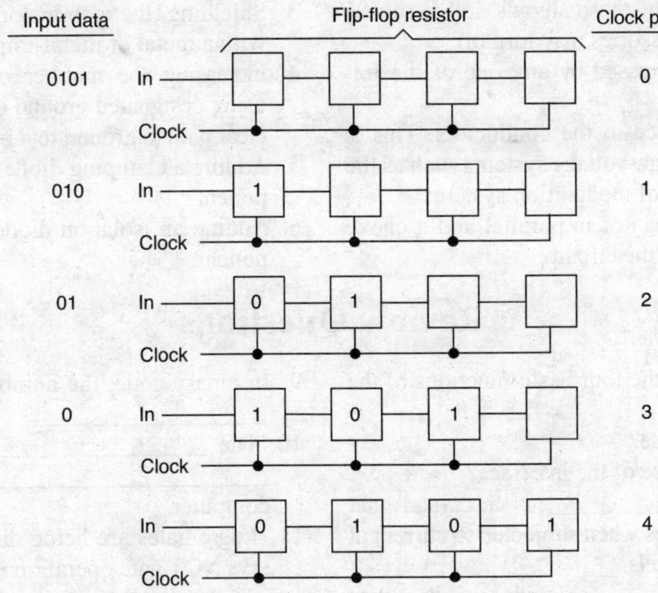


**Figure 24** *It takes four clock pulses to load 4 bits into the register.*

2. **Multiplexer (MUX).** The basic computer is not capable of looking at all of the inputs at the same time. A multiplexer is used to examine one of many inputs depending on a programmed priority rating **(Figure 20)**. This process is called sequential sampling. This means the computer will deal with all of the sensors and actuators one at a time.
3. **Demultiplexer (DEMUX).** DEMUX operates similarly to the MUX except that it controls the order of the outputs **(Figure 21)**.
4. **RS and clocked RS flip-flop circuits.** Logic circuits that remember previous inputs and do not change their outputs until they receive new input signals. **Figure 22** shows a basic RS flip-flop circuit. The "R" is "reset" when Q output is equal to 0 and $\bar{Q}$ is equal to 1. The "S" is "set" when Q is equal to 1 and $\bar{Q}$ is equal to 0. The clocked flip-flop circuit has an inverted clock signal as an input so that circuit operations occur in the proper order **(Figure 23)**. Flip-flop circuits are called **sequential logic circuits** because the output is determined by the sequence of inputs. A given input affects the output produced by the next input.
5. **Driver circuits.** A *driver* is a term used to describe a transistor device that controls the current in the output circuit. Drivers are controlled by a computer to operate such controls as fuel injectors, ignition coils, and many other high-current circuits. The currents handled by a driver are not really that high; they are just more than what a transistor typically handles. Several types of driver circuits are used on automobiles, such as quad, discrete, peak and hold, and saturated switch driver circuits.
6. **Registers.** A combination of flip-flops that transfer bits from one to another every time a clock pulse occurs **(Figure 24)**. It is used in the computer to store information temporarily.
7. **Accumulators.** Registers designed to store the results of logic operations that can become inputs to other modules.

## EMI Suppression

As manufacturers began to increase the number of electronic components and systems in their vehicles, the problem of **electromagnetic interference (EMI)** had to be controlled. EMI is an undesirable creation of electromagnetism whenever current is switched on and off. EMI is also caused by static electricity that is created by friction. The friction is a result of the tires and their contact with the road or the result of fan belts contacting the pulleys. The low-power ICs used on modem vehicles are sensitive to the signals produced as a result of EMI.

EMI can disrupt the vehicle's computer systems by inducing false messages to the computer. The computer requires messages to be sent over circuits in order to communicate with other computers, sensors, and actuators. If any of these signals are disrupted, the engine and/or accessories may turn off.

EMI can be suppressed by any one of the following methods:

1. Adding a resistance to the conductors. This is usually done to high-voltage systems such as the secondary circuit of the ignition system.
2. Connecting a capacitor in parallel and a choke coil in series with the circuit.
3. Shielding the conductor or load components with a metal or metal-impregnated plastic.
4. Increasing the number of paths to ground by using designated ground circuits. This provides a clear path to ground that is very low in resistance.
5. Adding a clamping diode in parallel to the component.
6. Adding an isolation diode in series to the component.

## Review Questions

1. List and describe the four basic functions of the computer.
2. What is binary code?
3. What is the purpose of the interface?
4. The ____________________ is a crystal that electrically vibrates when subjected to current at certain voltage levels.
5. ____________________ contains specific data that pertains to the exact vehicle in which the computer is installed.
6. Computer memory is being discussed: Technician A says ROM can be written to by the CPU. Technician B says RAM stores temporary information that can be read from or written to by the CPU. Who is correct?
   A. Technician A only
   B. Technician B only
   C. Both Technician A and Technician B
   D. Neither Technician A nor Technician B
7. Describe the basics of NOT, AND, and OR logic gate operation.
8. What is the difference between ROM, RAM, and PROM?
9. In binary code, the number 4 is represented by ____________________.
10. The ____________________ ____________________ ____________________ is the heart of the computer.
11. Logic gates are being discussed: Technician A says NOT gate operation is similar to that of two switches in series to a load. Technician B says an AND gate simply reverses binary 1s to 0s and vice versa. Who is correct?
    A. Technician A only
    B. Technician B only
    C. Both Technician A and Technician B
    D. Neither Technician A nor Technician B
12. Technician A says volatile RAM is erased when it is disconnected from its power source. Technician B says nonvolatile RAM will retain its memory if removed from its power source. Who is correct?
    A. Technician A only
    B. Technician B only
    C. Both Technician A and Technician B
    D. Neither Technician A nor Technician B

CHAPTER

# 6 Computer System Diagnosis

## Introduction

Because the power-train control module controls many of the functions of the engine operating systems, it is important for technicians to be able to properly diagnose problems with this system. The use of computers has expanded to include the functions of ignition, fuel, and emission systems. In addition, computers are used to control the operation of climate control, lighting circuits, cruise control, antilock braking, electronic suspension systems, electronic shift transmissions, and alternator regulation.

As discussed in Chapter 5, a computer processes the physical conditions that represent information (data). The operation of the computer is divided into four basic functions: input, processing, storage, and output. Understanding these four computer functions will help you organize the troubleshooting process. When a system is tested, you are attempting to isolate a problem with one of these functions.

In the process of controlling the various systems, the PCM continuously monitors operating conditions for possible system malfunctions. The computer compares system conditions against programmed parameters. If the conditions fall outside of these limits, the computer detects a malfunction. A **diagnostic trouble code (DTC)** is set to indicate the portion of the system at fault. DTCs are two- to-five digit characters displayed in the diagnostic display if the testing and failure requirements are both met. The technician can access this code for aid in troubleshooting.

If a malfunction results in improper system operation, the computer may minimize the effects by using **failsoft** or limp-in mode. This provides for system operation, but at a limited level of function. For example, if the engine coolant temperature sensor voltage input is considered invalid, the computer will provide a fixed value as its own input. This fixed value can be either a programmed value in the computer's memory, or it can be the last received signal from the sensor prior to failure. This allows the system to operate on a limited basis instead of shutting down completely.

There are several things you need to know before you learn how to access the computer's memory to gain information concerning system operation. You need to become familiar with what you are looking at, and you must follow proper precautions when servicing these systems.

## Electronic Service Precautions

The technician must take some precautions before servicing *any* computer or any of its controlled systems. Computers are designed to withstand normal current draws associated with normal operation. However, overloading any of the system circuits will result in damage to the computer. Follow these service precautions to prevent computer and circuit damage:

1. Do not ground or apply voltage to any controlled circuits unless the service manual instructs you to do so.
2. Use only a high **impedance** multimeter (10 megohm or greater) to test the circuits. Impedance is the combined opposition to current created by the resistance, capacitance, and inductance of the meter. Never use a test light unless specifically instructed to do so in the service manual.
3. Make sure the ignition switch is turned off before making or breaking electrical connections to the computer.
4. Unless instructed otherwise in the service manual, turn off the ignition switch before making or breaking any electrical connections to sensors or actuators.
5. Turn the ignition switch off whenever disconnecting or connecting the battery terminals. Also turn it off when pulling and replacing the fuse.
6. Do not connect any other electrical accessories to

the insulated or ground circuits of the computer-controlled systems.

7. Use only manufacturer's specific test and replacement procedures for the year and model of vehicle being serviced.
8. Wear an **electrostatic discharge (ESD) strap**. The ESD strap grounds your body to prevent static discharges that may damage electronic components.

**You Should Know**

*Static electricity can be 25,000 volts or higher.*

By following these precautions, plus those listed in the service manual, you can avoid having to replace expensive components.

### Electrostatic Discharge

Some manufacturers mark certain components and circuits with a code or symbol to warn technicians that the units are sensitive to electrostatic discharge **(Figure 1)**. Static electricity can destroy or render an electronic component useless.

When handling any electronic part, especially those that are static sensitive, follow the guidelines below to reduce the possibility of electrostatic buildup on your body and the inadvertent discharge to the electronic part. If you are not sure if a part is sensitive to static, treat it as if it is.

1. Always touch a known good ground before handling the part. This should be repeated while handling the part; repeat more frequently after sliding across a seat, sitting down from a standing position, or walking a distance.
2. Avoid touching the electrical terminals of the part unless you are instructed to do so in the written service procedures. It is good practice to keep your fingers off all electrical terminals as

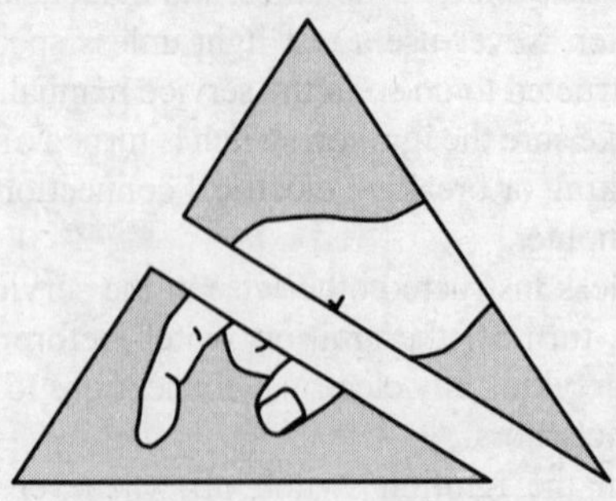

**Figure 1** *The electrostatic discharge (ESD) symbol warns the technician that the component or circuit is sensitive to static.*

the oil from your skin can cause corrosion.

3. When you are using a voltmeter, always connect the negative meter lead first.
4. Do not remove a part from its protective package until it is time to install the part.
5. Before removing the part from its package, ground yourself and the package to a known good ground on the vehicle.
6. When replacing a PROM, ground your body by putting a metal wire around your wrist and connect the wire to a good ground.

### Electromagnetic Interference

Electromagnetic interference or radio frequency interference can cause problems with the vehicle's on-board computer. Unfortunately, an automobile's spark plug wires, ignition coil, and generator all possess the ability to generate these radio waves. Under the right conditions, RFI can trigger sensors or actuators. The result may be an intermittent drivability problem that appears to be ignition system related.

To minimize the effects of RFI, make sure your visual inspection is thorough. Also check to make sure that sensor wires running to the computer are routed away from potential RFI sources. Rerouting a wire by no more than an inch or two (25 to 50 mm) may keep RFI from falsely triggering or interfering with computer operation. RFI can be present on a voltage signal or on a ground.

Most manufacturers shield their PCM from EMI and RFI. However, this shielding will only work if the PCM is properly grounded. Always confirm a good ground before condemning the PCM. Some PCMs may have up to five grounds.

## Powers and Grounds

For the control module to operate properly, it must have good power feeds and ground connections. Usually the module will have at least one terminal connected to a battery feed. This feed is used for keep alive memories. There may also be terminals that receive battery power when the ignition key is turned to the Run position. These feeds are usually used as "wake-up" signals, and to power the driver circuits of the module. All of these circuits must be free of resistance.

Most modules will use redundant grounds (more than one). These ground connections must be free of resistance. If the ground circuit is faulty, the control module may not power up. In addition to the ground terminals, there may also be a case ground. The case ground may be used to reduce RFI and EMI in the module. The case ground may also provide the ground for the driver circuits in the module. Often the sensors used by a module will have a common ground terminal into the module. This ground is commonly called "sensor return" or "signal return." Since this is a common ground for all sensors, resistance on this circuit will affect the values of every sensor. For example, if there is a 0.5 voltage drop on the sensor return circuit, every sensor will be 0.5 volt off.

> **You Should Know**
>
> *In light of the total number of vehicles being produced that use computer-controlled systems, very few customer complaints are actually the fault of the computer. The computer should only be replaced if all other possible causes have been checked and confirmed as operating properly. Always check the condition of the battery feed and ground circuit of the computer before condemning it.*

## Diagnosis of Computer Voltage Supply and Ground Circuits

> **You Should Know**
>
> *Never replace a computer unless the ground wires and voltage supply wires are proven to be in satisfactory condition.*

Like any other electrical or electronic component, a computer cannot operate properly unless it has satisfactory voltage supply at the required terminals and proper ground connections. A computer wiring diagram for the vehicle being tested must be available for these tests. First, measure the voltage across the battery terminals as a reference. Now, backprobe the battery feed terminal to the computer and connect the digital volt-ohm meter (DVOM) leads to this terminal and to ground. The voltage at the battery feed terminal should be 0.5 volt of battery voltage with the ignition switch off. If the proper voltage is not present at this terminal, check the computer fuse and related circuit. Turn the ignition switch to the Run position and check applied voltage to *all* battery and ignition feed terminals of the PCM. The voltage measured at these terminals should be within 0.5 volt of battery voltage. If the specified voltage is not available, test the voltage supply wires to these terminals. These terminals may be connected through fuses, fuse links, or relays. Always refer to the vehicle manufacturer's wiring diagram for the vehicle being tested.

> **You Should Know**
>
> *When diagnosing computer problems, it is usually helpful to ask the customer about service work that has been performed lately on the vehicle. If service work has been performed in the engine compartment, it is possible that a computer ground wire may be loose or disconnected.*

Computer ground wires usually extend from the computer to a ground connection on the engine or battery. Often there is more than one ground wire. In addition, the fasteners used to attach the computer to the vehicle chassis may be used as a ground. With the ignition switch in the Run position, perform a voltage drop test across the ground wires. Compare your results with specifications. A good circuit should not drop more than 0.2 volt on the ground circuit.

### Sensor Return Circuit

Many manufacturers use a single sensor return circuit for all of the engine sensors. This makes a good ground critical since a sensor return circuit that has high resistance will cause *all* of the sensors to indicate an erroneous voltage reading to the PCM. Consider a temperature-sensing circuit as an example **(Figure 2)**: (A) indicates the normal voltage sensed by the PCM with the thermistor at 10K ohms; (B) indicates the voltage sensed by the PCM with the thermistor still at 10K ohms, but with an additional 200 ohms of resistance in the sensor's return circuit. The voltage is now off by 25 millivolts. Since the 200 ohms of resistance also affects all of the other sensors that share this return circuit, they are all off as well. This may not seem like enough voltage to cause any problems, but remember that most systems are very sensitive to small voltage changes.

In addition to causing some drivability and emission problems, this type of fault may not set a fault code. The voltage values are within the normal operating parameters of the sensor, so continuity faults will not set. Also, since all sensors are off by the same amount, rationality faults are not detected. Drivability and emission problems may occur as a result of the computer making bad decisions based on erroneous information.

Poor grounds can also allow EMI or noise to be present on the reference voltage signal. This noise causes minute changes in the voltage going to the sensor: therefore, the output signal from the sensor will also have these voltage changes. The computer

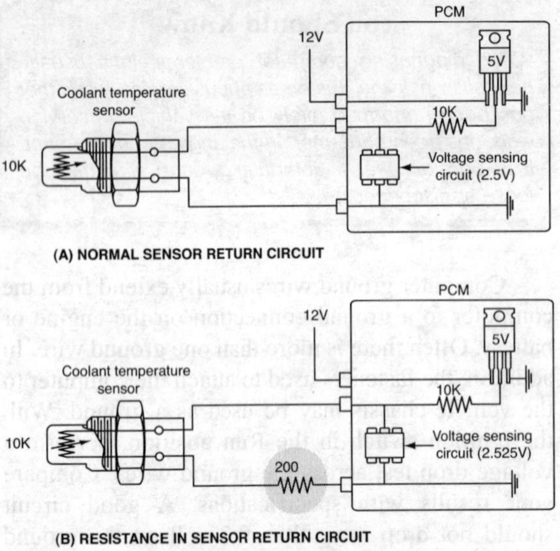


**Figure 2** *The effect of resistance on the sensor return circuit. (A) Normal sensor voltage. (B) Voltage reading with ETC at same resistance, but sensor return circuit has additional resistance.*

will try to respond to these changes, which can cause a drivability problem. The best way to check for noise is to use a lab scope.

Connect the lab scope between the 5 volt reference signal into the sensor and the ground. The trace on the scope should be flat **(Figure 3)**. If noise is present, move the scope's negative probe to a known good ground. If the noise disappears, the sensor's ground circuit is faulty or has resistance. If the noise is still present, the voltage feed circuit is faulty or there is EMI in the circuit from another source, such as the AC generator. Find and repair the cause of the noise.

Circuit noise may be present in either the positive or negative side of a circuit. It may also be evident by a popping noise from the radio or by an intermittent engine miss. However, noise can cause a variety of problems in any electrical circuit. The most common sources of noise are electric motors, relays and solenoids, AC generators, ignition systems, switches, and air conditioner compressor clutches. Typically, noise is the result of an electrical device being turned on and off. Sometimes the source of the noise is a defective suppression device. Some of the commonly used noise suppression devices are resistor-type secondary cables and spark plugs, shielded cables, capacitors, diodes, and resistors. If the source of the noise is not a poor ground or a defective component, check the suppression devices.

Diodes and resistors are also used for noise-suppression devices. Resistors do not eliminate the spikes, but they do limit their intensity. When testing a circuit that uses a resistor to limit noise, if the lab scope trace indicates large voltage spikes, the resistor may be bad. Clamping diodes are used to suppress noise and induced voltages in circuits such as AC compressor clutches and relays. If the diode is bad, a negative spike will result **(Figure 4)**. Capacitors or chokes are used to control noise from motors

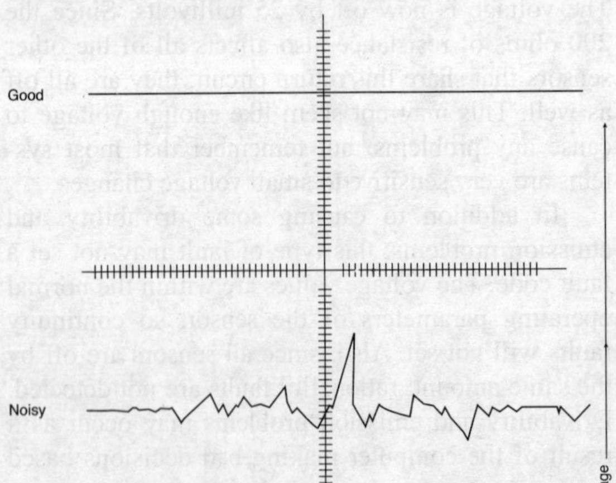


**Figure 3** *(Top) Voltage signal trace of a good circuit. (Bottom) Trace of a circuit with excess noise.*

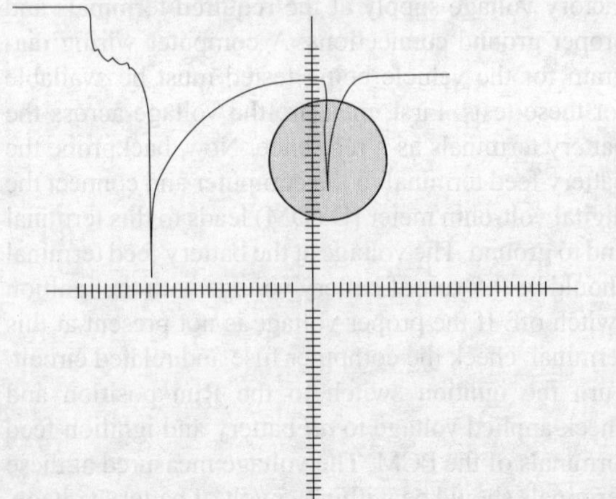


**Figure 4** *A voltage trace of air conditioner compressor clutch circuit with a bad diode.*

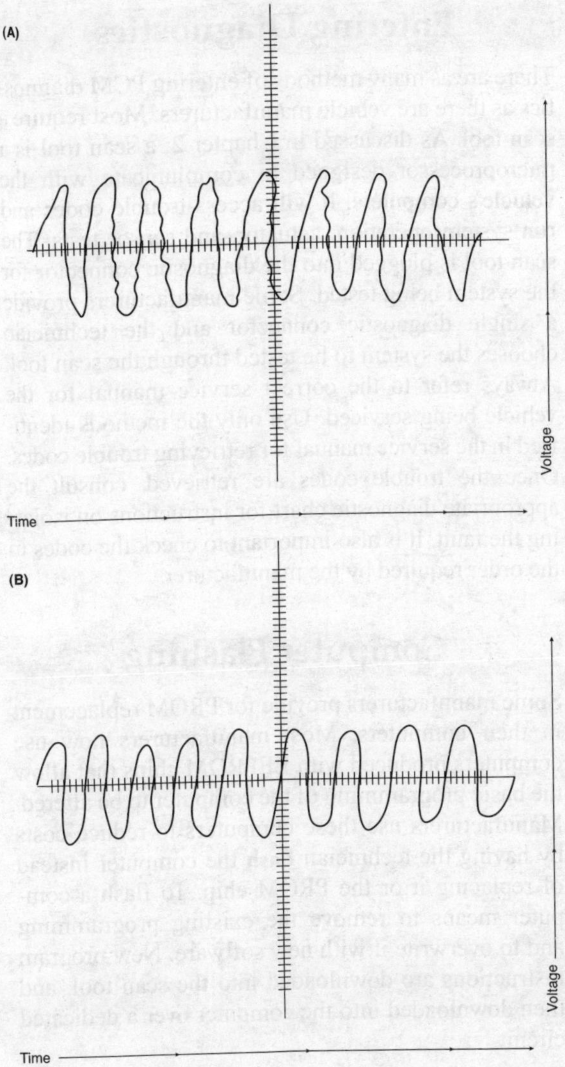


**Figure 5** *(A) Voltage trace of motor without a chock. (B) Voltage trace of a motor with a chock.*

and generators **(Figure 5)**. To avoid much frustration during diagnosis of a computer system (especially the inputs), check the integrity of the ground as one of the first steps.

## Trouble Codes

Most computers are capable of displaying the stored faults in memory. Fault codes can be displayed by the scan tool, and in early systems, by a method referred to as **flash codes**. These are codes that are displayed by flashing the **malfunction indicator light (MIL)** to provide the technician the DTC. The MIL is illuminated by the PCM if it detects a fault in a monitor system. The lamp is usually located in the instrument cluster and may be referred to as the Check Engine light.

The method used to retrieve the flash codes varies greatly, and the technician must refer to the correct service manual for the procedure. The flash code provides a means of retrieving DTCs when a scan tool is not available. Usually, the MIL is used to display the flash codes. The DTCs are read by counting the flash: for example, two quick flashes followed by a brief pause and three quick flashes indicate code number 23. By referencing the proper diagnostic information, the technician can then look up the correct procedure to diagnosis the displayed code. Manufacturers discontinued this practice of displaying DTCs without a scan tool beginning in 1996. Make sure that the vehicle you are working on supports this option.

Depending on system design, the computer may store codes for long periods of time; some lose the code when the ignition switch is turned off. Systems that do not retain the code when the ignition is turned off require the technician to test drive the vehicle and attempt to duplicate the fault. Once the fault is detected by the computer, the code must be retrieved before the ignition switch is turned off again.

Diagnostic trouble codes can be retrieved to assist the technician in identifying the *circuit* that has been detected as faulty by the PCM. Remember, the code displayed may not necessitate the replacement of the component. For example, if the fault code indicates a high TPS voltage, the fault may not be the throttle position sensor itself; it might be an open in the circuit. A low O2 sensor fault may not be a faulty O2 sensor; it might be a vacuum leak that the O2 sensor is sensing. To locate the problem, follow the diagnostic procedure in the service manual for the code received.

### Differences Between Hard and Intermittent Codes

Some computers will store trouble codes in their memory until they are erased by the technician, or until a set amount of engine starts have passed. Some computers will display two sets of fault codes. Usually, the first set of codes displayed represents all codes stored in memory, including both hard and intermittent codes. The second set of codes displayed is only **hard codes**. The codes displayed in the first set, but not in the second set, are **intermittent codes**. Hard codes are failures that are detected somewhere in the system at the time of the system self-test. These faults are also referred to as current or active faults. Intermittent codes are those in which a fault occurred in the past but is not present during the last test of the circuit. Intermittent codes are also called soft, memory, or history codes.

Most diagnostic charts cannot be used to locate causes of intermittent codes. This is because the testing at various points of the chart depends on the fault being present to locate the problem. If the fault is not present, the technician may be erroneously instructed to replace the PCM, even though it is not defective.

Many intermittent problems are the result of poor electrical connections. Diagnosis should start with a good visual inspection of the connectors involved with the code. Even on hard codes, visually inspect the circuit before conducting any other tests.

## Visual Inspection

Perhaps the most important check to be made before diagnosing a PCM-controlled system is a complete visual inspection. The visual inspection can identify faults that could cause the technician to spend wasted time in diagnostics. In addition, the problem can be pinpointed without any further steps.

Inspect the following:

1. All sensors and actuators for physical damage
2. Electrical connections into sensors, actuators, and control modules
3. All ground connections
4. Wiring for signs of burned or chaffed spots, pinched wires, or contact with sharp edges or hot exhaust manifolds
5. All vacuum hoses for pinches, cuts, or disconnects

The time spent performing a visual inspection is worthwhile. Put forth the effort to check wires and hoses that are hidden under other components.

## Entering Diagnostics

There are as many methods of entering PCM diagnostics as there are vehicle manufacturers. Most require a scan tool. As discussed in Chapter 2, a scan tool is a microprocessor designed to communicate with the vehicle's computers. It will access trouble codes and run system operation, actuator, and sensor tests. The scan tool is plugged into the diagnostic connector for the system being tested. Some manufacturers provide a single diagnostic connector and the technician chooses the system to be tested through the scan tool. Always refer to the correct service manual for the vehicle being serviced. Use only the methods identified in the service manual for retrieving trouble codes. Once the trouble codes are retrieved, consult the appropriate diagnostic chart for instructions on isolating the fault. It is also important to check the codes in the order required by the manufacturer.

## Computer Flashing

Some manufacturers provide for PROM replacement in their computers. Most manufacturers now use computers produced with EEPROM chips that allow the basic programming of the computer to be altered. Manufacturers use these computers to reduce costs by having the technician **flash** the computer instead of replacing it or the PROM chip. To flash a computer means to remove the existing programming and to overwrite it with new software. New program instructions are downloaded into the scan tool, and then downloaded into the computer over a dedicated circuit.

### You Should Know

*Sometimes removal and replacement of the PCM involves removal of covers and/or trim pieces. Be careful not to damage these components and to properly secure them upon installation. Keep the inside of the vehicle as clean as possible and remove any indications that you were ever in the vehicle before returning it to the customer.*

# Review Questions

1. The service precautions associated with servicing a control module are being discussed. Technician A says analog voltmeters must be used to test circuit voltage. Technician B says to make sure the ignition switch is turned off before making or breaking electrical connections to the module. Who is correct?
   A. Technician A only
   B. Technician B only
   C. Both Technician A and Technician B
   D. Neither Technician A nor Technician B
2. Technician A says that trouble codes will indicate the exact failure in the circuit. Technician B says that trouble codes will direct the technician to the circuit that has a fault in it. Who is correct?
   A. Technician A only
   B. Technician B only
   C. Both Technician A and Technician B
   D. Neither Technician A nor Technician B
3. Replacement of the PROM is being discussed. Technician A says when replacing the PROM chip, the notch in the PROM chip should be referenced to the smaller notch in the carrier. Technician B says the small notch of the carrier must be aligned with the small notch in the socket. Who is correct?
   A. Technician A only
   B. Technician B only
   C. Both Technician A and Technician B
   D. Neither Technician A nor Technician B
4. Technician A says hard code failures are those that have occurred in the past but were not present during the last test of the circuit. Technician B says intermittent codes are those that were detected the last time the module tested the circuit. Who is correct?
   A. Technician A only
   B. Technician B only
   C. Both Technician A and Technician B
   D. Neither Technician A nor Technician B
5. Retrieving trouble codes is being discussed. Technician A says the scan tool is the only method of pulling the codes. Technician B says to apply 12 volts to the diagnostic connector. Who is correct?
   A. Technician A only
   B. Technician B only
   C. Both Technician A and Technician B
   D. Neither Technician A nor Technician B
6. Technician A say installing a PROM chip backward will immediately destroy the chip. Technician B says electrostatic discharge will destroy the chip. Who is correct?
   A. Technician A only
   B. Technician B only
   C. Both Technician A and Technician B
   D. Neither Technician A nor Technician B

CHAPTER

# 7 Computer Inputs and Outputs

## Introduction

As discussed in Chapter 5, the CPU receives inputs that it checks with programmed values. Depending on the input, the computer will control the actuator(s) until the programmed results are obtained **(Figure 1).** The inputs can come from other computers, the driver, the technician, or through a variety of sensors.

Switches can be used as an input for any operation that only requires a yes/no, or on/off, condition. Other inputs include those supplied by means of a sensor and those signals returned to the computer in the form of feedback.

## Sensors

**Sensors** convert some measurement of vehicle operation into an electrical signal. There are many different designs of sensors. Some are nothing more than a switch that completes the circuit. Others are complex chemical-reaction devices that generate their own voltage under different conditions. Repeatability, accuracy, operating range, and **linearity** are all requirements of a sensor. Linearity refers to the sensor signal being as constantly proportional to the measured value as possible. It is an expression of the sensor's accuracy.

### Thermistors

A **thermistor** is used to sense engine coolant or ambient temperatures. A thermistor is a solid-state variable resistor made from a semiconductor material that changes resistance in relation to temperature changes. By monitoring the resistance of the thermistor, the computer can observe very small changes in temperature. The computer sends a reference voltage to the thermistor (usually 5 volts) through a fixed resistor. As the current flows through the thermistor resistance to ground, a voltage-sensing circuit measures the voltage after the fixed resistor **(Figure 2).** The voltage dropped over the fixed resistor will

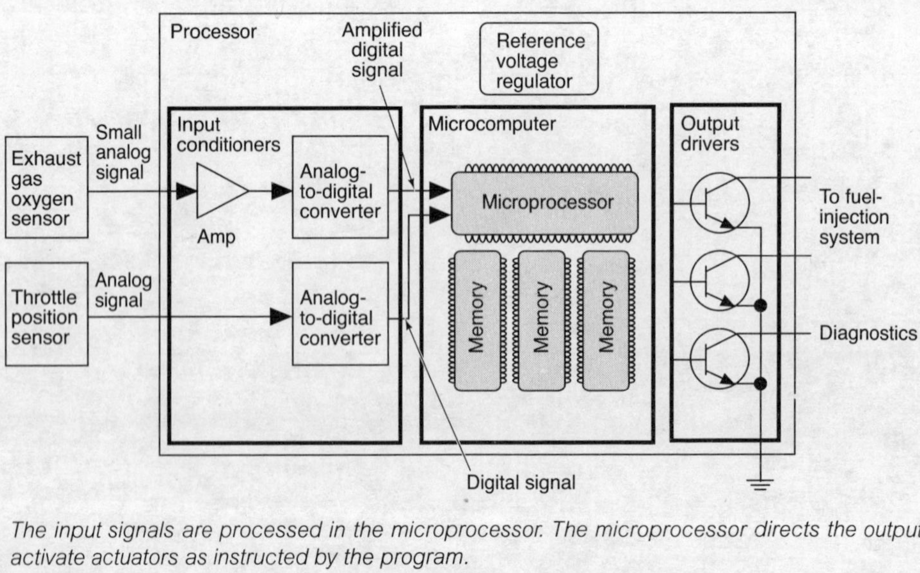


**Figure 1** *The input signals are processed in the microprocessor. The microprocessor directs the output drivers to activate actuators as instructed by the program.*

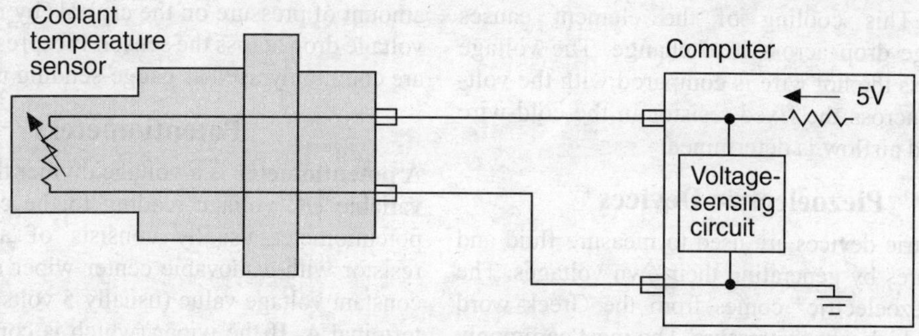


**Figure 2** *A thermistor is used to measure temperature. The sensing unit measures the resistance change and translates the data into temperature values.*

change as the resistance of the thermistor changes. Using its programmed values, the computer can translate the voltage drop into a temperature value.

There are two types of thermistors: **negative temperature coefficient (NTC)** thermistors and **positive temperature coefficient (PTC)** thermistors. NTC thermistors reduce their resistance as the temperature increases, while PTC thermistors increase their resistance as the temperature increases.

## Wheatstone Bridges

**Figure 3** shows the construction of the **Wheatstone bridge.** A Wheatstone bridge is nothing more than two simple series circuits connected in parallel across a power supply. Construction design varies among manufacturers, but usually three of the resistors are kept at exactly the same value and the fourth is the sensing resistor. When all four resistors have the same value, the bridge is balanced and the voltage sensor will indicate a value of 0 volt. The output from the amplifier acts as a voltmeter. Remember, since a voltmeter measures electrical pressure between two points, it will display this value. For example, if the reference voltage is 5 volts and the resistors have the same value, then the voltage drop over each resistor is 2.5 volts. Since the voltmeter is measuring the potential on the line between $R_S$ and $R_1$ and between $R_2$ and $R_3$, it will read 0 volt because both of these lines have 2.5 volts on them. If there is a change in the resistance value of the sense resistor, a change will occur in the circuit's balance. The sensing circuit will receive a voltage reading that is proportional to the amount of resistance change.

A common use of a Wheatstone bridge is as the hot-wire sensor in a mass airflow (MAF) sensor. The sensor consists of a hot-wire circuit, a cold-wire circuit, and an electronic signal processing area. The hot- and cold-wire circuits form the Wheatstone bridge. The cold-wire circuit is made of a fixed resistor and a thermistor. The temperature of the thermistor determines the amount of voltage dropped across the two resistors. The hot-wire circuit is made up of a fixed resistor and a variable resistance heat element (hot wire). The heat element generates heat in proportion to the amount of current flowing through it. This heat, in turn, changes its resistance. As air flows past the hot wire, it moves a small amount of heat from the

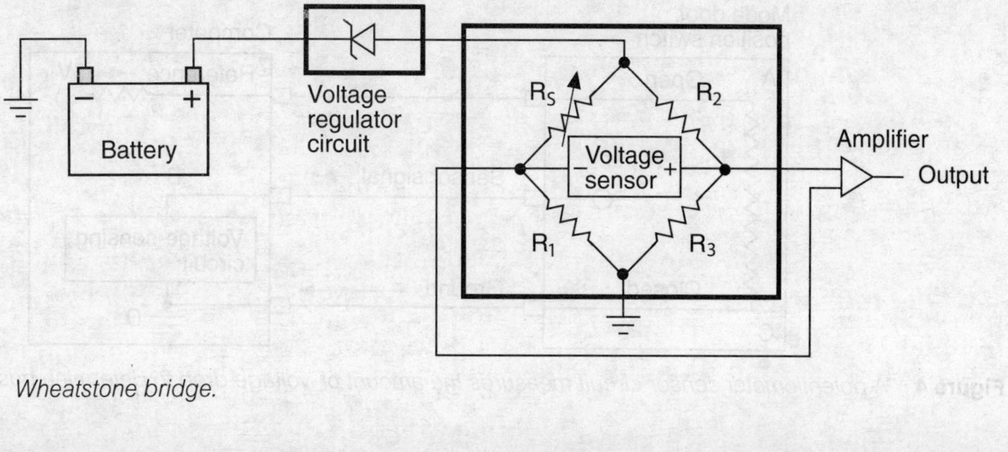


**Figure 3** *Wheatstone bridge.*

element. This cooling of the element causes the voltage drop across it to change. The voltage drop across the hot wire is compared with the voltage drop across the fixed resistor in the cold-wire circuit and airflow is determined.

## Piezoelectric Devices

Piezoelectric devices are used to measure fluid and air pressures by generating their own voltages. The word "piezoelectric" comes from the Greek word "piezo," which means pressure. The most commonly found piezoelectric device is the engine knock sensor. The knock sensor measures engine knock, or vibration, and converts the vibration into a voltage signal.

The sensor is a voltage generator and has a resistor connected in series with it. The resistor protects the sensor from excessive current flow in case the circuit becomes shorted. The voltage generator is a thin ceramic disc attached to a metal diaphragm. When engine knock occurs, the vibration of the noise puts pressure on the diaphragm. This puts pressure on the piezoelectric crystals in the ceramic disc. The disc generates a voltage that is proportional to the amount of pressure. The voltage generated ranges from 0 to 1 or more volts. Each time the engine knocks the sensor generates a voltage spike.

## Piezoresistive Devices

In construction, a piezoresistive device is similar to a piezoelectric one. However, it operates differently. The sensor acts like a variable resistor. Its resistance value changes as the pressure applied to the crystal changes. A voltage regulator supplies a constant voltage to the sensor. Since the amount of voltage dropped by the sensor will change with the change in resistance, the control module can determine the amount of pressure on the crystals by measuring the voltage drop across the sensor. Piezoresistive sensors are commonly used as gauge-sending units.

## Potentiometers

A **potentiometer** is a voltage divider that provides a variable DC voltage reading to the computer. The potentiometer usually consists of a wire-wound resistor with a movable center wiper **(Figure 4).** A constant voltage value (usually 5 volts) is applied to terminal A. If the wiper (which is connected to the shaft or movable component of the unit that is being monitored) is located close to this terminal, there will be low-voltage drop represented by a high-voltage signal back to the computer through terminal B. As the wiper is moved toward the C terminal, the sensor signal voltage to terminal B decreases. The computer interprets the different voltage values into different shaft positions. The potentiometer can measure linear or rotary movement. As the wiper is moved across the resistor, the computer can track the position of the unit.

Since applied voltage must flow through the entire resistance, temperature and other factors do not create false or inaccurate sensor signals to the computer.

## Magnetic Pulse Generators

**Magnetic pulse generators** use the principle of magnetic induction to produce a voltage signal. These sensors are commonly used to send data to the computer about the speed of the monitored component. The magnetic pulse generator is also used to inform the computer of the position of a monitored component. This is common in engine controls where the computer needs to know the position of the crankshaft in relation to rotational degrees.

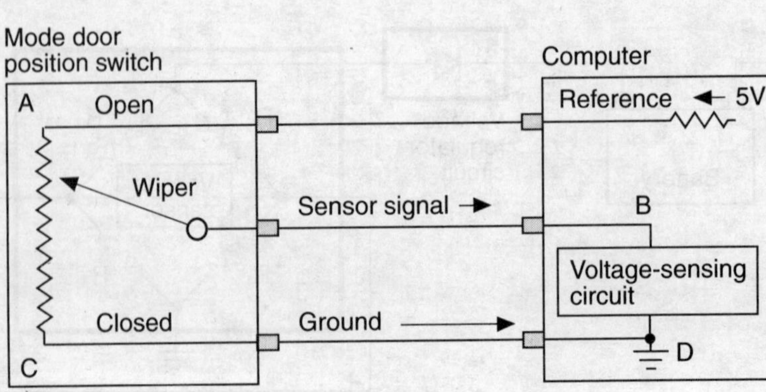


**Figure 4** *A potentiometer sensor circuit measures the amount of voltage drop to determine position.*

> **You Should Know**
>
> *Magnetic pulse generators are also called permanent magnet generators and variable reluctance sensors (VRS).*

The components of the pulse generator are:

1. A **timing disc** (reluctor) that is attached to the rotating shaft or cable. The number of teeth on the timing disc is determined by the manufacturer and depends on application. The teeth will cause a voltage generation that is constant per revolution of the shaft. For example, a vehicle speed sensor may be designed to deliver 4,000 pulses per mile. The number of pulses per mile remains constant regardless of speed. The computer calculates how fast the vehicle is going based on the frequency of the signal.

> **You Should Know**
>
> *The timing disc is known as an armature, reluctor, trigger wheel, pulse wheel, or timing core. It is used to conduct lines of magnetic force.*

2. A **pickup coil.** The coil consists of a permanent magnet that has fine wire wound around it. The changes of magnetic lines of force generate a small voltage signal in the coil.

> **You Should Know**
>
> *The **pickup coil** is also known as a stator, sensor, or pole piece. It remains stationary while the timing disc rotates in front of it.*

3. A magnet.
4. A pole piece.

An air gap is maintained between the timing disc and the pickup coil. As the timing disc rotates in front of the pickup coil, the generator sends an AC signal **(Figure 5).** As a tooth on the timing disc aligns with the core of the pickup coil, it repels the magnetic field. The magnetic field is forced to flow through the pickup coil and core **(Figure 6).** Since the magnetic field is not expanding, a voltage of 0 is induced in the pickup coil. As the tooth passes the core, the magnetic field is able to expand **(Figure 7).** The expanding magnetic field cuts across the windings of the pickup coil. This movement of the magnetic field induces a voltage in the windings. This action is repeated every time a tooth passes the core. The moving lines of magnetic force cut across the coil windings and induce a voltage signal.

When a tooth approaches the core, a positive current is produced as the magnetic field begins to concentrate around the coil **(Figure 8).** The voltage will continue to increase as long as the magnetic field is expandAs the tooth approaches the magnet, the magnetic field becomes smaller, causing the induced voltage to decrease and drop toward 0. When the tooth and core align, there is no more expansion or contraction of the magnetifield (thus no movement) and the voltage drops to 0 **(Figure 9).** When the tooth passes the core, the magnetic field expands and a negative current is produced **(Figure 10).** The resulting pulse signal is amplified, digitized, and sent to the microprocessor.

The toothed disc is mounted on the crankshaft, vibration damper, or distributor shaft in a very precise manner. When the disc teeth align with the pickup coil corresponds to the exact time certain pistons are nearing top dead center (TDC). This means the 0 voltage signal needed to trigger the secondary circuit occurs at precisely the correct time.

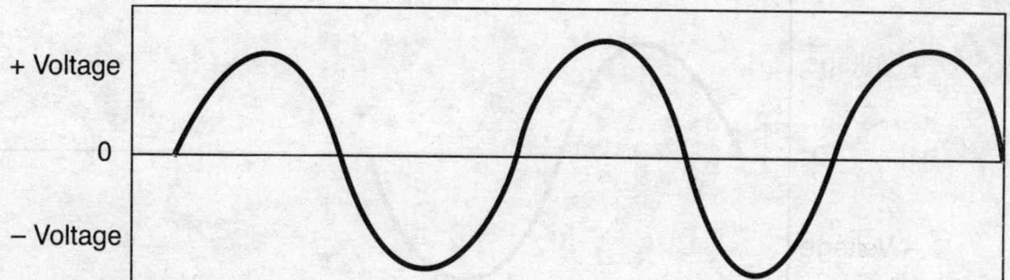


**Figure 5** *AC signal from a pulse generator.*

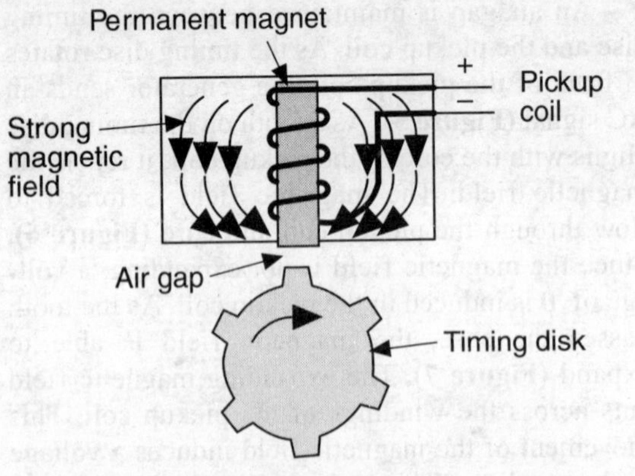


**Figure 6** *A strong magnetic field is produced in the pickup coil as the teeth align with the core.*

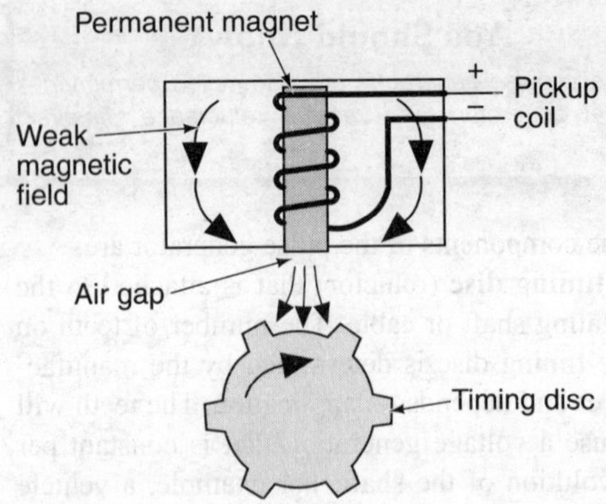


**Figure 7** *The magnetic field expands as the teeth pass the core.*

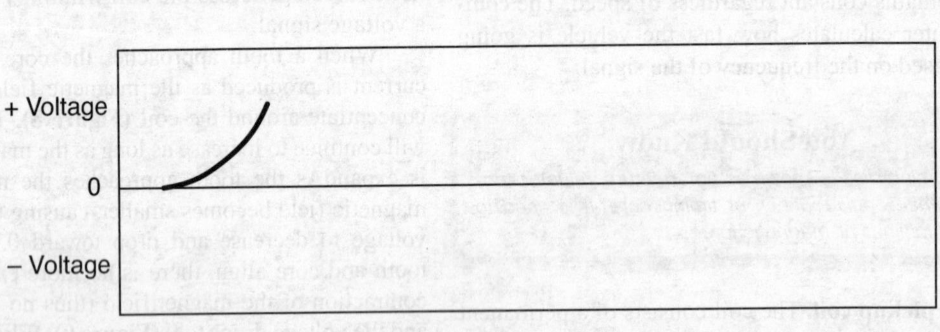


**Figure 8** *A positive voltage swing is produced as the tooth approaches the core.*

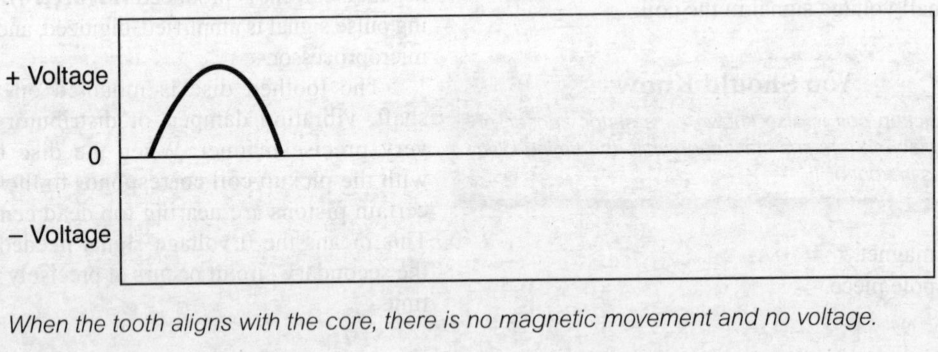


**Figure 9** *When the tooth aligns with the core, there is no magnetic movement and no voltage.*

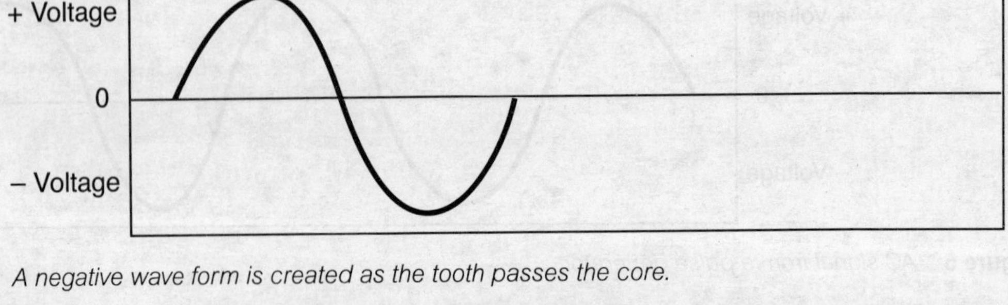


**Figure 10** *A negative wave form is created as the tooth passes the core.*

> **You Should Know**
>
> *The magnetic pulse generator operates on basic magnetic principles. Remember that a voltage can only be induced when a magnetic field is moved across a conductor. The magnetic field is provided by the pickup unit, and the rotating timing disc provides the movement of the magnetic field needed to induce voltage.*

## Metal Detection Sensors

**Metal detection sensors** are found on many early electronic ignition systems. They work much like a magnetic pulse generator with one major difference: the pickup coil of a metal detection sensor does not have a permanent magnet. Instead, the pickup coil is an electromagnet. An electronic control unit supplies a low level of current to the coil, inducing a weak magnetic field around the coil. As the reluctor on the distributor shaft rotates, the trigger teeth pass very close to the coil **(Figure 11).** As the teeth pass in and out of the coil's magnetic field, the magnetic field builds and collapses, producing a corresponding change in the coil's voltage. The voltage changes are monitored by the control unit to determine crankshaft position.

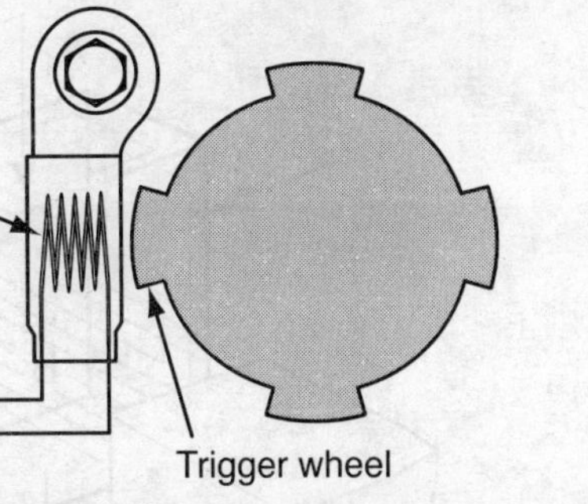


**Figure 11** *A metal detector sensor.*

## Hall-Effect Switches

The **Hall-effect switch** performs the same functions as the magnetic pulse generator; however, its operation is different. The Hall-effect switch operates on the principle that if a current is allowed to flow through thin conducting material that is exposed to a magnetic field, another voltage is produced. It contains a permanent magnet, a thin semiconductor layer made of gallium arsenate crystal (Hall layer), and a shutter wheel **(Figure 12).** The Hall layer has a negative and a positive terminal connected to it. Two additional terminals located on either side of the Hall layer are used for the output circuit.

The permanent magnet is located directly across from the Hall layer so that its lines of flux will bisect at right angles to the current flow. The permanent magnet is mounted so that a small air

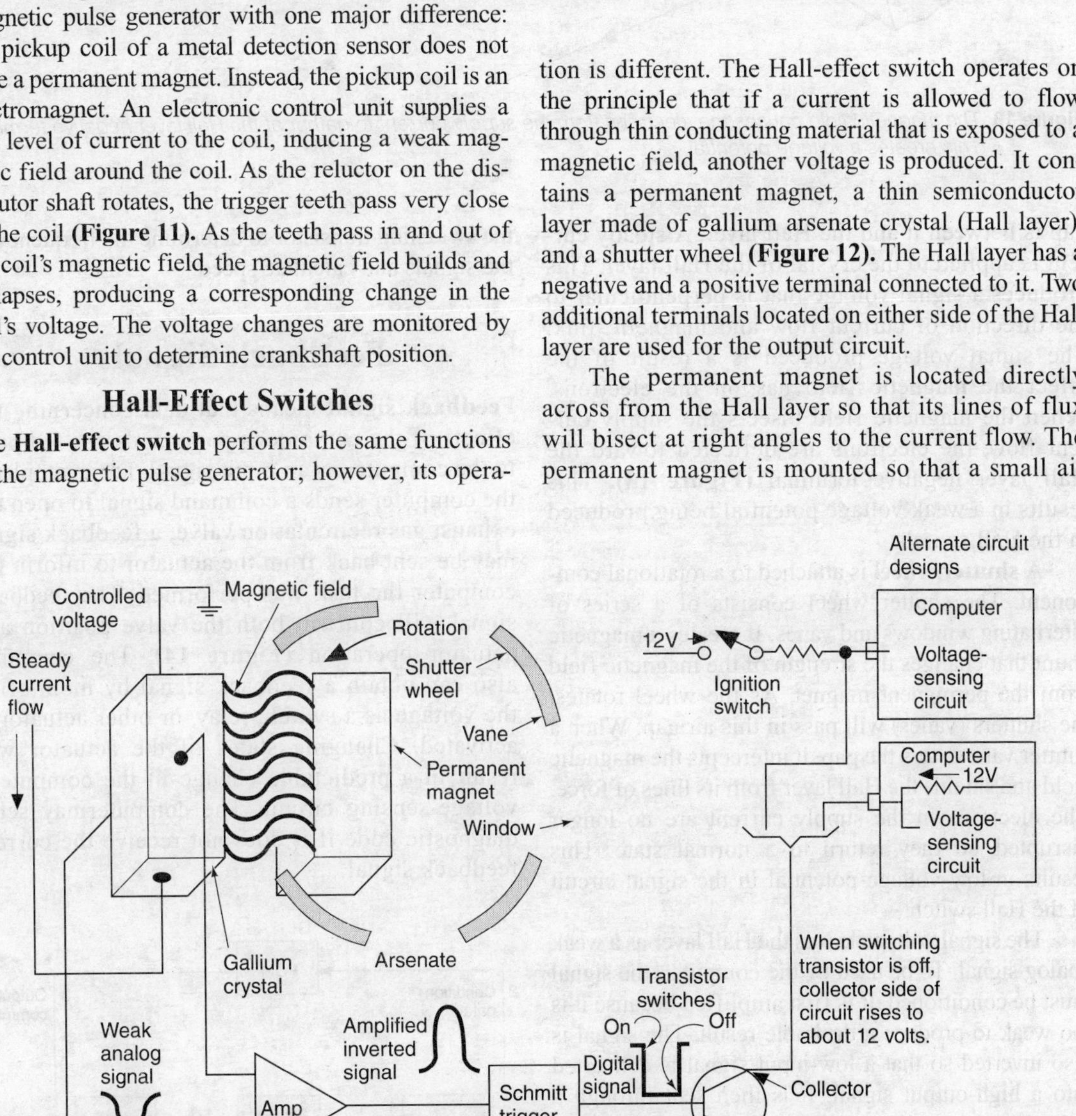


**Figure 12** *Typical circuit of a Hall-effect switch.*

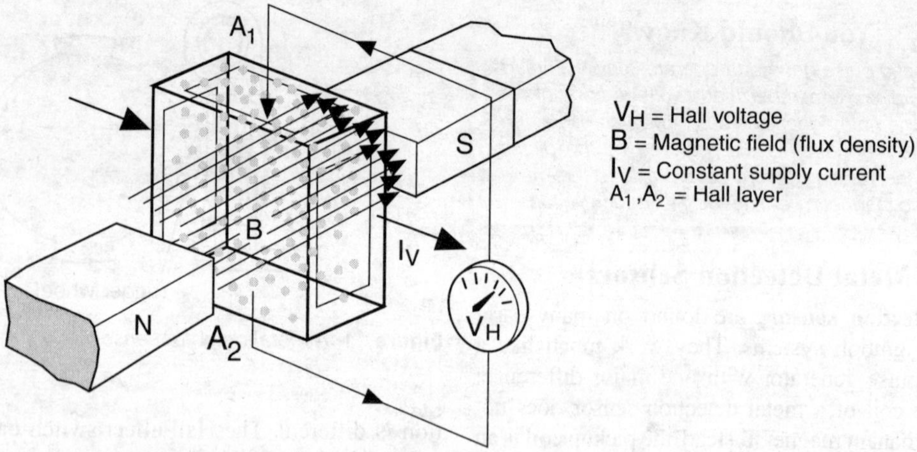


**Figure 13** *The magnetic field causes the electrons from the supply current to gather at the Hall layer negative terminal. This creates a voltage potential.*

gap is between it and the Hall layer. A steady current is applied to the crystal of the Hall layer. This produces a signal voltage that is perpendicular to the direction of current flow and magnetic flux. The signal voltage produced is a result of the effect the magnetic field has on the electrons. When the magnetic field bisects the supply current flow, the electrons are deflected toward the Hall layer negative terminal **(Figure 13).** This results in a weak voltage potential being produced in the Hall switch.

A **shutter wheel** is attached to a rotational component. The shutter wheel consists of a series of alternating windows and vanes. It creates a magnetic shunt that changes the strength of the magnetic field from the permanent magnet. As the wheel rotates, the shutters (vanes) will pass in this air gap. When a shutter vane enters the gap, it intercepts the magnetic field and shields the Hall layer from its lines of force. The electrons in the supply current are no longer disrupted, so they return to a normal state. This results in low-voltage potential in the signal circuit of the Hall switch.

The signal voltage leaves the Hall layer as a weak analog signal. To be used by the computer, the signal must be conditioned. It is first amplified because it is too weak to produce a desirable result. The signal is also inverted so that a low-input signal is converted into a high-output signal. It is then sent through a **Schmitt trigger** where it is digitized (converted from AC to DC) and conditioned into a clean square wave signal. The signal is finally sent to a switching transistor. The computer senses the turning on and off of the switching transistor to determine the frequency of the signals and calculate speed.

## Feedback Signals

**Feedback signal** means that data concerning the effects of the computer's commands are fed back to the computer as an input signal. For example, if the computer sends a command signal to open the exhaust gas recirculation valve, a feedback signal may be sent back from the actuator to inform the computer the task was performed. The feedback signal will confirm both the valve position and actuator operation **(Figure 14).** The computer also can obtain a feedback signal by monitoring the voltage as a switch, relay, or other actuator is activated. Changing states of the actuator will result in a predictable change in the computer's voltage-sensing circuit. The computermay set a diagnostic code if it does not receive the correct feedback signal.

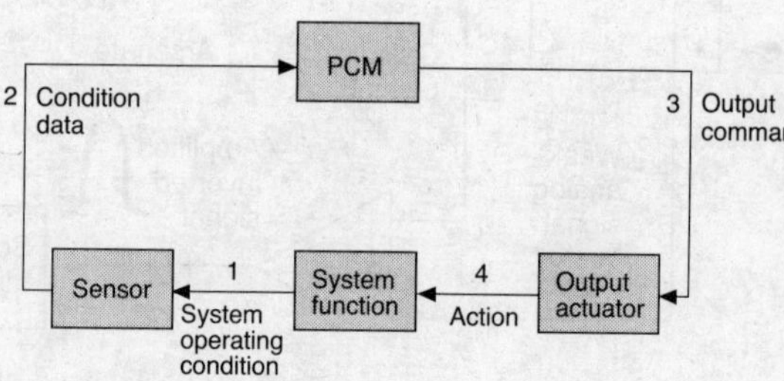


**Figure 14** *Principle of feedback signals.*

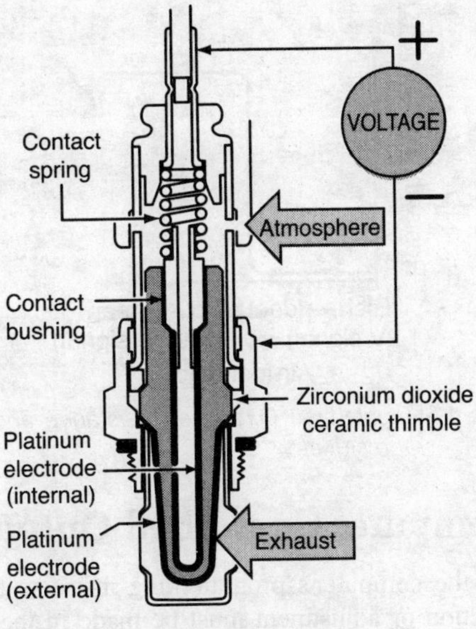


**Figure 15** *Oxygen ($O_2$) sensor design.*

## Oxygen ($O_2$) Sensors

One of the most commonly used feedback sensors is the $O_2$ sensor. The $O_2$ sensor is mounted in the exhaust gas stream and provides the PCM with a measurement of the oxygen in the engine's exhaust. The sensor is constructed of a zirconium dioxide ceramic thimble covered with a thin layer of platinum **(Figure 15).**

When the thimble is filled with oxygen-rich outside air and the outer surface of the thimble is exposed to oxygen-depleted exhaust gases, a chemical reaction in the sensor produces a voltage. The generation of voltage is similar to the same activity that takes place in a battery, except at much lower voltages. The voltage output varies with the level of oxygen present in the exhaust. As oxygen in the exhaust decreases, the voltage output increases. Likewise, as the oxygen level in the exhaust increases, the output voltage decreases **(Figure 16).**

Some $O_2$ sensors have a single wire that is connected from the oxygen-sensing element to the computer, and this wire acts as a signal wire. If an $O_2$ sensor has two wires, the second wire is a ground wire connected back to the computer.

Most $O_2$ sensors today use an internal heater element. Since the $O_2$ sensor does not produce a satisfactory signal until it reaches about 600ºF (315°C), the internal heater provides faster sensor warm-up time and helps to keep the sensor hot during prolonged idle operation. The internal $O_2$ sensor heater maintains higher sensor temperatures, which helps to burn deposits off the sensor. When the $O_2$ sensor has an internal heater, the sensor can be placed farther away from the engine in the exhaust stream, giving engineers more flexibility in sensor location. These $O_2$ sensors are identified by having three or four wires. The third wire is connected to an electric heating element in the sensor. Voltage is supplied from the ignition switch (or a relay that is energized when the engine is running) to this heater. In a four-wire sensor, there is an individual circuit for the signal wire, the heater wire, and the two ground wires. In these four-wire sensors, the heater and the sensing element have individual ground wires. A replacement $O_2$ sensor must have the same number of wires as the original sensor.

A lean air-fuel ratio provides excess quantities of oxygen in the exhaust stream because the mixture

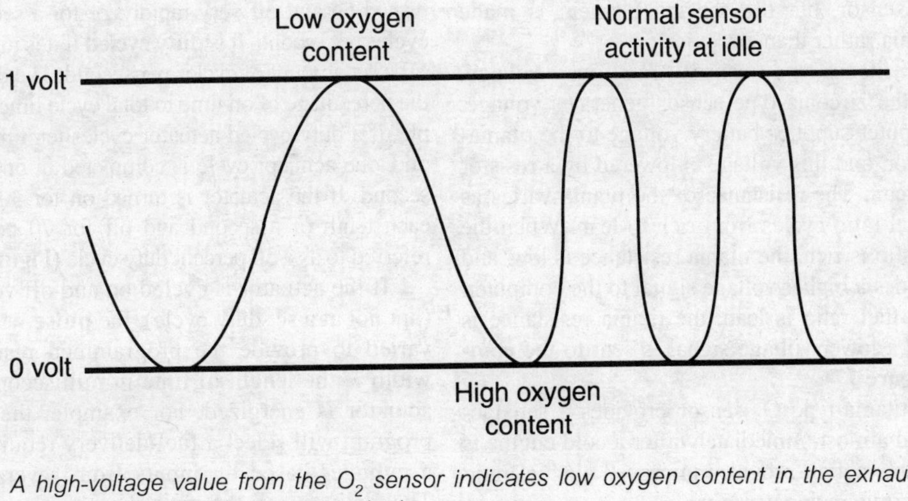


**Figure 16** *A high-voltage value from the $O_2$ sensor indicates low oxygen content in the exhaust. A low-voltage indicates high oxygen content.*

entering the cylinders has an excessive amount of air in relation to the amount of fuel. Therefore, air containing oxygen is left over after the combustion process. When the exhaust stream has high oxygen content, oxygen from the atmosphere is also present inside the $O_2$ sensor element. When oxygen is present on both sides of the sensor element, the sensor produces a low voltage.

A rich air-fuel ratio contains excessive fuel in relation to the amount of air entering the cylinders. A rich air-fuel mixture produces very little oxygen in the exhaust stream because the oxygen in the air is all mixed with fuel, and excess fuel is left over after combustion is completed. When the exhaust stream with very low oxygen content strikes the $O_2$ sensor, there is high oxygen content from the atmosphere inside the sensor element. With different oxygen levels on the inside and outside of the sensor element, the sensor produces up to 1 volt. As the air-fuel ratio cycles from lean to rich, the $O_2$ sensor voltage changes in a few milliseconds.

In a gasoline fuel system, the stoichiometric (ideal) air-fuel mixture is 14.7:1. At the stoichiometric air-fuel ratio, combustion is most efficient, and nearly all the oxygen in the air is mixed with fuel and burned in the combustion chambers. Computer-controlled fuel-injection systems maintain the air-fuel ratio at stoichiometric under most operating conditions. As the air-fuel ratio cycles slightly rich and lean from the stoichiometric ratio, the $O_2$ sensor voltage cycles from high to low. Once the engine is at normal operating temperature, the $O_2$ sensor voltage should vary between 0.3 volt and 0.8 volt if the air-fuel ratio is at, or near, stoichiometric.

Some vehicles are equipped with a titania-type $O_2$ sensor. This sensor design is similar to the zirconia-type sensor, but the sensing element is made from titania rather than zirconia.

The titania-type sensor modifies voltage, whereas the zirconia-type sensor generates voltage. The computer supplies battery voltage to the titania-type sensor, and this voltage is lowered by a resistor in the circuit. The resistance of the titania varies as the air-fuel ratio cycles from rich to lean. When the air-fuel ratio is rich, the titania resistance is low, and this provides a higher voltage signal to the computer. If the air-fuel ratio is lean, the titania resistance is high, and a lower voltage signal is sent to the computer **(Figure 17).**

The titania-type $O_2$ sensor provides a satisfactory signal almost immediately after a cold engine is started. This action provides improved air-fuel ratio control during engine warm-up.

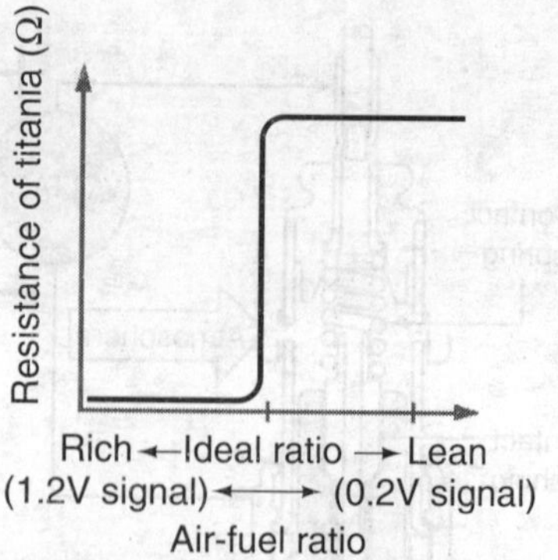


**Figure 17** *Titania-type $O_2$ sensor resistance and voltage signal.*

## Computer Controlled Outputs

Once the computer's programming instructs that a correction or adjustment must be made in the controlled system, an output signal is sent to an **actuator.** This involves translating the electronic signals into mechanical motion. Actuators perform the actual work commanded by the computer. They can be in the form of a motor, relay, switch, or solenoid.

An output driver is used within the computer to control the actuators. The driver usually applies the ground circuit of the actuator. The ground can be applied steadily if the actuator must be activated for a selected amount of time. For example, if the PCM inputs indicate that the air conditioner clutch relay needs to be energized, the driver keeps the relay energized steadily until inputs inform the PCM that air conditioner clutch operation is not needed. Then, the ground for the relay coil is removed.

Other systems require the actuator to either be turned on and off very rapidly, or for a set amount of cycles per second. It is **duty cycled** if it is turned on and off a set amount of cycles per second. The duty cycle is the percentage of on time to total cycle time. For example, if a duty-cycled actuator cycles ten times per second, one actuator cycle is completed in one-tenth of a second. If the actuator is turned on for 30 percent of each tenth of a second and off for 70 percent, it is referred to as a 30 percent duty cycle **(Figure 18).**

If the actuator is cycled on and off very rapidly (but not at a set duty cycle), the **pulse width** can be varied to provide the programmed results. Pulse width is the length of time in milliseconds that an actuator is energized. For example, the computer program will select a fuel-delivery requirement for a cylinder based on inputs from several sensors. The delivery of the fuel through the injector is

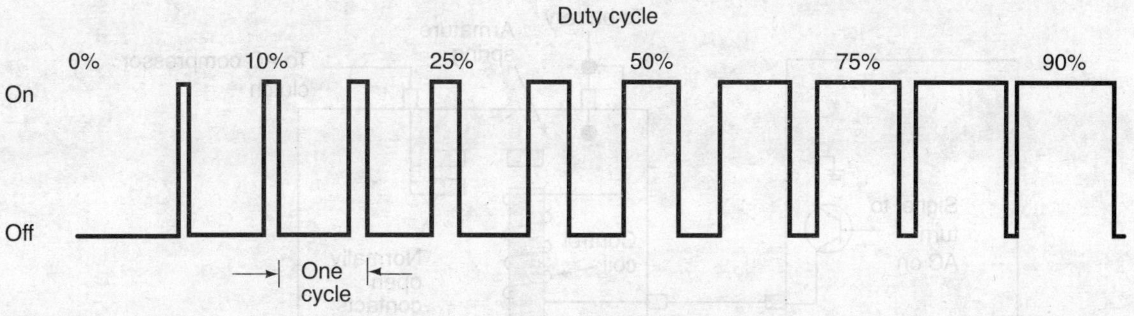


**Figure 18** *Duty cycle is the percentage of on time per cycle. Duty cycle can be changed; however, total cycle time remains constant.*

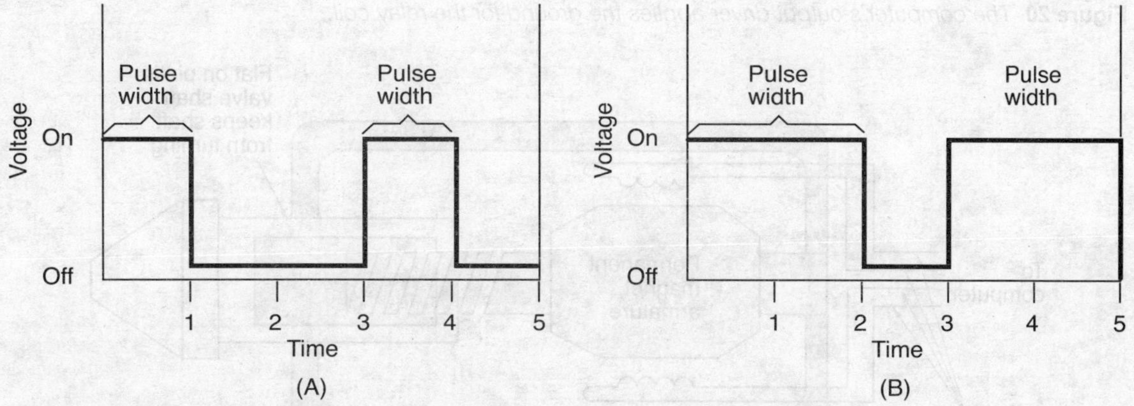


**Figure 19** *Pulse width is the duration of on time. (A) Pulse width modulation logic used to achieve fuel flow. (B) Pulse width modulation logic used to achieve increased fuel flow.*

achieved through pulse-width modulation of the injector driver. If more fuel is required, the pulse width is increased, which increases the length of on time. As the fuel quantity needs to be reduced, the pulse width is decreased **(Figure 19).**

## Actuators

Most computer-controlled actuators are electromechanical devices that convert the output commands from the computer into mechanical action. These actuators are used to open and close switches, control vacuum flow to other components, and operate doors or valves depending on the requirements of the system.

**Relays.** A relay allows control of a high-current draw circuit by a very low–current draw circuit. The computer usually controls the relay by providing the ground for the relay coil **(Figure 20).** The use of relays protects the computer by keeping the high current from passing through it. For example, the electromagnetic clutch used on an air conditioner compressor requires a high-current draw to operate. Instead of having the computer operate the clutch directly, it will energize a relay. With the relay energized, a direct circuit from the battery to the clutch is completed.

**Solenoids.** Computer control of the solenoid is usually provided by applying the ground through the output driver. A solenoid is commonly used as an actuator because it operates well under duty-cycling conditions. Many newer PCMs will use high-side drivers to control some solenoids. One of the most common uses of the solenoid is the fuel injector. Another common use is to control vacuum to other components.

**Motors.** Many computer-controlled systems use a **stepper motor** to move the controlled device to whatever location is desired **(Figure 21).** A stepper motor contains a permanent magnet armature with two, four, or more field coils. By applying voltage pulses to selected coils of the motor, the armature will turn a specific number of degrees. When the same voltage pulses are applied to the opposite coils, the armature will rotate the same number of degrees in the opposite direction.

Some applications require the use of a permanent magnet field servo motor **(Figure 22).** The polarity of

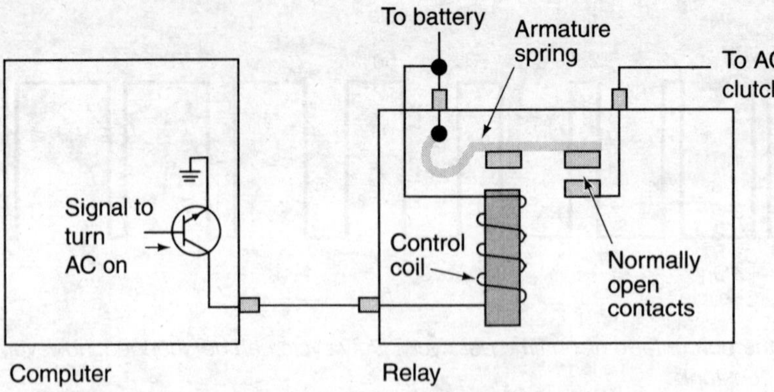


**Figure 20** *The computer's output driver applies the ground for the relay coil.*

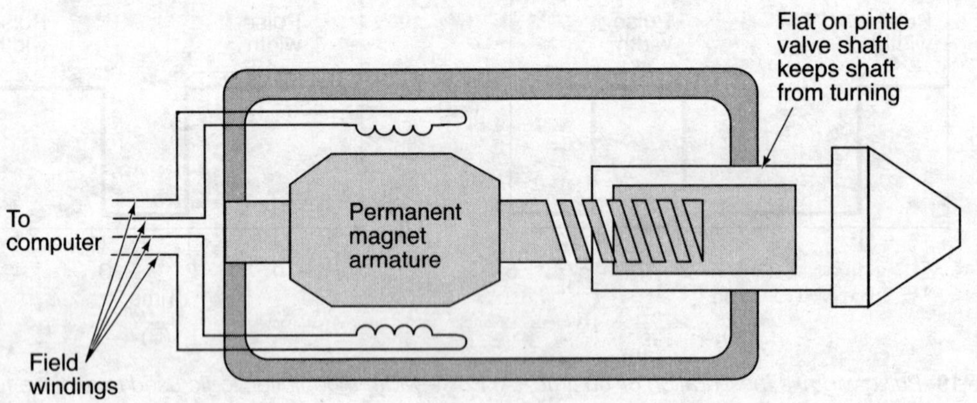


**Figure 21** *Typical stepper motor.*

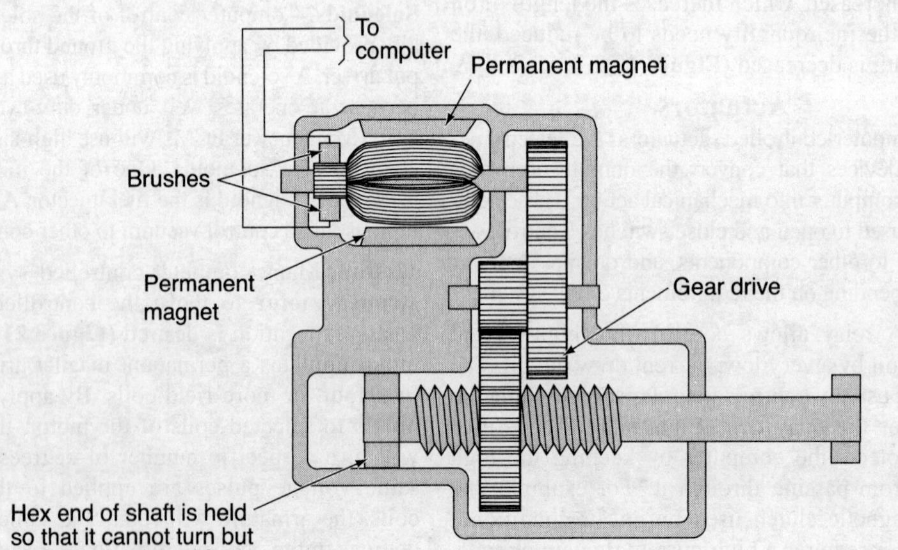


**Figure 22** *Reversible permanent magnet motor.*

the voltage applied to the armature windings determines the direction the motor rotates. The computer can apply a continuous voltage to the armature until the desired result is obtained.

## High-Side and Low-Side Drivers

Usually the computer will control an actuator by the use of **low-side drivers**. These drivers will complete the path to ground through a FET transistor to control the output device. The computer may monitor the voltage on this circuit to determine if the actuator operates when commanded **(Figure 23)**. Monitoring of the system can be done by either measuring voltage on the circuit or by measuring the current draw of the circuit.

Many newer vehicles are now using **high-side drivers** that control the output device by varying the positive (12 volt) side of the circuit. High-side drivers consist of a metal-oxide field-effect transistor (MOSFET) that is controlled by a bipolar transistor. The bipolar transistor is controlled by the microprocessor. The advantage of the high-side driver is that it can provide quick response self-diagnostics for shorts, opens, and thermal conditions. It also reduces vehicle wiring.

High-side driver diagnostic capabilities include the ability to determine a short-circuit or open-circuit condition. The high-side driver will take the place of a fuse in the event of a short-circuit condition. When it senses a high-current condition, it will turn off the power flow and then store a DTC in memory. The driver will automatically reset once the short-circuit condition is removed. In addition, the

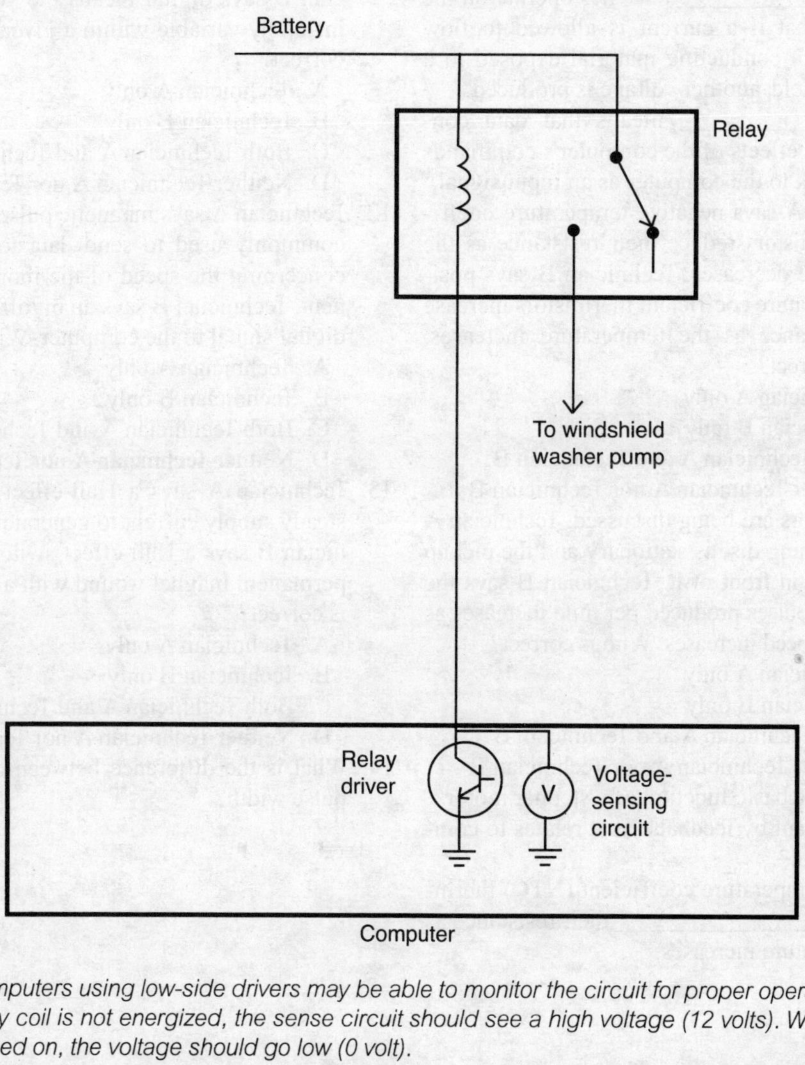


**Figure 23** *Computers using low-side drivers may be able to monitor the circuit for proper operation. When the relay coil is not energized, the sense circuit should see a high voltage (12 volts). When the relay is turned on, the voltage should go low (0 volt).*

high-side driver monitors its temperature. The driver reports the junction temperature to the microprocessor. If a slow-acting resistive short occurs in the circuit, the temperature will begin to climb. Once the temperature reaches 300ºF (150ºC), the driver will turn off and set a DTC.

The high-side driver can also detect an open circuit, even if the system is turned off. This is done by reading a feedback voltage to the microprocessor when the driver is off. A 5 volt, 50 μA current is fed through the circuit, which also has a resistor wired in parallel. Low voltage (less than 2.25 volts) will indicate a normal circuit. If the voltage is high (above 2.25 volts), a high resistance or open circuit is detected. If the open circuit is detected, a DTC is set.

## Review Questions

1. How does the Hall-effect switch generate a voltage signal?
2. ______________ convert some measurement of vehicle operation into an electrical signal.
3. ______________ switches operate on the principle that if a current is allowed to flow through thin conducting material exposed to a magnetic field, another voltage is produced.
4. ______________ means that data concerning the effects of the computer's commands are fed back to the computer as an input signal.
5. Technician A says negative temperature coefficient thermistors reduce their resistance as the temperature decreases. Technician B says positive temperature coefficient thermistors increase their resistance as the temperature increases. Who is correct?
   A. Technician A only
   B. Technician B only
   C. Both Technician A and Technician B
   D. Neither Technician A nor Technician B
6. Speed sensors are being discussed. Technician A says the timing disc is stationary and the pickup coil rotates in front of it. Technician B says the number of pulses produced per mile increases as rotational speed increases. Who is correct?
   A. Technician A only
   B. Technician B only
   C. Both Technician A and Technician B
   D. Neither Technician A nor Technician B
7. Describe the basic function of a stepper motor.
8. What is meant by feedback as it relates to computer control?
9. Negative temperature coefficient (NTC) thermistors ______________ their resistance as the temperature increases.
10. Magnetic pulse generators use the principle of ______________ to produce a voltage signal.
11. Technician A says analog means the voltage signal is either on/off, yes/no, or high/low. Technician B says digital means the voltage signal is infinitely variable within a given range. Who is correct?
    A. Technician A only
    B. Technician B only
    C. Both Technician A and Technician B
    D. Neither Technician A nor Technician B
12. Technician A says magnetic pulse generators are commonly used to send data to the computer concerning the speed of the monitored component. Technician B says an on/off switch sends a digital signal to the computer. Who is correct?
    A. Technician A only
    B. Technician B only
    C. Both Technician A and Technician B
    D. Neither Technician A nor Technician B
13. Technician A says a Hall-effect switch uses a steady supply current to generate a signal. Technician B says a Hall-effect switch consists of a permanent magnet wound with a wire coil. Who is correct?
    A. Technician A only
    B. Technician B only
    C. Both Technician A and Technician B
    D. Neither Technician A nor Technician B
14. What is the difference between duty cycle and pulse width?

CHAPTER

# 8 Input and Output Diagnostics

## Introduction

This chapter presents the testing of sensors and actuators to orient you to the basic diagnostic procedures. Specific procedures will be presented throughout this manual for individual systems and types of sensors.

There are many different designs of sensors, depending on the operation of the system they monitor. Some sensors are nothing more than a switch that completes the circuit. Others are complex chemical-reaction devices that generate their own voltage under different conditions.

Actuators can be simple relays while others are pulse width modulated solenoids. DC reversible stepper motors are also common actuators.

## Testing Sensors

Testing sensors can include the use of DMMs and lab scopes. Depending on the type of sensor, special test equipment may be required. Usually a sensor can be diagnosed with a scan tool. Common sensor testing procedures are included here.

### Potentiometer Testing

Potentiometers are tested by measuring the input voltage to the sensor and the feedback voltage to the computer. The feedback voltage to the computer should change smoothly as the resistance value of the sensor changes. To test these voltage signals, a series of jumper wires may be required (**Figure 1**). The jumper wires provide a method of gaining access to the terminals of weather-pack connectors without breaking the wire insulation.

An ohmmeter can be used to measure the changes in resistor values. Disconnect the sensor from the system. Connect the ohmmeter leads to the reference and ground terminals (**Figure 2**). Check

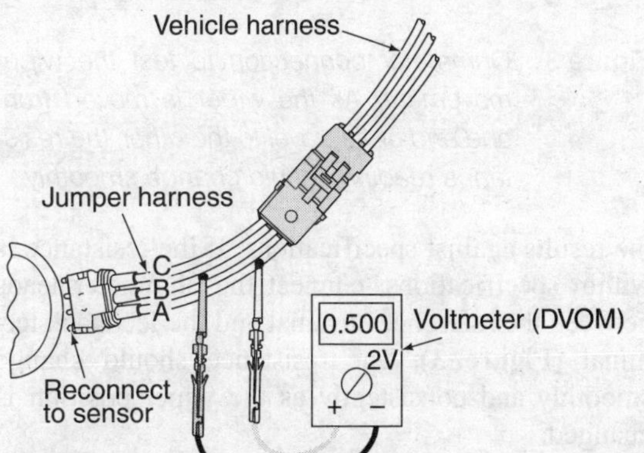


**Figure 1** *Jumper wires connected between the sensor and the harness allow the technician to probe for voltage or test resistance without damaging the wiring.*

**You Should Know**

*Do not use a test light to test for voltage. This may damage the system. Also, do not probe for voltage by sticking the wire insulation.*

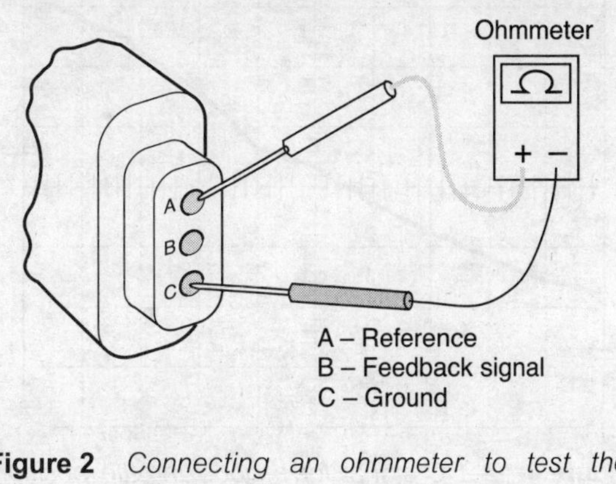


**Figure 2** *Connecting an ohmmeter to test the potentiometer. This would give the fixed resistance value.*

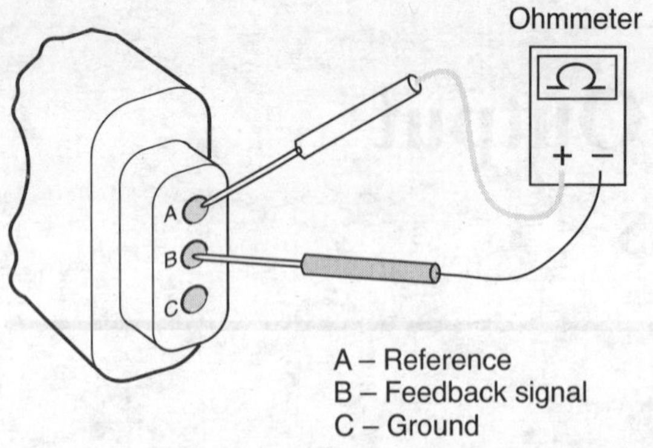


**Figure 3** *Ohmmeter connection to test the wiper movement. As the wiper is moved from one end of its travel to the other, the resistance reading should change smoothly.*

the results against specifications. If the resistance is within specifications, connect the ohmmeter leads between the reference terminal and the feedback terminal (**Figure 3**). The resistance should change smoothly and consistently as the wiper position is changed.

Potentiometer testing with a lab scope is a good way to watch the sweep of the resistor. The waveform (**Figure 4**) is an analog signal that moves up as the voltage increases. Most potentiometers in computer systems are fed a reference voltage of 5 volts; therefore, the voltage output of these sensors will range from about 500 millivolts (mV) to approximately 4.5 volts. The change in voltage should be smooth. Look for glitches in the signal. These can be caused by changes in resistance or an open.

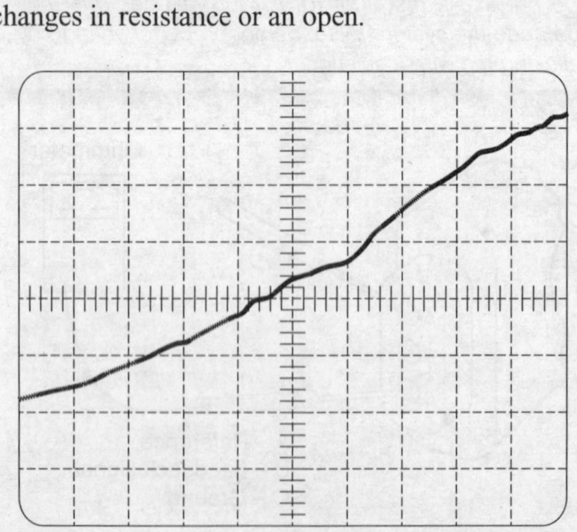

**Figure 4** *A DC signal for a good throttle position sensor measured from closed throttle to wide open throttle.*

## Thermistor Testing

Usually a thermistor circuit is a two-wire system with a signal wire and a return. An easy test of this system is to use a scan tool and monitor the voltage of the sensor. With the key on and engine off, record the voltage displayed. Next, disconnect the sensor connector and the voltage should change to the reference voltage (usually 5 volts). If it does not change, check the signal circuit for a short to ground. With the connector unplugged, jump a wire across the two terminals of the connector on the wire harness side. The scan tool should display 0 volt. If it does not, then test the signal wire and the return wire for opens. It the test results are correct, replace the sensor.

The thermistor can also be tested using an ohmmeter. You must first know if the thermistor is NTC or PTC. Measure the resistance of the thermistor by connecting the ohmmeter leads across the terminals. Compare your results with specifications. Next heat the sensor. This process depends on what the sensor is used for. If the sensor measures engine coolant temperature then it can be warmed in hot water. However, if the sensor measures ambient air temperature, then heat it by holding it in your hand or near a heat source. Compare the resistance value of the hearted sensor to specifications.

The lab scope is also a reliable method of testing the circuit and sensor. Connect the lab scope leads across the signal wire and a good ground. As the engine warms, the voltage value should change smoothly. Look for glitches in the signal. These can be caused by changes in resistance or an open.

## Testing Magnetic Pulse Generators

The pickup coil of a magnetic pulse generator or metal detection sensor is checked for proper resistance using an ohmmeter. Connect the ohmmeter to the pickup coil terminals to test the pickup coil for an open or a shorted condition. While the ohmmeter leads are connected, pull on the pickup leads and watch for a change in the readings. Any change would indicate an intermittent open in the pickup leads. Most pickup coils have 150 to 900 ohms of resistance, but always refer to the manufacturer's specifications. If the pickup coil is open, the ohmmeter will display an infinite reading. When the pickup coil is shorted, the ohmmeter will display a reading lower than specifications.

Connect the ohmmeter from one of the pickup leads to ground to test the pickup for a short to

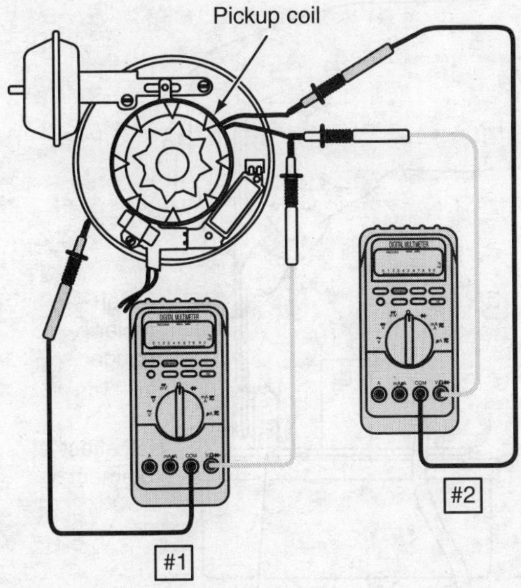


**Figure 5** *Ohmmeter tests on a pickup coil. Testing for opens and shorts to ground.*

ground. If there is no short to ground, the ohmmeter will give an infinite reading. In **Figure 5**, ohmmeter #1 is testing for a short to ground, while ohmmeter #2 is testing the resistance of the pickup coil.

A scope can be used to check pickup coil operation. Connect the scope's leads across the pickup coil and set the scope to a low AC voltage scale. When the distributor shaft is spun, an AC waveform should appear on the screen (**Figure 6**). The trace is not a true sine wave, but it should have both a positive and a negative pulse.

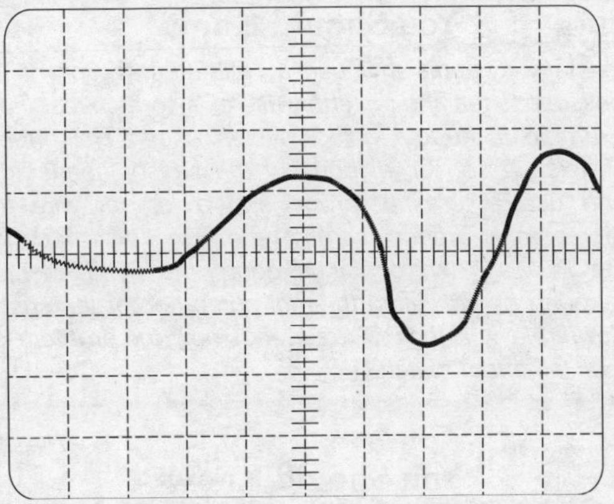

**Figure 6** *An AC voltage signal from a magnetic pulse generator used as a wheel speed sensor. The signal is changing over time because the speed of the wheel is accelerating quickly.*

Another method of measuring the pickup coil's AC signal is with a voltmeter set on its low AC voltage scale. The meter registers AC voltage during cranking. Measure this voltage as close to the control module as possible to account for any resistance in the connecting wire from the pickup coil to the module.

## Testing Hall-Effect Sensors

The best way to test Hall-effect inputs is to use a scope. A DVOM can be used to confirm that the proper voltages and grounds are supplied to the sensor, but it will not be able to indicate the quality of the signal. With a lab scope, the unit can be checked while the engine is running. Connect the positive lead of the scope to the signal wire by back probing the connector. The connector must be plugged into the Hall-effect switch. Since Hall-effect switches are commonly used to measure camshaft and crankshaft rotation, the engine will need to be rotated or started to check the signal pattern. The trace should show a clean 5-volt or 12-volt square wave pattern (based on design) that increases in frequency as shaft rpm increases and decreases (**Figure 7**). The voltage value should go to a full 5 or 12 volts. If it does not, then there is resistance on the signal circuit between the Hall-effect switch and the computer. The voltage should also return to 0 volt. If it does not, then there is resistance on the ground circuit.

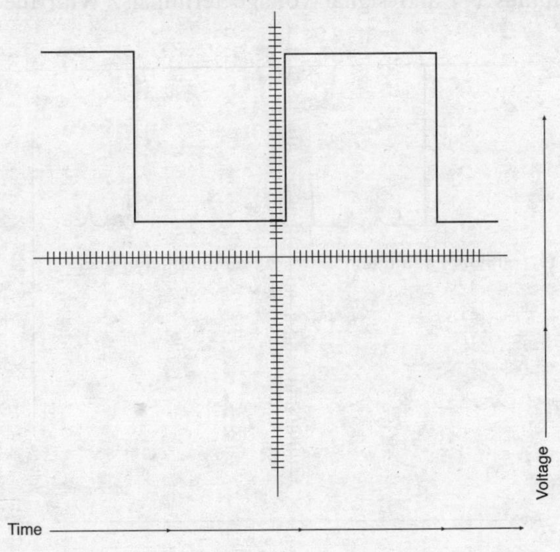


**Figure 7** *A good Hall-effect switch signal should be a square wave pattern.*

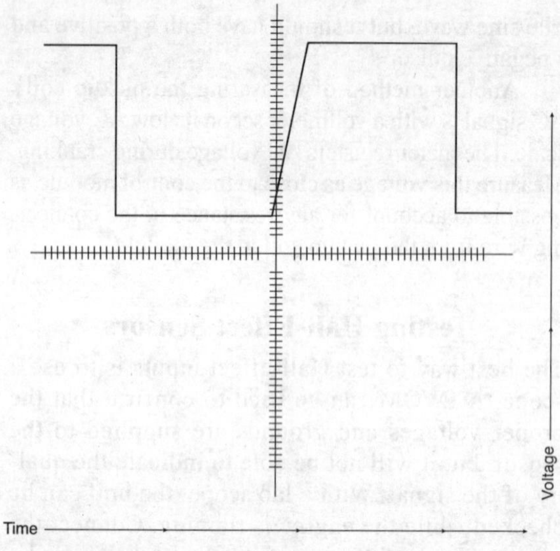


**Figure 8** *A sloping rising or falling line may indicate the air gap is out of adjustment or the transistor in the Hall-effect switch is faulty.*

In addition, a sloping rising and falling line may indicate that the tone wheel is too far away from the magnet of the Hall-effect switch or that the transistor is faulty (**Figure 8**). Check the trace for glitches and noise that may be the result of RFI or EMI (**Figure 9**).

To test a distributor-mounted Hall-effect sensor with a voltmeter, begin by connecting a 12-volt battery across the plus (+) and minus (–) voltage terminals of the Hall layer and the voltmeter across the minus (–) and signal voltage terminals. With the

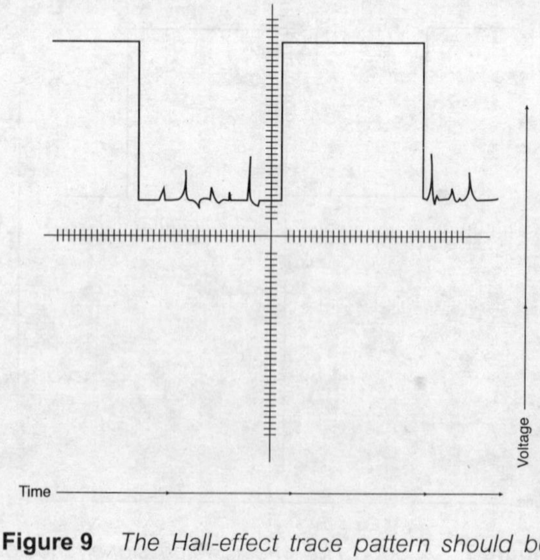


**Figure 9** *The Hall-effect trace pattern should be free of glitches and noise.*

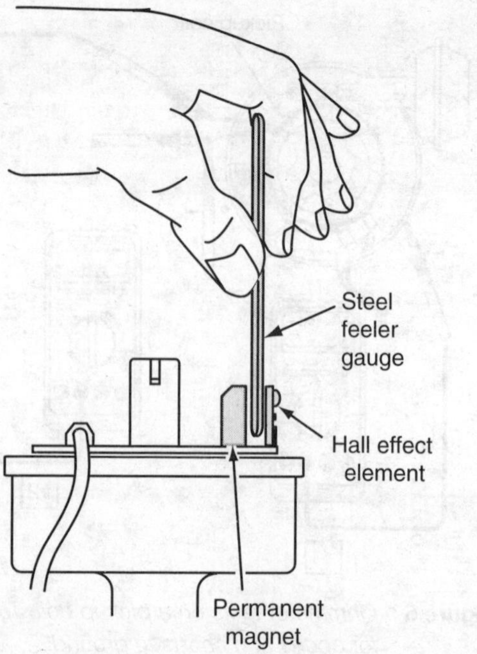


**Figure 10** *Testing a Hall-effect switch with a steel feeler gauge.*

voltmeter hooked up, insert a steel feeler gauge or knife blade between the Hall layer and magnet (**Figure 10**). If the sensor is good, the voltmeter should read within 0.5 volt of specifications when the feeler gauge or knife blade is inserted and touches the magnet. When the feeler gauge or blade is removed, the voltage should read less than 0.5 volt.

### You Should Know

*In many systems that use a Hall-effect switch, it is possible to test the switch by use of a scratch test. For example, if the fuel-ignition system uses a Hall-effect crankshaft sensor, disconnect the sensor connector and use a terminal probe and a jumper wire to momentarily ground (scratch) the signal wire back to the PCM. If the circuit is good, the PCM should respond by activating the fuel pump and/or the shutdown relays. If the PCM responds now, the problem is the Hall-effect switch.*

### You Should Know

*Since magnetic pulse generator and Hall-effect switches require movement to generate their respective signals, be careful not to place your hands and fingers around moving components. Also be sure your clothing does not hang loosely.*

**Figure 11** *The breakout box provides test points for voltmeter and ohmmeter connections.*

## Breakout Boxes

The vehicle's wiring harness creates special problems when it comes to diagnosing circuits. Some manufacturers use a breakout box (**Figure 11**) connected between the module and a wiring harness that will allow the technician to "see" the exact information the computer is receiving and sending. The breakout box is a diagnostic tool that connects in series between the PCM and the engine wiring harness to allow the technician to measure values at each terminal pin. Once the breakout box is connected, each PCM terminal is connected to a corresponding numbered breakout box terminal.

Once the breakout box is connected into the system, a DMM can be used to measure the voltage signals and resistance values of the circuit. The diagnostic manual provided with the breakout box will direct the technician through a series of test procedures. Comparing test results with specifications will lead the technician to the problem area.

The breakout box has the advantage that it taps directly into the sensor or actuator circuit. This provides the technician with the exact voltage signal being sent or received. The scan tool provides the technician with an interpretation of these values only. In addition, you can test hard-to-get-to components without having to disconnect them from the circuit.

## Testing Actuators

Testing of actuators is included here to orient you to the basic procedures. Specific procedures will be presented throughout this manual for individual systems and types of actuators.

Most computer-controlled actuators are electromechanical devices that convert the output commands from the computer into mechanical action. These actuators are used to open and close switches, control vacuum flow to other components, and operate doors or valves depending on the system requirements. Most systems allow for testing of the actuator through the scan tool. Actuators that are duty cycled by the computer are more accurately diagnosed through this method. This will allow the technician to activate selected actuators to test their operation.

If the actuator must be tested by means other than the scanner, follow the manufacturer's procedures very carefully. Because many of the actuators used by the PCM operate with 5 to 7 volts, do not connect a jumper from a 12-volt source unless the service manual directs you to do so. Some actuators are easily tested with a voltmeter by testing for input voltage to the actuator. If there is input voltage of the correct level, check for a good ground connection. If both of these are good, then the actuator is faulty. If you need to use an ohmmeter to measure the resistance of an actuator, disconnect the actuator from the circuit first.

### Testing Actuators with a Lab Scope

Since most actuators are electromechanical devices, when they fail it is because they are electrically faulty or mechanically faulty. By observing the action of an actuator on a lab scope, you will be able to watch its electrical activity. Normally, if there is a mechanical fault, it will affect its electrical activity as well. Therefore, you get a good sense of the actuator's condition by watching it on a lab scope.

To test an actuator, you need to know what it is. Most actuators are solenoids. The computer controls the action of the solenoid by controlling the pulse width of the control signal. By watching the control signal, you can see the turning on and off of the solenoid (**Figure 12**). The voltage spikes are caused by the discharge of the coil in the solenoid.

Some actuators are controlled pulse width modulated signals (**Figure 13**). These signals show a changing pulse width. These devices are controlled by varying the pulse width, signal frequency, and voltage levels.

Both waveforms should be checked for amplitude, time, and shape. You should also observe changes to the pulse width as operating conditions change. A bad waveform will have noise, glitches, or rounded corners. You should be able to see evidence that the actuator immediately turns on and off according to the computer's commands.

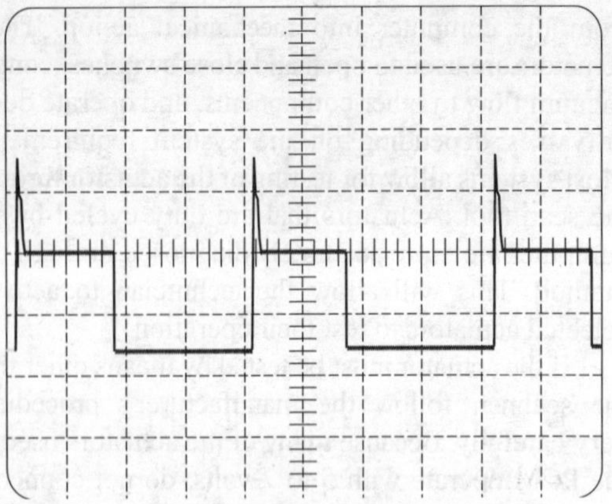

**Figure 12** *A typical solenoid control signal.*

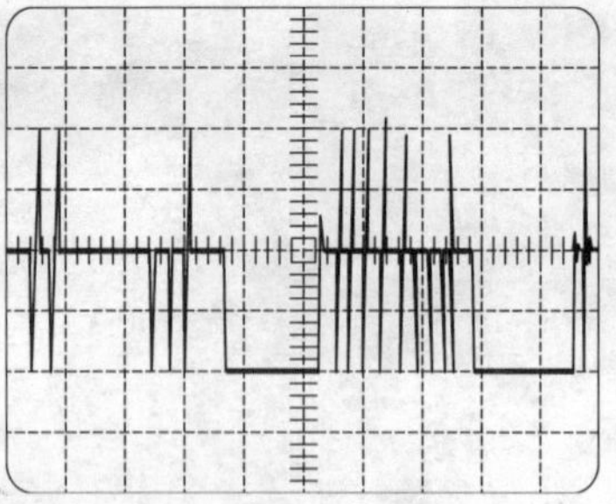

**Figure 13** *A typical pulse width modulated solenoid control signal.*

# Review Questions

1. The engine coolant temperature sensor is being tested. Technician A says an NTC-type engine coolant temperature (ECT) sensor voltage signal will drop in voltage as the coolant temperature increases. Technician B says that a NTC-type ECT sensor will increase resistance as coolant temperature increases. Who is correct?
   A. Technician A only
   B. Technician B only
   C. Both Technician A and Technician B
   D. Neither Technician A nor Technician B
2. Testing of sensors is being discussed. Technician A says sensors can only be tested with a scan tool. Technician B says a breakout box can be used to test sensor signals. Who is correct?
   A. Technician A only
   B. Technician B only
   C. Both Technician A and Technician B
   D. Neither Technician A nor Technician B
3. Technician A says the best way to test Hall-effect inputs is to use a scope. Technician B says a DVOM can be used to confirm that the proper voltages and grounds are supplied to the sensor. Who is correct?
   A. Technician A only
   B. Technician B only
   C. Both Technician A and Technician B
   D. Neither Technician A nor Technician B
4. Technician A says if the Hall-effect switch voltage does not indicate full signal voltage then there is resistance on the signal circuit between the Hall-effect switch and the computer. Technician B says if the voltage does not return to 0 volt then there is resistance on the ground circuit. Who is correct?
   A. Technician A only
   B. Technician B only
   C. Both Technician A and Technician B
   D. Neither Technician A nor Technician B
5. Technician A says the throttle position sensor voltage signal will increase as the throttle is opened. Technician B says the TPS voltage should decrease in steps as the throttle is closed. Who is correct?
   A. Technician A only
   B. Technician B only
   C. Both Technician A and Technician B
   D. Neither Technician A nor Technician B
6. Technician A says to use an ohmmeter to test the resistance value of the coil of the magnetic pulse generator. Technician B says the voltage generation of the magnetic pulse generator sensor can be tested by connecting a voltmeter across the sensor terminals. Who is correct?
   A. Technician A only
   B. Technician B only
   C. Both Technician A and Technician B
   D. Neither Technician A nor Technician B
7. Describe the lab scope pattern of a typical Hall-effect switch.
8. Describe common methods of testing actuators.

SECTION

# 3 AUTOMOTIVE FUELS

## Section Objectives

After you have read, studied, and practiced the contents of this section, you should be able to:

- Explain the refining process of petroleum-based fuels.
- Describe the combustion process of an internal combustion engine.
- Describe the four performance characteristics of gasoline.
- Explain how Reid vapor pressure is measured.
- Explain the difference between winter and summer blend gasoline.
- Explain how octane ratings are determined.
- Explain what the octane rating represents.
- Describe the common methods of determining the octane number.
- Detail the differences between regular and premium gasoline.
- Describe the various types of gasoline and additives.
- Explain the purpose of oxygenated fuels.
- List and describe the common oxygenated fuel blends.
- Describe the different alternative fuels.
- Detail the properties of diesel fuel.
- Explain the differences between diesel fuel grades.
- Explain the cetane rating of diesel fuel.
- Describe how cetane ratings affect power output.
- Explain what API gravity is and how it affects power output.
- Measure the Reid vapor pressure of a gasoline sample.
- Test gasoline for alcohol content.
- Test the specific gravity of diesel fuel.
- Perform the task of relieving fuel pressure safely.

### Interesting Fact

*Today there are individuals and companies that convert diesel engines to burn vegetable oils. This really is nothing new. Rudolph Diesel expected that his engine would be powered by vegetable oils and seed oils. At the 1900 World's Fair, Diesel ran his engines on peanut oil.*

CHAPTER

# 9 Automotive Fuels

## Introduction

Before studying the automotive fuel or emission systems, the properties of the most commonly used fuels in automobiles must be detailed. This chapter takes a look at the composition and qualities of gasoline, diesel fuel, and some of the alternate fuels being used by some engines. More advanced alternate fuels, such as fuel cells, will be discussed in Chapter 41.

## Gasoline

The term **gasoline** describes a complex mixture of various hydrocarbons refined from crude petroleum oil for use as a fuel in engines. **Gasoline** contains hydrogen and carbon molecules. The chemical symbol for this liquid is $C_8H_{15}$. Gasoline is a colorless liquid with excellent vaporization capabilities. Today, gasoline is the most commonly used fuel source in automotive engines. However, recent advancements in technology concerning alternate fuel strategies may mean that hydrocarbon fuels (such as gasoline and diesel) will soon be replaced or supplemented.

### Gasoline Composition

Crude oil is used to make such petroleum-based products as gasoline, diesel fuel, motor oil, solvents, and many others (**Figure 1**). Crude oil is a combination of large and small (light and heavy) **hydrocarbons** along with petroleum gas, hydrogen sulfide, and water. Hydrocarbons are organic compounds that contain only hydrogen and carbon.

There are many kinds of liquid hydrocarbons, consisting of approximately 12 percent hydrogen and 82 percent carbon. The refining process separates the hydrocarbons so they can be used. Refining breaks the crude oil down into different parts called fractions. Hydrogen is a light gas vapor, while carbon is a heavy black solid. These materials, as well as hydrogen, sulfide, and water, are removed during the refining process.

Crude oil is separated into many products at the oil refinery. During refining, the crude oil is heated by pumping it through pipes routed through hot furnaces and into a fractionating column (**Figure 2**). During refining, light hydrocarbon molecules are separated from heavier ones. Since crude oil is a

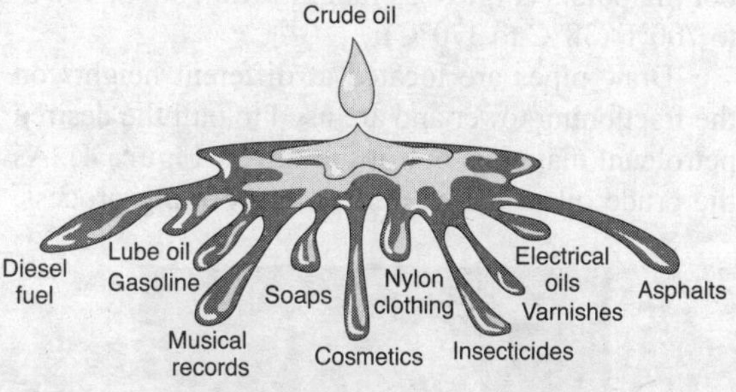


**Figure 1** *Some of the different products made from crude oil.*

**Figure 2** *The tall towers that are seen in the skyline are the fractionating towers of the refinery.*

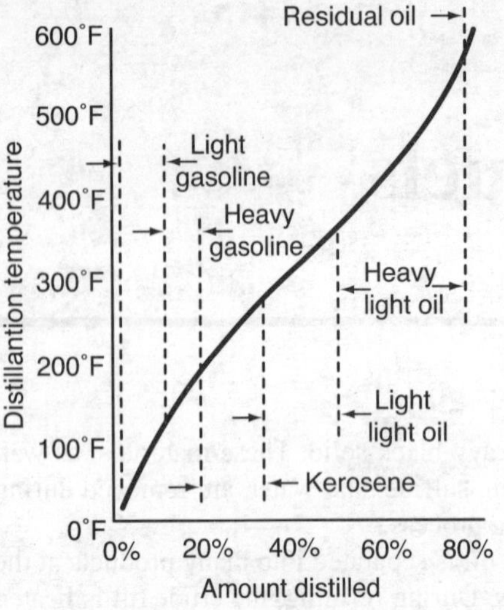


**Figure 3** *The compounds of crude oil have different boiling points.*

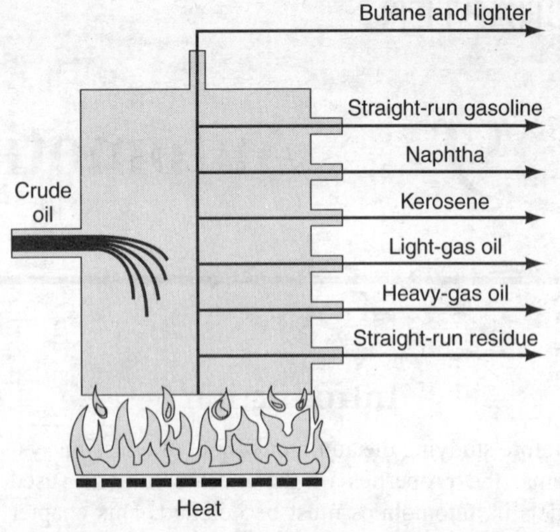


**Figure 5** *Illustration of the different materials that will be pulled off the tower.*

composite of different substances, it has different boiling points (**Figure 3**) ranging from about 100°F to 700°F (38°C to 370°C).

Draw pipes are located at different heights on the fractioning tower and are used to pull the desired petroleum materials out of the tower (**Figure 4**). As the crude oil is boiled during the refining process, the lightest products (butane) are taken from the top of the fractioning tower (**Figure 5**). The next draw pipe pulls straight-run gasoline from the tower. As the fractions go down the tower, the draw pipe pulls materials that have increasingly higher boiling points. The heaviest products boil last and are taken from the bottom where the temperature is the highest. The substance pulled from the bottom of the tower is a tar-like residue.

**Figure 4** *Draw pipes are located at different heights of the fractionating towers. The higher draw tubes pull the lightest materials, while the lower draw tubes pull progressively heavier materials.*

Some of the fractions are used as raw materials that will be blended with other materials. For example, the gasoline will be blended to create the desired octane, emissions, volatility, and storage life. Some of the lower fractions, such as kerosene or diesel, can be used directly. The last fraction can be used as fuel oil or as a basic stock for lubricants. Fractions may also be reformulated by heat or catalysts to produce more of the lighter stocks. The distillation process is repeated in other plants as the oil is further refined for its most efficient use (**Figure 6**).

Crude oil is measured and sold by the barrel, which is 42 U.S. gallons. **Figure 7** shows the percentage of various products produced from an average barrel of crude oil.

## Combustion Chemistry

Most internal combustion engines are designed to burn organic fuels to produce power. The term **organic** refers to a product from a source that originally was living. All petroleum products are

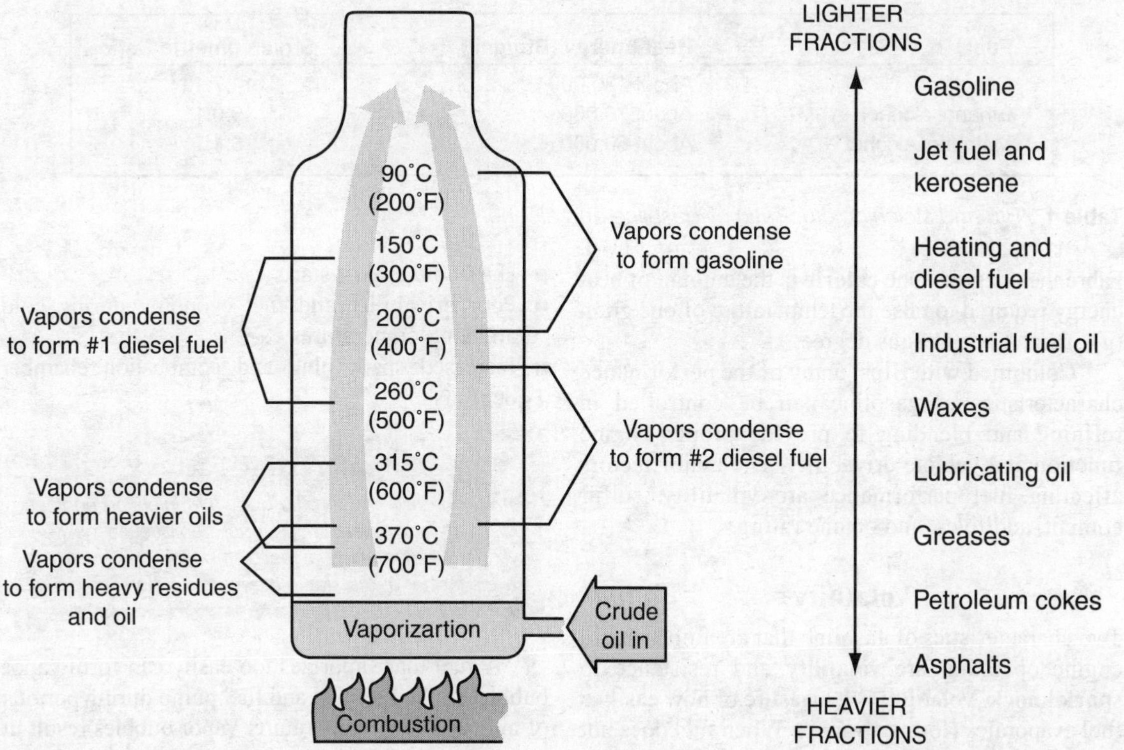


**Figure 6** *Oil-refining process.*

considered organic and are hydrocarbon fuels. The combustion process combines oxygen ($O_2$) from the air with the hydrogen (H) and carbon (C) from the fuel. In a gasoline engine, a spark starts the combustion process, which takes about 3 milliseconds (0.003 second) to be completed. The chemical reaction that takes place is basically:

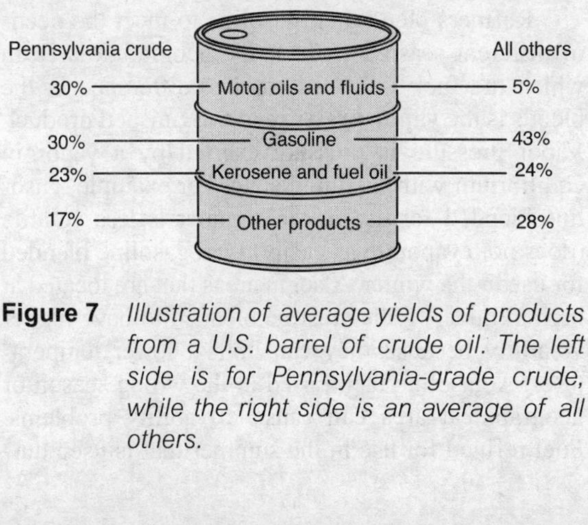


**Figure 7** *Illustration of average yields of products from a U.S. barrel of crude oil. The left side is for Pennsylvania-grade crude, while the right side is an average of all others.*

Hydrogen (H) + carbon (C) + oxygen ($O_2$) + nitrogen (N) + other chemicals + spark = Heat + water ($H_2O$) + carbon monoxide (CO) + carbon dioxide ($CO_2$) + hydrocarbons (HC) + oxides of nitrogen ($NO_x$) + other chemicals

If the combustion process is complete, all the hydrocarbons (HC) will be completely combined with all the available oxygen ($O_2$). The ratio of air to fuel that accomplishes this is referred to as **stoichiometric**. Stoichiometric, in simplest terms, means chemically correct. An air-fuel mixture is considered chemically correct when it is neither too rich nor too lean. The stoichiometric quantities for gasoline are 14.7 parts air for 1 part gasoline by weight. Different fuels have different stoichiometric ratios (**Table 1**).

## Gasoline Characteristics

The energy content of gasoline is between 109,000 and 125,000 **British thermal units (Btus)** per gallon. One British thermal unit is the amount of heat energy required to raise one pound of water one

| Fuel | Heat Energy (Btu/gal.) | Stoichiometric Ratio |
|---|---|---|
| Gasoline | About 130,000 | 14.7:1 |
| Ethanol alcohol | About 76,000 | 9.0:1 |
| Methanol alcohol | About 60,000 | 6.4:1 |

**Table 1** *Btu and stoichiometric ratio of gasoline and alcohol.*

Fahrenheit degree. One **calorie** is the amount of heat energy required to raise the temperature of one gram (g) of water one Celsius degree.

Combined with Btus, many of the performance characteristics of gasoline can be controlled in refining and blending to provide proper engine function and vehicle drivability. The major factors affecting fuel performance are volatility, sulfur content, additives, and octane rating.

## Volatility

Two characteristics of gasoline that are important to engine operation are **volatility** and resistance to **spark knock**. Volatility is a measure of how easily a fuel evaporates (forms a vapor). When fuel does not vaporize easily, this is called low volatility. Spark knock is a metallic noise an engine makes, usually during acceleration, resulting from abnormal or uncontrolled combustion inside the cylinder.

> **You Should Know**
>
> *Spark knock is also referred to as detonation or engine knock.*

Gasoline is very volatile; consequently, it readily evaporates. This volatility allows its vapors to adequately mix with air to promote combustion. This evaporation into a vapor is very important to the combustion process since only vaporized fuel supports complete combustion. Thus, to ensure complete combustion, complete vaporization must occur. For fuel to burn properly in the cylinder, its vaporization point should be near the temperature inside the intake manifold. If liquid fuel enters the cylinders, it will not burn. Unburned gasoline will form droplets that will break down the oil film on the cylinder wall. Oil breakdown can result in scuffing or scoring of the cylinder. The fuel will eventually enter the crankcase and mix with the oil, resulting in decreased lubrication properties of the oil.

Fuel with a low volatility can result in the following undesirable conditions:

- Hard cold engine starts
- Poor drivability and fuel economy during cold ambient temperatures
- Increased spark plug and combustion chamber deposits

> **Interesting Fact**
>
> *On carbureted and throttle body-fuel injection engines, low volatility can cause unequal combustion between cylinders due to unequal fuel distribution. Vaporized fuel travels farther and faster in the manifold than liquid fuel.*

A fuel that vaporizes too easily can form vapor bubbles in the fuel line and fuel pump during periods of high ambient temperature. Vapor bubbles result in **vapor lock** or loss of performance. Vapor lock is a result of fuel boiling in the fuel line or tank. Unlike liquid fuel, vapor is compressible. This means that the fuel cannot be pumped to the carburetor or injectors, so the engine stalls. After the fuel line cools sufficiently, the engine will run again. Fuel-injected engines use electric fuel pumps that keep the fuel under higher pressure. Higher pressure raises the fuel's boiling point so vapor lock is less likely to occur. Other problems that can result are:

- High evaporative emissions
- Poor drivability when the engine is hot
- Poor fuel economy
- Hard starting after a hot soak condition

Refiners blend their gasoline to meet the needs of different seasons and for the geographic area in which the fuel will be used. The difference in the blends is the **vapor pressure** of the finished product. Vapor pressure is pressure exerted by a vapor in equilibrium with its liquid state. For example, gasoline blended for use in the summer is less volatile (does not evaporate as easily) than gasoline blended for use in the winter. Also, in areas that are located at high altitudes, fuels must be blended to have higher volatility because they can boil at lower temperatures. A gasoline blend used in the wrong season or geographical area can cause drivability problems. Fuel refined for use in the summer that is used during the winter can cause hard starting. If winter fuel is used in the summer, vapor lock can result. Refiners attempt to schedule delivery of the different blends to be at retail outlets at the appropriate time; however, unseasonable weather can result in drivability problems due to the blend being used.

The definition of "volatility" assumes that the vapors will remain in the fuel tank or fuel line and will cause a certain pressure based on the temperature of the fuel. **ASTM International** (formerly known as the American Society for Testing and Materials or ASTM), provides a global forum for the development and publication of voluntary consensus standards for materials, products, systems, and services. ASTM standards serve as the basis for manufacturing, procurement, and regulatory activities. ASTM International has six volatility classes for gasoline: AA, A, B, C, D, and E. AA is the least volatile **(Table 2)**. For example, the 10 percent evaporated maximum standard for AA gasoline means that 10 percent of the fuel evaporates before it reaches a temperature of 158°F (70°C). For all volatility classes, gasoline evaporates at the endpoint maximum temperature of 437°F (225°C).

There are several methods of measuring the volatility of gasoline. The most common are **Reid vapor pressure (RVP)**, distillation, drivability index (DI), and vapor-liquid ratio.

The best known method of measuring volatility is the RVP test. Vapor pressure is the pressure exerted by vapor formed over a liquid in a closed container. The RVP test is performed by placing a sample of gasoline into a sealed metal chamber that has a pressure-measuring device attached. The chamber is submerged in water that is heated to 100°F (38˚C). As the fuel is heated by the water, it vaporizes. Remember, the more volatile a fuel is, the easier it will vaporize. As the fuel vaporizes, it creates vapor pressure within the chamber. More volatile fuels will create more pressure. The vapor pressure is measured in pounds per square inch (psi).

The distillation method is done by heating the fuel and measuring the temperature when a certain percentage of the sample evaporates. Using the temperature for 10 percent (T10), 50 percent (T50), and 90 percent (T90) evaporation, a correlation for cold engine start-up and warm-up performance is obtained. This correlation is known as the **drivability index (DI)**. The ASTM International specification ranges from 1,200 in the winter months to 1,250 in the summer months.

The DI is a mathematical expression of distillation properties that describe the influence of fuel volatility on drivability:

$$DI = 1.5\,T10 + 3\,T50 + T90 + 11 \text{ oxygenates}$$

### Interesting Fact

*Oxygenate level is the concentration of oxygenates measured in the percentage of weight.*

The higher the DI number, the less volatile the fuel. Most premium gasoline sold in the United States has a higher DI index than regular or midgrade gasoline. High DI gasoline causes drivability problems and higher emissions due to its lower volatility.

The vapor-liquid ratio method measures the volume of vapor formed from a given volume of gasoline at a specified temperature. The temperature at which the vapor liquid ratio equals 20, known as TV/L20,

ASTM D 4814 Gasoline volatility requirements

| Vapor pressure/ distillation class | Distillation temperatures 10% Evap. ˚F | 50% Evap. ˚F | 90% Evap. ˚F | End point Max. ˚F | Vapor pressure psi/Max. | Vapor lock protection class | Vapor-liquid ratio of 20 Min. ˚F |
|---|---|---|---|---|---|---|---|
| AA | 158 | 170–250 | 374 | 437 | 7.8 | 1 | 140 |
| A | 158 | 170–250 | 374 | 437 | 9.0 | 2 | 133 |
| B | 149 | 170–245 | 374 | 437 | 10.0 | 3 | 124 |
| C | 140 | 170–240 | 365 | 437 | 11.5 | 4 | 116 |
| D | 131 | 150–235 | 365 | 437 | 13.5 | 5 | 105 |
| E | 122 | 150–230 | 365 | 437 | 15.0 | 6 | 95 |

*(Courtesy of Changes in Gasoline III, Downstream Alternatives, Inc.)*

**Table 2** *Gasoline volatility requirements.*

has proven to protect against vapor lock. The TV/L20 specification is 105°F (40.5°C) in the winter 4 months and 140°F (60°C) in the summer months.

**Winter-blend gasoline**. As discussed, gasoline must be vaporized (mixed with air) in order to support combustion in the engine. Since cold temperatures reduce the normal vaporization of gasoline, refiners will increase the RVP of their gasoline to allow the engine to start easier in cold weather. The ASTM International standards for winter-blend gasoline allow for up to 15 psi RVP. The RVP for gasohol can be up to 1 psi higher.

> **You Should Know**
>
> *Gasoline mixed with alcohol is sometimes referred to as gasohol.*

If winter-blend gasoline is used in an engine during warm weather, the following problems may occur:

- Rough idle
- Stalling
- Vapor lock
- Hesitation on acceleration
- Surging
- Evaporative system damage
- Engine flooding during a hot soak

**Summer-blend gasoline.** Since warm ambient temperatures cause gasoline to vaporize easily, the RVP of summer-blend fuel is reduced. The vehicle's fuel system is designed to operate with liquid gasoline. If the RVP were not reduced, vapor bubbles would form and the engine would not run properly. According to ASTM International standards, the maximum RVP for summer blend is 10.5 psi. In the United States, the EPA requires gasoline that is to be sold in retail outlets between June 1 and September 15 to have a fuel vapor pressure (VP) below 9.0 psi. In areas of the country classified by the EPA as nonattainment areas, the VP must be lower than 7.8 psi.

## Sulfur Content

Since sulfur is a component of crude oil, gasoline can contain some sulfur. A sulfur content that is too high can cause corrosion in the engine and exhaust system. For this reason, the refining process removes most of the sulfur.

When gasoline is burned in the combustion chamber, one of the products of combustion that is formed is water. Due to the high temperatures of combustion, the water will leave the combustion chamber as steam; however, this steam can condense back into water as it passes through the exhaust system. In addition to water in the exhaust, the steam in the combustion chamber and in the crankcase blowby will condense when the engine is shut off and cools. When the sulfur that is in the gasoline is burned, it combines with oxygen and forms sulfur dioxide. When sulfur dioxide combines with water, it forms sulfuric acid. Sulfuric acid is highly corrosive. Corrosion caused by this acid may cause exhaust valve pitting and exhaust system deterioration. Sulfur dioxide can cause the obnoxious odor of rotten eggs as it passes through the catalytic converter and the exhaust system. To reduce corrosion caused by sulfuric acid, the sulfur content in gasoline is limited. In the United States, current requirements are that the sulfur content be less than 0.01 percent. In most European and Asian countries, the standards are stricter.

## Fuel Additives

For many years, lead compounds such as tetraethyl lead (TEL) were added to gasoline to improve octane ratings. However, since the mid-1970s, vehicles have been designed to run on unleaded gasoline only. Leaded fuels are no longer available as automotive fuels due to the addition of special antipollution devices, such as catalytic converters, on vehicles. These systems require the use of unleaded fuel in order to work properly.

Gasoline additives have different properties and a variety of uses. They are expensive and are added in limited quantities. Although the exact amount of additives a refinery adds to its gasoline is a closely guarded secret, it has been estimated that as little as 100 pounds of additive can be used to treat 20,000 gallons of gasoline. The following are examples of the additives that may be formulated into gasoline:

- **Detergents.** These are used to keep the fuel system clean of deposits. In the United States, deposit control additives have been required by law since 1995. Polyether amine is an added detergent that helps dissolve deposits and keeps injectors clean. However, the added detergents tend to create deposits on the intake valves.
- **Anti-icing or deicer.** Isopropyl alcohol is added seasonally to gasoline to prevent gas line freeze-up in cold weather.
- **Metal deactivators and rust inhibitors.** These additives are used to inhibit reactions between the

fuel and the metals used in the fuel system that can form abrasive substances.

- **Gum or oxidation inhibitors.** Some gasoline blends contain aromatic amines and phenols to prevent the formation of gum and varnish. During storage, gum deposits can form as the result of some gasoline components reacting with oxygen. Oxidation inhibitors promote gasoline stability.

## Octane Ratings

There are two important factors that affect the power and efficiency of a gasoline engine: the **compression** ratio and control of **detonation**. The compression ratio is the ratio of the volume in the cylinder above the piston when the piston is at bottom dead center (BDC) to the volume in the cylinder above the piston when the piston is at top dead center (TDC). It is an indication of how much air-fuel mixture can be brought into a cylinder and how tightly the mixture will be compressed. Detonation is the result of an amplification of pressure waves, such as sound waves, occurring during the combustion process when the piston is near TDC. The actual "knocking" sound of detonation is due to these pressure waves pounding against the insides of the combustion chamber and on the top of the piston.

Theoretically, the higher the compression ratio the greater the power output and efficiency of the engine. The better the efficiency, the less fuel will be consumed to produce a given power output. The problem is that the higher the compression ratio, the greater the tendency to have detonation. Most gasoline engines have a compression ratio between 8:1 and 10:1. High-performance engines may have higher compression ratios.

Normal combustion occurs smoothly and progresses across the combustion chamber from the point of ignition (**Figure 8**). Normal combustion propagation is between 50 to 250 meters per second. The speed of combustion propagation is dependent upon engine, load, air-fuel ratio, combustion chamber design (determining amount of turbulence), temperature, and **octane rating**. Measured in time, normal combustion of fuel occurs in a matter of milliseconds (thousandths of a second). Even though the speed of combustion is very fast, it is not an explosion. It is a controlled burn. The flame front advances smoothly across the combustion chamber until all the air-fuel mixture has been burned.

The octane rating is a measure of how easily the gasoline can be ignited. The higher the octane, the harder the fuel is to ignite and the slower it burns. Generally, the higher the octane, the greater the number of hydrocarbons containing larger numbers of carbon atoms. More carbon atoms per hydrocarbon means that more oxygen and more heat are needed to burn the fuel. The octane rating provides an indication of its antiknock qualities. The higher the octane number, the less tendency for knock.

During periods of detonation, the combustion speed increases by up to ten times and approaches the speed of sound. Detonation is the result of the pressure waves caused by two flame fronts, one front started by the spark plug and the other started by fuel self-ignition (**Figure 9**). When the spark plug is fired, the flame front is started and is carried outward from the spark plug by eddies, swirls, and flow patterns created by the combustion chamber turbulence. In front of the flame front is a pressure wave that is moving through the unburned air-fuel mixture. Normally the pressure rise in the combustion chamber during combustion is between 20 and 30 psi (138 and 207 kPa) per degree of crankshaft rotation. Once the pressure rises faster than about 35 psi (241 kPa) per degree, the engine will run rough. This is due to the mechanical vibration of the engine components caused by too great a pressure rise. The pressure wave can be strong enough to cause the remaining fuel to decompose and self-ignite. Self-

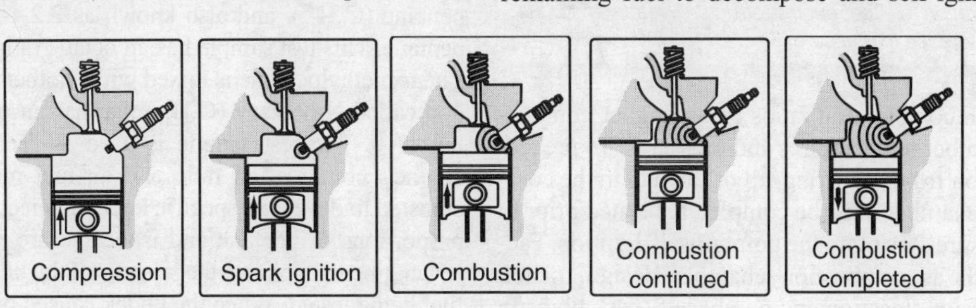


**Figure 8** *Normal combustion has a smooth, controlled burn across the combustion chamber.*

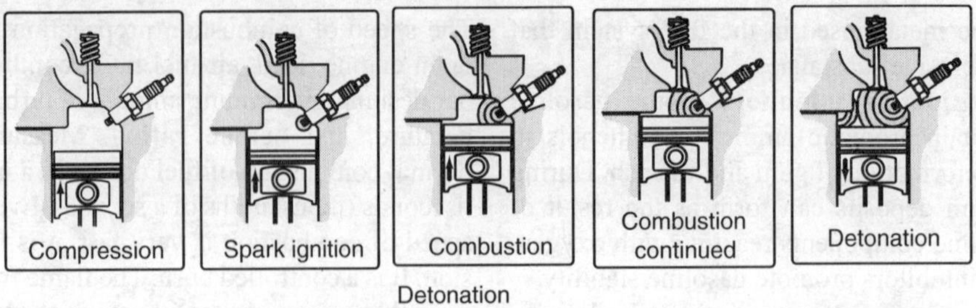


**Figure 9** *Detonation occurs when a second flame is started as a result of high temperatures.*

ignition results from the remaining fuel in the end gas lacking sufficient octane rating to withstand this combination of heat and pressure. The ignition of the end gases also causes a very high and rapid pressure increase in the combustion chamber. The pressure rise can be up to 8 bars per degree of crankshaft angle or up to 50,000 bars per second during detonation. This pressure increase is for a very short duration. Detonation usually happens at the edges of the piston crown. It is here that reflecting pressure waves from the piston or combustion chamber walls recombine to create a very high local pressure. If the speed of this pressure buildup is greater than the speed that the air-fuel mixture burns, both the pressure waves from the initial ignition at the spark plug and the self-ignited fuel will set off immediate explosions (instead of normal combustion) of the air-fuel mixture. The spike in pressure creates a force in the combustion chamber that causes the structure of the engine to ring, or resonate. Resonance, which is caused by the unequal pressures from the initial ignition and the second ignition, occurs at about 6,400 hertz (depending on engine design). The pinging (knocking) sound is the result of the structure of the engine reacting to the pressure spikes.

## You Should Know

*Often, detonation is explained as being the result of "two flame fronts colliding." The noise is actually the result of the vibration of the engine structure reacting to the sharp increase in pressure.*

High compression ratios create higher combustion chamber temperatures and pressures. To prevent detonation from occurring, all of the fuel in the combustion chamber must be completely burned prior to the pressure's rising to the point of self-ignition. Factors such as combustion chamber design, piston dome design, location and number of spark plugs in the cylinder, valve timing, and ignition timing all contribute to detonation control. Most electronically controlled ignition systems have a sensor to detect if knock is occurring so the computer can retard the ignition timing to prevent detonation. Another factor in preventing detonation is the octane rating of the gasoline being burned.

In the early 1920s, TEL was added to gasoline to reduce the tendency for spark knock. Shortly after the introduction of TEL as a fuel additive, the Society of Automotive Engineers (SAE) established standards for the antiknock rating of gasoline. The purpose for these standards was to allow the automobile manufacturers to produce engines that would operate without knock based on the quality of the available fuel.

Two commonly used methods for determining the octane number of motor gasoline are the motor octane number (MON) and the research octane number (RON). Both use a laboratory single-cylinder engine called a cooperative fuel research (CFR) engine. The MON method test engine is run at a constant 900 rpm with no load. The RON method uses the same test engine run placed at 600 rpm and under a load that is more closely related to actual driving conditions. The CFR engine is equipped with a variable head and knock meter to indicate knock intensity. The compression ratio of the CFR can be varied between 4:1 and 18:1. The test provides an octane rating by comparing knock intensity between the test sample of fuel and a known combination of knock-resistant hydrocarbon isooctane, chemically called trimethylpentane ($C_8H_{18}$), and also known as 2,2,4-trimethylpentane. This test sample has an octane rating of 100. The trimethylpentane is mixed with another hydrocarbon called N-heptane ($C_7H_{16}$) that has an octane rating of 0. The test sample is used as fuel, and the engine's compression ratio and air-fuel mixture are adjusted to develop a specific knock intensity. Various proportions of heptane and isooctane are run in the test engine to duplicate the severity of the knock of the fuel being tested. When the knock caused by the heptane and isooctane mixture is identical to the test fuel, the octane number is established by the percentage of isooctane in the mixture. If the test sample of fuel has the exact antiknock characteristics of isooctane, it is rated as 100 octane gasoline. If 85 percent isooctane and 15 percent N-heptane produce the same severity of knock as the fuel being tested, the fuel is assigned an octane number of 85.

The RON gives a higher reading for the same fuel as the motor method (typically eight to ten numbers higher). It affects low-speed knock and engine run-on. The MON gives a measurement of how much engine knock will be present under heavy loads such as when passing or climbing hills. For example, a fuel with a RON of 93 might have a MON of 85.

**You Should Know**

*Engine run-on is referred to as dieseling since the engine continues to run without the benefit of a spark plug firing.*

By law, U.S. gas pumps are labeled with a yellow sticker containing an Antiknock Index (AKI) number, which is the average of the RON and MON rating systems. The AKI index, also known as road octane number or pump octane number, is referred to as (R + M)/2. Using the previous examples of MON and RON ratings, the rating posted on the pumps would be:

$$(RON + MON)/2 = 93 + 85/2 = 178/2 = 89$$

**You Should Know**

*Gasoline is not mixed with octane. "High octane fuel" does not mean there is more octane in the fuel compared with "low octane fuel." The octane rating is a measure of the fuel's ability to control the burning process (to prevent detonation); it is not a function of burning "hotter" or "colder."*

As discussed previously, there are many factors that affect spark knock other than just the octane rating of the gasoline. To recap, these factors include:

- **Lean fuel mixture.** A lean mixture burns slower than a rich mixture. The heat of combustion is higher, which promotes the tendency for unburned fuel in front of the spark ignition flame to detonate.
- **Overadvanced ignition timing.** Advancing the ignition timing induces knock. Retarding ignition timing suppresses knock.
- **Compression ratio.** Higher compression ratios increase cylinder pressures.
- **Valve timing.** Valve timing that fills the cylinder with more air-fuel mixture promotes higher cylinder pressures, increasing the chances for detonation.
- **Turbocharging.** Turbocharging (or supercharging) forces additional fuel and air into the cylinder. This induces higher cylinder pressures and temperatures, which promote knock.
- **Coolant temperature.** Hot spots in the cylinder or combustion chamber due to inefficient cooling raise combustion chamber temperatures and promote knock.
- **Cylinder-to-cylinder distribution.** If an engine has poor distribution of the air-fuel mixture from cylinder to cylinder, the leaner cylinders could promote knock.
- **Excessive carbon deposits.** Carbon deposits on the piston, valves, and combustion chamber causes poor heat transfer from the combustion chamber. In addition, carbon accumulation also artificially increases the compression ratio.
- **Air inlet temperature.** The higher the air temperature entering the cylinder, the greater the tendency to knock.
- **Combustion chamber design.** The optimum combustion chamber design for reduced knocking is the hemispherical or pentroof design. Both use a spark plug located in the center of the chamber (**Figure 10**). These chambers promote faster combustion, allowing less time for detonation to occur ahead of the flame front.
- **Faulty components.** These include the knock sensor, advance delay valves, and EGR valves.
- **Humidity.** The drier the air, the greater the potential for knock.
- **Altitude.** The lower the altitude, the more potential for knock.

## Regular and Premium Fuels

Most gasoline stations offer three octane grades of unleaded gasoline: regular (87 octane), midgrade (89 octane), and premium (93 octane). The amount of octane required for a particular engine is called the engine octane number requirement (ONR). The

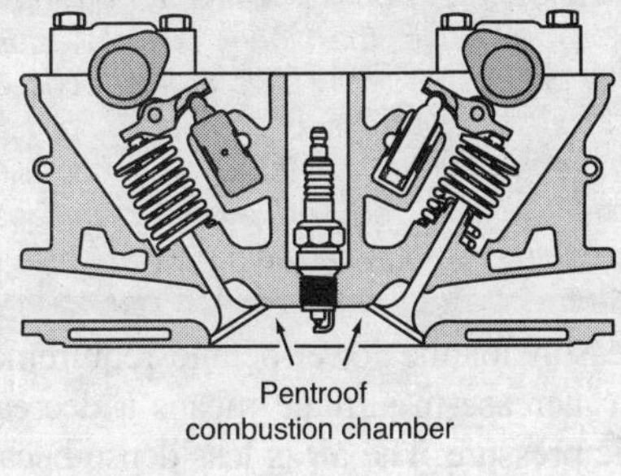


**Figure 10** *Cross-sectional view of a pentroof combustion chamber.*

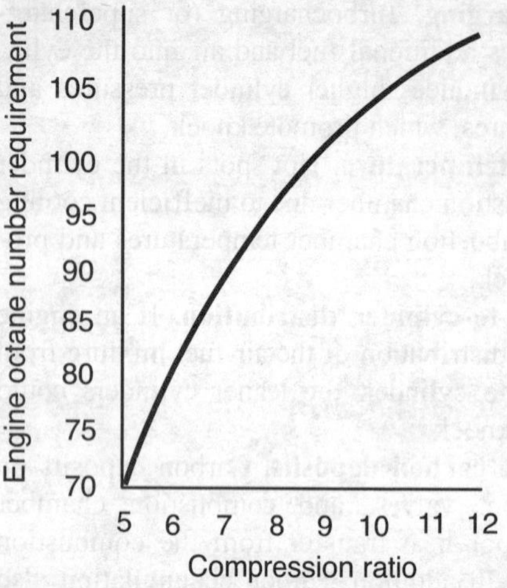


**Figure 11** *Graph indicating that as compression ratio increases, so does the octane requirement.*

octane requirements go up as the compression ratio increases (**Figure 11**).

To obtain the full benefit of a higher-octane fuel, the engine, ignition system, and fuel-control system must be designed and adjusted to take advantage of the higher-octane gasoline. Most modern engines are designed to operate efficiently with regular grade gasoline (87 octane) and do not require a high-octane premium grade.

Changes in outside air temperature, humidity, and pressure also affect octane requirements. For example, in the high altitude of the Colorado Rockies, all the grades of gasoline have lower-octane ratings than other parts of the country. Regular grade is usually rated at 85, midgrade is 87, and premium fuel has a rating of 91. As a rule, octane rating of gasoline is reduced about one octane number per 1,000 feet or 300 meters in altitude.

### You Should Know

*Using premium gasoline in a car designed to run on regular will not harm it, but the gains from using premium fuel do not offset the higher cost. In addition, since higher octane fuels may have greater concentrations of alcohol, it is possible to experience hard starting after the engine is hot. On the other hand, using a lower-octane fuel than required by the manufacturer can cause engine damage and void the warranty.*

The reason for the lower-octane requirement is because an increase in altitude means a decrease in atmospheric pressure. The air is less dense because a pound of air takes more volume. The octane rating of fuel does not need to be as high because the engine cannot take in as much air. This process will reduce the combustion (compression) pressures inside the engine. A secondary reason for the lowered octane requirement of engines running at higher altitudes is the normal enrichment of the air-fuel ratio and lower engine vacuum with the decreased air density.

### Octane Improvers

Because lead will cause the vehicle's catalytic converter to fail, gasoline is limited to a lead content of 0.06 gram per gallon. When federal EPA regulations required gasoline companies to remove tetraethyl lead from gasoline, other methods of controlling spark knock were developed. Octane improvers can be grouped into three categories:

1. Aromatic hydrocarbons (hydrocarbons containing the benzene ring) such as xylene and toluene
2. Alcohols such as ethanol, methanol, and tertiary butyl alcohol (TBA)
3. Metallic compounds such as methylcyclopentadienyl manganese tricarbonyl (MMT)

## Oxygenated Fuels

Some areas of the country require that oxygenated fuels be used during the winter months, and some gasoline stations sell blends of oxygenated fuels year-round. Oxygenated fuels are mixtures of conventional gasoline and one or more combustible liquids that contain oxygen (oxygenates). Oxygenates are compounds such as alcohols and ethers that contain oxygen in their molecular structure. The idea behind the use of oxygenated fuels in these areas is to reduce CO emissions. By blending oxygenates into the gasoline, the fuel requires less ambient oxygen for complete burning. Therefore, for the same carburetor or fuel-injector settings, oxygenated gasoline produces a leaner air-fuel mixture and generates less carbon monoxide. The extra oxygen in the fuel ensures that there is enough oxygen to convert CO into $CO_2$ during the combustion process, either in the engine or in the catalytic converter.

Examples of oxygenated fuels include methanol, ethanol, and ethyl tertiary butyl ether (ETBE). Ethyl alcohol, (ethanol) in up to a 10 percent concentration, is the most common oxygenate.

### Methanol

**Methanol** is a mixture of gasoline and methyl alcohol. Methanol can be distilled from coal, natural gas, oil shale, wood, or garbage. Most of what is used today is derived from natural gas. Most automotive manufacturers warn customers against using a fuel that contains more than 10 percent methanol and **cosolvents** by volume unless the vehicle was specifically produced to burn this fuel. Cosolvents are another substance (usually another alcohol) that is soluble in both methanol and gasoline and is used to reduce the tendency of the liquids to separate.

Methanol is recognized as being more corrosive to fuel system components than ethanol. Another disadvantage of methyl alcohol–mixed fuel is that it produces only about half the energy of gasoline. In order to use methyl alcohol, the fuel system needs to be adjusted to provide an air-fuel ratio of about 6.5:1. This is because methanol is 50 percent oxygen. Gasoline containing 5 percent methanol would have an oxygen content of 2.5 percent by weight.

## Ethanol

The most common alcohol-gasoline mixture is ethanol. Ethanol contains a mixture of about 10 percent **ethyl alcohol**. Ethyl alcohol is an alcohol made from grain. Ethanol was first used to extend gasoline supplies during the gasoline shortages of the 1970s. Ethanol has an oxygen content of approximately 35 percent, thus a 10 percent concentration adds about 3.5 percent oxygen to the mixture. Another benefit of ethanol is that it increases the octane rating of the fuel. A 10 percent ethanol mixture will raise an 87 octane fuel by at least 2.5 octane numbers **(Figure 12)**.

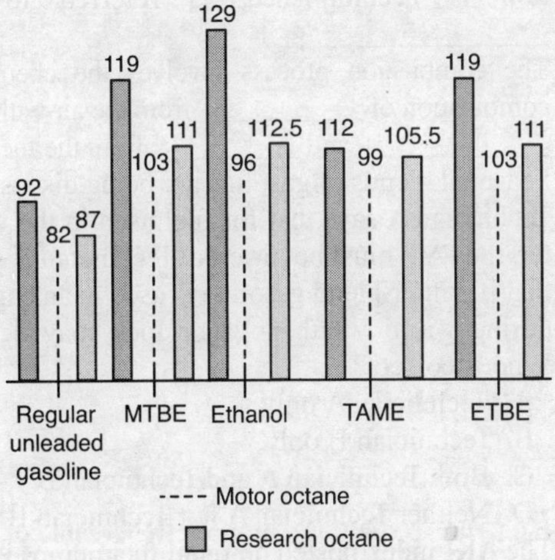


Exact octane values will vary based on octane and composition of gasoline to which the oxygenate is added

**Figure 12** *Comparison between the octane ratings of gasoline and oxygenated fuels.*

However, the alcohol added to the base gasoline also raises the volatility of the fuel about 0.5 psi (3.5 kilopascals, or kPa). Most automobile manufacturers permit up to 10 percent ethanol if drivability problems are not experienced.

Gasoline blended with 10 percent alcohol or less does not require changes to the fuel system. However, vehicles burning any amount of gasohol may require that the fuel filter be changed more often. This is due to the cleaning effect that alcohol has on the vehicle's fuel tank. Oxygenates suspend water in the fuel and tend to keep it from accumulating in the gas tank. One gallon of gasoline can hold only 0.5 teaspoon of water. As a result, the water separates and accumulates at the bottom of the tank. Comparative, one gallon of 10 percent ethanol can attract and hold 12 teaspoons of water.

## Ethyl Tertiary Butyl Ether

ETBE is derived from ethanol. The maximum allowable volume level is 17.2 percent. ETBE is made by combining isobutylene and ethanol. Like methyl tertiary butyl ether (MTBE) and methanol, ETBE does not have a significant amount of free ethanol and does not have ethanol properties.

## Advantages and Disadvantages of Alcohol Additives

The following is a list of the advantages of using alcohol as an additive to gasoline:

1. Absorbs moisture in the fuel tank.
2. Ten percent alcohol added to gasoline raises the octane rating, using the (R + M)/2 method, by nearly three points.
3. Alcohol cleans the fuel system.
4. The addition of alcohol reduces CO emissions.

The disadvantages of alcohol blends include:

1. The use of alcohol blends can result in the clogging of fuel filters.
2. Alcohol raises the volatility of fuel about 0.5 psi (3.5 kPa), resulting in possible hot weather drivability problems.
3. Since the heat energy of alcohol is between 60,000 to 75,000 Btus per gallon compared with about 120,000 Btus per gallon for gasoline, the addition of alcohol reduces the heat content of the resulting fuel mixture. Reformulated gasoline will yield 2 to 3 percent lower fuel mileage than nonoxygenated gasoline.
4. Alcohol absorbs water and then separates from the gasoline, especially as the temperature drops.

5. Separated alcohol and water on the bottom of the tank can cause hard starting during cold weather.
6. Alcohol does not vaporize easily at low temperatures.

## Reformulated Gasoline

The 1990 Clean Air Act required the EPA to issue regulations requiring gasoline to be "reformulated." The purpose of this fuel is to result in the reduction of ozone-forming and toxic air pollutants from vehicle emissions. This cleaner gasoline is called **reformulated gasoline (RFG)**. RFG is a general term for federally mandated gasoline that is specially processed and blended to reduce the emission of pollutants such as hydrocarbons, toxics, and nitrogen oxides. This fuel is required in the most severe ozone nonattainment areas of the United States. In addition, several other areas with ozone levels exceeding the public health standard have voluntarily chosen to use RFG. The main characteristics of RFG are:

- RFG has lower levels of benzene compounds that contribute to air pollution.
- RFG has a lower RVP and does not evaporate as easily.
- RFG contains oxygenates.

The RFG program has been implemented in two phases: Phase I ran from 1995 through 1999, and Phase II began in 2000. For Phase I, the Clean Air Act required that reformulated gasoline result in summertime emissions of volatile organic compounds (VOCs) and year-round emissions of air toxics that are 15 percent less than those that would occur from the use of conventional gasoline. Phase II required that VOCs and toxics be reduced by an additional 25 percent.

For gasoline to be considered reformulated, it must have an oxygen content of at least 2 percent by weight and a benzene content no greater than 1 percent by volume. It also must contain no heavy metals, including lead or manganese. According to EPA estimates, RFG reduces hydrocarbon emissions by at least 15 percent.

## Government Test Fuel

Because of the variation of commercially available fuel, the federal government uses a standardized fuel for testing engines for emissions and fuel economy. The fuel is indolene 30, and it is used as a standard replacement for regular unleaded gasoline during these tests.

## Review Questions

1. Explain the refining process of petroleum-based fuels.
2. List and briefly describe the four performance characteristics of gasoline.
3. Explain the purpose of oxygenated fuels.
4. The__________ __________ is a measure of how easily the gasoline can be ignited.
5. One __________ __________ __________ is the amount of heat required to raise 1 pound of water 1 Fahrenheit degree.
6. Detonation is described as:
   A. A smooth flame front moving across the combustion chamber, burning all of the air-fuel mixture
   B. A controlled burn during the combustion process
   C. Uncontrolled burn of the air-fuel mixture as a result of a second flame front
   D. None of the above
7. Explain how Reid vapor pressure is measured.
8. Describe the common methods of determining the octane number.
9. The ratio of air to fuel, where complete combustion is accomplished, is referred to as __________________.
10. The combustion process involves the chemical combination of __________ from the air with the __________ and __________ from the fuel.
11. Seasonal blends of gasoline are being discussed. Technician **A** says that for fuel used in the winter, the RVP must be lowered. Technician B says that if winter-blend gasoline is used in an engine during warm weather, vapor lock may occur. Who is correct?
    A. Technician A only
    B. Technician B only
    C. Both Technician A and Technician B
    D. Neither Technician A nor Technician B
12. The AKI index posted on retail fuel pumps is an average of what two methods of determining octane numbers of motor gasoline?
    A. The SG and Boil Point methods
    B. The API and the ATMI methods
    C. The AKI and CFR methods
    D. The MON and RON methods

CHAPTER

# 10 Alternative Fuels

## Introduction

Tighter federal, state, and local emissions regulations have led to a search for **alternative fuels**. This chapter looks at alternative fuels, which are those other than gasoline. Liquid petroleum lp gas and alcohol are currently in use in automobiles and trucks. Hydrogen is also another example of an alternative fuel. Many considerations determine the viability of an alternative fuel, including emissions, cost, fuel availability, fuel consumption, safety, engine life, fueling facilities, weight and space requirements of fuel tanks, and the range of a fully fueled vehicle. Currently, the major competing alternative fuels include ethanol, methanol, propane, and natural gas. Although diesel fuel has been in use for many years, its properties are included in this section. Ethanol and methanol were presented in Chapter 9.

## Liquid Petroleum Gas

**Liquid petroleum (LP)** gas is a by-product of crude oil refining, and it is also found in natural gas wells. LP gas is the term widely used to describe a family of light hydrocarbons called "gas liquids." The most prominent members of this family are propane ($C_3H_8$) and butane ($C_4H_{10}$).

> **You Should Know**
>
> *Because of its high propane content, many people simply refer to LP gas as propane.*

LP gas is similar chemically to gasoline; however, LP gas is mostly propane with a small percentage of butane (up to 8 percent). LP gas is a vapor above –40°F (–40°C). It is called liquid petroleum because it is stored as a liquid in a pressurized bottle. The pressure increases the boiling point of the gas and prevents it from vaporizing.

Because LP gas burns very clean and creates lower emissions than gasoline, it is often used in fleets such as utility companies, taxis, buses, and for indoor machinery such as forklifts. LP gas has been used for a number of years on farm tractors as an alternative to regular gasoline. Even though LP gas has lower heat energy per volume than gasoline (about 84,000 Btu), it does have a higher octane rating (104). This allows LP gas to be used in engines with higher compression ratios.

LP gas burns clean in the engine because it vaporizes at atmospheric temperatures and pressures. Since it vaporizes so readily, it does not puddle in the intake manifold. This means it emits fewer hydrocarbons and less carbon monoxide without added emission control devices.

## Natural Gas

One of the cleanest burning fuels available is **compressed natural gas (CNG)**. CNG is a mixture of hydrocarbons, mainly methane ($CH_4$), and is produced either from gas wells or in conjunction with crude oil production. It has 99 percent less carbon monoxide and 85 percent less reactive hydrocarbon emissions than gasoline, with no particulates and almost no sulfur dioxide. As a result of its low emissions and clean burn, oil change intervals can be increased and spark plugs last longer. A big benefit of natural gas is that it is available in vast quantities in North America.

Vehicles have been designed with gasoline/CNG, diesel/CNG, and dedicated CNG engine applications. CNG vehicles offer several advantages over gasoline:

- The fuel costs less.
- It is the cleanest alternative hydrocarbon fuel.
- It is lighter than air, so it dissipates quickly.
- It has a higher ignition temperature.
- The fuel tanks on natural gas vehicles are safer than those used on gasoline-fueled vehicles. The

fuel tanks used for CNG are aluminum or steel cylinders with walls that are ½ to ¾ inch (13 to 19 mm) thick. They can withstand severe crash tests, direct gunfire, dynamite explosions, and burning beyond any standard sheet metal gasoline tank.

- It generally reduces vehicle maintenance since it burns cleanly.
- Natural gas is abundant and readily available in the United States.

At this time, the largest disadvantage of CNG is the lack of an infrastructure for refueling the vehicle. Fuel facilities are needed in greater numbers than are currently in existence due to the relatively shorter range of CNG vehicles. Lack of range is partly due to CNG having about 33,000 to 38,000 Btu at 3,000 psi. However, CNG has an octane rating of 120+. Many utility companies provide CNG filling stations, and some companies sell kits to allow filling from the natural gas line into your house; however, this is a low-pressure tap and it takes a long time to fill the tanks. The space taken by the CNG cylinders and their weight, about 300 pounds (136 kilograms), would also be considered disadvantages in most applications.

## Hydrogen

Hydrogen ($H_2$) is one of the most abundant elements in the universe and is a highly promising fuel. It is an ideal fuel since it is highly flammable and produces no polluting or poisonous emissions. The only by-products of  hydrogen that is burned are water and carbon dioxide. The sun is an example of burning hydrogen. Hydrogen has an energy content of 113,000 to 134,000 Btu per gallon and an octane rating of over 130. Hydrogen has been used effectively in some engines when mixed with natural gas. Blending hydrogen and CNG at a ratio of 20 percent $H_2$ and 80 percent CNG results in lower emissions than similar engines that are fueled solely by CNG. Most impressive is that the blended fuel results in a 50 percent reduction in $NO_X$. Besides producing lower emissions, hydrogen-fueled engines are 60 percent efficient.

Transportation of hydrogen from the plant to the consumer is the biggest obstacle to its use. Hydrogen is transported by pipeline or by over-the-road tankers. The use of pipelines is limited to those areas where large hydrogen refineries and chemical plants are located. Over-the-road transportation is limited to a range of about 200 miles. For distances of up to 1,000 miles, hydrogen is transported as a liquid and is then vaporized after delivery.

## E85

Ethanol, or grain alcohol, is produced by fermenting biomass. Ethanol is a renewable resource and contributes nothing in itself to the greenhouse gas loading of the atmosphere. As a motor fuel, E85 is blended in a mixture of 85 percent ethanol and 15 percent unleaded gasoline. However, it can be blended with any amount of gasoline in the tank of a flex-fuel vehicle. E85 has a per-gallon Btu rating of about 80,000 and an octane rating of 100.

## Diesel Fuel

Diesel engines are designed to use diesel oil as fuel. **Diesel fuel** is a light oil that is refined as part of the same process that makes gasoline. It has several properties that make it useful as a fuel. At the refinery, the petroleum is separated into three major components: gasoline, middle distillates, and all remaining substances. Diesel fuel comes from the middle distillate group, which has properties and characteristics different from gasoline.

A diesel engine is called a **compression ignition engine** since it uses the heat generated by compressing air to ignite the fuel charge. These engines do not use a spark plug to start the combustion process. A four-stroke diesel engine operates in a similar manner to a gasoline engine. The difference is that ignition in a diesel is controlled by the injection of fuel into the cylinder. A diesel engine has a very high compression ratio, usually between 16:1 and 20:1 (**Figure 1**). The pressure of compression (400 to 700 psi or 2,800 to 4,800 kPa) generates temperatures of 1,200°F to 1,600°F (700°C to 900°C). If the diesel fuel is first drawn into the cylinder and then

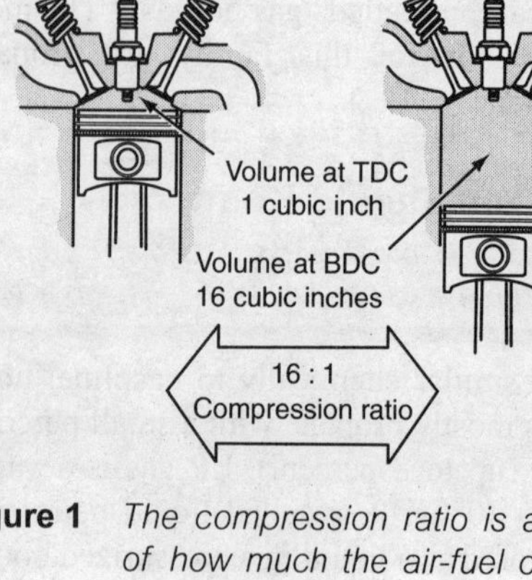


**Figure 1** *The compression ratio is an expression of how much the air-fuel charge will be compressed. This is an example of a compression ratio of 16:1.*

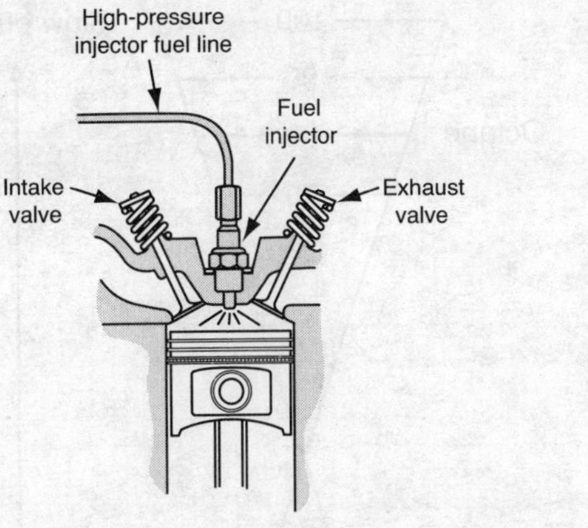


**Figure 2** *The diesel engine uses an injector to spray the fuel into the combustion chamber at the instant combustion is required.*

compressed (as it is in a gasoline engine), the fuel will self-ignite before the piston can reach TDC. In a diesel, the fuel is injected by a high-pressure injector at the instant that ignition is desired (**Figure 2**). The high pressures and temperature in the cylinder speed the preflame reaction to start the ignition of the fuel injected into the cylinder. The compressed, hot air in the cylinder causes the fuel to vaporize and burn.

The shape of the fuel spray, turbulence in the combustion chamber, beginning and duration of injection, and chemical properties of the diesel fuel all affect the power output of the diesel engine. Diesel fuel must meet an entirely different set of standards from gasoline. All diesel fuel must be clean, able to flow at low temperatures, and be of the proper **cetane rating**. This rating determines, to a great extent, the diesel fuel's ability to start the engine at low temperatures and to provide smooth warm-up and even combustion. The cetane number of a diesel fuel describes how easily the fuel will ignite. The significant chemical properties of diesel fuel are described in the following sections.

## Density

Diesel engines have a predictable variation in power output depending on the **density** of the fuel used. Density is the relative mass of a material. Relative density of fuels is usually expressed in degrees of American Petroleum Institute (API) gravity. A fuel with a high density will have a low API gravity, and vice versa. The formula for determining API gravity is:

API gravity (deg.) = (141.5/Specific gravity @ 60°F) – 131.5

Specific gravity is the ratio of the weight of a certain volume of fuel at 60°F (15.5°C) to the weight of the same volume of water at 60°F (15.5°C). Engines using fuels with a high density (low API gravity) will produce more power than those using fuels with lower densities (higher API gravities) because the thermal energy content of the fuel is higher. That is, low API gravity fuels have higher energy content than high API gravity fuels.

A general rule of thumb is that there is approximately a 3 percent to 5 percent decrease in the thermal energy (Btu) content of fuel for every 10-degree increase in API gravity. In addition, there is a 0.7 degree API gravity increase with an increase in temperature. The loss in the energy content of the fuel results in approximately the same percentage of loss in engine power output.

> **You Should Know**
>
> *The use of fuels with higher API gravity will cause an increase in fuel consumption. This is due to the lower thermal energy content per gallon of fuel.*

## Viscosity

In addition to power variations due to fuel density, variations in power output can be due to differences in fuel **viscosity**. Viscosity is the measure of a liquid's resistance to flow. Diesel fuel must be able to flow easily through the fuel system and be sprayed by the injectors. This requires the diesel fuel to have a low viscosity. The viscosity of diesel fuel directly affects the spray pattern of the fuel into the combustion chamber (**Figure 3**). Viscosity must be just high

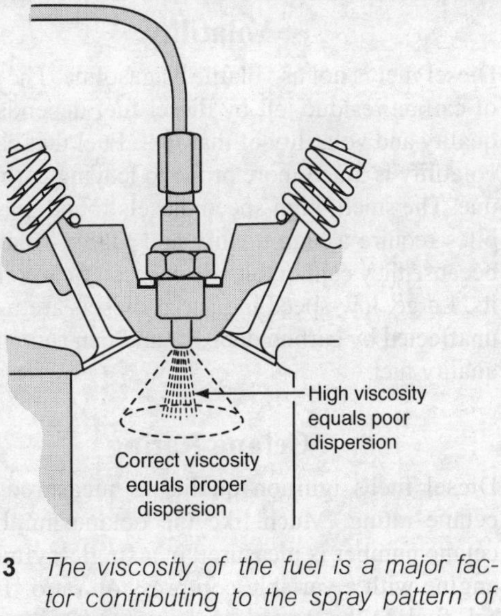


**Figure 3** *The viscosity of the fuel is a major factor contributing to the spray pattern of the fuel.*

enough to provide lubrication to the fuel injectors and fuel pump, but it must not be so high as to produce large droplets that are hard to burn. On the other hand, fuel with viscosity that is too low can create the following problems:

- Power variations due to changes in density with temperature changes. Viscosity deteriorates as temperatures increase
- Hard restart with a hot engine
- Rough idle
- White exhaust smoke
- Loss of torque due to the fine mist igniting too rapidly
- Inadequate lubrication and cooling of the injection pump and nozzles

## Cloud Point and Pour Point

Since diesel fuels contain a wax base called paraffin, ambient temperatures affect diesel fuel more than they do gasoline. As ambient temperatures drop to the wax appearance point (WAP) or **cloud point** of the fuel, the wax crystals grow larger. This is called the cloud point because the fuel appears cloudy when the wax separates out. If cold enough, the crystals will enlarge enough to restrict the flow of fuel through the filters and lines. Eventually, the fuel will stop flowing as a result of the wax crystals plugging the fuel filter or lines. The better the quality of the fuel, the lower the cloud point. As ambient temperatures continue to drop, the fuel will reach its **pour point**, where it solidifies and no longer flows.

## Volatility

Diesel fuel is not as volatile as gasoline. The amount of carbon residue left by diesel fuel depends on the quality and volatility of that fuel. Fuel that has a low volatility is much more prone to leaving carbon residue. The small, high-speed diesels found in automobiles require a high quality and highly volatile fuel because they cannot tolerate excessive carbon deposits. Large, low-speed industrial diesels are relatively unaffected by carbon deposits and can run on lower-quality fuel.

## Cetane Rating

Diesel fuel's ignition quality is measured by the cetane rating. Much like the octane number, the cetane number is measured in a single-cylinder test engine with a variable compression ratio. The diesel fuel to be tested is compared with **cetane**.

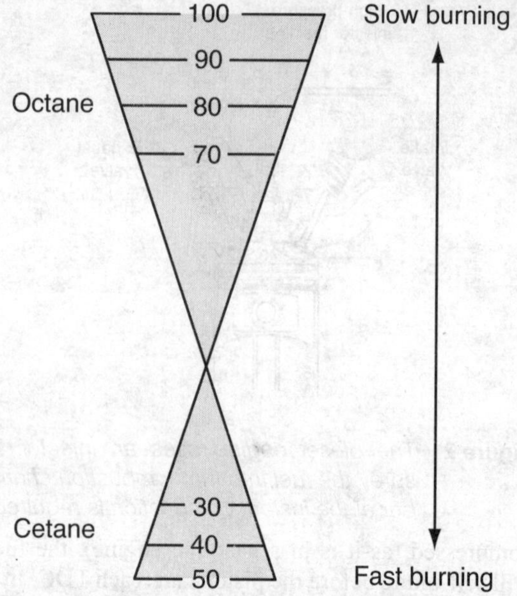


**Figure 4** *Difference between the octane rating of gasoline and the cetane rating of diesel fuel.*

Cetane is a colorless, liquid hydrocarbon that has excellent ignition qualities. Cetane is rated at 100. The higher the cetane number assigned to the diesel fuel, the shorter the ignition lag time (delay time) from the point the fuel enters the combustion chamber until it ignites. The cetane rating for diesel fuel and the octane rating for gasoline are not the same. The higher the cetane rating, the more easily the fuel is ignited, whereas the higher the octane rating, the more slowly the fuel is ignited (**Figure 4**).

In fuels that are readily available, the cetane number ranges from 40 to 55, with values of 40 to 50 being the most common. These cetane values are satisfactory for medium-speed engines whose rated speeds are from 500 to 1,200 rpm and for high-speed engines rated over 1,200 rpm. Low-speed engines rated below 500 rpm can use fuels in the above 30 cetane number range. The cetane number improves with the addition of certain compounds such as ethyl nitrate, acetone peroxide, and amyl nitrate. Amyl nitrate is commercially available for this purpose.

Fuels with a high cetane number will burn as soon as they are injected, so no knock occurs. The lower the cetane number, the higher the temperature required to ignite the fuel. If the cetane number is too low for the conditions, the fuel will take too long to light and will build up in the cylinder. Engine

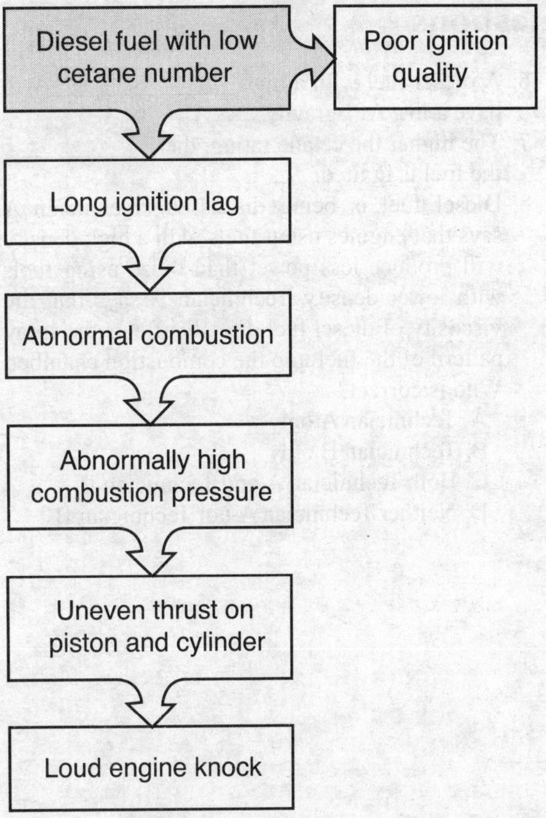


**Figure 5** *Possible problems resulting from use of a fuel with a too-low cetane rating.*

knocking can occur under this condition when the fuel is burned in too much quantity and out of control. **Figure 5** lists some of the problems caused by fuel with a too-low cetane rating.

## Flash Point

The **flash point** is the temperature at which the vapors on the surface of the fuel will ignite if exposed to an open flame. Although flash point does not affect diesel engine operation, a lower-than-normal flash point could indicate contamination of the diesel fuel with gasoline or a similar substance.

## Sulfur Content

On October 1, 1993, the U.S. Clean Air Act required that all diesel fuels sold for use in highway vehicles contain no more than 0.05 percent sulfur. The sulfur content of diesel fuel is very important to the life of the engine. Sulfur in the fuel creates sulfuric acid during the combustion process, which can damage engine components and cause piston ring wear.

## Diesel Fuel Grades

ASTM International has set minimum quality standards for diesel fuel grades. The commonly used fuel grades are described as follows:

**Grade 1.** Grade 1 (Number 1) diesel fuel is thinner and more volatile, and it is used at very low temperatures. It has the lowest boiling point and the lowest cloud and pour points. As a result, Grade 1 is suitable for use during low-temperature operation. Grade 1 may be specified for use in diesel engines involved in frequent changes in load and speed, such as those found in city buses and delivery trucks. The API gravity of Number 1 diesel is between 48 and 34 degrees; thus, it has a Btu rating per gallon of 134,637.

**Grade 2.** Grade 2 (Number 2) diesel fuel has a lower volatility than Number 1 fuel. Number 2 diesel is the most commonly specified fuel to be used for most driving conditions. Number 2 fuel has a higher boiling range, cloud point, and pour point as compared with Number 1. The API gravity of Number 2 diesel is between 40 and 24 degrees; thus, it has a Btu rating per gallon of 139,457.

In colder climates, Number 1 diesel fuel is blended with Number 2 to improve starting. In moderately cold climates, the blend may be 90 percent Number 2 to 10 percent Number 1. In very cold climates, the blend may be as high as 50/50. Diesel fuel economy can be expected to drop off during the winter months due to the use of Number 1 diesel in the fuel blend.

**Grade 4.** Grade 4 (Number 4) has a very high boiling range and is designed for use in low-speed engines running under a constant load. Grade 4 is used mainly in stationary power plants and ships and not as an automotive fuel.

### You Should Know

*Number 3 diesel fuel is no longer refined.*

## Cleanliness

It is imperative that the fuel used in a diesel engine be clean and free from water. Unlike gasoline engines, the fuel is the lubricant and coolant for the diesel injection pump and injectors. Good-quality diesel fuel contains additives such as oxidation inhibitors, detergents, dispersants, rust preventives, and metal deactivators.

## Review Questions

1. Describe the different alternative fuels discussed in this chapter.
2. Explain the cetane rating of diesel fuel.
3. The viscosity of diesel fuel directly affects the ____________ of the fuel into the combustion chamber.
4. Technician A says that LP gas burns clean in the engine because it vaporizes at atmospheric temperatures and pressures. Technician B says that the disadvantage of CNG is its high output of carbon monoxide. Who is correct?
   A. Technician A only
   B. Technician B only
   C. Both Technician A and Technician B
   D. Neither Technician A nor Technician B
5. Explain the differences among diesel fuel grades.
6. A diesel fuel with a high ____________ will have a low API gravity.
7. The higher the cetane rating, the ____________ the fuel is ignited.
8. Diesel fuel is being discussed. Technician A says that engines using fuels with a high density will produce less power than those using fuels with lower density. Technician B says that the viscosity of diesel fuel directly affects the spray pattern of the fuel into the combustion chamber. Who is correct?
   A. Technician A only
   B. Technician B only
   C. Both Technician A and Technician B
   D. Neither Technician A nor Technician B

CHAPTER

# 11 Basic Fuel Testing

## Introduction

As you progress through the following chapters, you will be exposed to several different types of test procedures relating to the fuels and emissions system. The following test procedures set the ground work for performing a complete diagnostic routine on these systems. Knowledge gathered here is a building block to understanding the procedures that follow in later chapters.

## Fuel Quality Testing

There are a couple of tests that can be performed in the shop to help the technician determine the quality of fuel. These include testing the Reid vapor pressure (RVP) and the alcohol content of gasoline. A specific gravity check can be used to help determine the quality of diesel fuel.

### Testing the RVP of Gasoline

The RVP is a measure of the volatility of gasoline. More volatile fuels vaporize more easily, creating more pressure. Increasing the RVP of a gasoline permits the engine to start easier in cold weather. The RVP of winter-blend fuel is about 9 psi. If winter-blend gasoline is used in an engine during warm weather, the following problems may occur:

1. Rough idle
2. Stalling
3. Hesitation on acceleration
4. Surging

Since warm ambient temperatures cause gasoline to vaporize easily, the RVP of summer-blend fuel is reduced. If the RVP of a fuel is too high for the ambient temperature, vapor will form and the engine will not run properly. The volatility of summer-grade gasoline should be about 7 psi RVP.

A special fuel vapor pressure tester is needed to test the RVP of gasoline (**Figure 1**). For accurate results, the vapor pressure test must be performed on chilled gasoline. If the fuel being drawn from the fuel tank is warm, it must be poured into a chilled container and securely capped as soon as it is removed from the fuel tank. It is important that the container be securely capped and the fuel be sealed as soon as possible after being removed from the tank to prevent loss of light vapors. The fuel sample then must be chilled prior to testing.

In preparation for testing the fuel vapor pressure, separate the aluminum fuel cup from the gauge and rinse the inside of both the fuel cup and the gauge's pressure chamber with hot water. Use low-pressure air or a siphon bulb, then dry the fuel cup and pressure chamber. Place the aluminum fuel cup into a container of chilled water. Finally, fill the insulated cup with hot water to within ¾ inch

**Figure 1** *Typical gasoline vapor pressure tester kit.*

(19 mm) of full. Fit the gauge and chamber assembly over the insulated cup with the open-ended air chamber immersed in the hot water. Let stand for three to five minutes to condition the gauge assembly and allow it to be close to the equilibrium when the temperature drops to 105°F (40°C).

> **You Should Know**
>
> *To prevent damage to the gauge assembly, do not direct high-pressure air into the cavity of the gauge.*

At a temperature of 105°F (40°C), the test should begin. Do not allow the temperature to drop below 105°F before the beginning of the test. If necessary, add hot water to prevent the temperature from dropping too low.

The following is a typical procedure to test the RVP of gasoline. These steps must be performed as quickly as possible:

1. Remove the aluminum fuel cup from the chilled water, shake loose any water, and place the fuel cup in the holder.
2. Using a chilled dropper (or syringe), withdraw chilled gasoline from the sample bottle and fill the fuel cup to within ⅛ inch (3 mm) of the top. Wipe off any fuel that may have spilled over the top of the fuel cup.
3. Remove the gauge from the insulated cup, holding it vertically at all times. Gently shake loose any water from the gauge's chamber and place the gauge over the fuel cup. Quickly attach the cup by screwing it on finger tight. During this operation, the gauge or chamber assembly must be held vertical at all times. Do not allow any gasoline to spill.
4. Place the gauge and fuel cup assembly and the cup cover over the insulated cup and let stand for at least two minutes. Place a thermometer into the small hole of the cup cover to monitor water temperature.
5. After two minutes, remove the gauge and fuel cup assembly from the cup cover and tilt the assembly with the gauge pointing downward at about 45 degrees and shake vigorously for a few seconds. Shake along the direction that the unit is held.
6. Return the assembly to the large hole in the cup cover and let it stand in the water bath for three minutes. After three minutes, read the pressure and temperature.

Unless the pressure reading was taken at 100°F (38°C), the reading will need to be corrected. Refer to **Table 1** to obtain the gasoline vapor pressure corrected to 100°F (38°C).

## Testing Gasoline for Alcohol Content

Excessive alcohol content can cause hard starting, poor fuel economy, and drivability problems. The

| Temp. | Correction Factor | Temp. | Correction Factor |
|---|---|---|---|
| 32°F (0°C) | −1.6 | 97°F (36.1°C) | 5.5 |
| 46°F (7.8°C) | −1.0 | 100°F (37.8°C) | 6.0 |
| 60°F (15.5°C) | 0 | 102°F (38.9°C) | 6.5 |
| 64°F (17.8°C) | 0.5 | 105°F (40.6°C) | 7.0 |
| 68°F (20°C) | 1.0 | 107°F (41.7°C) | 7.5 |
| 72°F (22.2°C) | 1.5 | 109°F (42.8°C) | 8.0 |
| 75°F (23.9°C) | 2.0 | 112°F (44.4°C) | 8.5 |
| 79°F (26.1°C) | 2.5 | 114°F (45.5°C) | 9.0 |
| 83°F (28.3°C) | 3.0 | 116°F (46.7°C) | 9.5 |
| 86°F (30°C) | 3.5 | 118°F (47.8°C) | 10.0 |
| 89°F (31.7°C) | 4.0 | 122°F (50°C) | 11.0 |
| 92°F (33.3°C) | 4.5 | 126°F (52.2°C) | 12.0 |
| 95°F (35°C) | 5.0 | 130°F (54.4°C) | 13.0 |

**Table 1** *Temperature conversion chart for vapor pressure testing.*

following is one method of determining the alcohol content of a gasoline sample:

1. Pour suspect gasoline into a small clean beaker or glass container.

**You Should Know**

*Do not smoke or run the test around sources of ignition.*

2. Carefully fill the graduated cylinder to the 100 milliliter (ml) mark.
3. Add 2 ml of water to the graduated cylinder by counting the number of drops from the eye dropper. (Before performing the test, the eye dropper must be calibrated to determine how many drops equal 2 ml).
4. Put the stopper in the cylinder and shake vigorously for one minute. Relieve built-up pressure by occasionally removing the stopper. Alcohol dissolves in water and will drop to the bottom of the cylinder.
5. Place the cylinder on a flat surface and let it stand for two minutes.
6. Take a reading near the bottom of the cylinder at the boundary between the two liquids.
7. For percent of alcohol in gasoline, subtract 2 from the reading and multiply by 10.
   For example, if the reading is 3.1 ml:

   3.1 − 2 = 1.1 × 10 = 11% alcohol.

   If the reading is 2.0 ml:

   2 − 2 = 0 × 10 = 0% alcohol (no alcohol).

If the increase in volume is 0.2 percent or less, it may be assumed that the test gasoline contains no alcohol.

## Diesel Fuel Testing

The **specific gravity** of diesel fuel is the ratio of the density of the fuel to the density of water. The density of the fuel affects the spray penetration as the fuel is injected into the cylinder. By measuring the specific gravity of the fuel it is possible to determine its API gravity also. Engines using fuels with a high density (low API gravity) will produce more power than those using fuels with lower densities (higher API gravities) because the thermal energy content of the fuel is higher.

The specific gravity reading is a strong indication of fuel quality. The specific gravity of Number 1 diesel is between 0.788 and 0.855 at 60°F (15°C). The specific gravity of number 2 diesel should be between 0.825 and 0.910 at 60°F (15°C). The higher the specific gravity number, the greater the power. For example, a fuel with specific gravity of 0.876 contains 3.5 percent more energy per gallon than a fuel with a specific gravity of 0.815.

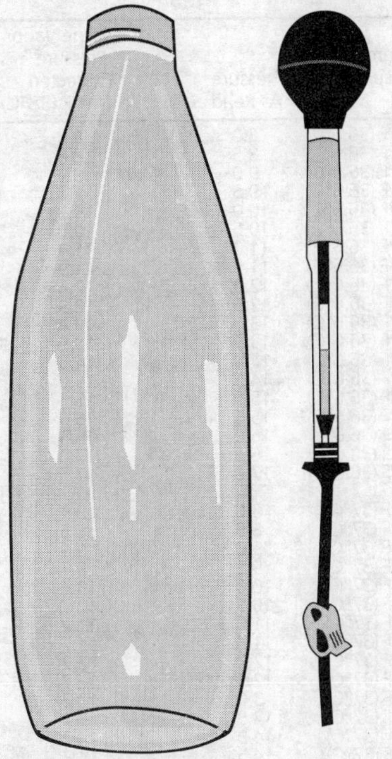

**Figure 2** *Hydrometer used to measure the specific gravity of diesel fuel.*

The specific gravity of the fuel can be tested by filling a beaker with fuel from the vehicle's tank and using a hydrometer (**Figure 2**). Fill the beaker with enough sample of the fuel to ensure the hydrometer will be able to float without touching the bottom of the beaker. Place the hydrometer into the beaker, making sure it is free to float in the fuel. With the beaker level, read the scale at the point the hydrometer touches the top of the fuel sample. Since temperature affects the specific gravity reading, the obtained reading may need to be corrected (**Table 2**).

With the specific gravity now known, use the following formula to determine its API gravity:

API gravity (degrees) =
(141.5/Specific gravity @ 60°F) − 131.5

Remember that a fuel with a high density will have a low API gravity, and vice versa.

| Temperature When Pressure Is Read | Pressure As Read | Gasoline Vapor Pressure Corrected To 100°F (38°C) | Temperature When Pressure Is Read | Pressure As Read | Gasoline Vapor Pressure Corrected To 100°F (38°C) |
|---|---|---|---|---|---|
| 98.0°F (36.6°C) | 8.0 | 8.3 | 102.0°F (38.9°C) | 8.0 | 7.7 |
| 98.0°F (36.6°C) | 8.5 | 8.8 | 102.0°F (38.9°C) | 8.5 | 8.2 |
| 98.0°F (36.6°C) | 9.0 | 9.3 | 102.0°F (38.9°C) | 9.0 | 8.7 |
| 98.0°F (36.6°C) | 9.5 | 9.8 | 102.0°F (38.9°C) | 9.5 | 9.2 |
| 98.0°F (36.6°C) | 10.0 | 10.3 | 102.0°F (38.9°C) | 10.0 | 9.7 |
| 98.0°F (36.6°C) | 10.5 | 10.8 | 102.0°F (38.9°C) | 10.5 | 10.2 |
| 98.0°F (36.6°C) | 11.0 | 11.4 | 102.0°F (38.9°C) | 11.0 | 10.6 |
| 98.0°F (36.6°C) | 11.5 | 11.9 | 102.0°F (38.9°C) | 11.5 | 11.1 |
| 98.0°F (36.6°C) | 12.0 | 12.4 | 102.0°F (38.9°C) | 12.0 | 11.6 |
| 98.0°F (36.6°C) | 12.5 | 12.9 | 102.0°F (38.9°C) | 12.5 | 12.1 |
| 98.0°F (36.6°C) | 13.0 | 13.4 | 102.0°F (38.9°C) | 13.0 | 12.6 |
| 98.0°F (36.6°C) | 13.5 | 13.9 | 102.0°F (38.9°C) | 13.5 | 13.1 |
| 98.0°F (36.6°C) | 14.0 | 14.5 | 102.0°F (38.9°C) | 14.0 | 13.5 |
| 98.0°F (36.6°C) | 14.5 | 15.0 | 102.0°F (38.9°C) | 14.5 | 14.0 |
| 98.0°F (36.6°C) | 15.0 | 15.5 | 102.0°F (38.9°C) | 15.0 | 14.5 |
| 98.0°F (36.6°C) | 15.5 | 16.0 | 102.0°F (38.9°C) | 15.5 | 15.0 |
| 98.0°F (36.6°C) | 16.0 | 16.5 | 102.0°F (38.9°C) | 16.0 | 15.5 |
| 98.0°F (36.6°C) | 16.5 | 17.0 | 102.0°F (38.9°C) | 16.5 | 16.0 |
| 98.0°F (36.6°C) | 17.0 | 17.5 | 102.0°F (38.9°C) | 17.0 | 16.5 |
| 99.0°F (37°C) | 8.0 | 8.1 | 103.0°F (39.4°C) | 8.0 | 7.6 |
| 99.0°F (37°C) | 8.5 | 8.6 | 103.0°F (39.4°C) | 8.5 | 8.1 |
| 99.0°F (37°C) | 9.0 | 9.1 | 103.0°F (39.4°C) | 9.0 | 8.6 |
| 99.0°F (37°C) | 9.5 | 9.7 | 103.0°F (39.4°C) | 9.5 | 9.0 |
| 99.0°F (37°C) | 10.0 | 10.2 | 103.0°F (39.4°C) | 10.0 | 9.5 |
| 99.0°F (37°C) | 10.5 | 10.7 | 103.0°F (39.4°C) | 10.5 | 10.0 |
| 99.0°F (37°C) | 11.0 | 11.2 | 103.0°F (39.4°C) | 11.0 | 10.5 |
| 99.0°F (37°C) | 11.5 | 11.7 | 103.0°F (39.4°C) | 11.5 | 10.9 |
| 99.0°F (37°C) | 12.0 | 12.2 | 103.0°F (39.4°C) | 12.0 | 11.4 |
| 99.0°F (37°C) | 12.5 | 12.7 | 103.0°F (39.4°C) | 12.5 | 11.9 |
| 99.0°F (37°C) | 13.0 | 13.2 | 103.0°F (39.4°C) | 13.0 | 12.4 |
| 99.0°F (37°C) | 13.5 | 13.7 | 103.0°F (39.4°C) | 13.5 | 12.8 |
| 99.0°F (37°C) | 14.0 | 14.2 | 103.0°F (39.4°C) | 14.0 | 13.3 |
| 99.0°F (37°C) | 14.5 | 14.7 | 103.0°F (39.4°C) | 14.5 | 13.8 |
| 99.0°F (37°C) | 15.0 | 15.2 | 103.0°F (39.4°C) | 15.0 | 14.3 |
| 99.0°F (37°C) | 15.5 | 15.7 | 103.0°F (39.4°C) | 15.5 | 14.8 |
| 99.0°F (37°C) | 16.0 | 16.3 | 103.0°F (39.4°C) | 16.0 | 15.2 |
| 99.0°F (37°C) | 16.5 | 16.8 | 103.0°F (39.4°C) | 16.5 | 15.7 |
| 99.0°F (37°C) | 17.0 | 17.3 | 103.0°F (39.4°C) | 17.0 | 16.2 |
| 100.0°F (38°C) | 8.0 | 8.0 | 104.0°F (40°C) | 8.0 | 7.5 |
| 100.0°F (38°C) | 8.5 | 8.5 | 104.0°F (40°C) | 8.5 | 8.0 |
| 100.0°F (38°C) | 9.0 | 9.0 | 104.0°F (40°C) | 9.0 | 8.4 |
| 100.0°F (38°C) | 9.5 | 9.5 | 104.0°F (40°C) | 9.5 | 8.9 |
| 100.0°F (38°C) | 10.0 | 10.0 | 104.0°F (40°C) | 10.0 | 9.4 |
| 100.0°F (38°C) | 10.5 | 10.5 | 104.0°F (40°C) | 10.5 | 9.8 |
| 100.0°F (38°C) | 11.0 | 11.0 | 104.0°F (40°C) | 11.0 | 10.3 |
| 100.0°F (38°C) | 11.5 | 11.5 | 104.0°F (40°C) | 11.5 | 10.8 |
| 100.0°F (38°C) | 12.0 | 12.0 | 104.0°F (40°C) | 12.0 | 11.2 |
| 100.0°F (38°C) | 12.5 | 12.5 | 104.0°F (40°C) | 12.5 | 11.7 |
| 100.0°F (38°C) | 13.0 | 13.0 | 104.0°F (40°C) | 13.0 | 12.2 |
| 100.0°F (38°C) | 13.5 | 13.5 | 104.0°F (40°C) | 13.5 | 12.6 |
| 100.0°F (38°C) | 14.0 | 14.0 | 104.0°F (40°C) | 14.0 | 13.1 |
| 100.0°F (38°C) | 14.5 | 14.5 | 104.0°F (40°C) | 14.5 | 13.6 |
| 100.0°F (38°C) | 15.0 | 15.0 | 104.0°F (40°C) | 15.0 | 14.0 |
| 100.0°F (38°C) | 15.5 | 15.5 | 104.0°F (40°C) | 15.5 | 14.5 |
| 100.0°F (38°C) | 16.0 | 16.0 | 104.0°F (40°C) | 16.0 | 15.0 |
| 100.0°F (38°C) | 16.5 | 16.5 | 104.0°F (40°C) | 16.5 | 15.4 |
| 100.0°F (38°C) | 17.0 | 17.0 | 104.0°F (40°C) | 17.0 | 15.9 |
| 101.0°F (38.3°C) | 8.0 | 7.9 | 105.0°F (40.6°C) | 8.0 | 7.4 |
| 101.0°F (38.3°C) | 8.5 | 8.4 | 105.0°F (40.6°C) | 8.5 | 7.8 |
| 101.0°F (38.3°C) | 9.0 | 8.9 | 105.0°F (40.6°C) | 9.0 | 8.3 |
| 101.0°F (38.3°C) | 9.5 | 9.3 | 105.0°F (40.6°C) | 9.5 | 8.7 |
| 101.0°F (38.3°C) | 10.0 | 9.8 | 105.0°F (40.6°C) | 10.0 | 9.2 |
| 101.0°F (38.3°C) | 10.5 | 10.3 | 105.0°F (40.6°C) | 10.5 | 9.7 |
| 101.0°F (38.3°C) | 11.0 | 10.8 | 105.0°F (40.6°C) | 11.0 | 10.1 |
| 101.0°F (38.3°C) | 11.5 | 11.3 | 105.0°F (40.6°C) | 11.5 | 10.6 |
| 101.0°F (38.3°C) | 12.0 | 11.8 | 105.0°F (40.6°C) | 12.0 | 11.0 |
| 101.0°F (38.3°C) | 12.5 | 12.3 | 105.0°F (40.6°C) | 12.5 | 11.5 |
| 101.0°F (38.3°C) | 13.0 | 12.8 | 105.0°F (40.6°C) | 13.0 | 12.0 |
| 101.0°F (38.3°C) | 13.5 | 13.3 | 105.0°F (40.6°C) | 13.5 | 12.4 |
| 101.0°F (38.3°C) | 14.0 | 13.8 | 105.0°F (40.6°C) | 14.0 | 12.9 |
| 101.0°F (38.3°C) | 14.5 | 14.3 | 105.0°F (40.6°C) | 14.5 | 13.3 |
| 101.0°F (38.3°C) | 15.0 | 14.8 | 105.0°F (40.6°C) | 15.0 | 13.8 |
| 101.0°F (38.3°C) | 15.5 | 15.3 | 105.0°F (40.6°C) | 15.5 | 14.3 |
| 101.0°F (38.3°C) | 16.0 | 15.7 | 105.0°F (40.6°C) | 16.0 | 14.7 |
| 101.0°F (38.3°C) | 16.5 | 16.2 | 105.0°F (40.6°C) | 16.5 | 15.2 |
| 101.0°F (38.3°C) | 17.0 | 16.7 | 105.0°F (40.6°C) | 17.0 | 15.6 |

**Table 2** *Chart for correcting specific gravity based on temperature of the sample.*

# Typical Procedures for Relieving Fuel Pressure

Prior to performing many services or tests on a modern throttle body injection (TBI), multipoint fuel injection (MFI), or sequential fuel injection (SFI) system, the fuel pressure must be relieved. All of these systems retain residual pressures in the fuel lines to ensure easy starting of the engine. There are several methods that can be used to relieve fuel pressure. Always refer to the correct service manual to determine the manufacturer's recommended procedure.

> **You Should Know**
>
> *Failure to relieve the residual fuel pressures can cause gasoline to be sprayed into your eyes or onto your clothing.Both can result in serious injury.*

One method is to remove the fuel pump relay, turn the ignition switch to the Run position, and use the scan tool actuator tests to energize the injector. This action lifts the injector plunger and the fuel discharges from the injector to relieve the fuel pressure. If you use this method, it is a good practice to remove the spark plug from the cylinder and crank the engine over a couple of revolutions with the ignition and fuel system disabled to remove the fuel from the top of the cylinder.

> **You Should Know**
>
> *If you use this method of relieving the fuel pressure, it is recommended that an oil change be performed before returning the vehicle to its owner.The gasoline discharged into the cylinder may run into the crankcase and dilute the engine oil.*

On vehicles with a Schrader valve in the fuel line, the manufacturer may recommend a drain hose be connected to the valve with the open end of the hose placed into an approved container (**Figure 3**). Depending on manufacturer, this may be all that is required, while others may have you disable the fuel delivery system and then attempt to start the engine.

Another method is to disable the fuel pump by removing the fuel pump relay, disconnecting the fuel pump, or pulling the fuel pump fuse. With the fuel pump disabled, start the engine and allow it to die. Then crank the engine a few times to remove all fuel from the rails.

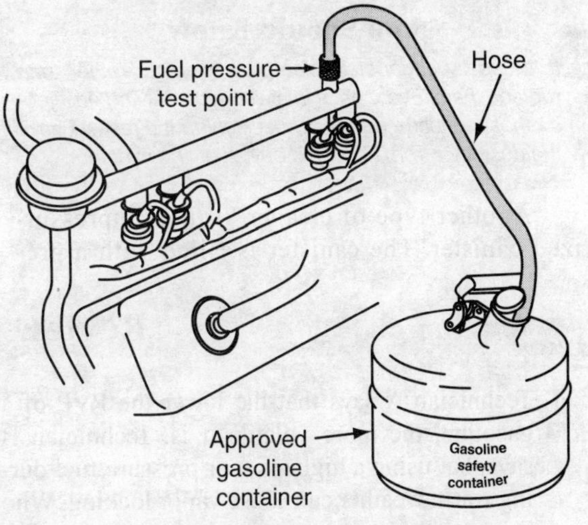


**Figure 2** *Relieving residual fuel pressure with a drain hose and approved container.*

Regardless of the method used, always remove the gas tank filler cap first. This will relieve any pressure in the tank and prevent siphoning.

# Fuel-Injector Cleaning

Fuel injectors spray a certain amount of fuel into the intake system. If the fuel pressure is low, not enough fuel will be sprayed. This is also true if the fuel injector is dirty. Normally clogged injectors are the result of inconsistencies in gasoline detergent levels and the high sulfur content of gasoline. When these sensitive fuel injectors become partially clogged, fuel flow is restricted. Spray patterns are altered, causing poor performance and reduced fuel economy.

From the standpoint of cost savings, it is usually more desirable to clean sulfated and/or plugged fuel injectors instead of replacing them. There are different kinds of fuel-injector cleaners available. One is a pressure tank that holds a mixture of solvent and unleaded gasoline. Follow the manufacturer's instructions for mixing, quantity, and safe handling of the mixture. The vehicle's fuel pump must be disabled, and on some vehicles the fuel line must be blocked between the pressure regulator and the return line. Then, connect the hose on the pressure tank to the service port in the fuel system. Pump the tank to build internal pressure. Partially open the in-line valve and start the engine. It should run at approximately 2,000 rpm for about 10 minutes to clean the injectors thoroughly.

**You Should Know**

*It is usually cheaper to clean the injectors rather than replace them. Every once in a while you will find an injector that cannot be cleaned, but cleaning the rest will save the customer from having to buy all of the injectors.*

Another type of cleaner kit uses a pressurized canister. The canister is filled with a premixed solvent solution. Use of the canister-type cleaner is similar to the pressure tank system described above but does not require mixing or pumping.

Connect the canister to the injection system's servicing fitting, and open the valve on the canister. Start the engine and allow it to run until it dies.

## Review Questions

1. Technician A says that the lower the RVP of a gasoline, the more volatile it is. Technician B says that using a higher vapor pressure fuel during warm weather can cause vapor locking. Who is correct?
   A. Technician A only
   B. Technician B only
   C. Both Technician A and Technician B
   D. Neither Technician A nor Technician B
2. Testing the specific gravity of diesel fuel is being discussed. Technician A says that the specific gravity of number 2 diesel should be between 0.825 and 0.910 at 60°F (15°C). Technician B says that the higher the specific gravity number, the less energy from the fuel. Who is correct?
   A. Technician A only
   B. Technician B only
   C. Both Technician A and Technician B
   D. Neither Technician A nor Technician B
3. Technician A says that prior to performing many services or tests on a modern throttle body injection (TBI), multipoint fuel injection (MFI), or sequential fuel injection (SFI) system, the fuel pressure must be relieved. Technician B says that fuel pressure does not need to be relieved if the vehicle has not been started for one hour or more. Who is correct?
   A. Technician A only
   B. Technician B only
   C. Both Technician A and Technician B
   D. Neither Technician A nor Technician B
4. Testing for alcohol content of gasoline is being discussed. Technician A says that if the increase in volume is 0.2 percent or less, it may be assumed that the test gasoline contains no alcohol. Technician B says that if the reading obtained was 4, the alcohol content is 20 percent. Who is correct?
   A. Technician A only
   B. Technician B only
   C. Both Technician A and Technician B
   D. Neither Technician A nor Technician B
5. Technician A says that the Reid vapor pressure (RVP) is a measure of the specific gravity of gasoline. Technician B says that RVP readings are taken at, or corrected to, 60°F (15°C). Who is correct?
   A. Technician A only
   B. Technician B only
   C. Both Technician A and Technician B
   D. Neither Technician A nor Technician B

SECTION

# 4 FUEL DELIVERY

## Section Objectives

After you have read, studied, and practiced the contents of this section, you should be able to:

- Explain the purpose of the fuel-delivery system.
- List the commonly used components of the fuel-delivery system.
- Detail the function of the individual components of the fuel-delivery system.
- Explain how expansion, contraction, and overflow are controlled in the fuel tank.
- Explain the purpose of the check valves used in the fuel-delivery system.
- Describe the function of the rollover valve.
- Detail the emissions controls associated with the fuel-delivery system.
- Describe the operation of the electric fuel pumps.
- Explain how the fuel pressure regulator operates.
- Explain the differences between return-type and returnless-type fuel systems.
- Describe the types of materials and fittings used in common fuel lines.
- Detail the function of the intake manifold.
- Explain typical fuel pump circuit operation.
- Conduct a visual inspection of a fuel system and determine needed repairs.
- Remove and inspect fuel tanks.
- Perform fuel pressure and volume tests on an engine equipped with an electric fuel pump and determine needed repairs.
- Perform electrical tests on the electric fuel pump circuit and determine needed repairs.
- Remove, inspect, service, and replace electric fuel pumps and gauge-sending units.
- Remove and replace the fuel filter.
- Inspect and service fuel lines and tubing.
- Remove and replace the fuel rail.

### Interesting Fact

*During the early years of the automobile, many cars did not have fuel pumps. Some cars, like the Model A Ford, had fuel tanks mounted in the front of the instrument panel. The filler cap was positioned in front of the windshield. Since the gasoline tank was mounted higher than the carburetor, gravity caused the fuel to flow from the tank to the carburetor. Other cars in that era had vacuum tanks that used the engine vacuum to move the fuel from the tank to the carburetor. Since gravity was used to fill the carburetor bowl, it was not uncommon to see motorists driving up a hill backward so fuel from the tank would flow to the carburetor. The AC Spark Plug Company introduced the first successful mechanical fuel pump in 1927.*

CHAPTER

# 12 Fuel Tanks

## Introduction

The fuel-delivery system consists of all the components responsible for storing and delivering fuel to the engine's combustion chambers. In addition, this system must control the emissions associated with fuel storage. Since the fuel-delivery system can have a major impact on engine performance and emissions, today's technician must possess knowledge of the components used and their function. This chapter looks at fuel tanks. **Figure 1** illustrates a typical fuel-delivery system used in fuel-injected vehicles. Fuel is drawn from the fuel tank by an in-tank or chassis-mounted electric fuel pump. Before it reaches the injectors, the fuel passes through a filter that removes dirt and impurities. A fuel line pressure regulator maintains a constant fuel line pressure in the system. This fuel pressure generates the spraying force needed to inject the fuel. On some systems, excess fuel returns to the fuel tank through a fuel return line. A typical fuel-delivery system includes the following components:

1. Fuel tank
2. Filler cap
3. Fuel pump
4. Fuel lines
5. Fuel filter
6. Check valves
7. Pressure regulator
8. Liquid-vapor separator
9. Pressure and vacuum controls
10. Fuel rail
11. Injectors
12. Intake manifold

## Fuel Tank

The main function of the **fuel tank** is to store the fuel for the engine. The expanded function of the fuel tank is to prevent gasoline vapors (HC) from escaping into the atmosphere. Although the fuel tank appears to be a simple device, there is much research and development that goes into its design.

The fuel tank in most passenger cars and sport utility vehicles (SUVs) is located at the rear of the vehicle, situated between the frame and above the rear suspension cross-member **(Figure 2)** or in front of the rear axle under the rear seat floor. On light-duty and medium-duty trucks, the fuel tank may be located under the bed and behind the cab. Straps or brackets support the fuel tank in the vehicle. Fuel tanks are normally mounted with insulators between the top of the tank and the chassis to protect the tank and to prevent noise from transferring into the passenger compartment **(Figure 3)**.

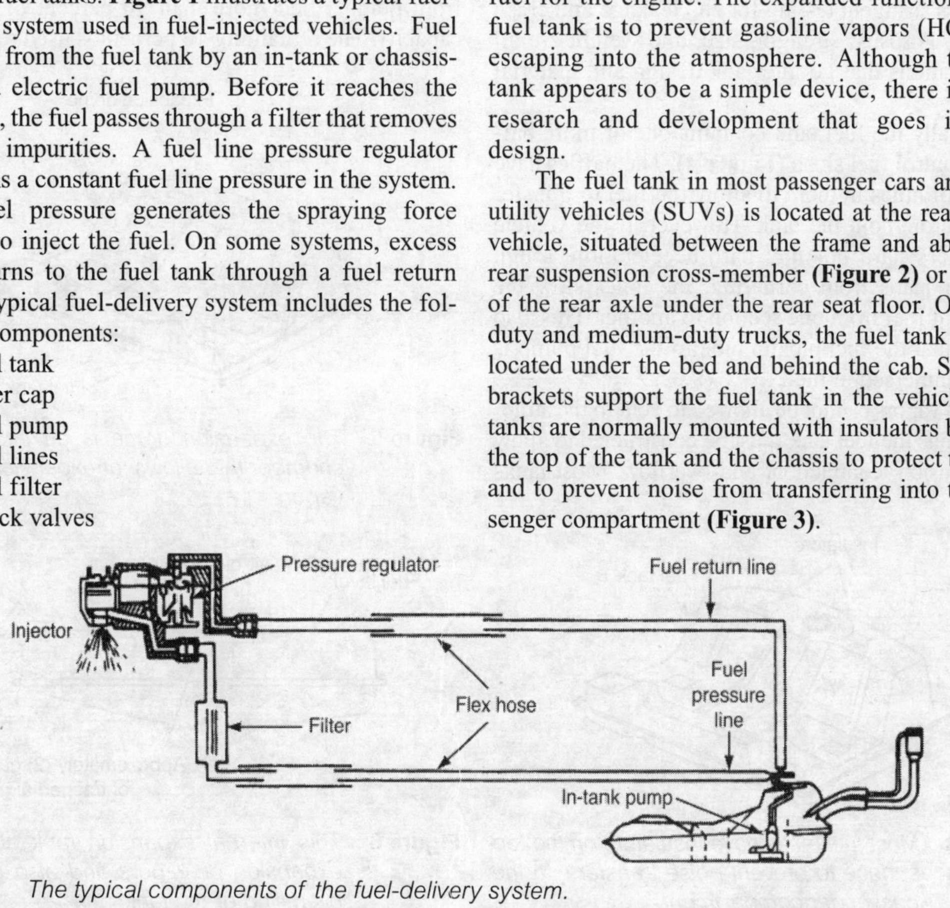


**Figure 1** *The typical components of the fuel-delivery system.*

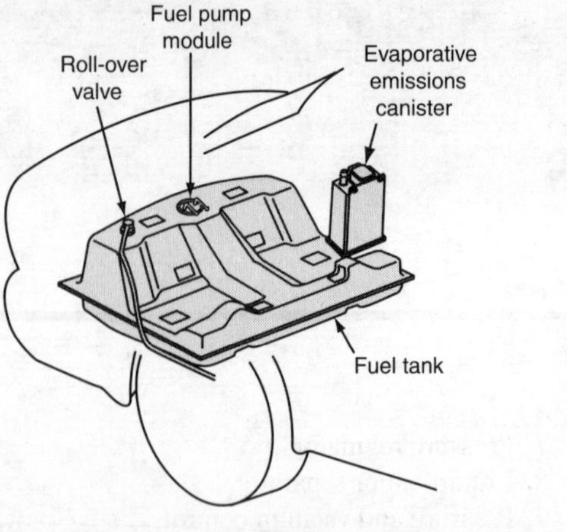


**Figure 2** *The fuel tank is usually located near the rear of the vehicle in most passenger cars and SUVs.*

Fuel tanks can be made from corrosion-resistant stamped steel, aluminum, or high-density polyethylene (HDPE). Prior to its use in a vehicle, the tank must pass stringent testing for HC leakage and crash integrity. Porosity, strength, size, and weight are all considerations that go into tank design and material selection.

Usually the fuel tank contains one or more baffles to control fuel slosh **(Figure 4)**. The baffles have a series of holes in them to allow the fuel to transfer evenly throughout the tank. However, if the vehicle experiences hard braking, hard acceleration, rough roads, or leans from cornering, the holes slow the transfer of fuel from one section to another. This also ensures that the fuel pickup tube to the fuel pump is always immersed in fuel.

Fuel vapors cannot be allowed to vent to the atmosphere, thus the fuel tank must be constructed to allow for expansion, contraction, and overflow. Most tanks have an internal air expansion chamber dome **(Figure 5)** or an expansion tank **(Figure 6)**. The expansion tank is a chamber of the fuel tank that allows for fuel expansion as ambient temperature increases.

Fuel tank overfill is controlled by the tank design. This can be accomplished by locating the filler tube lower in the tank **(Figure 7)** and/or by using vapor vent lines to the filler tube **(Figure 8)**. The filler tube is the pipe through which the fuel flows to fill the tank. These designs prevent the tank from being filled more than 90 percent of its total capacity. The remaining 10 percent is used to contain

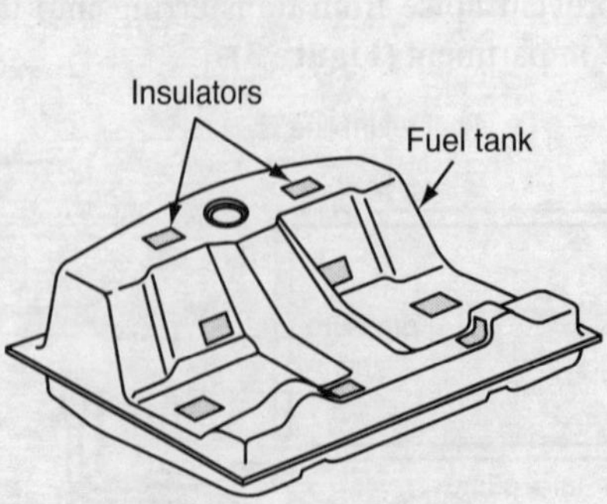


**Figure 3** *Most fuel tanks have insulators on the top surface to prevent noise transfer into the passenger compartment.*

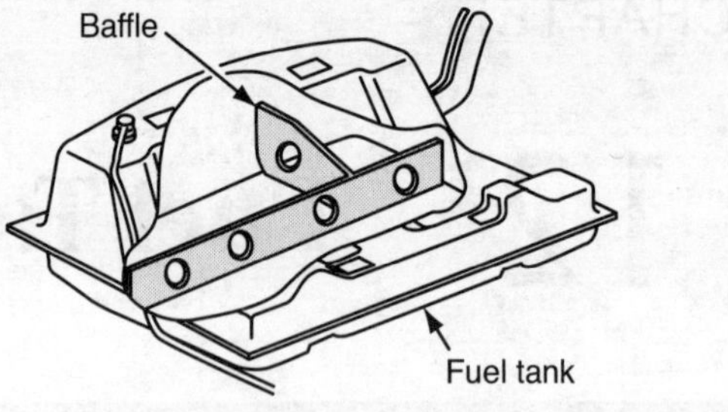


**Figure 4** *Internal baffles prevent fuel from rapidly moving from one side of the tank to the other.*

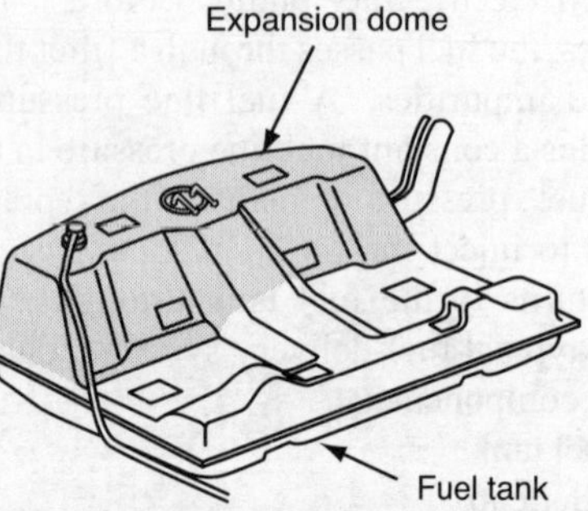


**Figure 5** *The expansion dome is an internal air chamber that allows for expansion of fuel vapors.*

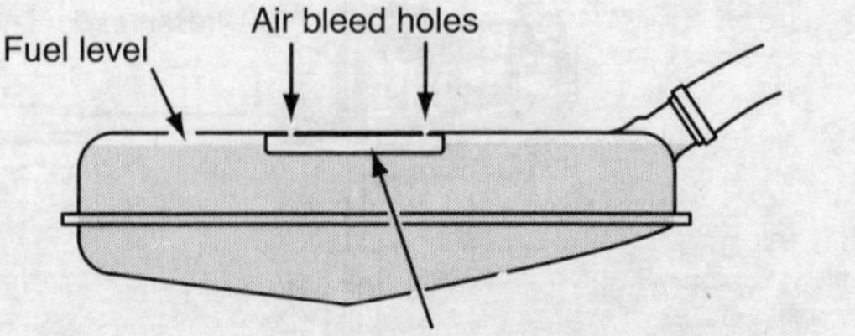


**Figure 6** *This internal expansion tank allows for expansion of vapors and also prevents overfilling of the tank.*

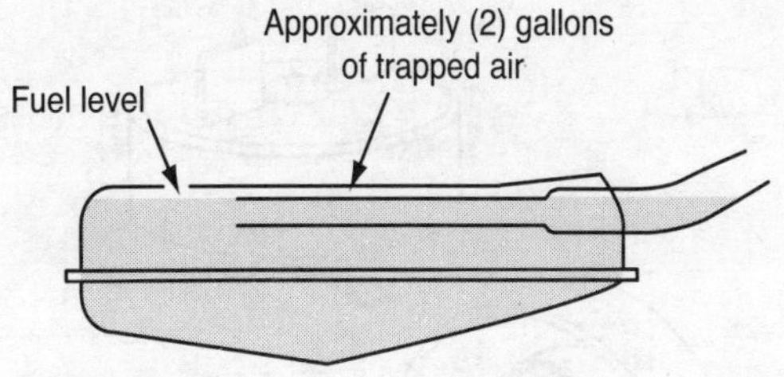


**Figure 7** *Locating the filler tube lower in the side of the tank prevents overfilling.*

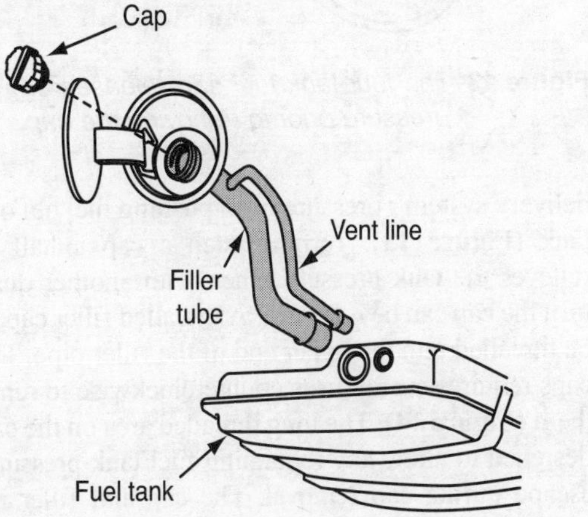


**Figure 8** *The use of vent lines controls fill volume of the tank.*

fuel vapors and to allow for expansion. Some fuel tanks may also contain a check valve at the bottom of the filler tube that closes when the tank is full, preventing overfilling.

Some form of liquid-vapor separator is incorporated into the system to prevent liquid fuel or bubbles from reaching the vapor storage canister or the engine crank-case. The separator may be located inside the tank, on the tank **(Figure 9)**, in fuel vent lines, or near the fuel pump. Other forms of control include the use of two or three pressure relief/rollover valves, one or two at the top of the tank and the other in the fuel filler tube. These valves prevent fuel flow through the vent valve hose serving the evaporative canister.

Most of today's vehicles have the **fuel pump module (Figure 10)** attached to the top or side of the tank. The fuel pump module assembly incorporates the fuel pump, fuel filter, fuel-level-sending unit, and fuel temperature sensor (if used) in a single assembly. The fuel-level sending unit includes a pickup tube and a float-operated fuel gauge **(Figure 11)**. A fuel line connects the fuel tank pickup tube to the fuel

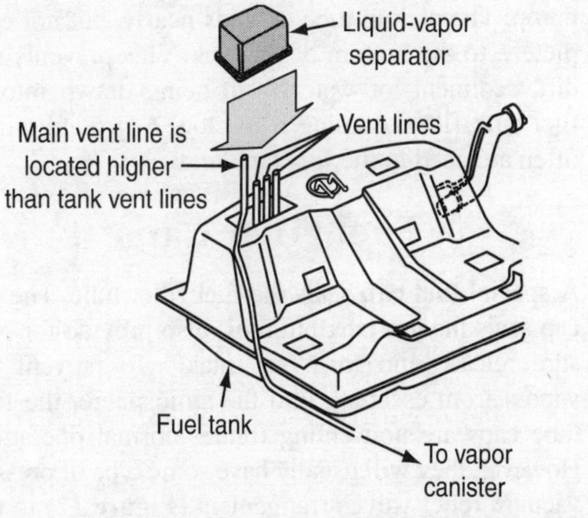


**Figure 9** *The liquid-vapor separator allows fuel vapors to be condensed back to fuel and returned to the tank. It also prevent raw fuel from entering the evaporative control system.*

**Figure 10** *Fuel pump module.*

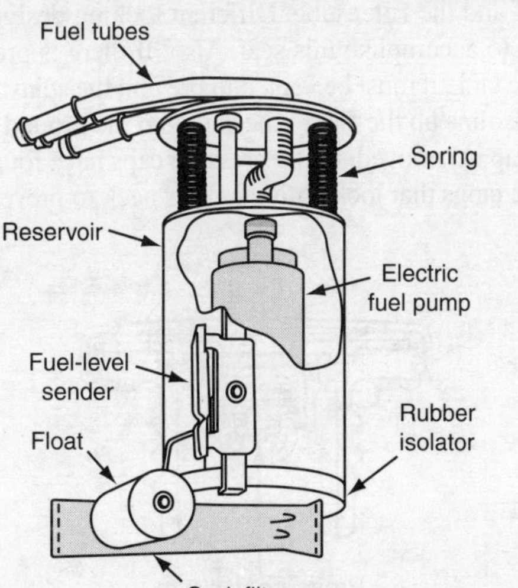


**Figure 11** *The fuel pump module includes the fuel pump, the fuel-gauge-sending unit, and the pickup filter.*

pump. The pickup tube extends nearly, but not completely, to the bottom of the tank. This prevents rust, dirt, sediment, or water from being drawn into the fuel tank filter, causing it to clog. A ground wire is often attached to the fuel tank unit.

## Fuel Filler Caps

A special **fuel cap** seals the fuel filler tube. The fuel cap seals the gas tank but must also provide for pressure release and fresh air intake. To prevent fuel vapors from escaping into the atmosphere, the filler tube caps are nonventing (under normal operation). However, they will usually have some type of pressure vacuum relief valve arrangement **(Figure 12)** to prevent fuel tank damage. Fresh air must be allowed to enter the tank and replace the fuel. Without air in the tank, the fuel pump would create a vacuum in the tank that could cause it to collapse. However, as the fuel temperature increases, a pressure is created inside the tank. If the pressure surpasses the calibrated set point of the valve (approximately 1 psi, or 6.89 kPa), the valve will open. The fuel vapors are not allowed to vent into the atmosphere; rather, they are vented to the vehicle's emissions system. Once the pressure or vacuum has been relieved, the valve closes.

Manufacturers may also use a cap design that does not contain a pressure/vacuum valve. To control tank pressures, a three-way valve is located in the vapor line between the tank and the carbon canister. The carbon canister stores the fuel vapors.

The cap must provide a positive seal between itself and the filler tube. Different locking designs are used to accomplish this seal. Also, if there is pressure in the tank, it must be vented to prevent the splash back of gasoline up the filler tube and onto the ground when the cap is removed. Many pressure caps have four antisurge tangs that lock onto the filler neck to prevent the

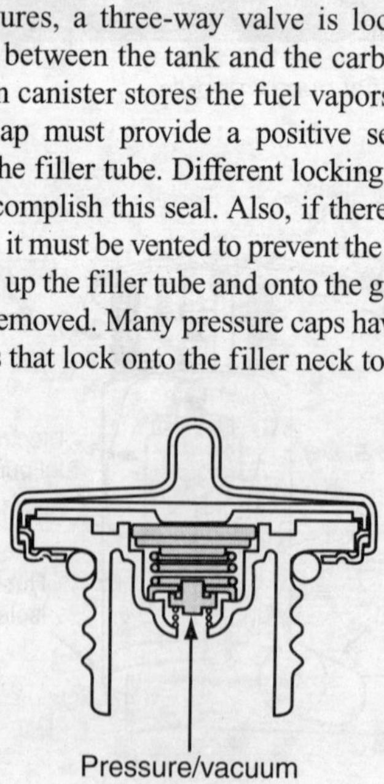


**Figure 12** *Filler cap with pressure/vacuum relief valve.*

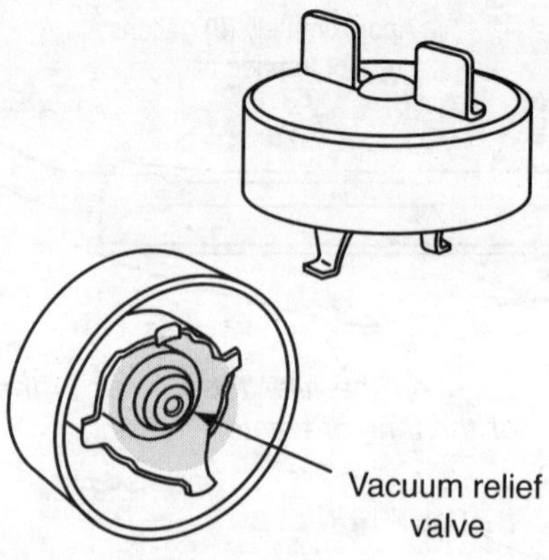


**Figure 13** *The four-tank fuel cap releases any tank pressure prior to removing the cap.*

delivery system's pressure from pushing fuel out of the tank **(Figure 13)**. Turning such a cap a half turn relieves the tank pressure. Then with another quarter turn the cap can be removed. A threaded filler cap may be threaded into the upper end of the filler pipe. These caps require several turns counterclockwise to remove them **(Figure 14)**. The long threaded area on the cap is designed to allow any remaining fuel tank pressure to escape during cap removal. The cap and filler neck have a ratchet-type, torque-limiting design to prevent overtightening. When the cap is installed, a clicking noise will be heard as the ratchet releases, indicating the cap is installed tightly and fully sealed. A third design is a variation of the threaded cap; however, it only requires a quarter turn to seal.

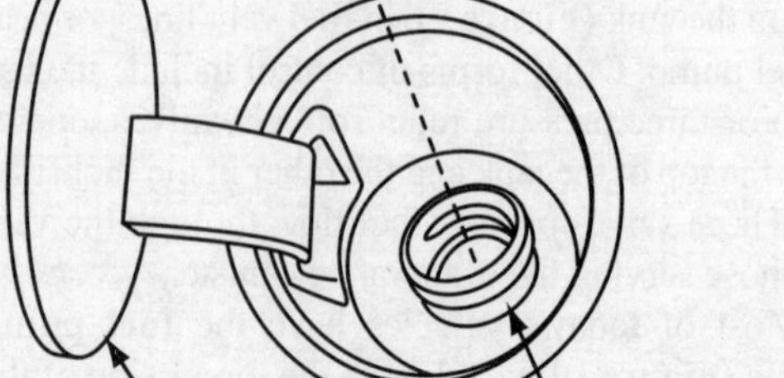


**Figure 14** *The threaded fuel cap releases pressure slowly as the cap is removed. This cap also has a ratchet system to indicate it is properly tightened.*

## Rollover Check Valve

Starting with the 1976 model year, a Federal Motor Vehicle Safety Standard (FMVSS 301) required manufacturers to install a control system to prevent gasoline leakage from passenger cars, certain light trucks, and buses after they were subjected to barrier impacts and rollovers. Tests conducted under these severe conditions showed the most common gasoline leak path was the gasoline supply line from the fuel tank to the carburetor. Although the carburetor system is no longer used, manufacturers must still prevent leakage between the fuel tank and the evaporative charcoal canister. One method of doing this is to use a **rollover check valve** located at the top of the fuel tank **(Figure 15)**. The rollover check valve is a safety valve that closes the fuel flow in the event of a rollover. It may also be part of the evaporative emissions controls. In addition, the rollover valve can be used to vent the fuel tank to the charcoal canister. During normal operation, the weight of the valve combined with the vapor pressure overpowers the force of the spring and opens the passage for fuel vapor. When the fuel level is high (or when driving around sharp corners), the liquid fuel causes the float to rise. This causes the float to close the port and prevents liquid fuel from entering the evaporative canister line. If the vehicle overturns, gravitational pull on the float along with spring pressure will close the port. With the port closed, liquid fuel will not be able to escape through the evaporative canister line. In addition, if the float sticks, causing the valve to remain closed, a relief valve will provide a means through which fuel vapor can be routed to the evaporative canister.

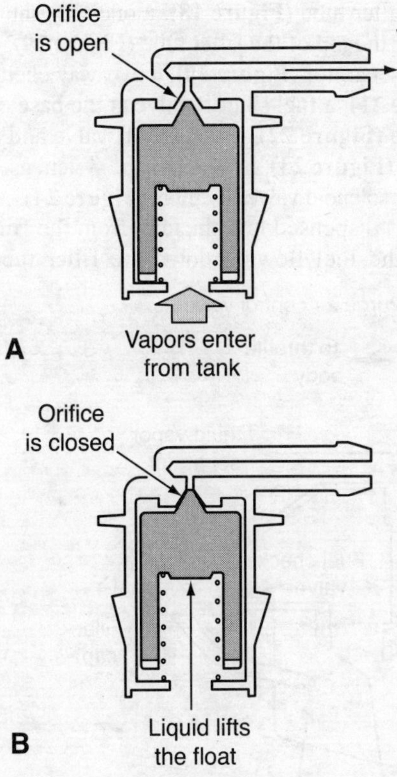


**Figure 15** *The rollover check valve will close the passage to the canister if the valve floats to the top or if the vehicle is involved in a rollover. The valve can also be used to prevent raw fuel from entering the evaporative control system.*

> **You Should Know**
>
> *The rollover valve is also referred to as the fuel cutoff valve.*

## Onboard Refueling Vapor Recovery

The vehicle's evaporative control system prevents hydrocarbon vapors that occur within the fuel tank from entering the atmosphere. This system (discussed in a later chapter) consists of a carbon-filled canister that stores the fuel vapors until the engine is started. Once the engine is running and other operating conditions are met, the vapor in the canister is allowed to enter the intake manifold and it is burned as part of the air-fuel mixture. However, during the refueling process, the vehicle's fuel tank hydrocarbons can escape into the atmosphere. Regulations require that vehicles have a method to prevent this. Phase in of this regulation began in the 1998 model year. The resulting control system is commonly referred to as **onboard refueling vapor recovery (ORVR) system**. The purpose of this system is to prevent hydrocarbon vapors from escaping into the atmosphere during refueling by drawing them into the fuel tank. Manufacturers can develop any means of refueling vapor control. The following are two examples of these systems.

One system uses a tapered filler tube, a one-way check valve at the base of the filler tube, a liquid-vapor separator, and a vapor control valve **(Figure 16)**. As the fuel tank is being filled, the fuel from the nozzle travels down the filler tube and opens the one-way check valve. This allows the fuel to enter the fuel tank. The filler tube is tapered to create a venturi vacuum as the fuel flows down it. This vacuum draws the

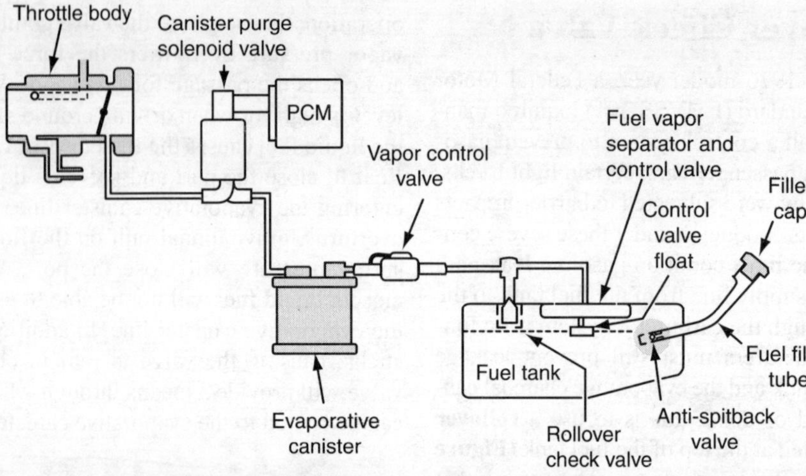


**Figure 16** *Onboard refueling vapor recovery system.*

hydrocarbon vapors that would escape to the atmosphere down the tube with the fuel and to the top of the tank. The vapors then flow through the vapor control valve and the liquid-vapor separator and they are stored in the charcoal canister. This action vents the tank. As the fuel tank is filled, the float of the vapor control valve rises. Once the tank is full, this float closes the passage to the canister. Since the tank is no longer venting, tank pressure increases and turns off the fuel nozzle. When the fuel flow down the filler tube stops, the one-way check valve closes to prevent a sudden surge of fuel up the filler tube.

Another type of ORVR system uses several check valves and additional tubing to prevent the escape of vapors **(Figure 17)**. This system consists of a venturi filler tube **(Figure 18)**. a one-way shutoff valve near the top of the filler tube **(Figure 19)**, a liquid-vapor separator **(Figure 20)**, a two-way check valve **(Figure 21)**, a fuel shutoff valve at the base of the filler tube **(Figure 22)**, a fuel cutoff valve and a leveling valve **(Figure 23)**. In addition, the system uses a canister vent solenoid valve assembly **(Figure 24)**.

As fuel is dispensed into the tank from the filling nozzle, the fuel flowing down the filler tube

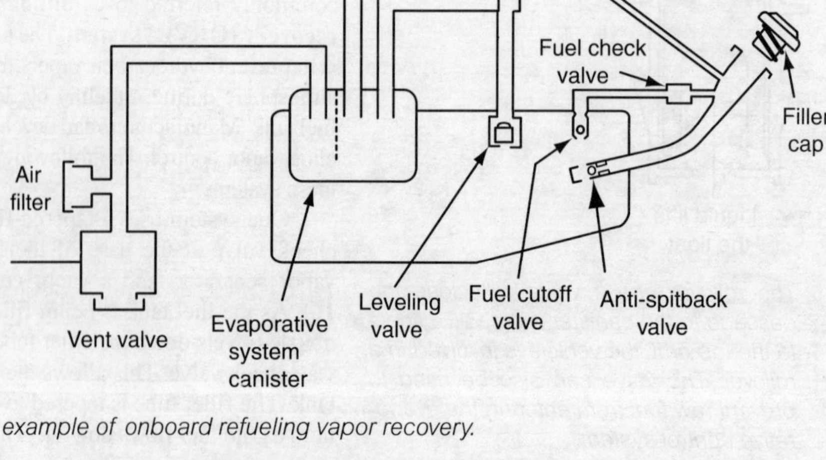


**Figure 17** *Another example of onboard refueling vapor recovery.*

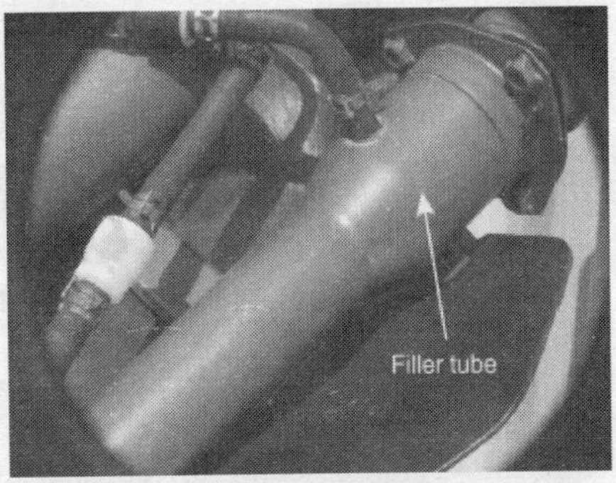


**Figure 18** *Venturi-shaped filler tube.*

**Figure 19** *One-way shutoff valve.*

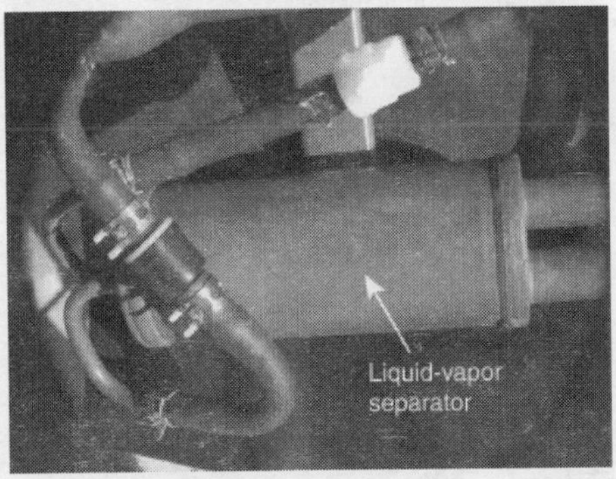


**Figure 20** *Liquid-vapor separator.*

## You Should Know

*The one-way check valve is also referred to as the anti-spitback valve.*

creates a venturi vacuum to draw the vapors into the tank. The fuel shutoff valve is open to allow the fuel and vapors to enter the tank **(Figure 25)**. The vapors flow past the leveling valve and into the liquid-

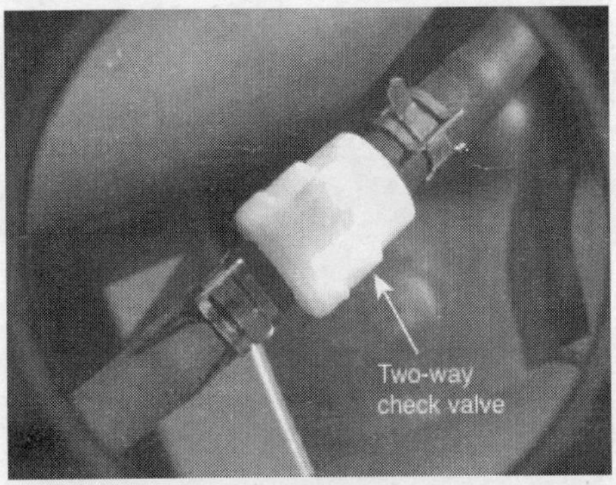


**Figure 21** *Two-way check valve.*

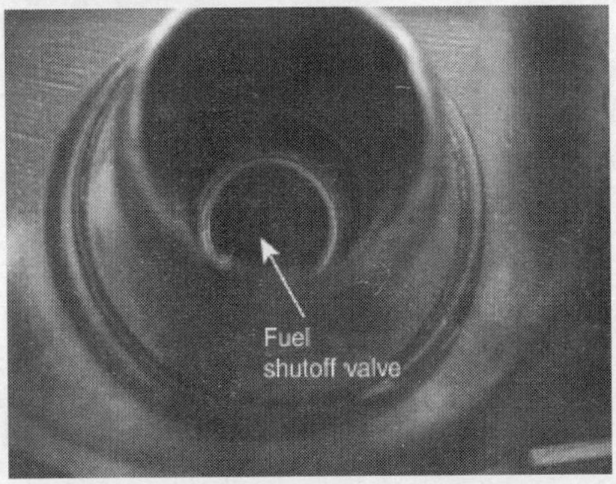


**Figure 22** *Fuel shutoff valve located at the bottom of the filler tube.*

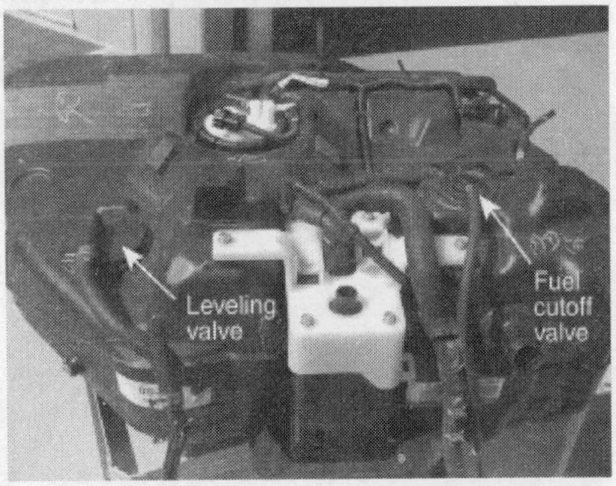


**Figure 23** *Fuel cutoff and leveling valve location.*

vapor separator. Liquid fuel returns to the tank past the leveling valve while vapors are routed to the canister for storage. Once the tank is filled, the leveling valve closes the passage to the canister and causes tank pressure to increase, shutting off the fuel nozzle.

With the vehicle parked, pressures inside the tank can increase if ambient temperature increases. Depending on the fuel level, the vapors in the tank

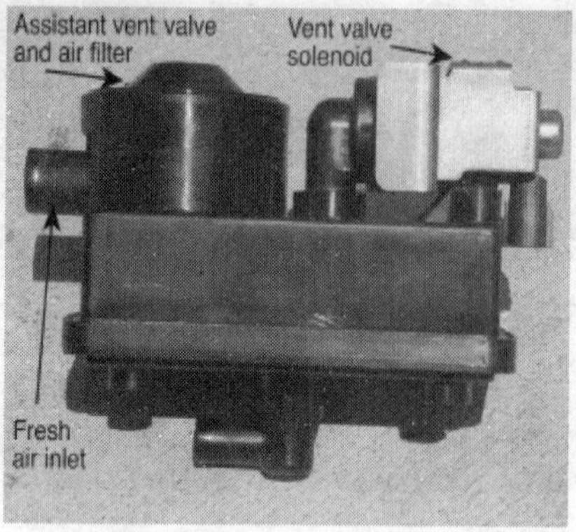


**Figure 24** *Vent solenoid valve assembly.*

can flow past the fuel cutoff valve or the leveling valve into the canister. Vapors passing the fuel cutoff valve enter the top of the filler tube through the two-way fuel check valve, then flow through the separator and into the canister. Vapors that flow past the leveling valve enter directly into the separator and flow into the canister. The canister vent solenoid valve and the assist vent valve are turned off to allow vapors in the canister to be cleaned by the filters and to vent the tank **(Figure 26)**.

While the vehicle is being driven, the level of the tank slowly lowers. Fresh air fills the space that was once occupied by vapors **(Figure 27)**. The canis-

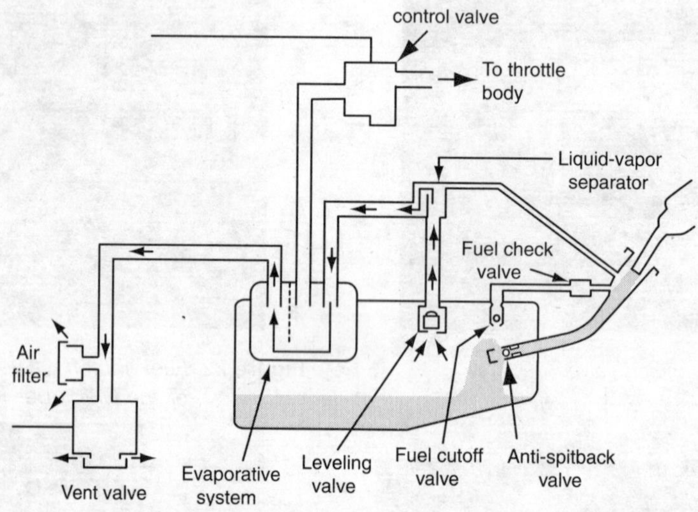


**Figure 25** *Vapor flow while the tank is being filled.*

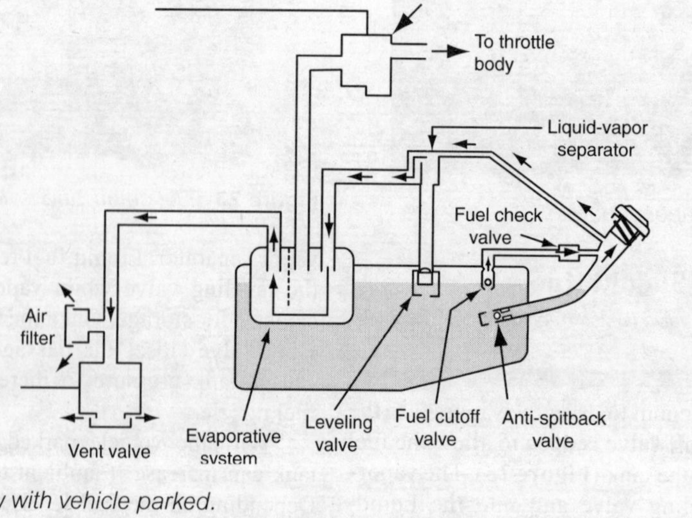


**Figure 26** *Vapor flow with vehicle parked.*

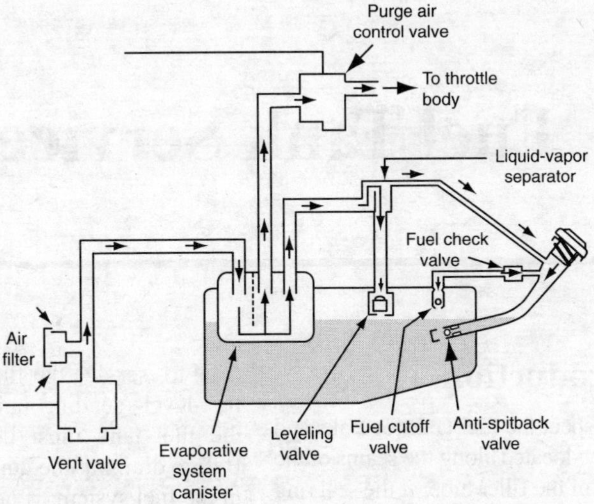


**Figure 27** *Vapor flow while the vehicle is being driven.*

ter vent solenoid valve and the assist vent valve are turned on. Fresh air passes through the air filter and enters the canister. The atmospheric pressure is used to "push" the vapors from the canister into the intake manifold when the purge air control valve allows this. Also, fresh air passes through the canister backward through the separator, past the leveling valve and into the tank. The atmospheric pressure in the tank is used to "push" fuel into the fuel pump module and to prevent a vacuum in the tank.

## Review Questions

1. Explain how expansion, contraction, and overflow are controlled in the fuel tank.
2. Describe the purpose of the onboard refueling vapor recovery (ORVR) system.
3. The main function of the fuel tank is to store the fuel for the engine. The expanded function of the fuel tank is to prevent ________ ________ from escaping into the atmosphere.
4. Technician A says that the threaded filler cap should be tightened until it clicks. Technician B says that the location of the filler tube to the tank is used to control fill level. Who is correct?
   A. Technician A only
   B. Technician B only
   C. Both Technician A and Technician B
   D. Neither Technician A nor Technician B
5. Most ORVR systems work using:
   A. the fuel pump to draw vapors down the filler tube.
   B. a venturi vacuum and a series of check valves.
   C. a vacuum internal to the filling nozzle.
   D. positive pressure to move the vapors back into the filling nozzle.
6. Describe the function of the rollover check valve.
7. Some fuel tank filler caps contain a pressure valve and a ________ valve.
8. The ORVR system prevents ________ vapors from escaping into the atmosphere while the fuel tank is being filled.
9. Technician A says that the one-way check valve prevents fuel flow from the underhood fuel system components into the fuel pump and the tank when the engine is shut off. Technician B says that the one-way check valve prevents fuel flow from the pump to the fuel filter and the fuel system if the engine stalls and the ignition switch is on. Who is correct?
   A. Technician A only
   B. Technician B only
   C. Both Technician A and Technician B
   D. Neither Technician A nor Technician B
10. The function of the fuel tank is:
    A. to store the fuel to be used by the engine.
    B. to prevent gasoline vapors (HC) from escaping into the atmosphere.
    C. Both A and B
    D. Neither A nor B

CHAPTER

# 13 Fuel Tank Service

## Introduction

A technician should inspect the fuel tank for leaks and damage. Leaks are often located along the seams of the tank, at the connection of the filler tube, at the seal ring of the fuel pump module or sending unit, and at the vent hose connections. Metal tanks should also be inspected for corrosion and rust. The technician should look for loose mounting bolts and damaged mounting straps. Leaks in the fuel tank, lines, or filter may cause gasoline odor in and around the vehicle, especially during low-speed driving and idling. Fuel leaks are not only costly for the customer, but they are also very dangerous. In most cases, the fuel tank must be removed for servicing. This chapter discusses fuel tank service.

Metal tanks that have seam leakage or rust should be replaced. Minor road damage leaks may be repaired by removing the tank and steam cleaning it or boiling it in a caustic solution to remove the gasoline residue. After the tank is completely cleaned, the leak can be soldered or brazed by a specialty shop equipped to do this. Small holes can sometimes be fixed by installing a sheet-metal screw into the hole with a neoprene washer.

Plastic tanks that have cracks or large holes should be replaced. Small holes in plastic tanks can sometimes be repaired by using a special tank repair kit. Be sure to follow the manufacturer's instructions when doing the repair.

## Fuel Tank Draining

In addition to removing the fuel tank from the vehicle to repair or replace it, the technician may often

> **You Should Know**
>
> *Hard-to-identify fuel tank and fuel line leaks can be located by using a "smoke machine" that induces smoke into the tank under a slight pressure. Watch for the escaping smoke to locate the leak.*

need to service the fuel pump, inlet strainer, or the fuel-level- sending unit. Regardless of the reason, the fuel tank must be drained prior to removal. Before draining the tank, always relieve the pressure in the fuel system. If needed, refer to Chapter 11 for how to perform this task. Also, always disconnect the negative battery cable.

> **You Should Know**
>
> *Some fuel pump modules are accessed by a door or panel located under the rear seat of the vehicle or under the trunk carpet.*

If the tank has a drain bolt, remove the bolt and drain the fuel into an approved container. If the fuel tank does not have a drain bolt, use commercially available gasoline-handling equipment **(Figure 1)**. This equipment can be used to siphon fuel up through the filler tube. If the filler tube has a check ball or screen that prevents the siphon hose from entering the fuel tank, the equipment can be connected to the tank's outlet line.

> **You Should Know**
>
> *Always drain gasoline into an approved container and use a funnel to avoid spills.*

If this equipment is not available, a hand-operated or air-operated pump can be used by installing the pump hose through the filler tube into the fuel tank. If the filler tube design does not allow the hose to enter the tank, a safe method of draining the fuel tank is as follows:

1. Remove the fuel cap to relieve any pressure in the tank.
2. Relieve the residual fuel pressure from the lines and the rail.
3. Remove the negative battery cable.
4. Raise the vehicle on a hoist.

**Figure 1** *The gasoline cart provides an easy and safe way to drain the fuel tank.*

5. Locate the fuel tank outlet line and disconnect it at the easiest location.
6. Install the appropriate adapter onto the tank side of the outlet line and connect the intake hose from a hand-operated or air-operated pump to this adapter.
7. Install the discharge hose from the hand-operated or air-operated pump into an approved gasoline container and operate the pump until all the fuel is removed from the tank.

> **You Should Know**
>
> *When servicing fuel system components, always place a Class B fire extinguisher near the work area.*

## Fuel Tank Removal

The following is a typical procedure for removing the fuel tank; however, the fuel tank removal procedure varies depending on the vehicle make and year. Always follow the procedure in the vehicle manufacturer's service manual.

> **You Should Know**
>
> *Abide by local laws for the disposal of contaminated fuels. Wear eye protection when working under the vehicle.*

1. Remove the filler cap to relieve any pressure in the tank.
2. Relieve the fuel system pressure in the lines and the rail.
3. Disconnect the negative battery cable.
4. Drain the fuel tank.
5. Raise the vehicle on a hoist or lift the vehicle with a floor jack and lower the chassis onto jack stands.
6. Use compressed air to blow dirt from the fuel line fittings and wiring connectors.
7. Remove the fuel tank wiring harness connector from the body harness connector.
8. If used, remove the ground wire retaining screw from the chassis.
9. Disconnect the fuel lines from the fuel tank. If these lines have quick-connect fittings, follow the manufacturer's recommended removal procedure in the service manual. Some quick-connect fittings are hand releasable and others require the use of a special tool. See Chapter 15 on fuel hose service for methods used to disconnect fittings.
10. Wipe the filler pipe and vent pipe hose connections with a shop towel and then disconnect the hoses from the filler pipe and vent pipe to the fuel tank.
11. Unfasten the filler tube from the tank **(Figure 2)**. If it is a rigid one-piece tube, remove the screws around the outside of the filler neck near the filler cap. If it is a three-piece unit, remove the neoprene hoses after loosening the clamp.

> **You Should Know**
>
> *Do not heat the bolts on the fuel tank straps in order to loosen them. The heat could ignite the fumes.*

12. If the fuel tank is supported by straps, loosen the bolts holding the straps to the vehicle **(Figure 3)** until they are about two threads from the end. Otherwise, evenly remove the bolts that attach the tank to the chassis until the tank is free.

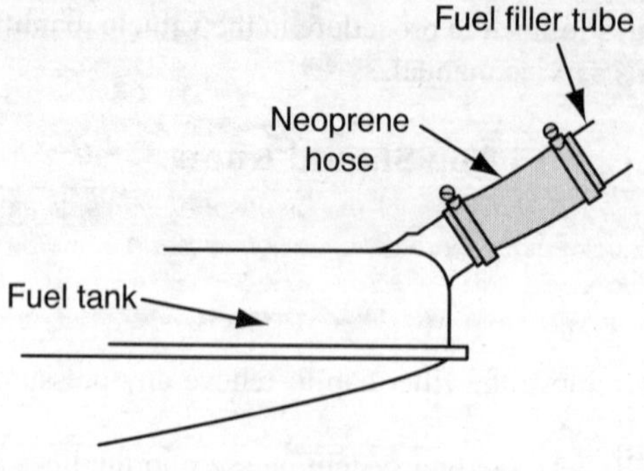


**Figure 2** *Disconnecting the filler tube.*

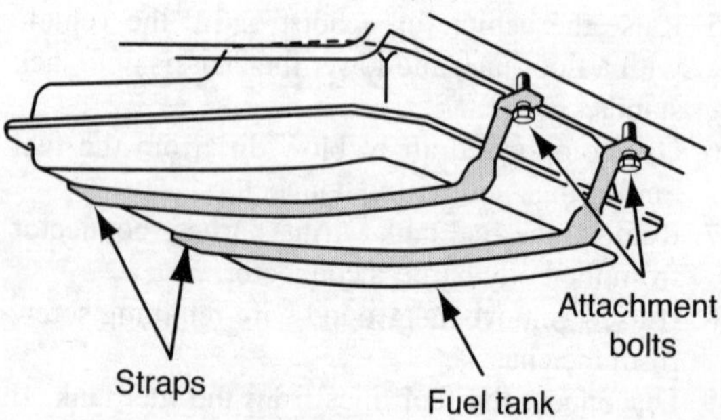


**Figure 3** *Many tanks are supported by straps.*

13. Holding the tank securely against the underchassis of the vehicle with one hand, remove the strap bolts and lower the tank to the ground. Make sure all wires and tubes are disconnected.

> **You Should Know**
>
> *Some fuel may remain in the tank after draining. If the tank is tipped to one side, the rushing fuel may cause the tank to become hard to handle, resulting in spills or dropping of the tank.*

> **You Should Know**
>
> *Use a drain pan to catch the fuel. Clean any spills immediately.*

14. To reinstall the new or repaired fuel tank, reverse the removal procedure. Be sure that all the rubber or felt tank insulators are in place. Then with the tank straps in place position the tank. Loosely fit the tank straps around the tank but do not tighten them.
15. Make sure that the hoses, wires, and vent tubes are connected properly. Check the filler neck for alignment and for insertion into the tank.
16. Tighten the strap bolts and secure the tank to the car. Install all the tank accessories (vent line, sending unit wires, ground wire, and filler tube). Fill the tank with fuel and check it for leaks, especially around the filler neck and the pickup assembly.
17. Reconnect the battery and check the fuel gauge for proper operation.
18. Start the engine and inspect the tank for any leaks.

> **You Should Know**
>
> *A transmission jack fitted with a square metal plate can be used to support the fuel tank during removal or installation.*

## Fuel Filler Cap Testing

Sealing the fuel system from the atmosphere can control fuel vapor emissions. However, air must enter into the fuel tank to take the place of the fuel as it is consumed. Without fresh air entering the tank, the fuel tank would collapse as the fuel pump continued to create a vacuum within it. Fuel flow would also be restricted to the engine. On many vehicles, the fuel tank is usually vented by the fuel filler tube cap. Without a vent, fuel could not flow and the air-fuel mixture would become lean. In fact, if the cap check valve became restricted, the vehicle could surge and then the engine would stop running after a few miles. Loosening the gas cap results in a large rush of air and enables the engine to run again.

Inspection of the filler cap includes the locking mechanism (either tangs or threads) and the gasket or seal. A leaking or missing fuel cap can allow up to 30 gallons (113.5 liters) of gasoline per year to evaporate into the atmosphere. If the cap is damaged, it should be replaced. If a cap is suspect, the easiest test is to replace it with a good cap. Filler tube cap testers are available **(Figure 4)**.

> **You Should Know**
>
> *When replacing a gasoline tank filler cap, use the same type of cap as the original. Otherwise, a malfunction could occur in the filling and venting system, resulting in higher emissions levels and the escape of dangerous hydrocarbons.*

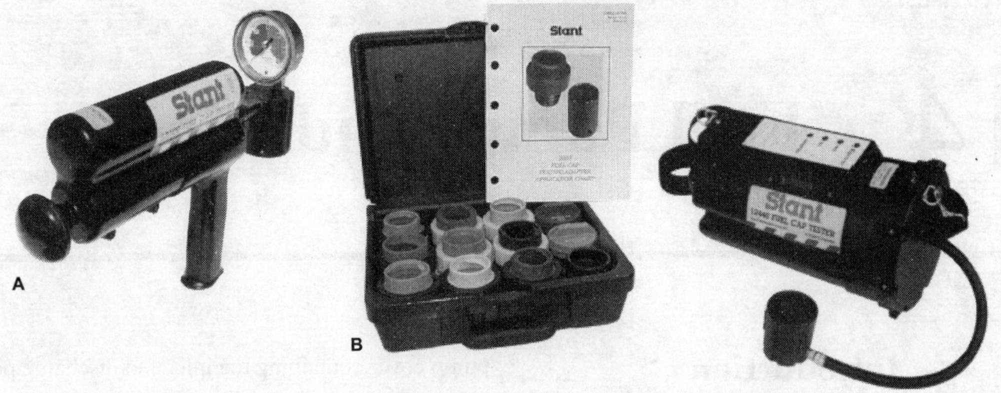


A

B

**Figure 4** *(A) Mechanical cap tester. (B) Communicating cap tester used with I/M240 testing.*

## Review Questions

1. The fuel tank needs to be removed from the vehicle. Technician A says that the fuel pressure in the system must first be relieved. Technician B says that to use a transmission jack to support the tank so the fuel does not need to be drained first. Who is correct?
   A. Technician A only
   B. Technician B only
   C. Both Technician A and Technician B
   D. Neither Technician A nor Technician B
2. Describe a typical procedure for draining a fuel tank without a drain plug.
3. List the common locations of fuel leaks around the tank.
4. The customer complains that the vehicle surges and dies after being driven a few miles. Technician A says that this can be caused by a restricted filler cap check valve. Technician B says that this can be caused by a loose filler cap. Who is correct?
   A. Technician A only
   B. Technician B only
   C. Both Technician A and Technician B
   D. Neither Technician A nor Technician B
5. Technician A says that steel tanks that have a leak at the seam should be welded, then the area should be covered with an epoxy. Technician B says that plastic tanks that have a small hole can be repaired by using a special tank repair kit. Who is correct?
   A. Technician A only
   B. Technician B only
   C. Both Technician A and Technician B
   D. Neither Technician A nor Technician B

CHAPTER

# 14 Fuel Pumps and Lines

## Introduction

From the fuel tank, fuel is pressurized and sent to the fuel rail and injectors. Proper operation of the fuel pump, lines, and connections is mandatory for proper engine performance. This chapter discusses the function of the fuel pump and related fuel-routing lines.

## Electric Fuel Pumps

The **fuel pump** is the device that draws the fuel from the fuel tank through the fuel lines and supplies pressurized fuel to the engine's carburetor or injectors. Electronic fuel-injection systems utilize an electric fuel pump that uses a DC motor to drive a pump gear or roller set to draw fuel.

Electric fuel pumps offer several advantages over mechanical fuel pumps. First, they maintain constant fuel pressure, which aids in starting and reduces the potential for vapor lock. In addition, they can be located either inside or outside the fuel tank. The reasons for locating the fuel pump inside the fuel tank are twofold: to keep the fuel pump cool while it is operating, and to keep the entire fuel line pressurized to prevent premature fuel evaporation.

The in-tank pump is usually a **positive displacement** design pump. This means that the same volume of fuel is delivered every rotation of the pump, regardless of speed. The pump can be a roller-vane-type design with a permanent magnet electric motor, a two-stage gerotor design, or a design that has an impeller connected to the end of the armature shaft. The pump and its electric motor are fitted in a common housing in the fuel tank and are continuously surrounded by fuel. The fuel pump assembly consists of the pump, noise dampener, fuel house, intake strainer (filter), and mounting plates.

The impeller design in-tank electric fuel pump contains a small DC electric motor with an impeller mounted on the end of the motor shaft **(Figure 1)**. A pump cover containing the inlet and discharge ports is mounted over the impeller. When the armature and impeller rotate, fuel moves from the tank to the inlet port and the impeller grooves pick up the fuel and force it around the impeller cover and out the discharge port. Fuel moves from the discharge port through the inside of the motor and out the check valve and outlet connection, which is connected to the fuel line.

> **You Should Know**
>
> *Although it is dangerous to have a spark near gasoline, the in-tank fuel pump is safe because there is no oxygen to support combustion in the tank.*

The fuel pump/sending unit assembly is installed inside the fuel tank through a large flange opening on the top of the fuel tank. This flange also provides the necessary electrical and fuel line connections. When installed, the flange seats against the synthetic rubber seal and is held in position by a lock ring.

### Reservoir Chamber

The purpose of the reservoir chamber is to provide fuel at the pump inlet during all operating conditions, especially when the fuel level is low. The reservoir chamber prevents the pump from losing prime while the vehicle is going around corners or when the vehicle experiences hard braking or acceleration with low fuel levels.

### Pressure Relief Valve

The fuel pump usually has two internal **check valves**. One is located on the inlet side of the pump. This check valve is used to regulate maximum fuel pump output. If the fuel-supply line is restricted and pump pressure becomes too high, the **pressure relief valve** (a safety valve) opens to prevent damage to the fuel-delivery system. When the pressure relief valve opens, fuel returns through this valve to the pump

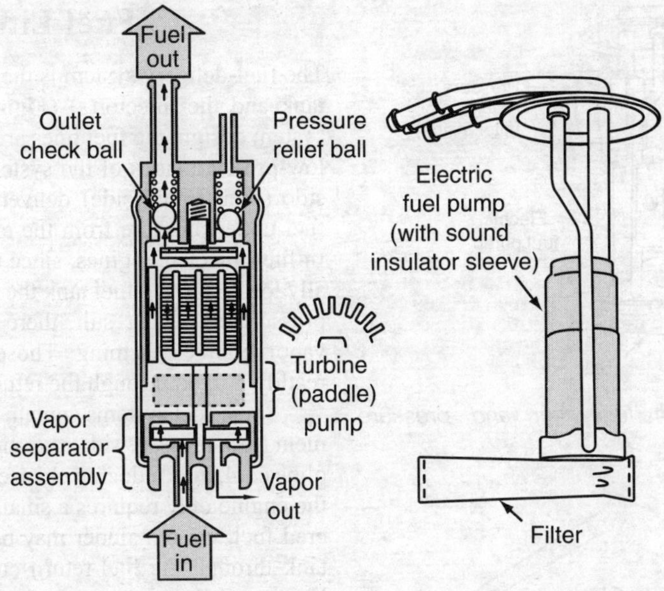


**Figure 1** *Electric fuel pump components.*

inlet. This action protects fuel system components from excessive fuel pressure. Different fuel-injection systems will have varying specifications for relief pressure. Many fuel systems used today have a relief pressure of approximately 120 psi (8 bar).

### One-Way Check Valve

A second check valve is used to prevent the fuel in the lines from returning to the tank after the fuel pump shuts off. It also prevents fuel from entering the fuel lines with the pump off in the event of an accident. The one-way check valve is located on the outlet side of the fuel pump. The valve traps a volume of fluid between the pump and injectors to help ensure faster engine starting.

> **You Should Know**
>
> *Remember that the purpose of the check valve is to maintain a volume of fluid in the supply line. It does not maintain pressure over long periods of time. It is normal for fuel pressures to drop as the fuel cools. However, pressure should not drop rapidly after the engine is shut off.*

### Return Line Check Valve

The return line check valve is hot staked in the fuel tank swirl reservoir assembly. The valve's only purpose is to prevent fuel flow in the reverse direction. Whenever the fuel tank is rotated from the normal position (a rollover accident, for example), the pressure in the fuel tank is greater than the pressure in the return line, so the valve closes to prevent fuel from draining out of the tank into the fuel line.

## Fuel Filters

**Fuel filters** are elements made from pleated paper, ceramic, or bronze. They are used to remove contamination in the fuel-delivery system. The fuel-delivery system usually consists of two filters and a screen. The first filter is located in the fuel tank on the fuel pump inlet. This filter is actually a strainer located on the fuel pickup tube ahead of the fuel pump. The strainer is made of a finely woven fabric. The purpose of this strainer is to prevent large contaminant particles from entering the fuel system, causing excessive fuel pump wear or plugging fuel-metering devices. It also helps to prevent passage of any water that might be present in the tank. The second filter is the main fuel filter, which is discussed next. The screen is installed in each injector's inlet port.

There are two different types of fuel filters. One type is contained in the same housing as the fuel pressure regulator and attaches to the fuel pump module (Figure 2). The other type is mounted outside the tank in the fuel supply line between the fuel tank and the engine (Figure 3). Most fuel filters contain a pleated paper element mounted in the filter

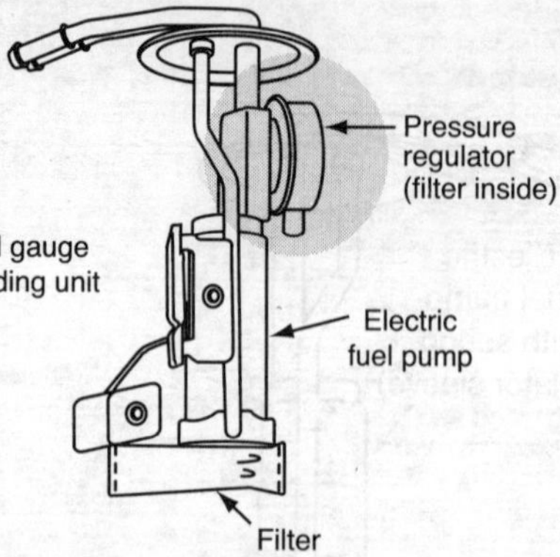


**Figure 2** *Combination fuel filter and pressure regulator.*

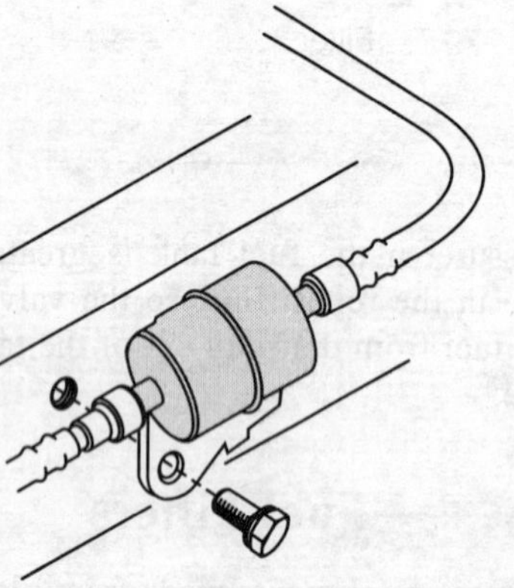

**Figure 3** *Electronic fuel-injection (EFI) fuel filter mounted on the frame rail.*

housing, which may be made from metal or plastic. Fuel filters on fuel-injected vehicles usually have a metal case. On many fuel filters, the inlet and outlet fittings are identified and the filter must be installed properly. An arrow on some filter housings indicates the direction of fuel flow through the filter.

## Fuel Lines

The fuel-delivery system is the link between the fuel tank and the injector(s) **(Figure 4)**. Depending on system design, the fuel lines are part of the high- and low-pressure sides of the system. The high-pressure side (fuel-supply side) delivers a constant flow of fuel under pressure from the tank to the injector(s). In fuel-injected engines, since the fuel pump is usually located in the fuel tank the entire delivery side is pressurized As a result, there is less likelihood of vapor bubbles forming. Those that may form are readily purged through the return circuit.

Some fuel systems contain a fuel-return arrangement (low-pressure side) that aids in keeping the gasoline cool, thus reducing chances of vapor lock. Since the engine only requires a small portion of the delivered fuel, the remainder may be returned to the fuel tank through the fuel-return circuit. The fuel-return line generally runs next to the fuel-supply line. The fuel-return system allows a metered amount of cool fuel to circulate through the tank and fuel pump, thus reducing vapor bubbles caused by overheated fuel in the tank. The return circuit is not pressurized.

Fuel lines can be made of metal tubing, flexible nylon, or synthetic rubber hose. Usually the fuel lines are constructed using all of these materials in different locations.

Metal lines are generally used where the supply lines and return lines follow the frame along the underchassis of the vehicle. These metal lines extend from near the tank to a point near the fuel-injector rail. Clips secure the fuel lines to the chassis to prevent movement and damage **(Figure 5)**. To absorb vibrations, lengths of high-pressure flexible hose join the metal line to the fuel pump and fuel rail.

The flexible high-pressure line, unlike filler neck or vent hoses, must work under pressure. Because of this, the flexible synthetic hoses must be

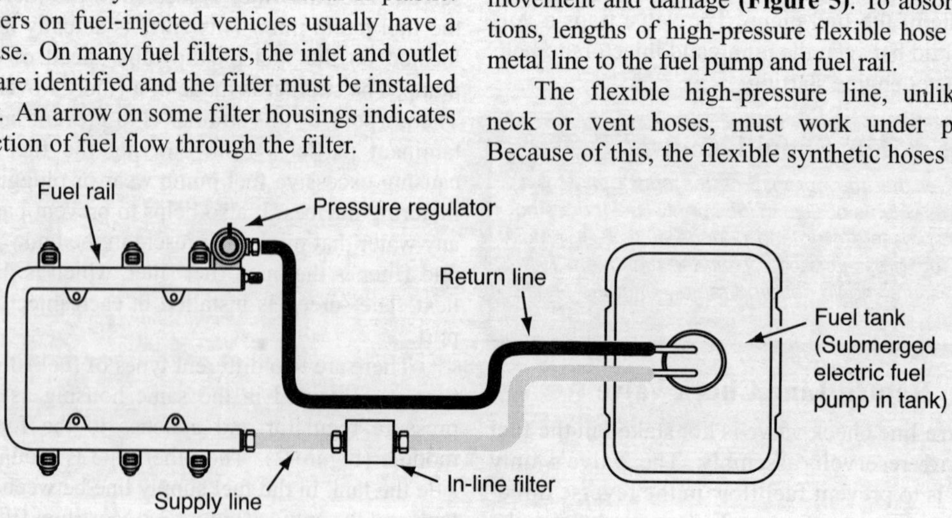


**Figure 4** *The fuel lines connect the fuel tank to the engine.*

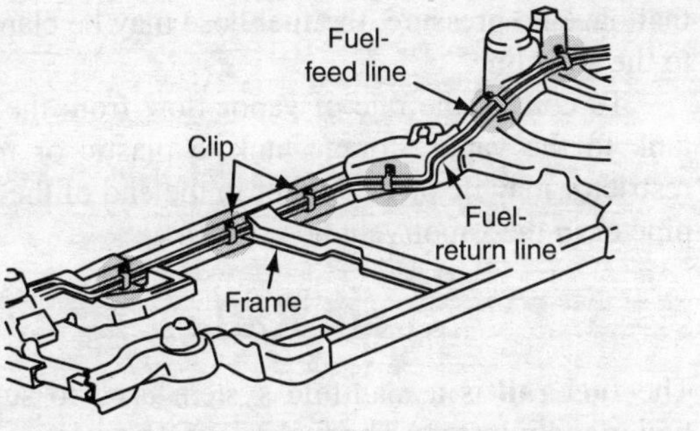


**Figure 5** *The fuel line must be properly supported along the frame.*

stronger. This is especially true for the hoses in fuel-injection systems, where pressures reach 58 psi (400 kPa) or more. For this reason, flexible fuel line hose must also have special resistance properties.

Any synthetic rubber hose used in the fuel-delivery or evaporative control systems must be able to resist gasoline. It must also be nonpermeable, so gas and gas vapors cannot evaporate through it. Ordinary rubber hoses, such as those used for vacuum lines, deteriorate when exposed to gasoline. Only hoses made for fuel systems should be used for replacement. Similarly, vapor vent lines must be made of materials that resist attack by fuel vapors. Replacement vent hoses are usually marked with the designation "EVAP" to indicate their intended use.

## Fuel Line Connections

There are several methods for making connections to the fuel lines. If the fuel filter is located in the fuel line, it is usually attached by means of a **banjo fitting (Figure 6).** The banjo fitting is a metal connection using a circular filling with a hollow bolt and two steel sealing rings. The banjo fitting provides a positive connection by using two metal gaskets compressed between the two ends of the connection.

Sections of fuel line are assembled together with special fittings. A common fitting is the quick-connect design **(Figure 7)**. The quick-connect fuel

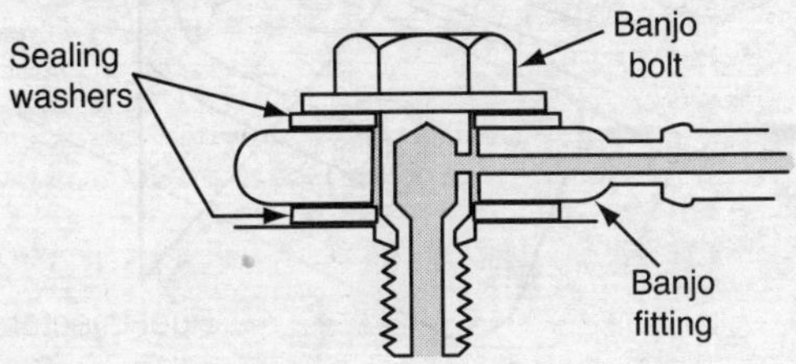


**Figure 6** *The banjo fitting uses a hollow bolt and two sealing washers.*

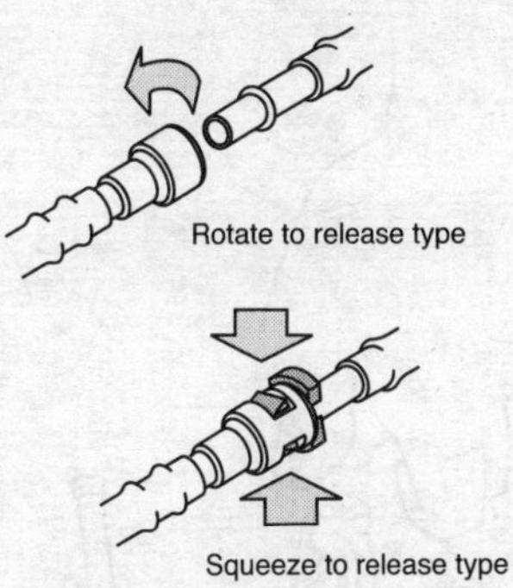


**Figure 7** *Example of quick-connect fittings.*

coupler consists of a plastic retainer, two O-rings, and a plastic spacer. The retainer engages and locks into position on a raised bead on the fuel line. The dual O-ring arrangement and the plastic spacer prevent leakage through the connection.

Another type of fitting is threaded. The two most common threaded-type tube fittings are the **compression fitting** and the **double-flare fitting (Figure 8)**. The compression fitting compresses the end of a sleeve to provide a seal. The double-flare fitting uses a procedure that folds the ends of the tube over itself, which doubles the thickness of the tube end and creates two sealing surfaces. The double-flare fitting is

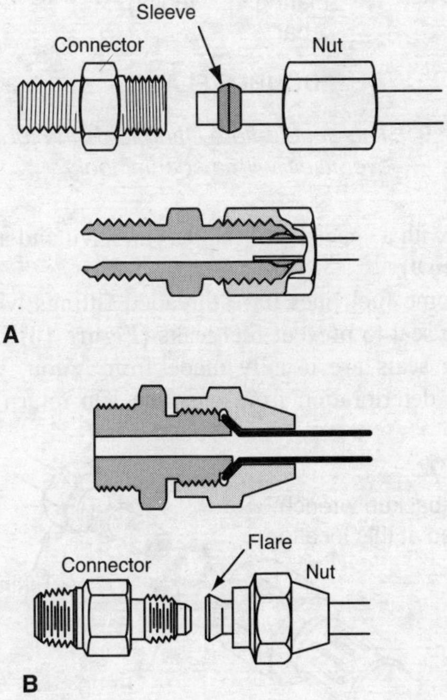


**Figure 8** *(A) Compression fitting. (B) Double-flare fitting.*

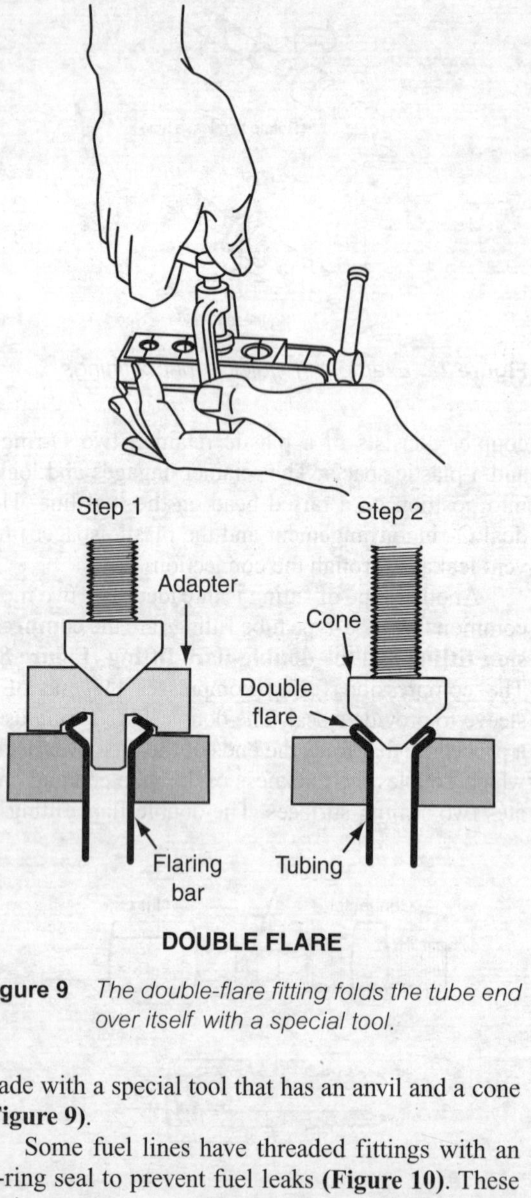


**Figure 9** *The double-flare fitting folds the tube end over itself with a special tool.*

made with a special tool that has an anvil and a cone **(Figure 9)**.

Some fuel lines have threaded fittings with an O-ring seal to prevent fuel leaks **(Figure 10)**. These O-ring seals are usually made from Viton, which resists deterioration from gasoline. On return lines that have no pressure, the fuel hose may be clamped to the steel line.

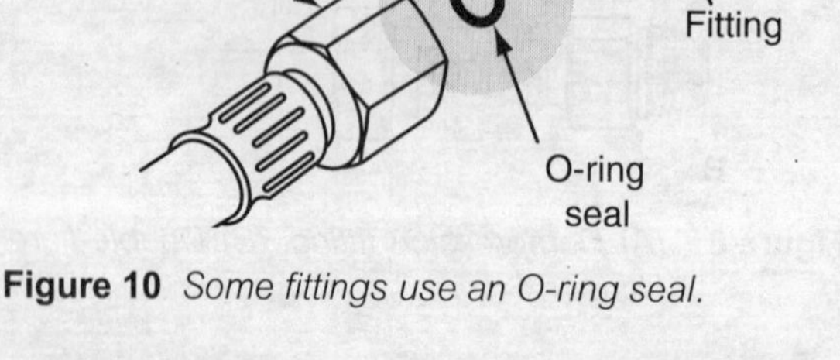


**Figure 10** *Some fittings use an O-ring seal.*

To control the rate of vapor flow from the fuel tank to the vapor storage tank, a plastic or metal restrictor may be placed either in the end of the vent pipe or in the vapor/vent hose itself.

## Fuel Rail

The **fuel rail** is a manifold system used to supply fuel to the injectors. The fuel rail can be mounted on the intake manifold **(Figure 11)** or it can be an integral part of the intake manifold. The fuel rail can be constructed of metal or composite materials. The injectors mount to the fuel rail with push-on retaining clips and use O-rings to prevent leakage between the injectors and the fuel rail.

## Fuel Pressure Regulation

Since the fuel pump is a positive displacement design, a **pressure regulator** is needed to ensure a constant pressure difference at the injector(s). The same pressure difference is programmed into the computer because the fuel pump produces more pressure and volume than the engine needs. A pressure regulator is needed to lower the pressure to the value programmed into the computer.

There are two basic types of regulation used depending on fuel-delivery system classification:

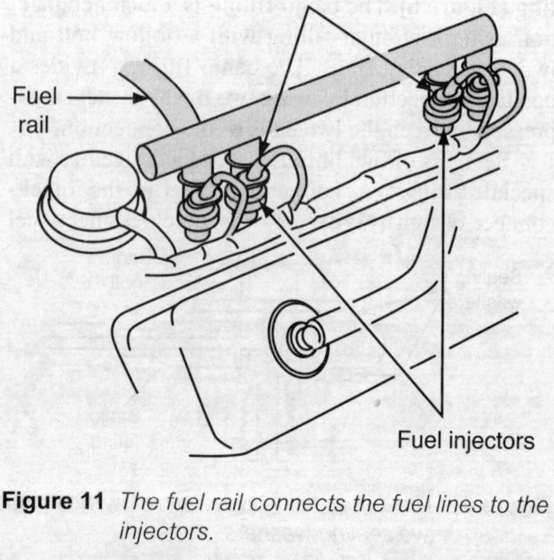


**Figure 11** *The fuel rail connects the fuel lines to the injectors.*

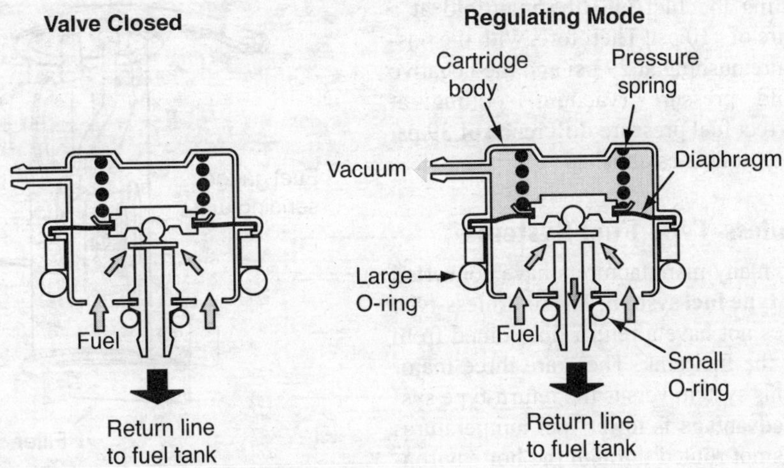


**Figure 12** *Fuel pressure regulator operation.*

return type and returnless type. Although regulation works similarly in both systems, there are some major differences.

## Return-Type Systems

On most return-type fuel systems, a fuel pressure regulator mounted on the fuel rail **(Figure 12)** controls the pressure. On this type of fuel system, all fuel is routed through the fuel rail; the remaining fuel that the engine does not use is returned to the fuel tank. The pressure regulator is a mechanical device that is not controlled by the power-train control module. The regulator contains a calibrated spring and a diaphragm that activates the regulator valve. Fuel pressure operates on one side of the diaphragm, while spring pressure operates on the other side. The diaphragm opens the valve to the return port, allowing fuel to be dumped back into the fuel tank. System fuel pressure reflects the amount of fuel pressure required to open the port. The spring on the opposite side of the diaphragm attempts to close the valve, causing an increase of pressure on the fuel as it travels to the fuel rail.

On port fuel-injection (PFI) systems, a constant pressure difference between the tip of the injector and the intake manifold must be maintained under all driving conditions. To accomplish this, manifold vacuum is applied to the spring side of the diaphragm. The holding pressure of the spring will vary with the amount of assist it receives from the intake manifold vacuum. At idle, vacuum being applied to the regulator lessens the effective spring rate, allowing fuel to be returned to the tank when pressure reaches desired specifications (approximately 38 psi [262 kPa], for example). Because less fuel is used at idle, a greater amount of fuel must be bypassed to the tank to maintain the correct fuel pressure differential. Also, there is less fuel pressure at idle because a high manifold vacuum helps pull fuel from the injector.

With a low vacuum, such as at full throttle, very little vacuum assist is available and the full force of the spring is exerted to seal the outlet, thus raising positive fuel pressure. Under high speed and load conditions, a greater volume of fuel is required. With a limited vacuum assist on the diaphragm, the flow back to the tank is restricted to prevent a drop in pressure and volume. The fuel pressure must increase since there is less manifold vacuum pulling the fuel from the injector. With a reduction of vacuum on the injector, less fuel is delivered. The increase in fuel pressure will overcome this.

In short, the pressure regulator maintains a constant pressure differential by controlling the amount of fuel entering the fuel-return line. As the intake manifold vacuum changes, the differential pressure between the positive pressure in the fuel rail and the negative pressure (vacuum) in the intake manifold also changes. The following is an example of this function: When the intake manifold vacuum is a steady 20 in. Hg (inches of mecury), or –10 psi, the fuel pressure regulator must maintain a positive 29 psi fuel pressure. Under these conditions the fuel system (regulated by the pressure regulator) is pushing fuel through the injector and into the intake manifold at 29 psi. In addition, 20 in. Hg (–10 psi) of intake manifold vacuum is assisting the fuel-supply system by pulling the fuel into the manifold at a negative pressure of –10 psi. Therefore, with the positive fuel pressure pushing at 29 psi and the negative intake manifold pressure (vacuum) pulling at –10 psi, the correct fuel pressure difference of 39 psi in the intake manifold is established.

## Returnless-Type Fuel Systems

In resent years, many manufacturers have converted to a **returnless-type fuel system**. The returnless-type fuel system does not have a return line routed from the fuel rail to the fuel tank. There are three major advantages to this system versus the return-type system. The first advantage is lower fuel temperatures since the fuel is not routed through the hot environment of the engine compartment and then returned to the tank. This reduces evaporative emissions, resulting in less evaporative canister purging. Second, since fuel passes through the fuel filter just one time before it is used by the engine, the fuel filter lasts longer. The third advantage is reduced cost since fewer components are used.

Like the return-type system, the pressure regulator on the returnless-type system is the mechanical device containing a calibrated spring and a diaphragm that activates the regulator valve **(Figure 13)**. Fuel pressure operates on one side of the diaphragm while spring pressure operates on the other. The diaphragm opens the valve to the return port, allowing fuel to be dumped back into the fuel tank. System fuel pressure reflects the amount of fuel pressure required to open the port. The spring on the opposite side of the diaphragm attempts to close the valve, causing an increase of pressure on the fuel as it travels to the fuel rail. Usually the pressure regulator is part of the fuel pump module **(Figure 14)**. On some vehicles it is part of a filter/regulator assembly and on others it is a separate component **(Figure 15)**.

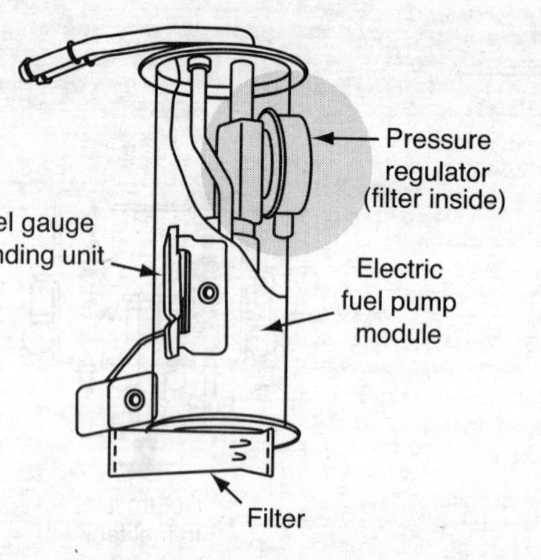


**Figure 14** *Returnless-type fuel system pressure regulator attached to the fuel pump module.*

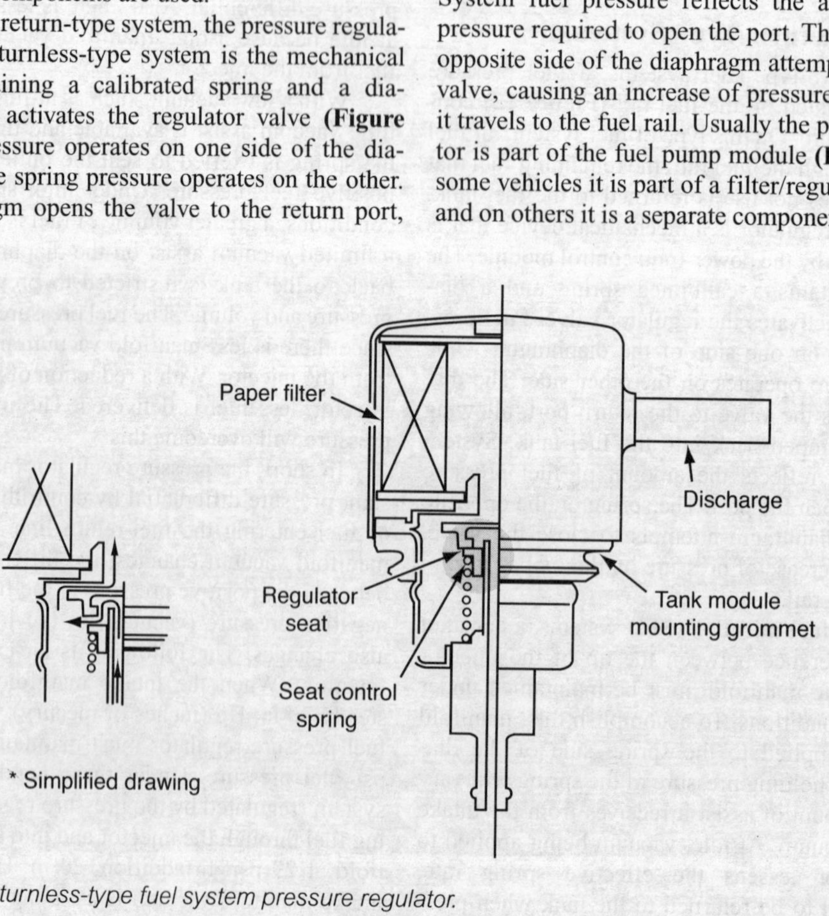


**Figure 13** *Returnless-type fuel system pressure regulator.*

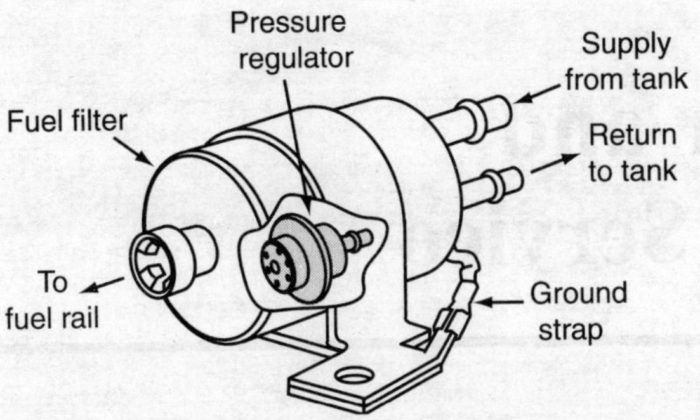


**Figure 15** *Returnless-type fuel system with remote mounted pressure regulator.*

Returnless-type systems do not use an engine vacuum like the return-type systems. With the regulator mounted on the tank, constant fuel pressure is always supplied to the injectors. The PCM uses a special formula that calculates the pressure differential across the injector and then adjusts injector pulse width accordingly.

## Review Questions

1. Explain how the vacuum-assisted fuel pressure regulator operates.
2. Explain the purpose of the check valves used in an electric fuel pump.
3. The electric in-tank pump is a ____________ displacement design pump.
4. The fuel pressure regulator contains a ____________ ____________ and a ____________ that activates the regulator valve.
5. Technician A says that banjo fittings may be used to connect the fuel filter to the supply circuit. Technician B says that a quick-connect fitting may require special tools. Who is correct?
   A. Technician A only
   B. Technician B only
   C. Both Technician A and Technician B
   D. Neither Technician A nor Technician B
6. Explain the differences between return-type and returnless-type fuel systems.
7. In an electric fuel pump, the relief valve opens if the fuel line becomes ____________.
8. The ____________ connects the fuel lines to the individual injectors.
9. Which of the following filters may be found within the fuel-delivery system?
   A. Pickup strainer
   B. Screen in the injector
   C. Filter in the supply circuit
   D. All of the above
10. Technician A says that return-type fuel systems typically use a fuel pressure regulator that does not sense engine vacuum. Technician B says that returnless-type fuel systems typically use pressure regulators that have a vacuum line connected to them to control pressures under different operating conditions. Who is correct?
    A. Technician A only
    B. Technician B only
    C. Both Technician A and Technician B
    D. Neither Technician A nor Technician B

CHAPTER 15

# Fuel Filter and Fuel Line Service

## Introduction

This chapter discusses fuel filter and fuel line service. The fuel filter (or element) is serviced by replacement only. Some vehicle manufacturers recommend fuel filter replacement at 30,000 miles (48,000 kilometers). Always replace the fuel filter at the vehicle manufacturer's recommended mileage. If contaminated fuel is placed in the fuel tank or the vehicle owner lives on a dusty road, the filter may require replacing before the recommended mileage. A plugged fuel filter may cause the engine to surge and cut out at high speed or hesitate on acceleration. A restricted fuel filter causes low fuel pump pressure.

**Figure 1** *Connecting the fuel drain hose to the Schrader valve.*

**Figure 2** *Relieving pressure from the fuel rail.*

## In-Line Filter Replacement

The in-line type of fuel filter is installed in the fuel line. It is usually located near the pump or it is integral to the fuel pressure regulator. Fuel filter replacement procedures vary depending on the make and year of the vehicle and the type of fuel system. Always follow the filter replacement procedure in the appropriate service manual. A typical procedure for removing a fuel filter begins by disconnecting the negative battery cable. Next, remove the filler cap to release any pressures in the tank. Remove the **Schrader valve** cap and wrap a shop towel around the valve (**Figure 1**). The Schrader valve resembles a tire valve stem and has a threaded fitting to allow easy connection of the pressure gauge to the fuel rail. Connect the pressure gauge or drain hose to the Schrader valve. Next, place the end of the drain hose into an approved container and open the drain valve (**Figure 2**). Lift the vehicle to a comfortable height and be sure to locate the lift pads at the specified locations. Flush the area around the fuel filter and the connectors with water and blow dry the area with compressed air. Now, disconnect the inlet connector to the filter (**Figure 3**). Plug the tank side of the line to prevent fuel from flowing out of the hose. Finally, disconnect the outlet connector and plug the line. With the connectors removed, the fuel filter retainer bolt can be removed and the filter replaced.

When replacing the in-line filter, it is best to first disconnect the inlet connector. Plug the line to prevent fuel from pouring from the tank. The fuel lines may be connected to the filter with quick-connect fittings. To release most of these fittings, grasp the large connector collar and twist in both directions while pulling the connector off the filter (**Figure 4**). Some

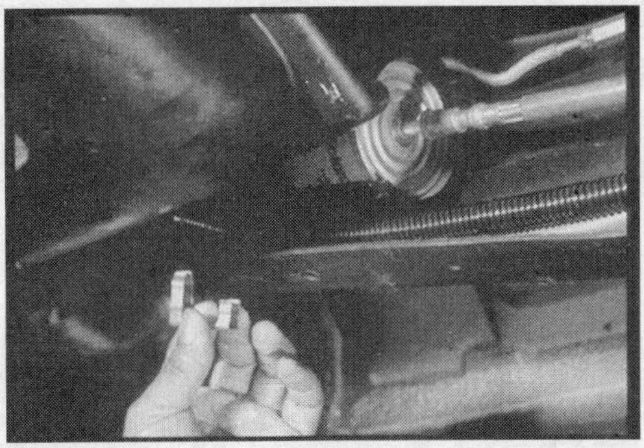

**Figure 3** *Disconnecting the inlet connector to the filter.*

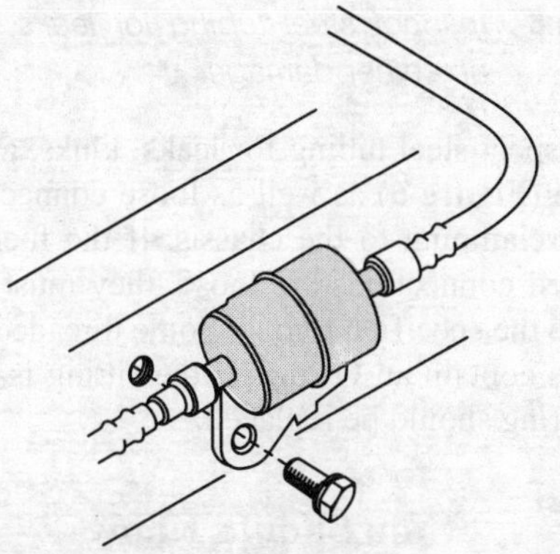

**Figure 4** *Fuel filter with quick-connect fittings.*

quick-connect fittings require special tools. These types of fittings are discussed later in this chapter. Also, the lines may be connected with banjo fittings. Always replace the metal gaskets when disconnecting these types of fittings. In addition, some filters may be connected with a threaded fitting and require an O-ring seal. Replace the O-ring any time the fitting is disconnected.

When installing the new filter, apply a few drops of clean engine oil to the male tube ends of the filter prior to engaging the quick-connect fitting. Check the quick connectors to ensure the large collar on each connector has rotated back to the original position. The springs must be visible on the inside diameter of each quick connector. Install the connectors onto the filter tubes and press the connectors firmly in place until the spring snaps into position. Grasp the fuel line and try to pull this line from the filter to ensure the quick connector is locked in place. Also, make sure the filter is installed in the correct direction. The filter is usually marked with an arrow or other indication of correct fuel flow. After installation is complete, start the engine and check for fuel leaks at the filter.

## Fuel Line Service

Fuel lines can be made of metal tubing, flexible nylon, or synthetic rubber (neoprene) hose. Neoprene hose must be able to resist gasoline and be nonpermeable to prevent gas and gas vapors from evaporating through it. Ordinary rubber hose such as that used for vacuum lines deteriorates when exposed to gasoline. When replacing hoses, only use those made for fuel systems.

**You Should Know**

*All fuel line hoses are made of special high-pressure material (MS-EA-235). No other hose material should be used on electronic fuel-injection (EFI) vehicles. To prevent any cuts in the hose lining, only use fuel hose clamps specified for EFI usage.*

The fuel line must work under pressure or vacuum. This is especially true for the hoses in fuel-injection systems, where pressures reach 50 psi (344 kPa) or more. For this reason, flexible fuel line hoses used in these systems must have special resistance properties. Many auto manufacturers recommend that only flexible hose be used for delivery to the fuel-metering unit in fuel-injection systems. However, flexible hose should not be used on the pressure side of the injector systems. This application requires a special high-pressure hose.

### Fuel Line Inspection

All fuel lines should periodically be inspected for holes, cracks, leaks, kinks, or dents. Many fuel system troubles that occur in the lines are blamed on the fuel pump. For instance, a small hole in the fuel line between the tank and the fuel pump will emit air, but there may not be any signs of leakage under the car. Air pressure that can now enter the fuel line pushes fuel back into the tank. When the driver attempts to start the engine, the fuel pump sucks air through the hole in the fuel line instead of drawing fuel from the tank. The correct diagnostic procedure will lead the technician to perform a fuel pump pressure and volume test. Since pressure and volume will both be low, the technician can be fooled into thinking the fuel pump is faulty. Remember to confirm powers and grounds before replacing the electric fuel pump! In this case, the pump will run and electrical tests will indicate no problems. Before replacing the pump, check for air leaks in the fuel line. This can be done by disconnecting the line at the tank and pressurizing the line with air. The leaking air is usually easy to locate.

Another example is when the fuel line between the pump and the injectors is under pressure. This makes locating a fuel leak on this side of the pump easier since the fuel will be pushed out the hole. Replace a damaged fuel line with one of similar construction; steel with steel, and the flexible type with nylon or synthetic rubber. Always use new clamps when installing flexible tubing. Because the old clamps lose some of their tension when they are removed, they will not provide an effective seal on the new line.

### You Should Know

*Do not substitute aluminum or copper tubing for steel tubing. Never use hose within 4 inches of any hot engine or exhaust system component. A metal line must be installed.*

Fuel-supply lines from the tank to the engine are usually routed to follow the frame along the underchassis of vehicles. Sometimes the manufacturer may route the lines inside the frame. Generally, rigid lines are used along the frame and are joined to other sections of line by short lengths of flexible hose (**Figure 5**). The flexible hose absorbs engine and chassis vibrations.

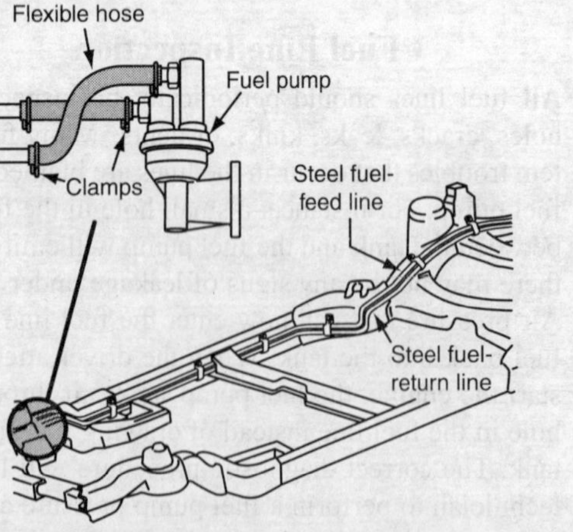


**Figure 5** *The fuel lines usually follow along the frame or under the floor board. Flexible hose attaches the steel line to the pump and fuel rail.*

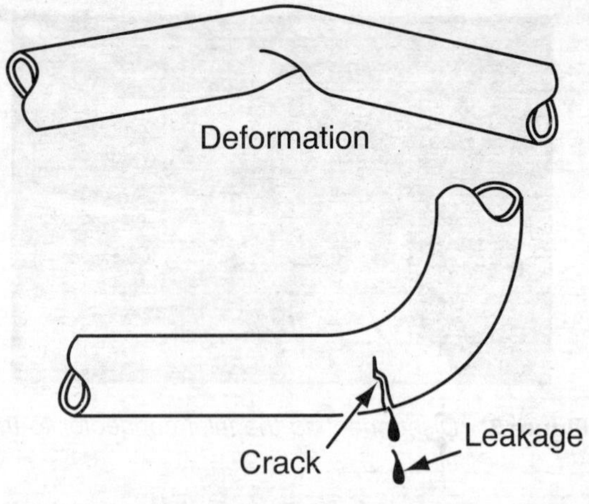


**Figure 6** *Inspect steel tubing for leaks, kinks, or any other damage.*

Inspect steel tubing for leaks, kinks, and deformation (**Figure 6**) as well as loose connections and proper clamping to the chassis. If the fuel tubing's threaded connections are loose, they must be tightened to the specified torque. Some threaded fuel line fittings contain an O-ring. If the fitting is removed, the O-ring should be replaced.

### You Should Know

*O-rings in fuel line fittings are usually made from fuel-resistant Viton. Other types of O-rings must not be substituted for fuel line fitting O-rings.*

## Steel Fuel Line Repair

If a fuel line is damaged or leaking, it must be repaired. Depending on the manufacturer's requirements and shop policy, the technician may need to replace the entire section of line or just the damaged portion. If a new line is to be fabricated, select the correct tube and fitting dimension and start with a length that is slightly longer than the old line. With the old line as a reference, use a tubing bender to form the same bends in the new line as those that exist in the old line. When bending tubing, remember that too sharp a bend will result in a kink or restriction. It is best to use a tubing bender

### You Should Know

*If a tubing bender is not available, bend the tube by holding it over a large piece of pipe and slowly forming it. When possible, use the tubing that is being replaced as a guide.*

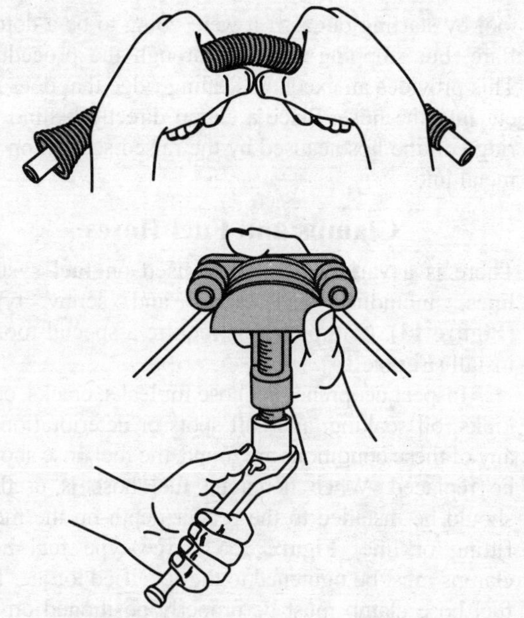

**Figure 7** *The best way to form a steel line is to use a tubing bender.*

(**Figure 7**). For gentle curves, the steel tubing can be bent by hand. Be careful not to bend the tubing too sharply.

If the end of the tubing will be close to a bend, it is better to install the fittings and flare the ends before bending the tubing. If the bend is too close to the flare, the flaring tool will not be able to clamp to the line.

### You Should Know

*Do not use a hacksaw to cut the tubing. The hacksaw will leave a rough edge that might not be square. When the end of the tube is dressed off with a file, metal chips can get into it. If the chips are not removed thoroughly serious damage to fuel system components can result.*

Use a tubing cutter to cut the tube to the desired length (**Figure 8**). Be sure the cut is square. First tighten the tubing cutter against the tubing and then roll it around the tubing. As the cutter rotates around the tubing, tighten the handle to advance the cutting blade. Do not overtighten the cutting blade or the cutter will cut through too soon and damage the tubing.

A burr usually remains on the end of the tubing after it is cut. Remove the burr using the cutter's reamer blade. After completing the cut, remove any chips from the end of the tubing.

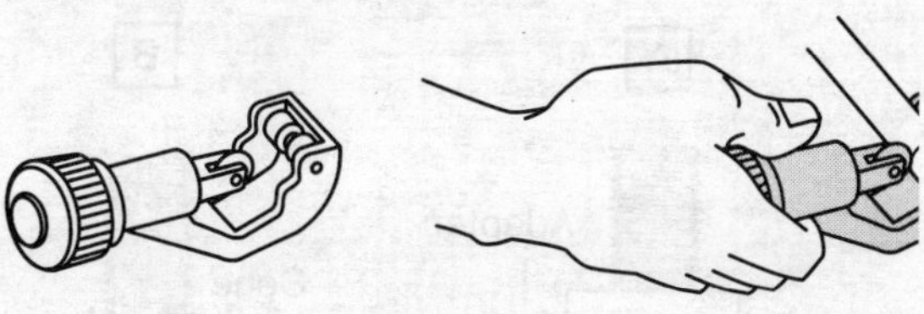

**Figure 8** *Use a tube cutter to ensure square ends.*

**Figure 9** *Secure the tube end so it protrudes the width of the adapter.*

Two common tubing fittings are the compression and the double flare. The double flare is made with a special tool that has an anvil and a cone. To make a double flare, begin by first sliding the fitting onto the line. Select the correct size hole in the flaring tool bar and clamp the line in the flaring tool bar. The end of the tube should extend out of the bar by the width of the flaring tool adapter (**Figure 9**). A threaded flaring cone and clamp are used to form the end of the tubing (**Figure 10**). Insert the adapter in the end of the tubing and tighten down on the flar-

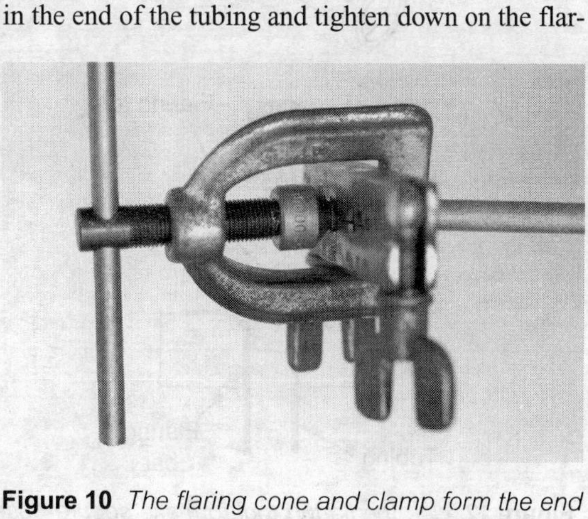

**Figure 10** *The flaring cone and clamp form the end of the tubing.*

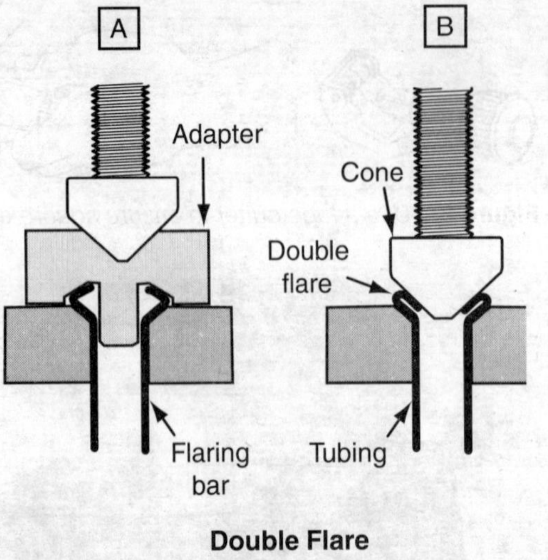


**Figure 11** *(A) To make a double flare, the adapter is used first. (B) The cone is then used to complete the flare.*

ing cone until it bottoms out (**Figure 11-A).** Remove the adapter and tighten the flaring tool against the line again to complete the flare (**Figure 11-B).**

The flare tool can also be used to secure nylon and synthetic rubber hose connections. By putting a partial double-lip flare on the end of the tubing (**Figure 12),** there will be less chance of the hose coming off when the hose is slid over and clamped in place. This can be done quickly with the proper flaring tool by starting out as if it were going to be a double flare, but stopping halfway through the procedure. This provides an excellent sealing ridge that does not cut into the hose. Place a clamp directly behind the ridge on the hose caused by the raised section on the metal line.

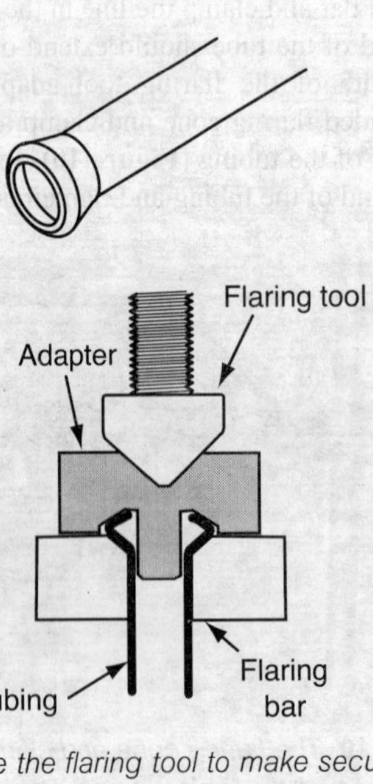


**Figure 12** *Use the flaring tool to make secure hose connections.*

## Clamps and Fuel Hoses

There is a variety of clamps used on fuel system lines, including the spring and screw types (**Figure** 13). Crimp clamps require a special tool to install (**Figure 14**).

Inspect neoprene fuel hose for leaks, cracks, cuts, kinks, oil soaking, and soft spots or deterioration. If any of these conditions are found, the fuel hose should be replaced. When neoprene fuel hose is used, it should be installed to the proper depth on the metal fitting or line (**Figure** 15). Screw-type fuel hose clamps must be tightened to the specified torque. The fuel hose clamp must be properly positioned on the hose in relation to the steel fitting or line.

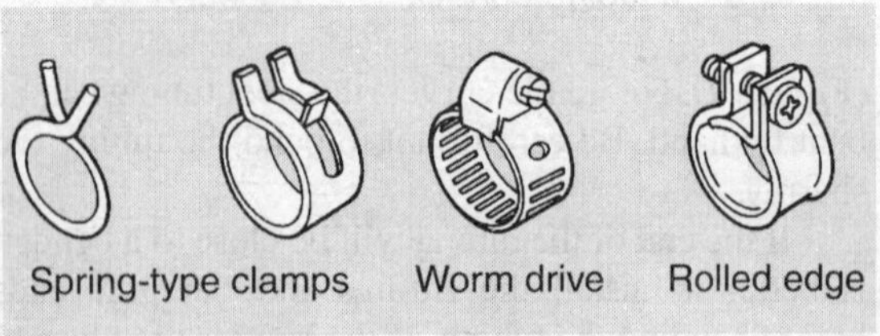


**Figure 13** *Different types of clamps used on fuel lines.*

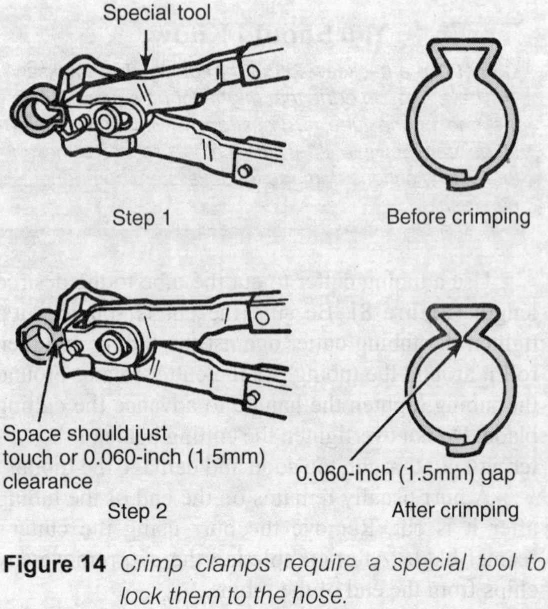


**Figure 14** *Crimp clamps require a special tool to lock them to the hose.*

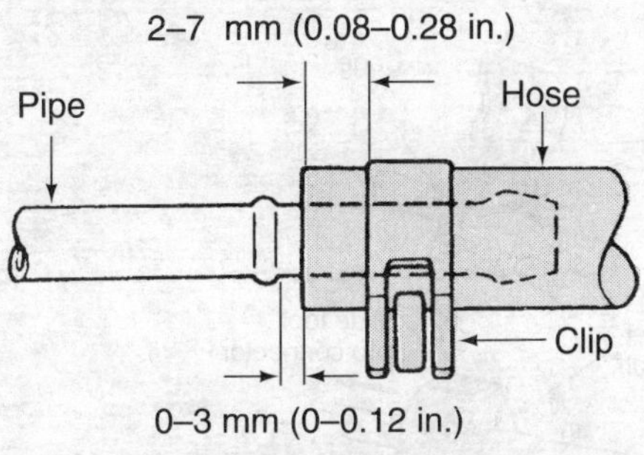


**Figure 15** *Proper rubber hose installation onto a steel line.*

To control the rate of vapor flow from the fuel tank to the vapor storage tank, a plastic or metal restrictor may be placed either in the end of the vent pipe or in the vapor vent hose itself. When the latter hose must be replaced, the restrictor must be removed from the old vent hose and installed in the new one.

> **You Should Know**
>
> *Always cover a nylon fuel hose with a wet shop towel before using a torch or other heat source near the line. Failure to observe this precaution may result in fuel leaks, personal injury, and property damage.*

## Nylon Hose Repair

Some manufacturers use nylon fuel hoses. Nylon fuel hoses provide a certain amount of flexibility and can be formed around gradual curves under the vehicle. Inspect nylon fuel hoses for leaks, nicks, scratches, cuts, kinks, melting, and loose fittings. Kinked or damaged hoses must be replaced. Nylon fuel hoses must be secured to the chassis at regular intervals to prevent fuel hose wear and vibration.

A hairpin push connector, a duckbill connector, or a spring lock may connect the nylon hose to the steel tubing. To remove a hairpin push connector **(Figure 16),** separate the clip legs by about ⅛ in. (3 mm) and pull the triangular tab to remove the clip. Next, firmly and gently pull the nylon line from the steel line. To remove the duckbill connector (**Figure 17),** use a pair of narrow-jawed pliers to compress both retaining clips on the side of the push connector at the same time. The retaining clips will remain on the metal tube when the connector is removed.

The spring-lock connector (**Figure 18)** is held together by the flared end on the female fitting lodged behind the garter spring inside the cage on the male fitting. There are two fluoroelastomer O-rings on the male coupling. Always use the correct O-rings when

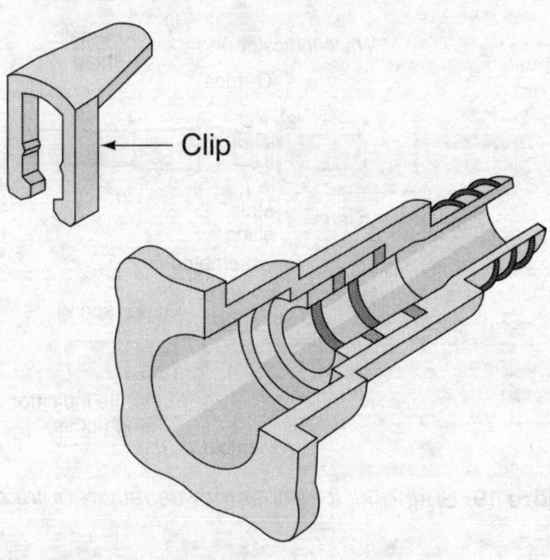


**Figure 16** *Hairpin push connector.*

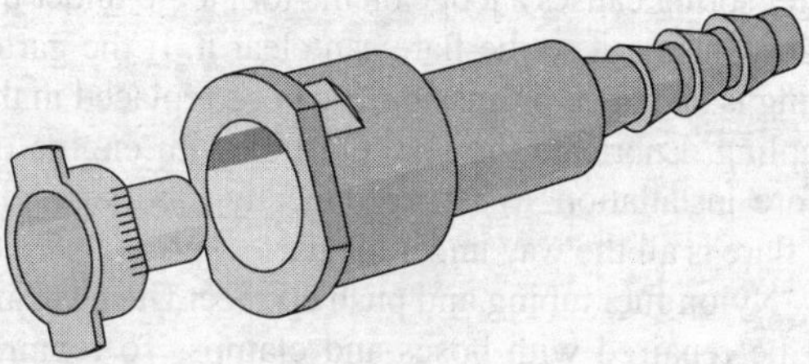

**Figure 17** *Duckbill connector.*

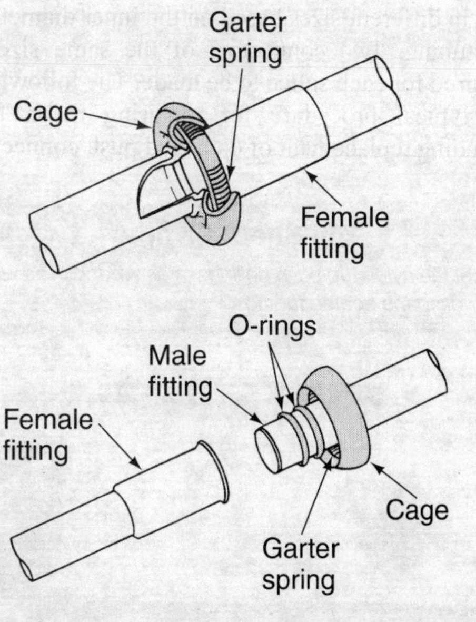


**Figure 18** *Spring-lock connector.*

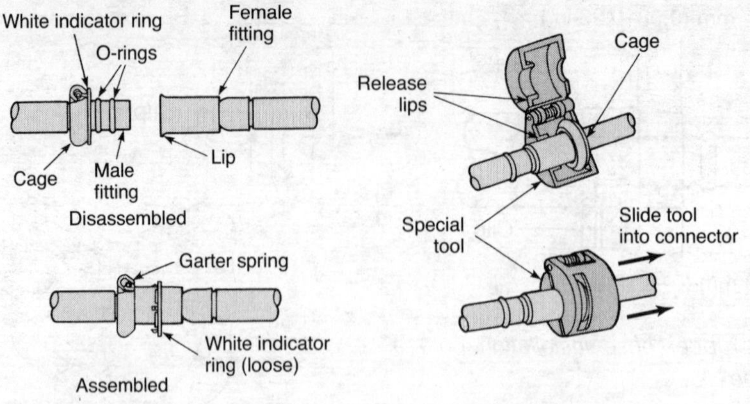


**Figure 19** *Spring-lock connectors usually require a special tool to release them.*

replacing them. A special tool is used to release the garter spring so that the two halves of the coupling can be released (**Figure 19**). Sliding the tool toward the garter spring causes a ledge on the tool to go under the spring, raising it so the flare can clear it. If the garter spring is damaged or missing, it can be replaced in the coupling. Lubricate the new O-rings with engine oil before installation. When reconnecting, be sure that the flare is all the way under the garter spring.

Nylon fuel tubing and push connect fittings cannot be repaired with hoses and clamps. To repair a nylon hose, use manufacturer-approved service parts to correct faults safely (**Figure 20**). Replacement tubing and the required barbed connectors are available in different sizes based on the inner diameter of the tubing. Two connectors of the same size are required for each splice to be made. The following is the typical procedure for repairing nylon hose, including replacement of damaged push connectors:

> **You Should Know**
>
> *Heat the nylon tubing in boiling water. Wear gloves when handling the heated tubing.*

**Figure 20** *Nylon fuel hose repair kit.*

1. Relieve the system pressure as described in Chapter 11.
2. Disconnect the negative battery cable.
3. Raise the vehicle on a hoist or jack stands.
4. Remove any fuel line retaining clips on both sides of the damaged section of tubing to provide enough slack to work easily with the line.
5. Cut out the damaged section of tubing.
6. Use the section cut from the fuel line as a guide for cutting the same length of replacement tubing.
7. Insert both ends of the replacement tubing into a container of boiling water for 2 minutes.
8. Remove the replacement section from the boiling water and immediately insert a barbed connector as far as it will go into each end of the tubing.
9. Insert one end of the original tubing to be spliced into the container of boiling water for 2 minutes.
10. Remove the tubing end from the water and immediately insert one barbed connector into one end of the tubing splice.
11. Repeat Steps 9 and 10 to complete the other end of the splice.
12. Reinstall any removed fuel line retaining clips.
13. Start the engine and check for leakage at the repaired area.

The connection of the original hose to the barbed connector may differ based on the system and hose size. Always refer to proper service information for the correct procedure. Often the 5⁄16 in. (8 mm) hose used on central fuel-injection (CFI) or EFI systems uses a crimp clamp on both ends of the connector. If the line is 5⁄16 in. (8 mm) and is not being used on CFI or EFI systems, then the original hose is installed over two barbs and crimp clamped. A tubing splice using ¼ in. (6 mm) hose does not require a clamp.

## Quick-Connect Fuel Line O-Ring Replacement

> **You Should Know**
>
> *Always lube the O-rings inside the quick-connect fittings with engine oil before reassembling the fuel line connections at the fuel pump module, fuel filter, fuel lines, and fuel rail.*

To disconnect the fittings, the retainer has two tabs that must be squeezed against the fuel tube and then pulled outward to disconnect the fuel tube from the quick-connect coupling. The retainer will stay on the fuel tube when the tube is disconnected. The O-rings and spacer will remain in the connector.

> **You Should Know**
>
> *Fuel system pressure must be released before disconnecting the fuel line couplings.*

A repair kit consisting of replacement O-rings, spacer, and retainer may be available. The replacement parts are installed on a disposable plastic plug. The following is a typical procedure:

1. Push the disposable plug with the kit installed on it into the quick-connect fitting until you hear a "click" (**Figure 21).**
2. Grasp the end of the disposable plug and pull outward to remove it from the fitting.
3. Push the fuel tube into the quick-connect fitting until you hear a "click" (**Figure 22).**
4. Verify that the connection is secure by firmly pulling back on the fuel tube. The tube should be locked in place.
5. Inspect all hoses and clamp connections for completeness and ensure that no leaks are present.

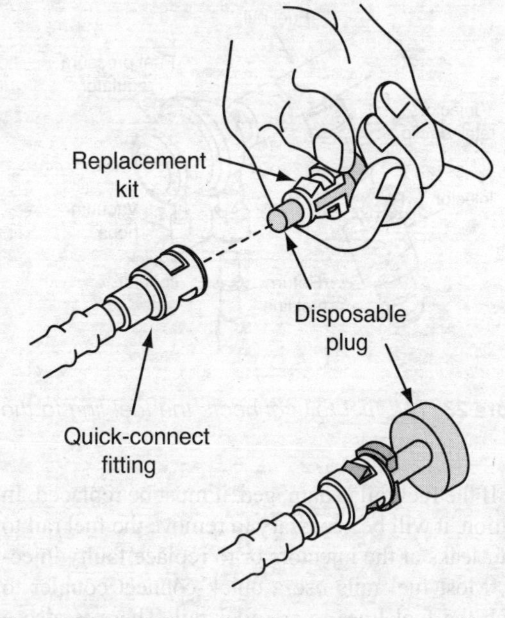


**Figure 21** *Installing the disposable plug with a kit installed into the fitting.*

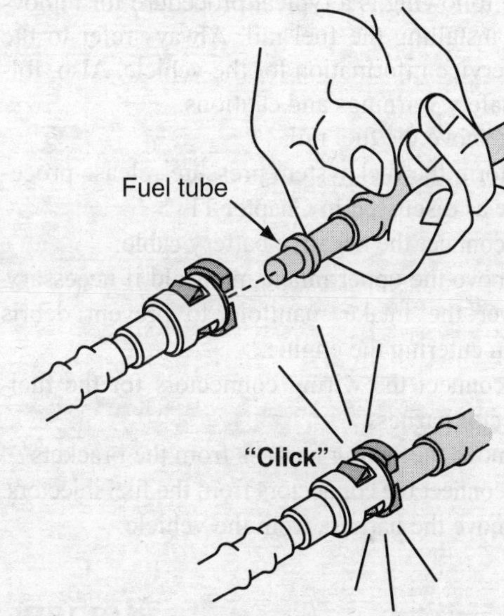


**Figure 22** *Connecting the fuel lines is confirmed by a "click."*

# Fuel Rail Service

Because the fuel rail (**Figure 23)** supplies fuel to the injectors, an adequate volume of fuel under a constant pressure differential must be maintained in the fuel rail. The fuel rail should be inspected for damage and leaks. Also, since the injectors are installed in separate fittings that extend from the rail and are held in position with a push-on spring clip, the O-rings that prevent fuel leakage between the injector and the intake manifold should be inspected.

> **You Should Know**
>
> *The injector O-rings may have different colors to distinguish their position on the fuel injector. To prevent leakage, the proper O-rings must be installed on the intake manifold end of the injector and on the fuel rail end of the injector.*

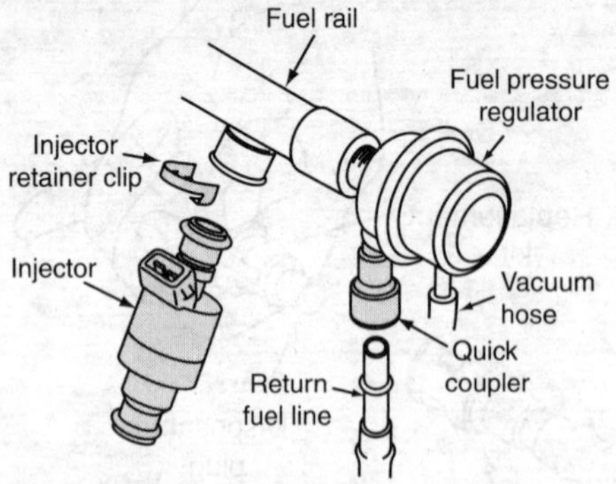


**Figure 23** *The fuel rail connects the fuel line to the injectors.*

If the fuel rail is damaged, it must be replaced. In addition, it will be necessary to remove the fuel rail to repair leaks at the injectors or to replace faulty injectors. Most fuel rails use a quick-connect coupler to attach the fuel lines to the fuel rail. There is also a quick-connect coupler that connects the fuel-return line flexible hose to the metal return line.

The following is a typical procedure for removing and installing the fuel rail. Always refer to the proper service information for the vehicle. Also, follow all safety warnings and cautions.

To remove the fuel rail:

1. Perform the fuel system pressure-release procedure as discussed in Chapter 11.
2. Disconnect the negative battery cable.
3. Remove the upper intake manifold if necessary. Cover the intake manifold to prevent debris from entering the engine.
4. Disconnect the wiring connectors for the fuel-injector harness.
5. Remove the wiring harness from the brackets.
6. Disconnect the connectors from the fuel injectors.
7. Remove the harness from the vehicle.
8. Disconnect the quick-connect fitting from the chassis tube to the fuel rail.

**You Should Know**

*Wrap a shop towel around the hoses to absorb any gasoline spillage.*

9. Remove the fuel rail attaching bolts.
10. Remove the fuel rail, being careful not to damage the injector O-rings when removing them from their ports.

To install the fuel rail:

1. Ensure that the injector holes are clean.
2. Replace any damaged O-rings.
3. Lubricate the injector O-rings with a drop of clean engine oil to aid in installation.
4. Put the tip of each injector into its port. Push the assembly into place until the injectors are seated in the ports.
5. Install the fuel rail mounting bolts. Tighten the bolts to the proper torque.
6. Connect the fuel hose quick-connect fitting. Push the fitting together until it clicks into place. Pull on the fitting to ensure complete insertion.
7. If the upper intake manifold was removed, clean the surfaces of the manifold and install the upper intake manifold. Torque the fasteners to specifications.
8. Connect the connectors to the fuel injectors.
9. Install the wiring harness to the brackets.
10. Connect the wiring connectors to the fuel-injector harness.
11. Connect the negative battery cable.
12. Jump across the fuel pump relay to pressurize the fuel system and check for leaks.

## Review Questions

1. Quick-disconnect fuel line fittings are being discussed. Technician A says some quick-connect fittings may be disconnected with a pair of snap ring pliers. Technician B says that some quick-connect fittings are hand releasable. Who is correct?
   A. Technician A only
   B. Technician B only
   C. Both Technician A and Technician B
   D. Neither Technician A nor Technician B
2. Describe a typical procedure for replacing an in-line fuel filter.
3. Explain how a leak in the fuel line between the tank and the fuel pump can lead to a false diagnosis of a fuel pump failure.
4. Explain the removal procedure for the three common types of connectors used with nylon hose.
5. Quick-connect fuel line fittings are being discussed. Technician A says that on some hand-

releasable, quick-connect fittings, the fitting may be removed by pulling on the fuel line. Technician B says that some hand-releasable, quick-connect fittings may be disconnected by twisting the large connector collar in both directions and pulling on the connector. Who is correct?
   A. Technician A only
   B. Technician B only
   C. Both Technician A and Technician B
   D. Neither Technician A nor Technician B

6. Technician A says that a plugged fuel filter may cause the engine to surge and cut out at high speed. Technician B says that a restricted fuel filter causes low fuel pressure. Who is correct?
   A. Technician A only
   B. Technician B only
   C. Both Technician A and Technician B
   D. Neither Technician A nor Technician B

7. Describe the procedure for making a double flare on a steel fuel tube.

8. Technician A says that nylon fuel line tubing and push connect fittings can be repaired by replacing the tube with a section of rubber hose and worm clamps. Technician B says that the two connectors that are used to splice a repair into a nylon tube need to be 1 mm larger than the tube. Who is correct?
   A. Technician A only
   B. Technician B only
   C. Both Technician A and Technician B
   D. Neither Technician A nor Technician B

CHAPTER

# 16 Electric Fuel Pump Circuits

## Introduction

Manufacturers use several methods to control the operation of electric fuel pumps used in fuel-injected engines. This chapter details some of the most common electric fuel pump circuits.

## Fuel Pump Relay

Most electric fuel pump systems use a relay that the PCM controls. Since the PCM controls when the relay will be energized, it in effect controls the fuel pump activation. The fuel pump relay is located inside the power distribution center (PDC) or junction box on most vehicles. The fuel pump relay is energized under the following conditions: to provide power to operate the fuel pump for approximately 0.7 to 1.5 seconds during the initial key-on cycle and while the PCM is receiving an rpm signal that exceeds a predetermined value.

Ignition voltage is provided to the fuel pump relay's coil any time the key is in the Run/Start position **(Figure 1)**. The PCM provides the ground control to energize the relay. The relay is energized when the key is cycled to Run in order to prime the fuel rail with liquid fuel, allowing for a quick start-up.

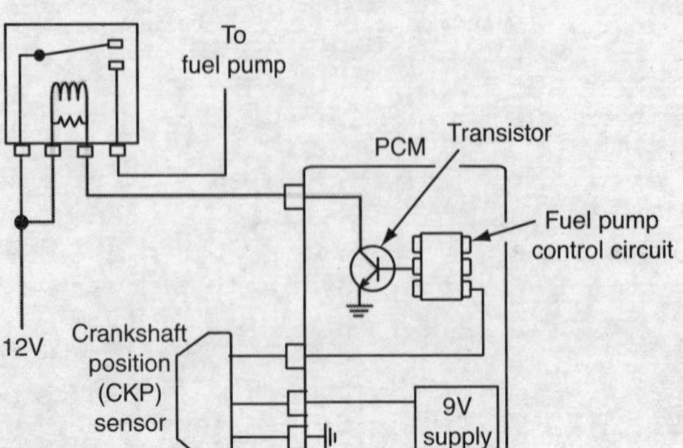


**Figure 1** *Fuel pump relay control circuit.*

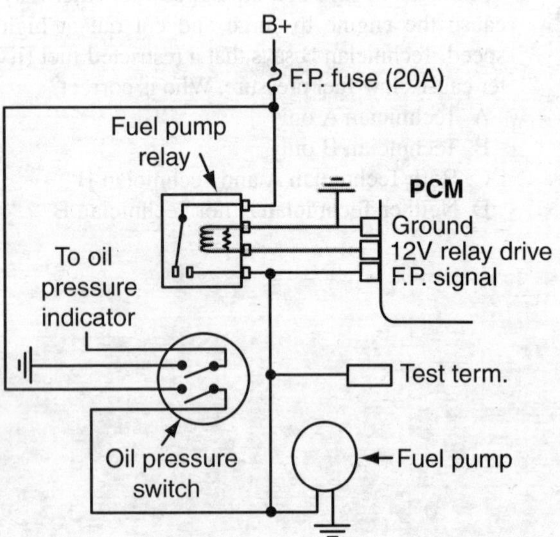


**Figure 2** *Fuel pump relay control using high-side switching.*

Any time the PCM receives an rpm signal that exceeds a predetermined value, the relay is energized to ensure proper fuel pressure and volume during engine cranking and running conditions. When the contacts in the relay close, battery voltage is supplied to the fuel pump. If the rpm signal is lost (engine has been shut off or the sensor indicates no rpm), the fuel pump relay is deenergized.

Some manufacturers supply voltage to the winding of the fuel pump relay instead of providing ground **(Figure 2)**. Except for the use of high-side switching, the operation of the fuel pump is similar to that discussed above.

## Inertia Switch

On Ford products, an **inertia switch** is connected in series in the fuel pump circuit. It is used to shut off the fuel pump in the event of an accident. The inertia

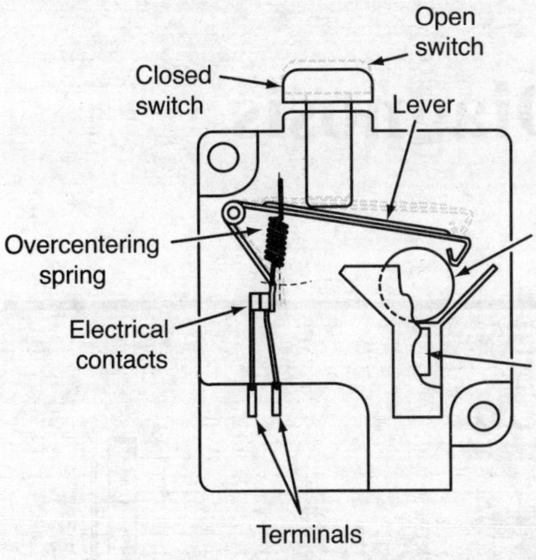


**Figure 3** *The inertia switch uses a weighted ball to trip open the circuit contacts if the vehicle is involved in an accident. The reset button closes the contacts.*

switch consists of a steel ball inside a conical ramp, a target plate, and a set of electrical contacts. The magnet holds the steel ball in the bottom of the conical ramp. In the event of a collision, the inertia of the ball causes it to break away from the magnet, roll up the conical ramp, and strike the target plate **(Figure 3)**. The force of the ball striking the plate causes the electrical contacts in the inertia switch to open, removing voltage to the fuel pump. If tripped, a reset button on top of the inertia switch must be pressed to close the switch and restore fuel pump operation.

## Circuit-Opening Relay

On many Toyota vehicles, the fuel pump relay is called a **circuit-opening relay**. The circuit-opening relay is

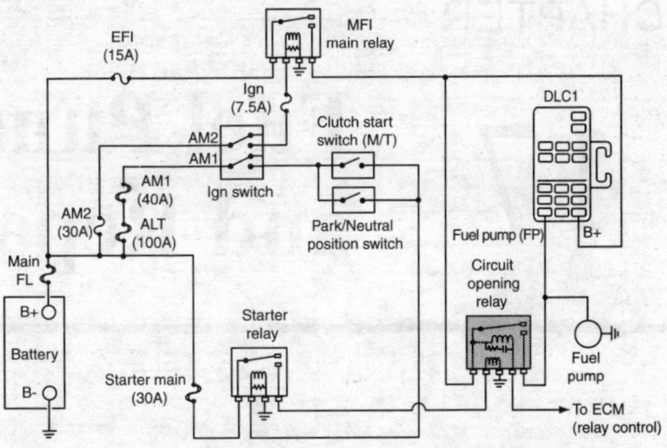


**Figure 4** *Circuit diagram of a circuit-opening relay fuel pump system.*

a dual-winding relay used to send current to the fuel pump while the engine is cranking and after it starts **(Figure 4)**. One winding is connected in a series with the starter relay contacts and ground. The second relay winding is connected from the positive battery terminal to the PCM. When the engine is cranking and the starter relay contacts are closed, current flows through the starter relay contacts to the circuit-opening relay winding and on to ground. This current flow creates a magnetic field around the circuit-opening relay winding that closes the relay contacts. When these contacts are closed, current flows to the fuel pump.

Once the engine starts, the starter relay is no longer energized and current stops flowing through these relay contacts to the first winding of the circuit-opening relay. However, the PCM grounds the other winding of the circuit-opening relay whenever the engine is running. This action keeps the relay contacts closed while the engine is running.

## Review Questions

1. Explain the operation of the circuit-opening relay.
2. An ____________ switch in the fuel pump circuit opens the fuel pump circuit immediately if the vehicle is involved in a collision.
3. Most electric fuel pump systems use a relay that is controlled by the ____________ .
4. Explain the operation of a typical fuel pump relay control circuit.
5. Explain the purpose and operation of the inertia switch.

CHAPTER

# 17 Fuel Pump Diagnosis and Repair

## Introduction

An engine must be supplied with the correct amount of fuel to run efficiently. Delivering the correct amount of fuel to the combustion chamber is the responsibility of the carburetor or fuel-injection system. However, there are many other components that are also responsible for the proper delivery of fuel. For everything to function as designed, the fuel must be stored, pumped out of storage, piped to the engine, and filtered. All this must be accomplished efficiently and safely.

Problems that require the fuel-delivery system to be inspected and serviced include fuel leaks, fuel odors, failure to start, hard starting, stalling, lack of power, poor fuel economy, and higher emissions. Fuel pressure that is not within specifications may cause too little or too much fuel to be delivered to the combustion chamber. Lean mixtures are often caused by insufficient amounts of fuel being drawn from the fuel tank. Various test results suggest lean mixtures, including high hydrocarbon (HC) readings on an exhaust analyzer and high firing lines on a scope.

## Fuel Pump Diagnosis and Service

A faulty fuel pump can cause several problems, including failure to start, hard starting, excessive emissions, surging, rough idle, lack of power, and poor fuel economy. Basically, there are two types of fuel pumps: mechanical and electrical. Electric fuel pumps are the most commonly used in today's vehicles.

### You Should Know

*Any time a vehicle is brought into the shop due to a no-start condition, always confirm there is fuel in the tank. Do not rely on the fuel gauge reading. Technicians who fail to do this simple step waste time performing unnecessary diagnosis and parts replacement.*

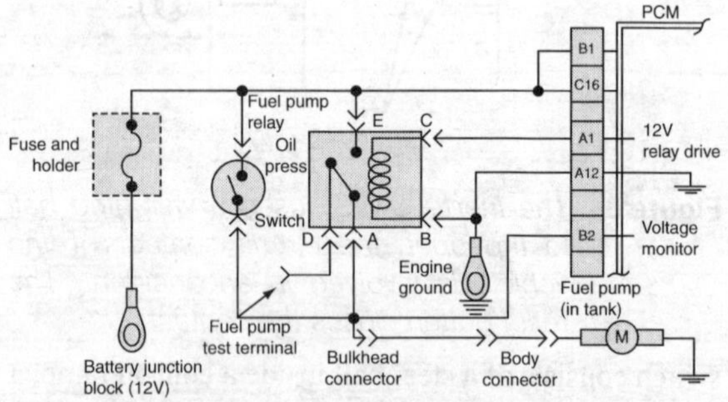


**Figure 1** *Wiring diagram of an electric fuel pump system.*

Electric fuel pump system failures can be caused by a problem in the electrical circuit **(Figure 1)**, a faulty pump motor, faulty check valves, a faulty regulator, leaking lines, or plugged filters or lines.

Usually a fuel pressure test will indicate or confirm a problem with the electric fuel pump system. It is a good practice to test the fuel pressure whenever a vehicle comes into the shop with a driveability problem. The following is a typical procedure for testing the fuel pressure on a fuel-injection engine. Always refer to the service information pertaining to the

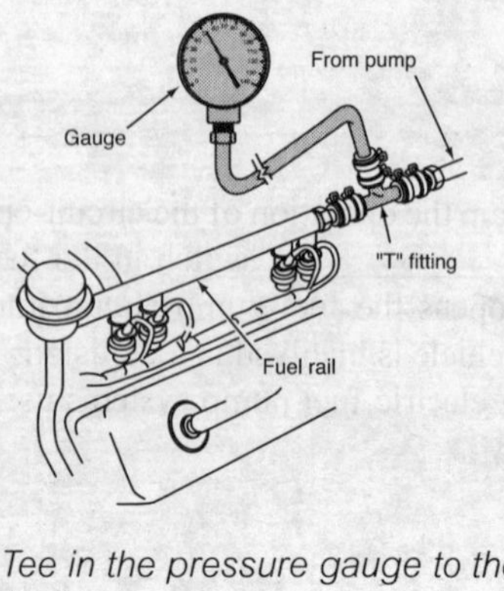


**Figure 2** *Tee in the pressure gauge to the fuel line if the engine is not equipped with a Schrader valve.*

**Figure 3** *Monitor the fuel pressure when the ignition switch is turned to Run.*

vehicle. Most systems have a Schrader valve located on the fuel rail to provide a connection point for the fuel pressure gauge. If the system does not have a Schrader valve, an adapter hose that connects the fuel rail and the fuel line will be required **(Figure 2)**. Prior to making any fuel pressure gauge or adapter connections, relieve the fuel rail of any pressure.

Begin the fuel pressure test with a thorough visual inspection of the components of the fuel-delivery system. Correct any faults before continuing. Be sure to inspect closely the connection between the fuel line and the fuel rail since this is a likely location of a leak. Locate the Schrader valve and remove its cover and relieve the fuel pressure from the system. After the pressure has been relieved, connect the pressure gauge to the Schrader valve. If the system uses a vacuum-controlled pressure regulator, connect a hand-operated vacuum pump to the fuel pressure regulator's vacuum sense connection. Turn the ignition switch to the Run position and observe the gauge reading **(Figure 3)**. Next, apply vacuum to the pressure regulator with the hand pump. As the vacuum increases, pressure should decrease. This may require a key cycle in order to run the pump again.

On some fuel-injection systems, a **pulsation damper** (or accumulator) is installed into the fuel rail. The pulsation damper is a spring-loaded accumulator used to reduce the effects of fuel pump pulsations and pressure changes due to injectors opening. Based on the manufacturer's recommendations, the connection point of the damper to the rail may be the specified location for performing fuel pressure testing. Remove the damper by covering it with a rag and loosening it with a proper size wrench **(Figure 4)**. Begin by loosening the damper one turn to relieve any pressure. Then the damper can be

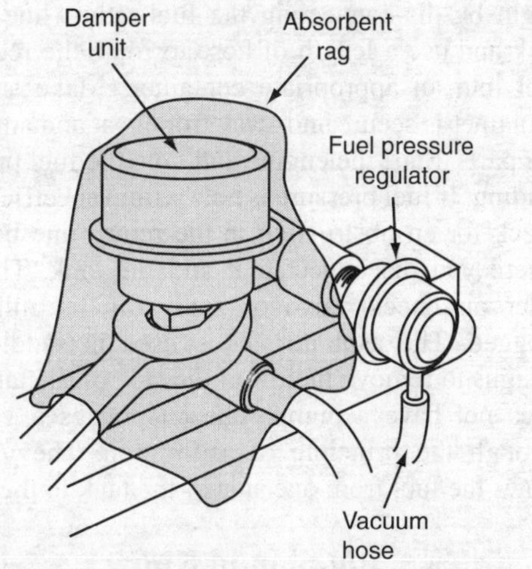


**Figure 4** *Removing the fuel pulsation damper.*

removed and the pressure gauge installed using the correct adapters **(Figure 5)**.

> **You Should Know**
>
> *The pulsation damper is also called a fuel rail accumulator.*

A higher-than-specified pressure reading is usually an indication of either a faulty pressure regulator or an obstructed return line. To isolate the cause,

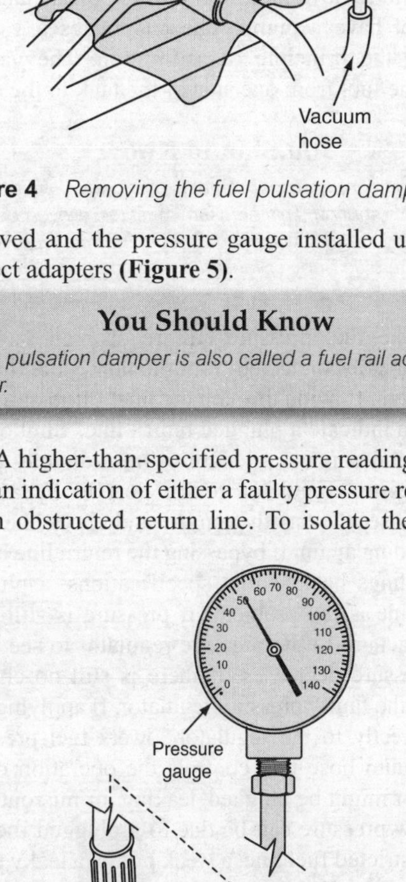


**Figure 5** *Connecting the pressure gauge to the pulsation damper fitting.*

begin by dis-connecting the fuel-return line at the tank and use a length of hose to route the returning fuel into an appropriate container. Make sure the container is secure and away from heat and any moving parts. Start the engine and note the fuel pressure reading. If fuel pressure is now within specifications, check for an obstruction in the return line between where you disconnected it and the tank. The fuel reservoir check valve or **aspirator jet** might be clogged. The aspirator jet is used in saddle tank designs to remove fuel from the side of the tank that does not have a pump. The jet creates a vacuum through the principle of as-piration. The vacuum draws the fuel from one side of the tank to the other.

**You Should Know**

*The fuel-supply lines remain pressurized after the engine is shut off. This pressure must be relieved before servicing the fuel system. A valve is normally provided on the throttle body or fuel rail for this purpose.*

If the fuel pressure still reads high with the return line disconnected from the tank, note the volume of fuel flowing through the line. Little or no fuel flow can indicate a plugged return line. Shut off the engine and connect a length of hose directly to the fuel pressure regulator return port to bypass the return hose. Restart the engine and check the pressure reading again. If bypassing the return line brings the readings back within specifications, a plugged return line is the problem. If pressure is still high, apply vacuum to the pressure regulator to see if the fuel pressure decreases. If there is still no change, replace the faulty pressure regulator. If applying vacuum directly to the regulator lowers fuel pressure, the vacuum hose that controls the operation of the regulator might be plugged, leaking, or misrouted.

Low pressure can be due to a plugged fuel filter, a restricted fuel line, a weak pump, a leaky pump check valve, a defective fuel pressure regulator, or a dirty filter sock in the tank. It may be possible to eliminate the filter and line restrictions as a cause of the problem by checking the pressure at the pump outlet. A higher reading at the pump outlet of at least 5 psi (34.5 kPa) as compared with the reading at the fuel rail means there is a restriction in the filter or line. If the reading at the pump outlet is unchanged, then the problem is a weak pump, a clogged pick-up screen, or a defect in the pump's electrical system (low amperage). All of these may require the tank to be dropped a few inches or removed in order to check it further.

**You Should Know**

*If the filter sock is clogged with dirt or debris, it is a good practice to clean out the tank along with replacing the filter.*

A leaking pump check valve may cause the engine to start hard. If the ignition key needs to be cycled from Off to Run several times before attempting to start the engine, this could be an indication of a leaking check valve. This action "primes" the fuel-delivery system and provides a volume of fuel to the injectors. The check valve can be tested by watching the fuel pressure gauge after the engine is shut off. Once the engine is shut off, pressure should remain within specifications for at least 10 minutes. If the pressure drops rapidly after the engine is shut off, the check valve may be leaking. Pinching a flexible fuel hose on the discharge line of the pump may help to determine if the check valve is leaking. If the gauge drops rapidly after the engine is shut off, start the engine again and then shut it off, and clamp the hose immediately. If the fuel pressure does not drop as fast as it did previously, the check valve is leaking. The fuel pump will need to be replaced since the check valve is internal to it.

**You Should Know**

*The purpose of the fuel pump check valve is to hold a volume of fuel in the line so it will be available to the injectors at key on. It is normal for the pressure in the line to drop to zero if the vehicle sits for a long period of time without being started. However, the pressure should return to normal as soon as the ignition key is turned to the Run position.*

Keep in mind that a leaking injector will also cause the fuel pressure to drop rapidly after the engine is shut off. If pinching the fuel hose does not stop the rapid pressure drop, then an injector is probably leaking. To identify a leaking injector, remove the spark plugs one at a time and use an exhaust analyzer to "sniff" each cylinder while observing the hydrocarbon meter. If an injector is leaking, the hydrocarbons will be excessive.

Depending on the type of pump used, the check valve can also prevent the reverse flow of fuel and relieve internal pressure to regulate maximum pump output. Check valves can stick or leak. So if a pump runs but does not pump fuel, a bad check valve is to blame. The check valve is usually an integral part of the pump assembly that is sealed at the factory. Therefore, if the check valve is causing trouble, the entire pump must be replaced.

Many manufacturers do not supply fuel volume specifications for engines with fuel injection. Typically the volume test is performed using the return line from the throttle body or the regulator. The fuel pump is run for the specified length of time while the volume of fuel is measured and the measurements are compared with the specifications.

> **You Should Know**
>
> *If a vehicle has a no-start condition, listen for pump noise when the key is in the Run position. The pump should buzz for a few seconds to build system pressure. If the pump does not run, perform electrical testing of the circuit.*

A lower-than-specified condition or no fuel pressure can be the result of an electrical defect in the pump circuit. Study the vehicle's wiring diagram and determine what type of controls it uses for the fuel pump. The different types of controls include an oil pressure switch, an inertia switch, and relays. Most late-model cars use the PCM to energize the pump relay when it receives a cranking signal from the distributor pickup or crankshaft sensor **(Figure 6)**. An oil pressure switch may still be included in the circuitry for safety purposes and to serve as a backup in case the relay or computer signal fails. Failure of the pump relay or the computer driver signal can cause slow starting. This is because the fuel pump does not come on until the engine cranks long enough to build up sufficient oil pressure to trip the oil pressure switch.

If pump pressure is below specification, check for the presence of proper voltage levels at the pump's electrical connectors. If the pump fails to make a buzzing sound when the key is turned to the Run position, check the fuses used in the circuit before continuing. If a fuse is blown, find the cause of the high current and repair the circuit. Replacing the fuse might restore power to the pump, but the problem is not solved until it is determined what caused the fuse to blow. The most likely causes of a blown fuse are a short in the wiring between the relay and the pump or a short inside the oil pressure switch or the relay.

> **You Should Know**
>
> *When testing an electric fuel pump, do not let fuel contact any electrical wiring. The smallest spark (electric arc) could ignite the fuel.*

The best electrical check is a voltage drop test. Remember, if you disconnect the connector and check for voltage, you are performing an open circuit test. Since the circuit is open, there will be no voltage drop because there is no current flow. This means that every stand in the wire between the fuel pump relay and the pump connector could be broken, except for one, and you will still have 12 volts at the connector. Testing the circuit under load provides the best indication of circuit integrity. To determine the amount of voltage being applied to the pump motor, leave the connector attached to the pump and back probe the feed wire with a pin. Connect the DVOM to the pin and ground

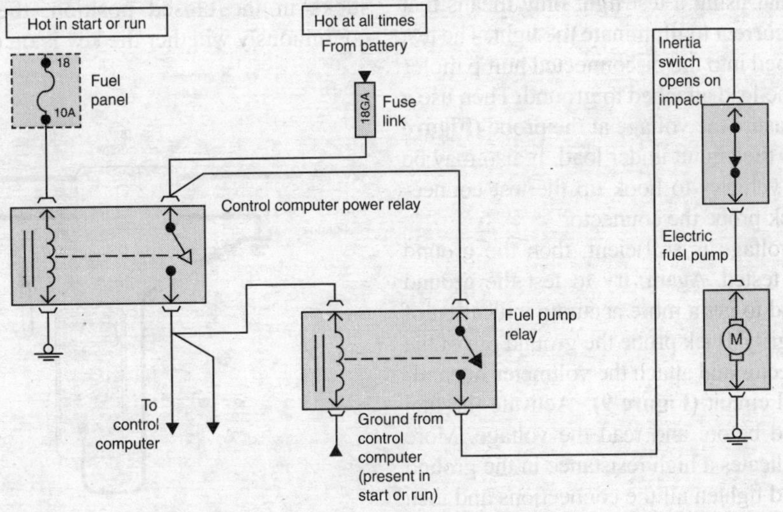


**Figure 6** *Wiring diagram of electric fuel pump circuit that is controlled by the computer and uses an inertia switch.*

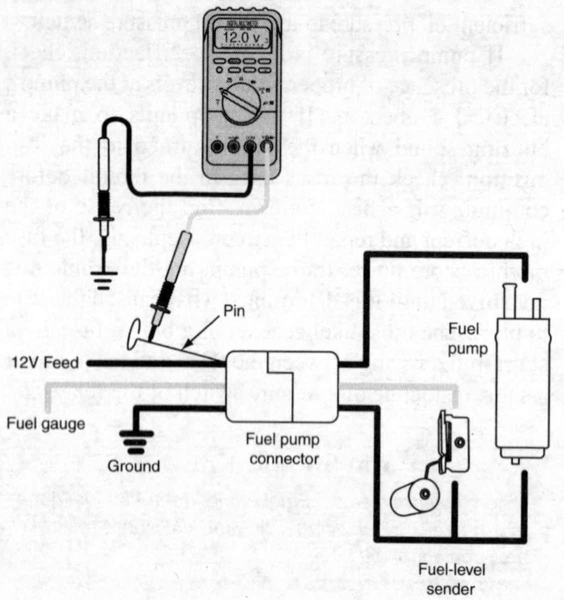


**Figure 7** *Testing for applied voltage on the feed wire at the fuel pump connector.*

**(Figure 7)**. Activate the fuel pump by turning the ignition key to the Run position or by jumping across the appropriate terminals of the fuel pump relay. The voltage reading should be close to battery voltage (usually most specifications require 10 volts or more). If the voltage is too low, work back up the circuit toward the battery until the location of the high resistance is found. If the voltage is zero, then work up the circuit until you find the location of the open.

A test light can also be used to check the circuit. Keep in mind that using a test light only means that there is enough current to illuminate the light. The test light can be probed into the disconnected pump motor connector and the lead attached to ground. Then use a voltmeter to measure the voltage at the probe **(Figure 8)**. This will test the circuit under load. It also may be easier on some vehicles to hook up the test connections than to back probe the connector.

If battery voltage is sufficient, then the ground circuit must be tested. Again, try to test the ground circuit under load to get a more accurate indication of the circuit's integrity. Back probe the ground pin of the fuel pump connector and attach the voltmeter in parallel to the ground circuit **(Figure 9)**. Activate the fuel pump as you did before and read the voltage. More than 500 mV indicates a high resistance in the ground circuit. Clean and tighten all the connections and then retest. To check the ground with a test light, disconnect the pump motor connector and attach the test light

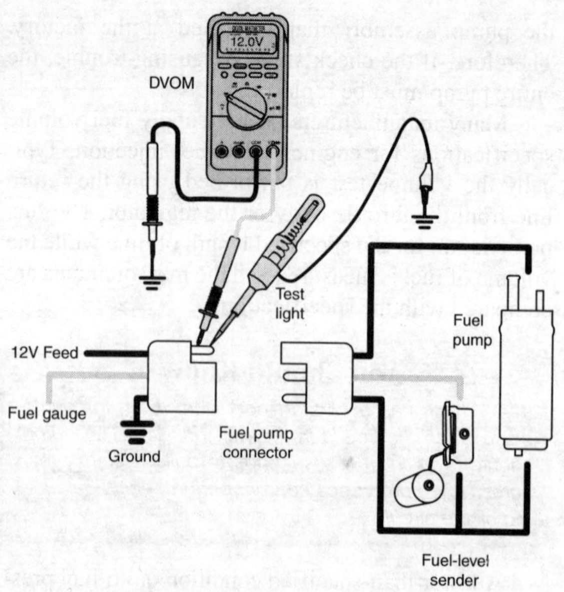


**Figure 8** *Testing for voltage under load using a test light and DVOM.*

across the ground and feed wires **(Figure 10)**. If the electrical circuit checks out but the pump does not run, the pump is probably bad and should be replaced.

No voltage at the pump terminal when the key is on and the engine is cranking indicates serveral possibilities: a faulty oil pressure switch, inertia switch, pump relay, or relay drive circuit in the computer, or a wiring problem. A faulty oil pressure switch can be checked by bypassing it with a jumper wire. If this restores power to the pump and the engine starts, replace the switch. If an oil pressure switch or relay sticks in the closed position, the pump can run continuously whether the key is on or off, depending

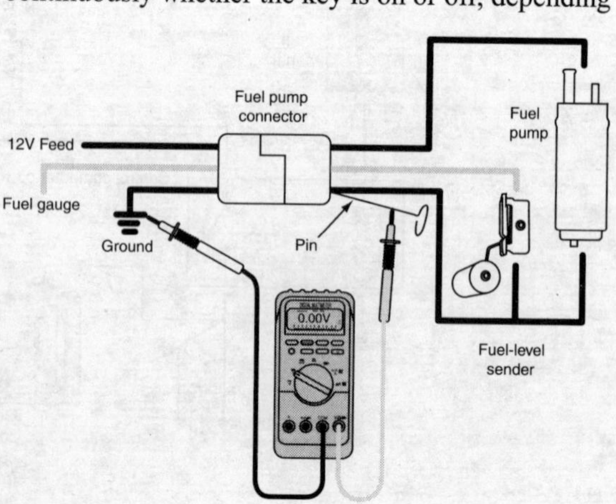


**Figure 9** *Performing a voltage drop test of the ground circuit of the fuel pump.*

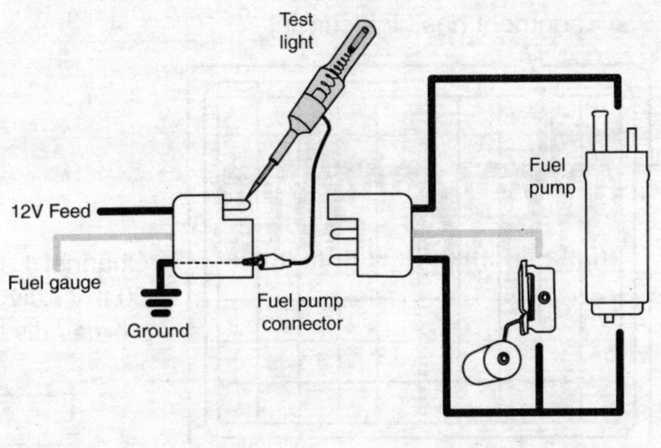


**Figure 10** *Testing the ground circuit with a test light.*

on how the circuit is wired.

The easiest way to test a fuel pump relay is to replace it with a known good relay. Usually there are several relays of the same type located within the power distribution center or junction box. Simply swapping these relays and testing the system operation may quickly indicate the cause of the problem. If swapping the relay does not result in normal fuel pump operation, further testing is needed. Remove the relay and use a test light or DVOM to check for voltage at the battery terminal of the relay's output circuit, usually relay pin 30 **(Figure 11)**. Also check for voltage at the relay's control circuit terminal, usually relay pin 86. If voltage is not present at either terminal, inspect the affected circuit for an open. If voltage is present at both terminals, then connect the test light or DVOM across the relay's control circuit terminals, usually relay pins 86 and 85 **(Figure 12)**. Turn the ignition key to the Run position while

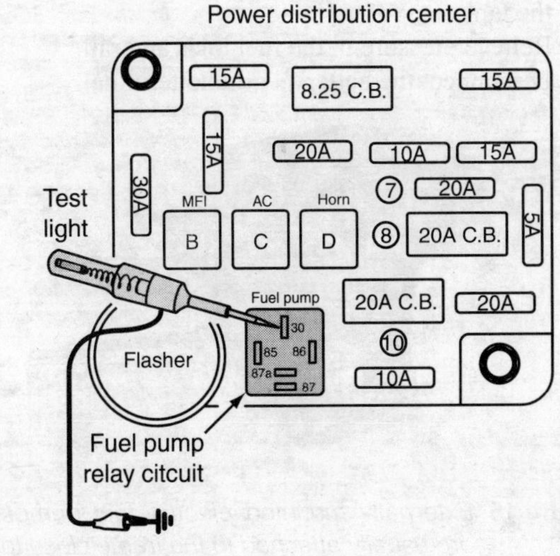


**Figure 11** *Checking for voltage to the fuel pump relay.*

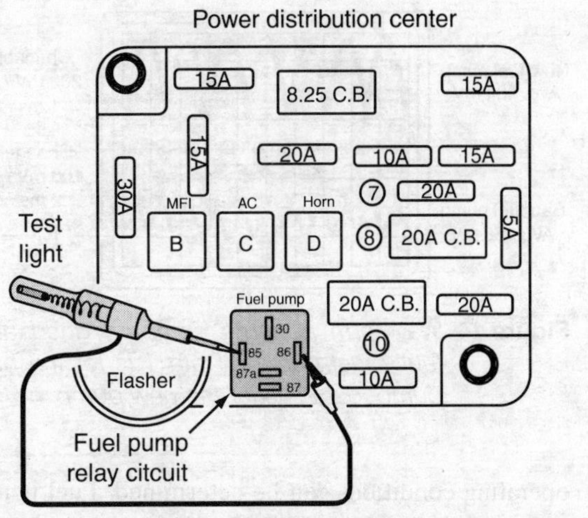


**Figure 12** *Checking the control side of the relay circuit for power and ground.*

observing the light. The light should come on for a few seconds, then go out. If this occurs, the PCM is performing its function of providing ground for the relay coil. If the light does not come on, the problem is an open in the ground circuit or a faulty PCM.

Another way to test the fuel pump circuit is to use an ampere probe and a lab scope. This will provide a waveform of the current in the fuel pump circuit **(Figure 13)**. New roller-vane-type automotive fuel pumps usually draw between 4 and 8 amperes at 5,000 to 6,000 rpms (approximately) when they are running loaded. Over a period of time, the fuel pump will become less efficient and eventually stop running. Fuel pumps usually begin to fail in one of three areas: worn brushes, armature, or bushing. When the brush material wears away, the spring tension lessens, increasing the resistance at the brush contact point on the armature. The increased resistance lowers the current flow, slowing the fuel pump speed. Unfortunately, the fuel pump may actually pass a pressure test at this point.

The value of using the lab scope and inductive ampere probe is that by clamping around a single wire feed to the fuel pump and operating the pump, the actual loaded condition of the pump under

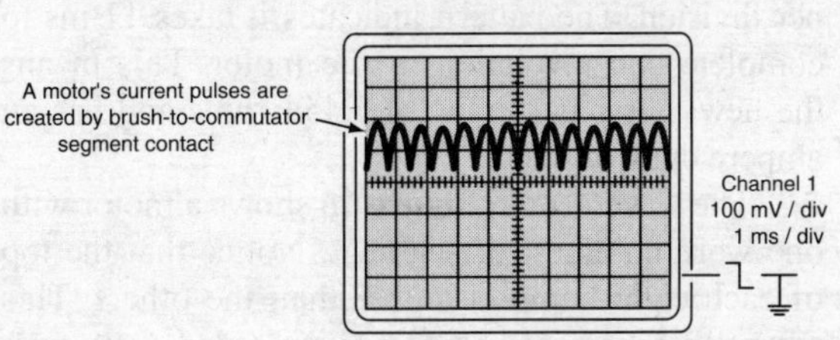


**Figure 13** *Lab scope waveform indicating a good fuel pump.*

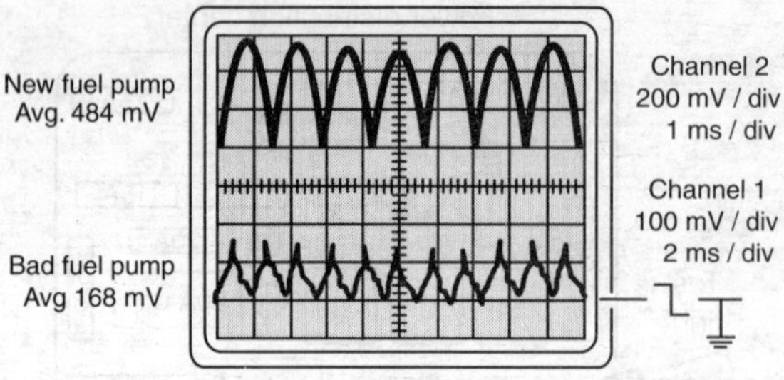


**Figure 14** *Waveform patterns compare good and faulty fuel pumps. Data 1 is for the bad pump; data 2 is for the new pump.*

operating conditions can be determined. Fuel pump ampere draw, operating rpm, brush contact, and wear can be determined by the waveform. The best setting for the lab scope is to use a voltage setting of 100 mV per division and a time setting of 1 or 2 ms. **Figure 14** shows a comparison between a bad fuel pump (channel 1) and a good fuel pump (channel 2).

When looking at the lab scope, the voltage reading can be translated into the amperage. Referring to Figure 14, the bad fuel pump had an average voltage reading of 168 mV. This means the amperage reading is 1.68 amperes. Current draw will increase with load and decrease with lack of load. Also, poor brush contact will lower current draw. Brush spring tension is critical to the motor operation.

Most fuel pumps use a total of eight (8) commutator segments. By using this number and the time base of the scope, the actual operational rpm of the pump motor can be determined. For example, in Figure 14 the division time is set at 2 ms. The bad pump took 9 divisions to complete one revolution, as shown between the time lines. This means it took 18 ms (9 divisions x 2 ms per division) for the pump to turn one revolution. Since 1 minute is equal to 60,000 ms, divide 60,000 by 18 to determine the rpm of 3,333. For the bad fuel pump, both the amperage and rpms were too low. The timing division for the new pump on channel 2 of Figure 14 is set at 1 ms per division. The pattern indicates it takes 11 ms to complete one revolution of the motor. This means the new motor is turning at 5,454 rpms and has an ampere draw of 4.84.

The waveform in **Figure 15** shows a motor with one worn commutator segment. Notice that the top of each eight-hump is higher than the others. This means that increased amperage is needed every time that commutator segment passes the brush. This pump would probably pass the fuel pressure test.

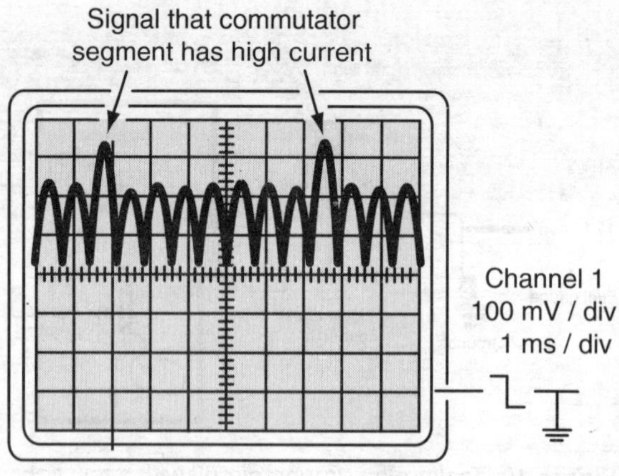


**Figure 15** *Waveform pattern indicates a bad commutator segment. Notice the high amperage every eight cycles.*

## Replacing Electric Fuel Pumps

Procedures for replacing the electric fuel pump differ based on location. The fuel pump may be external or internal to the fuel tank. Regardless of its location, be sure the replacement unit meets the vehicle's minimum requirements of pressure and volume. This information can be found in the service manual.

External electric fuel pumps are mounted on bushings to reduce noise and vibration. They are usually located underneath the vehicle on a frame rail near the fuel tank **(Figure 16)**. The bushings should be inspected and worn or damaged bushings should be replaced. Replace a typical electric fuel pump mounted externally to the fuel tank as follows:

1. Remove the filler cap to relieve any pressure in the tank.
2. Relieve pressure in the fuel lines and rail.
3. Disconnect the battery's ground terminal.

**Figure 16** *Externally mounted electric fuel pumps are usually attached to the frame close to the fuel tank.*

4. Raise the vehicle on a hoist or support it with safety stands.
5. Note that some vehicles have a protective covering over the pump housing that must be removed to access the pump fasteners and connections.
6. Disconnect the electrical connectors on the electric fuel pump. Label the wires to aid in connecting them to the pump. Reversing polarity on most pumps will destroy the unit.
7. Remove fuel and vapor lines from the pump. Label the lines to aid in connecting them to the new pump.
8. Remove the fasteners that hold the pump to the chassis.
9. Compare the replacement pump with the old one. If necessary, transfer any fuel line fittings from the old pump to the new one.
10. Replace any worn insulator bushings.

### You Should Know

*Avoid the temptation to test the new pump by energizing it with jumper wires before reinstalling the fuel tank. The pump relies on fuel for lubrication and cooling so running it dry can damage it.*

11. Locate the pump to the chassis and torque the retainer.
12. Reconnect the fuel lines using new gaskets or seals and torque to specifications.
13. Reconnect the electrical connectors.
14. Replace the cover if equipped.
15. Reconnect the battery's ground terminal.
16. Start the engine and check all connections for fuel leaks.

Fuel pumps that are mounted internally to the tank

### You Should Know

*Never turn on the ignition switch or crank the engine if the fuel line is disconnected. This action will result in gasoline discharge from the disconnected line, which could cause a fire, or personal injury and/or property damage.*

may require the fuel tank be removed. However, some vehicles have access panels to allow servicing the fuel pump without the need of removing the fuel tank **(Figure 17)**. Therefore, always check with the proper service information concerning fuel pump removal. Regardless of how access to the pump module is gained, most fuel pumps are removed by turning the retaining cam ring from the pump module with a spanner wrench **(Figure 18)**. Other methods of module-to-tank attachments include retaining bolts **(Figure 19)**.

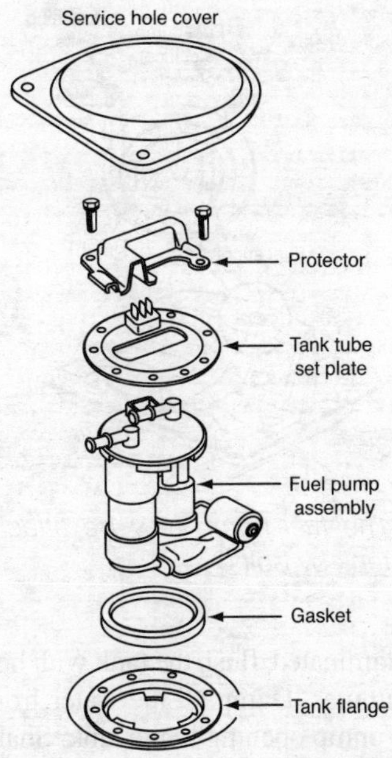


**Figure 17** *Some fuel pump modules can be serviced through an access panel in the floor of the passenger compartment or in the trunk.*

### You Should Know

*It is a good practice to replace any rubber hoses and clamps that attach to the pump whenever the pump unit is removed from the tank. The fuel-soaked rubber hoses may turn brittle after being exposed to air and then installed back into the tank.*

Prior to removing the retaining ring or bolts, clean the surrounding area to remove any accumulated dirt that might otherwise fall into the tank. Remove the pump assembly from the fuel tank. In most cases, the entire pump, filter, and gauge-sending unit is replaced with a new assembly. If the pump inlet

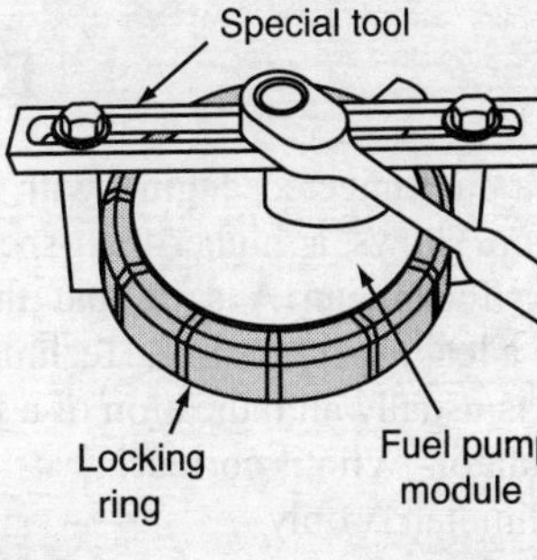


**Figure 18** *Most fuel pump modules are attached to the tank with a locking ring that requires a special spanner wrench to remove and install.*

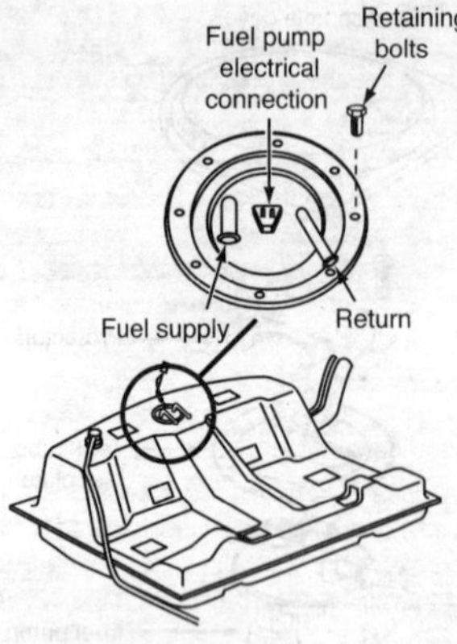


**Figure 19** *The fuel pump may be attached to the tank by bolts.*

filter is contaminated, flush the tank with hot water for at least 5 minutes. Dump all the water from the tank through the pump opening in the tank. Shake the tank to be sure all the water is removed. Also, check all fuel hoses and tubing on the fuel pump assembly. Replace fuel hoses that are cracked, deteriorated, or kinked. When fuel tubing on the pump assembly is damaged, replace either the tubing or the pump.

When installing the replacement pump, always install new gaskets, seals, O-rings, and fuel pump filter if applicable. Be sure the sound insulator sleeve is in place on the electric fuel pump and check the position of the sound insulator on the bottom of the pump. If the pump filter is serviced separately, note the position of the filter sock on the old pump to aid with installing the new one in the same relative position. When inserting the new pump into the sending unit bracket, be careful not to bend the bracket. Tighten the pump module retainer according to the manufacturer's specifications.

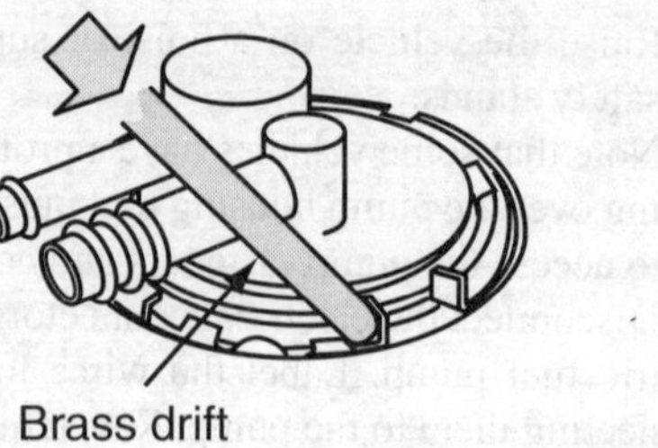


**Figure 20** *The metal lock ring used to attach the sending unit to the tank can be released or tightened with a brass punch.*

## Fuel-Level Sending Unit Replacement

Some vehicles have separate fuel pumps and gauge-sending units. These units are held in the tank by either a retaining ring or screws. The easiest way to remove a gauge-sending unit's retaining ring is to use a special tool designed for this purpose. This tool fits over the metal tabs on the retaining ring. A quarter turn will loosen the ring and the sender unit can be removed. If the special tool is not available, a drift punch or worn screwdriver and ball peen hammer will usually do the job **(Figure 20)**.

When removing the sending unit from the tank, be very careful not to damage the float arm, the float, or the fuel gauge sender. Check the unit carefully for any damaged components. Shake the float. If fuel can be heard inside the float, replace it. Make sure the float arm is not bent. It is usually wise to replace the filter and O-ring before replacing the unit. Check the fuel gauge by following the procedure in the service manual. When reinstalling the pickup pipe-sending unit, be very careful not to damage any of the components.

## Review Questions

1. Testing a fuel-injected engine with an electric fuel pump shows a higher-than-specified fuel pressure. Technician A says that the problem could be a leaking return line. Technician B says that this is usually an indication of a faulty pressure regulator. Who is correct?
   A. Technician A only
   B. Technician B only
   C. Both Technician A and Technician B
   D. Neither Technician A nor Technician B
2. The pressure regulator is being tested. Technician A says that the fuel rail pressure should decrease if the vacuum sensing hose to the regulator is disconnected. Technician B says that if the fuel rail pressure does not change when the vacuum sensing hose is removed, the hose may have a leak. Who is correct?
   A. Technician A only
   B. Technician B only
   C. Both Technician A and Technician B
   D. Neither Technician A nor Technician B

3. Fuel pressure testing shows normal pressure when the engine is running, but when the engine is shut off the pressure drops quickly. Technician A says that a faulty pump check valve is the problem. Technician B says that a faulty vent valve in the filler cap could be the problem. Who is correct?
   A. Technician A only
   B. Technician B only
   C. Both Technician A and Technician B
   D. Neither Technician A nor Technician B
4. A vehicle is hard to start after resting overnight, but it starts easily after short shutdown periods. Technician A says that the fuel pump check valve may be weak. Technician B says that the electric pump relay's bypass circuit could be open. Who is correct?
   A. Technician A only
   B. Technician B only
   C. Both Technician A and Technician B
   D. Neither Technician A nor Technician B
5. The fuel pressure of a fuel-injected engine is too low when tested at the fuel rail, but it is within specification when tested at the connection from the pump to the fuel line. Technician A says that the fuel line has a restriction. Technician B says that the fuel filter may be plugged. Who is correct?
   A. Technician A only
   B. Technician B only
   C. Both Technician A and Technician B
   D. Neither Technician A nor Technician B
6. A vehicle will not start due to low fuel pressure. When the fuel pump feed wire is back probed at the pump connector and the ignition key is turned to the Run position, the DVOM reads 11.8 volts. Moving the DVOM leads in parallel to the ground wire at the connecter and turning on the pump results in a DVOM reading of 8 mV. Technician A says that there is high resistance in the feed wire. Technician B says that there is an open in the ground wire. Who is correct?
   A. Technician A only
   B. Technician B only
   C. Both Technician A and Technician B
   D. Neither Technician A nor Technician B

SECTION

# 5 FUEL SYSTEM PRINCIPLES

## Section Objectives

After you have read, studied, and practiced the contents of this section, you should be able to:

- Describe the importance of the air-fuel ratio to engine performance and emissions levels.
- Describe the importance of the stoichiometric ratio as it relates to emissions and performance.
- Explain the difference between open-loop and closed-loop fuel strategies.
- Explain the operation of the speed density electronic fuel-injection (EFI) system.
- Explain the operation of the mass airflow (MAF) EFI system.
- Explain the operation of the MAF sensor.
- Describe the function of adaptive memory.
- Explain the fuel strategies used during different modes of operation.
- Describe the operation of the throttle body injection (TBI) system.
- Describe the difference between the sequential fuel-injection (SFI) system and the multipoint injection (MPI) system.
- Describe the meaning of pulse width.
- Describe the function of the central port injection (CPI) system.
- Describe the function of the gasoline direct-injection (GDI) system.
- Perform a preliminary diagnostic procedure on a TBI, MPI, or SFI system.
- Test the fuel injector and determine needed repairs.
- Properly perform injector cleaning.
- Remove and replace an injector on a port fuel-injection (PFI) engine.
- Diagnose and determine needed repairs of common input sensors used in fuel-injection systems.
- Perform a minimum idle speed adjustment.
- Perform scan tester diagnoses on various vehicles.
- Properly interpret data recordings.
- Determine the root cause of computer system failures.
- Determine the root cause of repeated component failures.
- Determine the root cause of multiple component failures.

### Interesting Fact

*Since the 1970s, fuel system technology has evolved quickly from carburetors to computer-controlled carburetors and then to throttle body injection systems. Many of the throttle body injection (TBI) systems have been replaced with multipoint injection (MPI) systems and a significant number of MPI systems have been changed to sequential fuel-injection (SFI) systems.*

CHAPTER

# 18 The Importance of the Air-Fuel Ratio

## Introduction

This chapter discusses the importance of the air-fuel ratio. The internal combustion engine converts the energy found in fuel into heat energy, which is changed to the mechanical energy that is used to move a vehicle. The conversion of energy results from the mechanical design of the engine and the burning process of combustion. During combustion, a chemical reaction takes place between the air and the fuel that enter the engine's cylinders. During this chemical reaction, heat energy is released. The release of heat results in an increase in the pressure of the air in the cylinders. This high pressure pushes on the engine's piston, forcing it to move downward. The movement of the piston is converted to rotational movement in order to move the vehicle.

The energy resulting from the movement of the piston depends on the amount of pressure on the piston. The amount of pressure on the piston depends on the amount of heat generated by the combustion process. Complete combustion of the air and the fuel will provide for the maximum amount of heat from that amount of air and fuel.

In order to have complete combustion, four things must be present: the correct amount of air (1) must be mixed with the correct amount of fuel (2) in a sealed container (3) and this mixture must be shocked by the correct amount of heat at the correct time (4). Although other factors can affect combustion, these are the most important ones.

No engine can achieve complete combustion during all operating conditions, but total combustion is striven for in the design of engine and fuel systems. The result of complete combustion is the generation of large amounts of heat energy and the conversion of all the cylinder's air and fuel into water and oxygen. When complete combustion does not occur, full conversion of the air and fuel does not take place. This results in the release of pollutants and reduced power from the engine.

The correct **air-fuel ratio** is required for an engine to provide good driveability, reduce vehicle emissions, and to prevent internal damage to engine parts. The engine's air-fuel ratio is measured by weight in pounds. The ideal ratio is about 15 pounds of air to 1 pound of fuel. A 15:1 air-fuel ratio is 9,000 gallons of air to 1 gallon of fuel. However, the requirements change based on driving conditions. A 15:1 ratio would not produce enough energy to satisfactorily power the vehicle up a hill or to accelerate away from a stop sign. A normal air-fuel mixture

### Interesting Fact

*Electronic fuel injection had its beginnings in Italy when engineer Ottavio Fuscaldo incorporated an electrical solenoid as a means to control fuel flow. The progress leading up to fuel-injection systems began as early as 1883 with Edward Butler, the Deutz Company and other pioneers. During World War II, Germany further pursued fuel-injection by bringing the Robert Bosch Company in to develop fuel-injection systems for the aviation field. During that time, Great Britain and the United States combined efforts to build a system to use in the Patton tank. In 1949, an Indy race featured a fuel-injected Offenhauser engine. The system was developed by Stuart Hillborn and featured an indirect-injection system. Later, Chevrolet in 1957 introduced the Rochester Ramjet full-injection system. It was also used in the 1957 Pontiac Bonneville. This system used a lot of the features designed by Hillborn. However, the system was not popular with the general public and it was dropped after 1959. The Corvette, though, used it as an option until 1965. In 1975, General Motors introduced the first mass-produced domestic fuel-injection system in the 1976 model Cadillac Seville. It was based on the Bosch "D" Jetronic system that had been used by European manufacturers since 1968. The system consisted of a throttle body, eight fuel injectors mounted on a fuel rail directing fuel into the intake, and a crude analog computer and various sensors, all on a modified intake using an Oldsmobile 350 V8. It had been developed by Bendix, Bosch, and General Motors.*

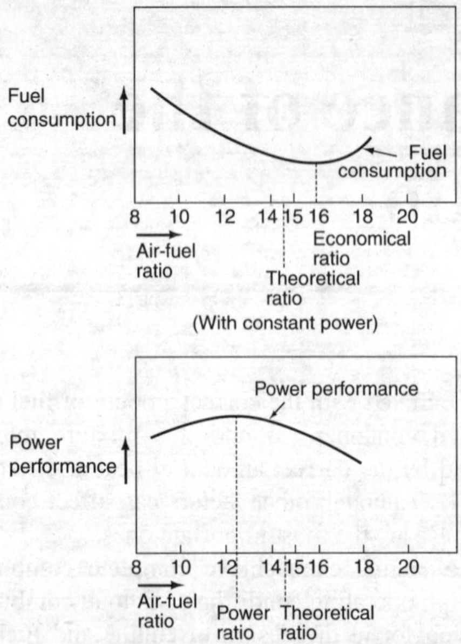


**Figure 1** *The air-fuel ratio must change to meet the requirements of various operating conditions of the engine.*

for high power is about 12 parts of air to 1 part of fuel **(Figure 1)**. The ratio for maximum economy is 15:1–16:1, while maximum power occurs from an air-fuel ratio of 12:1–12.5:1 **(Figure 2)**.

Not only is the air-fuel ratio a major factor in engine performance, but it is also a common element in the formation of each of the three main pollutant gases: hydrocarbons (HC), carbon monoxide (CO), and nitrogen oxide ($NO_x$). Each of these emissions can be caused by improper air-fuel ratio during the combustion process **(Figure 3)**. Maintaining the correct air-fuel ratio has therefore become one of the main goals of emissions-control system design and has ultimately resulted in today's computer-controlled fuel-injection systems.

| Engine Operating Condition | Air-Fuel Ratio (Air:Fuel) |
|---|---|
| Starting (Air temperature approx. 0°C) | Approx. 1:1 |
| Starting (Air temperature approx. 20°C) | Approx. 5:1 |
| Idling | Approx. 11:1 |
| Running slow | 12–13:1 |
| Accelerating | Approx. 8:1 |
| Max. output (full load) | 12–13:1 |
| Running at medium (economical) speed | 16–18:1 |

**Figure 2** *Different air-fuel ratios for different operating conditions.*

An air-fuel mixture that is too **lean** can result in excess levels of HC and $NO_x$ (depending on other circumstances). A lean mixture results if excess oxygen is mixed with the gasoline. An excessively lean mixture can also cause a **lean-burn misfire**. The lean-burn misfire results when a given volume of fuel mixture simply does not contain enough gasoline to ignite. Since there is too little fuel in the mixture to ignite, the fuel in the combustion chamber passes through the exhaust system unburned. This results in a cylinder that produces reduced power output and high HC levels. In addition, when the fuel mixture is lean, but not too lean to burn, the excess oxygen produces such a hot flame that combustion temperatures become excessive. These high temperatures cause nitrogen in the intake air to combine with oxygen to form harmful $NO_x$.

A **rich** mixture may be desirable under certain operating conditions. A rich air-fuel ratio contains excessive fuel in relation to the amount of oxygen entering the cylinders. When the engine is under load, the mixture must be enriched (a 12:1 ratio, for example) in order to produce enough energy to move the vehicle. In addition, when the engine starts it requires a richer air-fuel mixture. The colder the engine temperature, the richer the mixture must be.

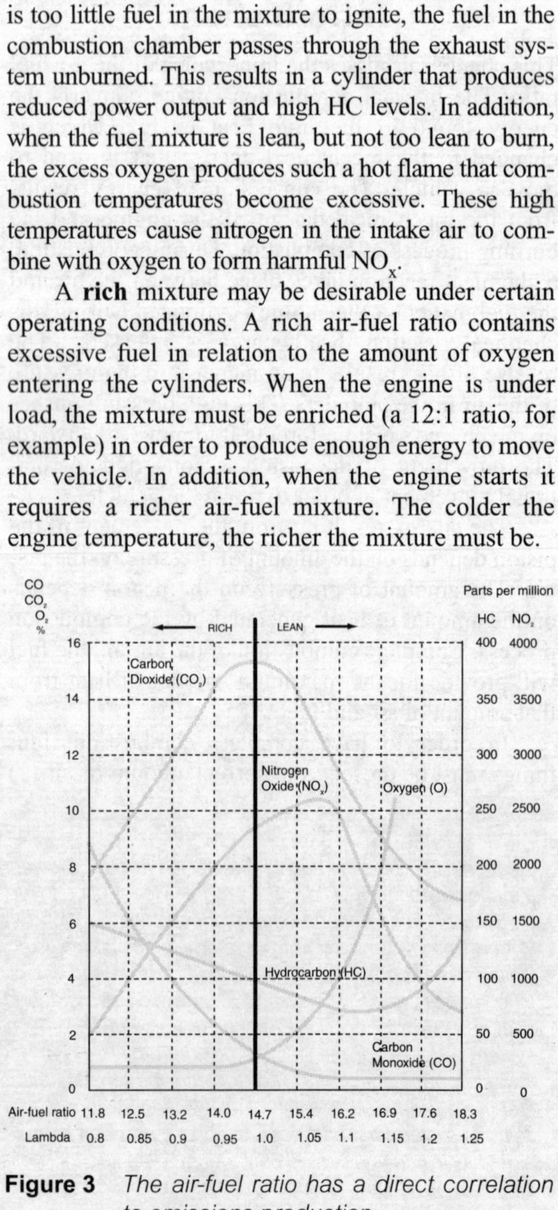


**Figure 3** *The air-fuel ratio has a direct correlation to emissions production.*

On EFI systems, the mixture is enriched for cold engine operation by increasing the injector pulse width. The amount of pulse width increase is based on the input from the coolant temperature sensor signal to the computer. The PCM supplies the proper air-fuel ratio and engine rpm for starting a cold engine. This also eliminates the need for the driver to depress the accelerator pedal while starting the engine.

> **You Should Know**
>
> *When an EFI-equipped engine is cold, the computer provides a very rich air-fuel ratio for faster starting. However, if the engine does not start because of an ignition defect, the engine floods quickly. Under this condition, excessive fuel may run past the piston rings into the crankcase. Therefore, when a cold EFI engine does not start, periods of long cranking should be avoided.*

If the driver suspects that the air-fuel ratio is extremely rich, the driver may depress the accelerator pedal to the wide-open position while cranking the engine. Under these conditions, the computer program provides a very lean air-fuel ratio. Most manufacturers have programmed their PCMs to shut down injector operation altogether during this time. This mode is referred to as a **clear flood**. However, under normal conditions, the driver should not push on the accelerator pedal when starting an engine with EFI. When the engine is decelerated, the computer reduces injector pulse width to provide a lean air-fuel ratio, which reduces emissions and improves fuel economy.

> **You Should Know**
>
> *Starting a fuel-injected engine without depressing the accelerator pedal is called no-touch starting.*

Although there are circumstances that require a richer air-fuel mixture, a mixture that is too rich (9:1, for example) can lead to reduced fuel economy and increased levels of HC and CO. However, $NO_x$ emissions will generally be low. Interestingly, HC emissions can be caused both by an excessively lean and an excessively rich mixture. In the case of a lean mixture, there is not enough gasoline to ignite, and it passes unburned through the exhaust system. In the case of a very rich mixture, there may not be enough oxygen present to support combustion. The end result is the same—unburned HC goes out the tailpipe.

High CO levels typically indicate a rich fuel mixture, especially if high HC levels are also present. In a rich mixture, there is insufficient oxygen to combine with the carbon atoms from the gasoline to form harmless carbon dioxide ($CO_2$). Instead, each carbon atom combines with only one oxygen atom to form poisonous CO.

$NO_x$ levels will usually be low in a rich-mixture situation because the shortage of oxygen results in lower combustion temperatures. Remember, $NO_x$ is formed when nitrogen from the air combines with oxygen at high temperatures.

## The Stoichiometric Ratio

At this point it is obvious that there must be an optimum air-fuel mixture. The optimum air-fuel ratio is called the **stoichiometric** (pronounced stoy-kee-o-MET-ric) ratio. The name "stoichiometric" comes from Greek words meaning "measured element." When the air-fuel ratio is at the stoichiometric ratio, every fuel molecule combines with every oxygen molecule with nothing left over (theoretically, at least). In a gasoline fuel system, the stoichiometric ratio is an air-fuel mixture of 14.7:1. This indicates that for every 14.7 pounds of air entering the air intake, the fuel injectors supply 1 pound of fuel. Mixing air and fuel at the stoichiometric ratio of 14.7:1 is the single most important technique used to control emissions levels. At the stoichiometric air-fuel ratio, combustion is most efficient and nearly all the oxygen in the air is mixed with fuel and burned in the combustion chamber. Computer-controlled fuel-injection systems maintain the air-fuel ratio at the stoichiometric ratio under most operating conditions. As the air-fuel ratio cycles from slightly rich or slightly lean to the stoichiometric ratio, the oxygen sensor ($O_2S$) voltage cycles from high to low. The computers **feedback** fuel system tailors the air-fuel mixture to adapt to various engine conditions. Feedback, as it relates to fuel systems, means the system uses an $O_2S$ to provide feedback information to the computer concerning combustion quality.

The stoichiometric ratio is the point at which the three pollutant gases are at controllable levels. At this ratio, there are exactly enough oxygen atoms to combine with the fuel molecules in the combustion chamber.

In working with air-fuel ratios, the term lambda refers to a number that compares the actual air-fuel ratio with the stoichiometric ratio:

Lambda = Actual ratio/Stoichiometric ratio

The lambda value is derived from the stoichiometric air-fuel ratio. The stoichiometric ratio is 14.7:1 when expressed as an air-fuel ratio or 1 when expressed as a lambda value. When the mixture is rich, lambda is less than 1. When the mixture is lean, lambda is greater than 1.

As the air-fuel mixture moves away from stoichiometric (either too lean or too rich), a number of conditions could occur, including emissions test failures and engine problems **(Table 1)**.

The quest for a reliable method of achieving stoichiometry has led to the evolution of fuel-delivery systems from simple carburetion to today's computer-controlled multiport fuel injection. There have been interim developments along the way, such as electronic feedback carburetors and mechanical fuel injection. However, conditions inside the engine's combustion chamber are not ideal. Even with a stoichiometric ratio, the engine's exhaust gases contain a certain percentage of unburned fuel in the form of HC and CO. In addition, at higher combustion temperatures, some of the free oxygen and nitrogen gases combine, forming various oxides of nitrogen. Obtaining stoichiometric efficiency makes the vehicle's emissions system more efficient at controlling these leftover emissions.

## Open Loop versus Closed Loop

**Open loop** means the oxygen sensor is out of the fuel strategy loop. The $O_2S$ is ignored during open-loop operation. In a computer-controlled fuel system, an open-loop condition occurs when the engine coolant is cold or the $O_2S$ signal is ignored by the PCM. When the engine is first started, the $O_2S$ is too cold to produce a satisfactory signal and the computer program controls the air-fuel ratio without the $O_2S$ input.

Open loop is also the computer strategy used for some driving conditions. The engine will always use the open-loop fuel strategy when it is first started (regardless of coolant temperature). Open-loop strategy is also used during wide-open throttle operation to provide a rich air-fuel mixture to meet the load requirements. Also, open-loop strategy is used when the throttle is closed quickly. Under this condition, the injectors may be turned off briefly, causing a very lean air-fuel mixture to prevent unburned fuel from exiting the exhaust.

Once the $O_2S$ warms sufficiently from the heat of the exhaust gases (and by an internal heating element) and it is able to provide an accurate voltage signal, the computer enters the **closed-loop** mode. In closed-loop mode, the computer uses the $O_2S$ Signal to control the air-fuel ratio. Determination of closed-loop operation varies among manufacturers. Some manufactures base closed-loop operation on $O_2S$ activity along with throttle position sensor (TPS) and manifold absolute pressure (MAP) or mass airflow (MAF) inputs. Some manufacturers use a closed-loop timer that puts the PCM into closed loop after a certain amount of engine run time has elapsed. The length of time is based on the engine coolant temperature (ECT) sensor value at engine start-up. In addition, a few systems do not enter the closed-loop

| Air-Fuel Mixture | Effect on Vehicle |
|---|---|
| Too lean | Increased $NO_x$ emissions<br>Poor engine power<br>Misfiring at cruising speeds<br>Burned valves<br>Burned pistons<br>Scored cylinders<br>Spark knock or ping |
| Slightly lean | Low exhaust emissions<br>High gas mileage<br>Reduced engine power<br>Slight tendency to knock or ping |
| Stoichiometric | Best all-around performance and emissions levels |
| Slightly rich | Increased CO emissions<br>Increased HC emissions<br>Maximum engine power<br>Higher fuel consumption<br>Less tendency to knock or ping |
| Too rich | Increased CO emissions<br>Increased HC emissions<br>Poor fuel mileage<br>Misfiring<br>Oil contamination<br>Black exhaust |

**Table 1** *Effects of air-fuel ratio on emissions and driveability.*

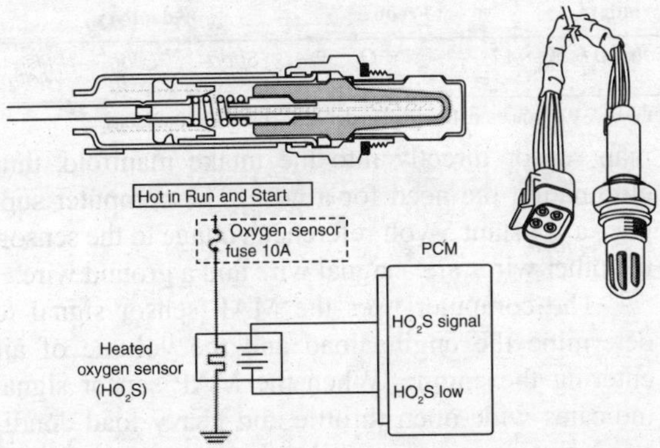


**Figure 4** *The heater element gets the $O_2S$ hotter faster and keeps the PCM in closed-loop mode during extended idle.*

mode until the ECT sensor signal informs the computer that the coolant temperature is approximately 175°F (79.4°C) or hotter and the $O_2S$ signal is valid.

Therefore, the ECT sensor signal is very important because it determines the open- or closed-loop status. For example, if the engine thermostat is defective and the coolant temperature never reaches 175°F (79.4°C), the computer-controlled fuel system never enters the closed-loop mode. When this open-loop condition occurs, the air-fuel ratio is rich, fuel economy is reduced, and emissions are increased. Some systems go back into open-loop mode during prolonged periods of idle operation when the $O_2S$ cools down. Also, most systems revert to the open-loop mode at or near wide-open throttle to provide a richer air-fuel ratio.

To keep the computer in closed-loop mode during idle and to heat the $O_2S$ faster, most (since 1993) oxygen sensors are heated electrically **(Figure 4)**. Battery voltage is sent to the heater element when the ignition switch is in the Run position. The heater element is a positive temperature coefficient (PTC) resistor that operates as a current regulator. When the $O_2S$ is hot, the PTC resistance increases thus, this resistance causes a reduction in current flow through the heater element.

## Computer-Controlled Air-Fuel Ratio Strategy

In an EFI system, the computer must know the amount (mass) of air entering the combustion chamber so it can determine the proper amount of fuel required to maintain the stoichiometric air-fuel ratio. Since the amount of air entering the combustion chamber is a constantly changing value, fast-responding onboard computer systems are needed. The computer uses an oxygen sensor to measure the oxygen content of the exhaust. This provides information concerning how well the computer did in calculating the correct air-fuel mixture for the previous burn. The computer then works to maintain driveability while maintaining the 14.7: 1 ratio as closely as possible.

Injector **pulse width** determines the amount of fuel that will be delivered to the combustion chamber. Pulse width is the length of time the computer energizes the injector. Under most closed-loop operating conditions, the computer provides the correct injector pulse width to maintain the stoichiometric air-fuel ratio. For example, the computer might ground the injector for 2 ms at idle speed and 7 ms at part throttle to provide the stoichiometric air-fuel ratio. There are two common strategies or methods manufacturers use to maintain the stoichiometric ratio: speed density and mass airflow (air density). Regardless of the method used, the computer must know the amount of air entering the engine in order to provide the proper pulse width.

## Speed-Density Systems

**Speed density** refers to the fuel-injection system that uses engine speed and **manifold absolute pressure (MAP) sensor** inputs to infer the mass of air entering the combustion chamber and respond based on programmed look-up tables in the PCM's memory chips. The speed-density fuel-injection system uses the rpm input as the most important sensor. Without the rpm input, the engine cannot run. The MAP sensor provides a varying voltage or frequency with engine load. The MAP sensor is a piezoresistive or capacitance discharge sensing device used to determine the load on the engine by sensing the difference between atmospheric pressure and engine vacuum. It is the second most important sensor in this type of system. Atmosphere pressure vacuum are used to calculate air density entering the combustion chamber. This calculated value is applied to a formula to determine the amount of fuel to be injected.

In addition, the rpm input determines the injection frequency (how many times the injector is fired) while the MAP sensor determines the greatest amount of fuel (pulse width). So the rpm signal determines how often the engine gets fuel and the MAP sensor determines how much fuel the engine gets. The other sensors/switches also have an effect on pulse width, but the MAP sensor determines the major amount (base pulse width). This type of EFI system is referred to as a speed-density system because the computer calculates the air intake flow from the engine rpm input and the density of the intake charge through the manifold vacuum input. Therefore, the computer must

| Load | Base Pulse Width | Feedback | Adaptives |
|---|---|---|---|
| *(RPM ÷ MAX RPM) × (MAP ÷ BARO)* | *× TPS × ECT × IAT × Sensed B off × LT** | *× $O_2$* | *× Short Term × Long Term = Pulse Width* |

* Long-term adaptive information is stored in memory and becomes part of the base PW equation. It is used under ALL operating conditions.

have accurate signals from these inputs to maintain the correct air-fuel ratio. The computer uses the other inputs to "fine tune" the air-fuel ratio. For example, if the TPS input indicates sudden acceleration, the computer will supply a richer air-fuel ratio. Following is the formula for a typical speed-density pulse-width calculation. The computer will assign a multiplicative value to the input signal and then determine the pulse width based on this value. The following is the pulse-width calculation used by the PCM:

The PCM requires crankshaft and camshaft position sensor inputs to determine when to fire the injectors. The crankshaft position sensor supplies engine speed information to the PCM. This information is required to determine if the engine is starting or is in the run mode. In addition, the PCM calculates the volume of fuel needed from the rpm signal (and other sensors). The rpm signal is used to determine how much air can enter the combustion chamber under that particular engine speed. All EFI systems use some method of determining the volume of airflow. Most systems will not start if the crankshaft speed sensor input is missing. However, some systems will start and run the engine as long as it receives a signal from either the crankshaft or the camshaft position sensor.

Most MAP sensors are a form of Wheatstone bridge stress gauge. The MAP sensor is usually mounted in the engine compartment. A hose may be connected from the intake manifold to a vacuum inlet on the MAP sensor. Three wires are connected from the sensor to the computer. Some systems insert the map sensor directly into the intake manifold, thus eliminating the need for a hose. The computer supplies a constant 5-volt reference voltage to the sensor; the other wires are a signal wire and a ground wire.

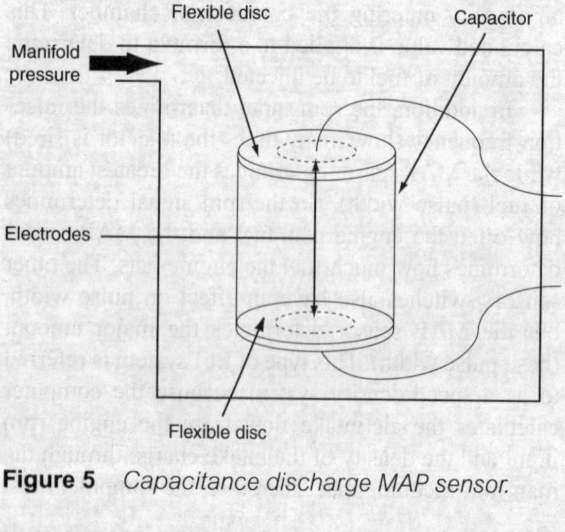


**Figure 5** *Capacitance discharge MAP sensor.*

The computer uses the MAP sensor signal to determine the engine load and the volume of air entering the engine. When the MAP sensor signal indicates wide-open throttle and heavy load conditions, the computer provides a richer air-fuel ratio. The computer will supply a leaner air-fuel ratio if the MAP sensor signal indicates a light load and moderate cruising speed conditions.

Many MAP sensors also perform the function of a barometric pressure (BARO) sensor. When the ignition switch is turned on, but before the engine is started, the MAP sensor signal informs the computer regarding atmospheric pressure. Atmospheric pressures vary in relation to altitude and atmospheric conditions such as humidity. Some applications have a separate barometric pressure sensor.

Many manufacturers use MAP sensors that contain a silicon diaphragm. When the manifold vacuum stresses the silicon diaphragm, an analog voltage signal is sent from the sensor to the computer in relation to the amount of vacuum. On a typical MAP sensor, the signal voltage changes from 1 volt to 1.5 volts at idle to 4.5 volts at wide-open throttle.

Another variation of the MAP sensor is a capacitance discharge MAP sensor. Instead of using a silicon diaphragm, this sensor uses a variable capacitor. In the capacitor capsule-type MAP sensor, two flexible alumina plates are separated by an insulating washer **(Figure 5)**. A film electrode is deposited on the inside surface of each plate and a connecting lead is extended for external connections. The result of this construction is a parallel plate capacitor. This capsule is placed inside a sealed housing that is connected to intake manifold pressure. As manifold pressure increases (goes toward atmospheric), the alumina plates deflect inward, resulting in a decrease in the distance between the electrodes. Since the formula for capacitance is:

$$C = \frac{\epsilon_0 A}{d}$$

Where

$\epsilon_0$ is the dielectric constant of air
$A$ is the area of the film electrodes
$d$ is the distance between electrodes

Since every value in the formula is constant except for the distance between the electrodes, a measurement of capacitance constitutes a measurement of pressure.

Ford MAP sensors contain a pressure-sensitive disc capacitor **(Figure 6)**. This type of MAP sensor changes manifold pressure to a varying frequency digital voltage signal. The MAP sensor actually senses the difference between atmospheric pressure and manifold vacuum. When the engine is idling, high intake manifold vacuum is about 18 in. Hg. Under this condition, the MAP sensor signal is approximately 95 Hz.

> **You Should Know**
>
> *Ford refers to this condition as low MAP because there is a greater difference between atmospheric pressure and manifold vacuum.*

If the engine is operating at or near wide-open throttle, the manifold vacuum may be 2 in. Hg and the MAP sensor signal will be approximately 160 Hz.

> **You Should Know**
>
> *This condition may be called high MAP because the manifold vacuum is closer to atmospheric pressure.*

Most manufacturers will use a limp-in value if the PCM considers the MAP sensor signal invalid. Limp-in values are used to keep the engine running. However, engine performance suffers greatly under these conditions. Usually the predetermined value is based on TPS and engine speed inputs.

## Mass Airflow Systems

EFI systems that use a **mass airflow (MAF) sensor** to determine air volume are referred to as **mass airflow (MAF)** systems. The MAF sensor is located in the air intake system. It measures the temperature and mass of air entering the engine as well as the altitude for the PCM. The mass of a given amount of air is calculated by multiplying its volume by its density and is expressed as grams per second. The denser the air, the more oxygen it contains. From a measurement of mass, the PCM modifies injector pulse width for the oxygen content in a given volume of air.

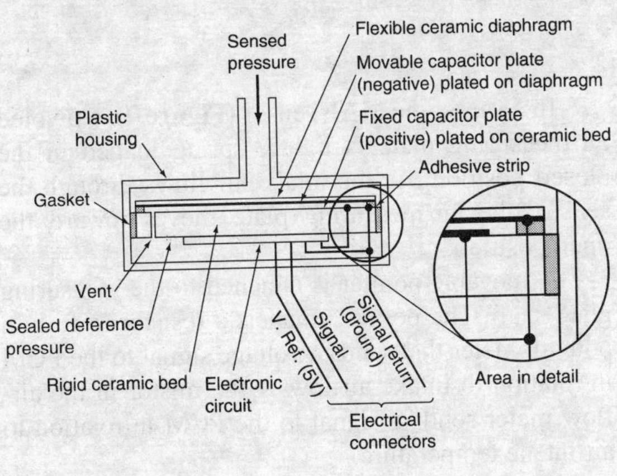


**Figure 6** *Typical variable capacitor sensor construction.*

Like the speed-density system, the MAF system uses the rpm input, the most important sensor. The MAF is the second most important sensor in this system.

The following is a list of typical sensors used in the air-density system and their order of importance:

> **You Should Know**
>
> *The MAF system is also referred to as an air-density system.*

> **You Should Know**
>
> *Each manufacturer's engine package is unique. The values provided in the following explanation of inputs are offered as examples based on the actual strategies used. The amount of influence an input has on the air-fuel mixture and ignition timing will vary.*

1. **Crank angle sensor (rpm).** In most cases, the engine cannot run without this sensor. First, this sensor is used for an rpm input. With that information, the PCM can determine ignition coil dwell and timing. Second, the PCM uses this sensor to start the injector firing.
2. **Mass airflow sensor.** This sensor is the load sensor. It "counts" the air going into the engine. The PCM then adds fuel based on the MAF sensor's output. A higher airflow will result in more fuel, while a lower airflow will result in less fuel. The MAF sensor affects the major amount of fuel (base pulse width). It is important to note that the MAF sensor's output also has a major influence on ignition timing. If the load is high (a large amount of airflow), the PCM will give the engine less advance than if there is a light load on the engine. Load is based on MAF, TPS, and rpm.
3. **Throttle position sensor.** Based on the TPS input, the PCM can increase fuel by 500 percent (quickly increasing TPS voltage) or decrease fuel by 70 percent (during deceleration). Also,

based on the TPS input, the PCM can lower the timing advance as the TPS voltage increases. If the TPS voltage is steady, it will have no effect on fuel modification.

4. **Engine coolant temperature sensor.** Based on the ECT input, the PCM can increase fuel by 60 percent at –22°F (–30°C) and will have no effect on fuel at 176°F (80°C). The ETC input can increase the timing by 14 degrees at –22°F (–30°C) and will have no effect on timing at 95°F (35°C).
5. **Barometric pressure sensor.** Based on the BARO input, the PCM can increase fuel by 12.5 percent (below sea level) or decrease fuel by 50 percent when above 8,000 ft. (2.4 km) in altitude. The BARO sensor input can also add 7 degrees of timing when above 7,000 ft. (2.1 km) in altitude and will have no effect on timing below 1,500 ft. (0.45 km).
6. **Intake air temperature (IAT) sensor.** Based on the IAT input, the PCM can increase fuel by 23 percent at 26°F (–3°C) air temperature or decrease fuel by 17 percent at 185°F (85°C) air temperature. This sensor will have no effect on fuel at 73°F (23°C) air temperature. This sensor has no effect on timing.
7. **Oxygen sensor.** The $O_2S$ can increase fuel by 17 percent (lean condition/too much oxygen) or decrease fuel by 17 percent (rich condition/lack of oxygen). This sensor has no effect on timing. The $O_2S$'s output is only used when the engine is in closed-loop mode.
8. **Adaptive memory.** This value can add or take away fuel and it is always used in the pulse-width formula. It comes from the $O_2S$ switching activity over a period of time.

The following is a typical formula the computer uses to calculate pulse width in an air-density system:

$$\text{Base pulse width} = \frac{\text{Airflow Hz}}{\text{Engine rpm}} \times \text{TPS} \times \text{coolant} \times \text{BARO} \times \text{Intake air temp} \times O_2 \text{ (closed loop)} \times \text{Air fuel commpensator (Open loop)} \times \text{Adaptive memory} \times \text{Detonation sensor (turbo)}$$

$$\text{Total pulse width} = \text{Base pulse Width} \times \text{TPS (accel)} + \text{injector lag time}$$

Some manufacturers use both MAF and MAP sensors. In such cases, the MAP sensor is used mainly as a backup if the MAF sensor fails. It is also used to determine faults. In some four-cylinder applications using the MAF strategy, the TPS input is unique in that it does not affect the fuel injector's base pulse width. Instead, the PCM (based on TPS input) will add four additional injector firings based on the rate of TPS increase. If the TPS input identifies a rapid opening of the throttle plates, the PCM will add four additional injector pulses (one to each cylinder). The extra pulses occur every 10 ms as long as the TPS input increases at a rapid rate. The ECT inputs determine the amount of fuel delivered by the extra pulses. However, the maximum pulse width is 4 ms. The effect of TPS inputs on fuel strategy is listed below:

| TPS Voltage | Multiplicative Value |
|---|---|
| Steady | 1.0 (no effect) |
| Rapidly increasing | 5.0 (increase pulse width by 500 percent) |
| Rapidly decreasing | 0.3 (decrease pulse width by 70 percent) |

Usually the BARO sensor input has no effect on fuel strategies at sea level. The authority of a typical BARO sensor over fuel ranges from reducing pulse width by about 50 percent to increasing it by 12.5 percent. This means at higher altitudes this sensor input can reduce the calculated base pulse width by half. If the engine is operating below sea level, more fuel will be added.

**Mass Airflow Sensors.** The MAF sensors may be classified as vane, heated grid, hot wire, pressure, or ultrasonic. The MAF sensor is usually mounted in the hose between the air cleaner and the throttle body so that all the intake air must flow through the sensor.

**You Should Know**

*Some vane-type MAF sensors are referred to as volume airflow meters.*

In a vane-type MAF sensor **(Figure 7)**, a pivoted air-measuring plate is lightly spring loaded in the closed position. As the intake air flows through the sensor, the air-measuring plate moves toward the open position.

A movable pointer is attached to the measuring plate shaft. The pointer contacts a resistor to form a potentiometer that sends a voltage signal to the PCM in relation to intake airflow. A thermistor in the airflow meter sends a signal to the PCM in relation to air intake temperature.

The heated resistor-type MAF sensor has a heated resistor mounted in the center of the air

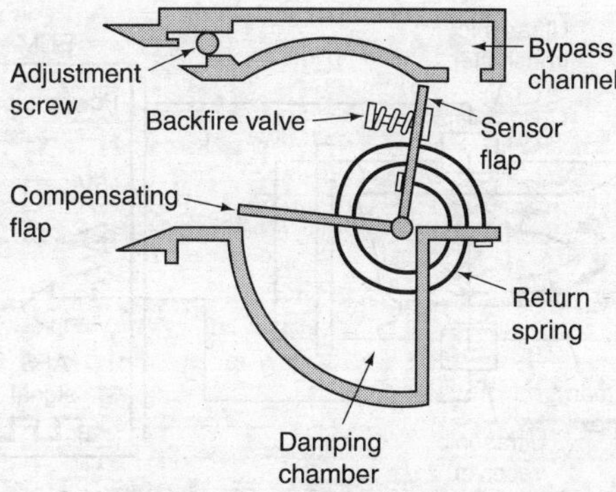


**Figure 7** *A vane-type mass airflow sensor used to measure the volume of air entering the engine.*

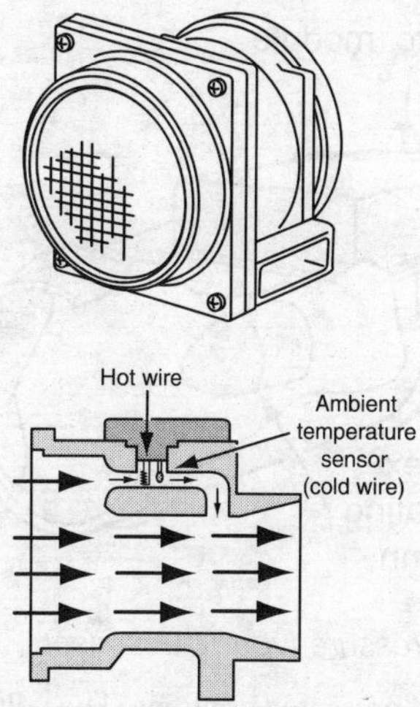


**Figure 8** *Hot-wire-type MAF sensor.*

passage. In addition, an electronic module is mounted on the side of the sensor. When the ignition switch is turned on, voltage is supplied to the module. The module sends current through the resistor to maintain a specific resistor temperature.

If an engine is accelerated suddenly, the rush of air tries to cool the resistor. Under this condition, the module supplies more current to maintain the resistor temperature. The module sends the increasing current signal (proportional to the airflow entering the engine) to the PCM. When the PCM receives this increasing current flow signal, it increases the amount of fuel delivery to go with the additional airflow entering the engine. The MAF module reacts in a few milliseconds to maintain the resistor temperature. Some MAF sensors have an electric grid in place of the resistor.

In a hot-wire-type MAF sensor **(Figure 8)**, a hot wire is positioned in the air stream through the sensor and an ambient temperature sensor wire is located beside the hot wire. The ambient temperature wire senses intake air temperature and may be referred to as a cold wire .

When the ignition switch is turned on, the MAF module sends enough current through the hot wire to maintain the temperature of this wire at 392°F (200°C) above the ambient temperature sensed by the cold wire. If the engine is accelerated suddenly, the rush of air tries to cool the hot wire. The module immediately sends more current through the wire to maintain the temperature of the wire. The module sends the increasing current signal(proportional to the intake airflow to the PCM). When this signal is received, the PCM supplies more fuel to go with the increased intake airflow. Some MAF sensors have a burn-off relay and related circuit. When the ignition switch is turned off, after the engine has been running for a specific length of time, the computer closes the burn-off relay. This relay activates a burn-off circuit in the MAF that heats the hot wire to a very high temperature (1,832°F, 1,000°C) for a short time period. This action burns contaminants off the hot wire.

Another version of the hot-wire-type MAF is the hot film. A nickel foil sensor is maintained at 147°F (75°C) above ambient temperature. This type of sensor does not require a burn-off period.

The pressure-type MAF sensor uses the Karman vortex phenomenon to determine the volume of air entering the engine. A vortex-generating column is located in the path of the airflow **(Figure 9)**. Vortices are generated in proportion to airflow speed (air volume). The Karman vortex occurs when a stream of air encounters an obstruction and separates as it moves around the object. In the MAF sensor the flow path is obstructed by a strut (vortex generating column) that creates vortex swirls that are formed alternately on either side of the strut. The swirls of air create a local increase in pressure and a local decrease in velocity on one side of the strut. At the same time, a local decrease in pressure and a local increase in velocity occurs on the other side of the strut. After shedding a swirl from one side, the process is reversed and a vortex or a swirl is shed from the other side. The frequency of this alternating shedding process is proportional to the velocity of the air as it flows past the strut.

A pressure inlet is positioned downstream from the vortex-generating column **(Figure 10)**. The pressure inlet is connected to a pressure sensor (stress gauge). The vortices generated by the column result in pressure variations that the pressure sensor detects. As the airflow through the MAF sensor increases, so does

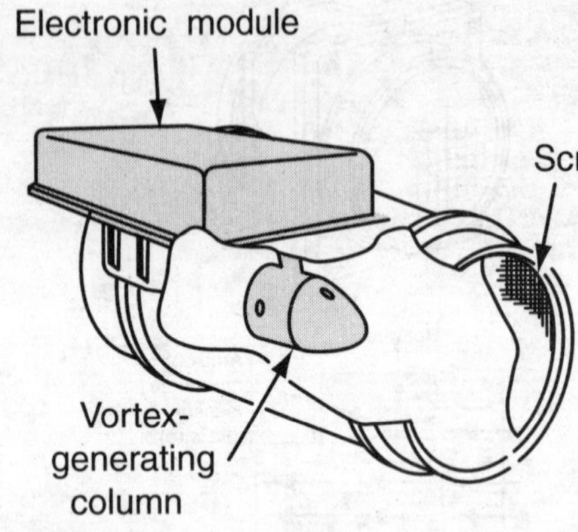


**Figure 9** *Pressure-type MAF sensor.*

the number of pressure variations. The voltage changes resulting from the pressure variations are changed to a pulsed digital signal that is proportional to intake airflow. The digital signal is then sent to the PCM.

The ultrasonic MAF sensor also uses the Karman vortex phenomenon. The difference is this sensor uses ultrasonic waves to measure airflow. An ultrasonic transmitter is positioned downstream of the vortex-generating column. The transmitter is located across from an ultrasonic receiver **(Figure 11)**. The sensor determines the volume of intake air by measuring the length of time required for the ultrasonic waves from the transmitter to reach the receiver.

When the ignition switch passes through the Run position on the way to the Start position, there is no airflow through the sensor. During this phase, a fixed reference time for the ultrasonic transmission is established. A clockwise-rotating vortex speeds up the ultrasonic waves and they reach the receiver in less time than the reference time. A counterclockwise-rotating vortex tends to slow the ultrasonic waves. The resulting signal is then converted into digital pulses and sent to the PCM. The greater the number of pulses, the greater the volume of air.

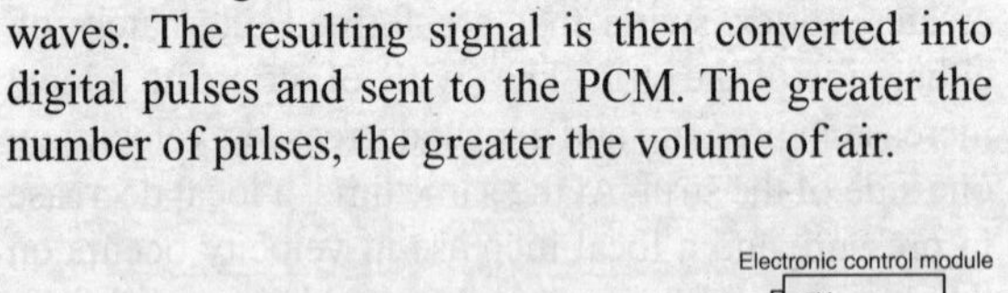


**Figure 10** *Pressure-type MAF sensor operation.*

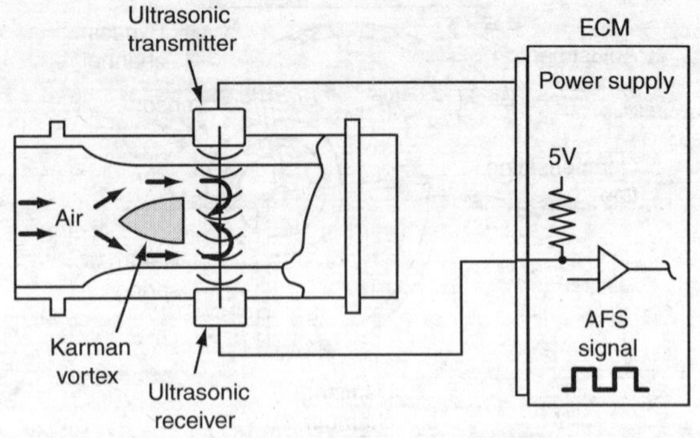


**Figure 11** *Ultrasonic mass airflow sensor.*

**Barometric Pressure Sensors.** Many systems use a BARO sensor to determine the amount of fuel delivery needed to start the engine and for spark-timing strategies. In some applications, the BARO sensor is part of the MAF sensor. As its name implies, this sensor measures atmospheric pressure. At lower pressures, the engine requires more air for efficient combustion. At higher pressures, it requires less air. Basically, the BARO sensor operates in the same way as the MAP sensor discussed earlier. The main difference is there is no vacuum source attached to the BARO sensor.

**Idle Air Control Motors.** Many TBI systems use a DC reversible motor to control idle speed. In this system, the pintle of the idle speed control (ISC) motor is located against a lever that is attached to the throttle plates. As the motor moves the pintle in and out, the throttle plates move. Moving the plate allows more or less air to enter the engine. For example, if idle speed needs to be increased, the PCM will activate the motor to move the pintle so the throttle plates will open slightly. At this time, more air is allowed to enter the engine. The $O_2S$ senses the high oxygen content and the PCM will increase the pulse width to

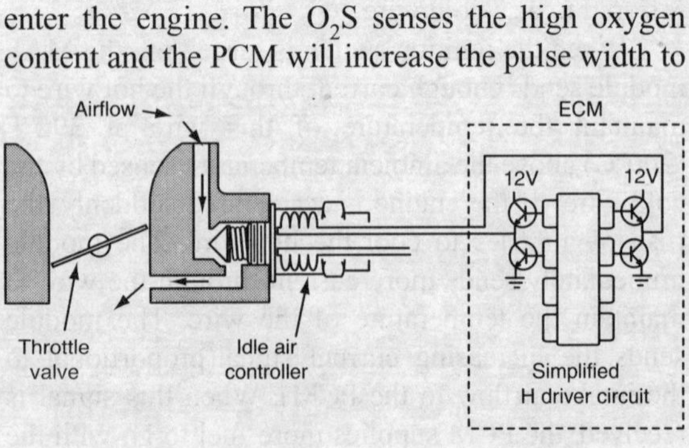


**Figure 12** *The H driver controls the direction of current, and thus the direction of motor rotation.*

maintain the stoichiometric ratio. With the increase of air and fuel, engine speed increases. The opposite will occur if idle speed needs to be decreased.

Newer EFI systems control idle speed by regulating airflow around the throttle plates. The idle air control (IAC) stepper motor is used to control the amount of air that bypasses the closed throttle plates.

Most IAC stepper motors use permanent magnet motors containing two to six electromagnets. Each command from the PCM results in one step movement. By switching which circuit supplies the voltage and which circuit supplies ground (through the use of an H driver that reverses polarity), the PCM can control the direction of motor rotation **(Figure 12)**. To decrease engine speed, the controller steps the motor in to close off the air bypass. If engine speed needs to be increased, the PCM drives the stepper motor pintle out to increase the size of the bypass.

Many manufacturers now use a solenoid to perform the function of idle air control. The solenoid is pulse-width modulated to control the air bypass. The use of the solenoid provides faster reaction time to obtain the desired results.

## EFI Modes of Operation

The following are typical modes of operation for most EFI systems. Variations of these modes are also used. For this reason, always reference the correct service manual when diagnosing a fuel system problem.

### Key On

Typically, when the ignition switch is turned to the On or Run position, the PCM receives a key "on" signal. The PCM then gathers the needed information in order to prepare the engine for starting. Usually the information required is:

- Barometric pressure
- Engine coolant temperature
- Throttle position
- Intake air temperature

Also during this time, the fuel pump is turned on for a short time to prime the system. This mode is known as initialization.

### Crank

When the ignition switch is in the Start position and the engine is cranking, the PCM receives crankshaft and camshaft position sensor signals. The PCM then turns on the fuel pump. Depending on the system, the PCM may also energize all of the injectors at the same time to provide the initial fuel to start the engine. This may be the only firing of the injectors until the rpm signal indicates the engine has started. The amount of fuel injected during this phase is based upon the ECT sensor input. Other systems may fire the injectors in order (sequentially) while the engine is being cranked. During the time the engine is being cranked, the PCM ignores the $O_2S$ input.

The TPS, MAP/MAF, and BARO sensors allow the PCM to monitor the amount of air delivered to the engine. Some EFI systems also receive an injector-sync signal at this point to inform the PCM which cylinder is approaching the compression stroke for proper fuel-injector sequencing.

### Open Loop

When the engine rpm increases to about 450 rpm, the PCM switches to "run" mode. Based on the input from the MAP or MAF sensor and engine speed inputs, the PCM determines base pulse width and generates ignition timing signals. Initially, the $O_2S$ input is ignored since the sensor is cold and a rich mixture was delivered during cranking to start the engine. The PCM delivers a predetermined air-fuel ratio and a higher idle speed to bring the engine up to operating temperature faster.

### Closed Loop

Once the PCM is satisfied that all required criteria have been met and it has determined That the $O_2S$ is warm enough to send valid readings, it switches to closed-loop operation. The PCM then adjusts the air-fuel mixture based on the $O_2S$ inputs.

### Acceleration

The rate of TPS voltage change is monitored to determine how much acceleration the driver is requesting. Based on the TPS input, along with input from the MAP or MAF sensor, the injection pulse width is increased to prevent hesitation and stumble. Based on the rate of acceleration, and how far the accelerator pedal is depressed, the system will stay in closed-loop operation.

### Wide-Open Throttle

The wide-open throttle (WOT) signal is received by the PCM from the TPS. MAP or MAF sensor inputs also identify a WOT condition. Upon recognition of a WOT condition, the PCM switches back to open-loop operation and ignores the $O_2S$ input. The main priority of the PCM during this mode is to increase the injector pulse width.

### Deceleration

Based on the response of the TPS when the accelerator pedal is released and on input from the MAP or MAF signals, the PCM may enter deceleration mode. Depending on the conditions and the programming, the PCM may turn off the injectors very briefly. The system will go into open-loop mode during this very lean condition.

## Review Questions

1. Describe the importance of the air-fuel ratio to engine performance and emissions levels.
2. Explain the difference between open-loop and closed-loop fuel strategies.
3. Explain the operation of the mass airflow (MAF) electronic fuel-injection (EFI) system.
4. The optimum air-fuel ratio is called the ____________ ratio.
5. The stoichiometric ratio of a gasoline engine is:
   A. 9.3:1
   B. 12.6:1
   C. 14.7:1
   D. None of the above
6. Technician A says that a rich air-fuel mixture will cause increased levels of hydrocarbons and carbon monoxide. Technician B says that a rich air-fuel mixture will increase nitrogen oxide emissions. Who is correct?
   A. Technician A only
   B. Technician B only
   C. Both Technician A and Technician B
   D. Neither Technician A nor Technician B
7. What is meant by stoichiometric?
8. Explain the operation of the speed-density fuel system.
9. Not only is the air-fuel ratio a major factor in engine performance, but it is also a common element in the formation of ________, ________, and ________.
10. Technician A says that the computer uses the TPS and ECT signals to determine the mass of air entering the engine in a speed-density system. Technician B says that the computer uses the TPS and $o_2$s signals to determine the mass of air entering the engine in an MAF system. Who is correct?
    A. Technician A only
    B. Technician B only
    C. Both Technician A and Technician B
    D. Neither Technician A nor Technician B
11. Technician A says that in an MAF system, the MAF sensor is used to measure the vacuum in the intake manifold to determine engine load. Technician B says that the rpm input is needed for the engine to start. Who is correct?
    A. Technician A only
    B. Technician B only
    C. Both Technician A and Technician B
    D. Neither Technician A nor Technician B

CHAPTER

# 19 Adaptive Memory

## Introduction

The PCM's goal is to use the input information and control outputs to provide a constant stoichiometric ratio. The operating modes discussed in Chapter 18 represent decision categories for the PCM. Within each of the modes of operation, smaller and more finite decisions are made. This chapter discusses the function of adaptive memory.

**Adaptive memory** is a component of the air-fuel ratio calculation. Adaptive memory is a function of the PCM to make corrections to the pulse-width calculation based on $O_2S$ activity. Within each operating cell, fuel pulse-width calculations occur. Usually a cell is defined by the MAP/MAF and rpm signals **(Figure 1)**. As discussed in Chapter 18, a set formula determines pulse width. If the $O_2S$ signals a rich or lean condition in a cell, the cell will update to aid in fuel control. Manufacturers use different methods to express the amount of correction. A common method is a percentage value change to the formula. For example, assume that the engine is operating in closed-loop mode (adaptive memory only updates in closed-loop mode) and the fuel pressure is a little low. The $O_2S$ reading will indicate a high oxygen content in the exhaust stream until the PCM increases the injector pulse width to return the air-fuel mixture to 14.7:1. The PCM records the percentage of pulse-width increase for that cell. Any time the same or different operating conditions reenter that cell, the PCM will use the percent of correction in its formula to determine pulse width.

| | Decel ← | | Cruise | | → Accel | | |
|---|---|---|---|---|---|---|---|
| Above 1,690 rpm | 1 | 3 | 5 | 7 | 9 | 11 | 13 |
| Below 1,690 rpm | 0 | 2 | 4 | 6 | 8 | 10 | 12 |
| MAP voltage | 1.38V | 2.00V | 2.64V | 3.26V | 3.90V | | Idle |

**Figure 1** *An example of memory cell construction based on engine rpm and MAP inputs.*

The following are examples of adaptive memory strategies used by various manufacturers. It is important to understand these principles since adaptive memory is used to run fuel system monitors in On-Board Diagnostic second generation (OBD-II) systems discussed later.

## Chrysler's Adaptive Strategies

When the fuel system enters closed-loop operation, there are two adaptive memory systems that begin to operate. The first system that becomes operational is the **short-term fuel trim**. The short-term fuel trim is an instantaneous correction to the pulse-width calculation. It is driven directly by the $O_2S$ activity **(Figure 2)**. As the $O_2S$ voltage switches in response to the air-fuel ratio changes, the PCM adjusts the amount of fuel delivered until the $O_2S$ reaches its switch point. When the switch point is reached, short-term adaptive begins with a quick kick. Then the adaptive correction slowly ramps up until the $O_2S$'s output voltage indicates the switch point in the opposite direction. Short-term adaptive will continue to increase and decrease the amount of fuel delivered based on the $O_2S$ input. For example, if the $O_2S$ output voltage goes toward 0 volt (high oxygen), short-term adaptive will start to add additional fuel until the $O_2S$ begins

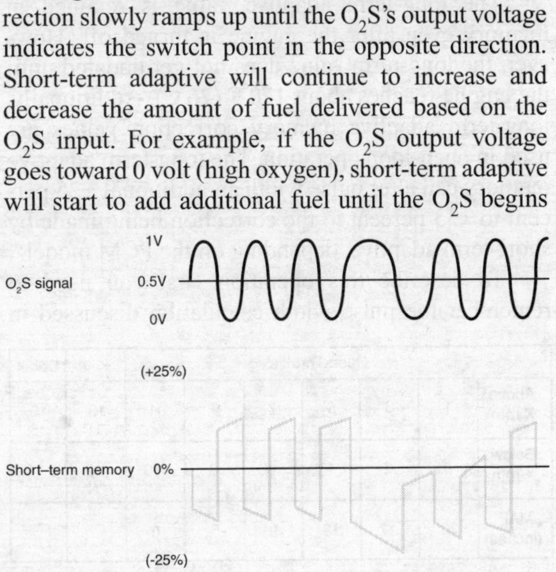


**Figure 2** *Short-term fuel trim is driven by $O_2S$ inputs.*

switching again. The maximum range of authority for short-term adaptive is ±25 percent to ±33 percent of base pulse width, depending on the PCM model.

Short-term adaptive memory resets to a value of zero when the engine is turned off. It will not update until the PCM enters closed loop. With normal short-term adaptive memory activity, the values jump back and forth crossing over zero.

The second system is called **long-term adaptive memory**. Long-term adaptive memory is a correction to the pulse-width calculation based on short-term activity. This memory uses a cell structure to better maintain correct emissions levels throughout all operating ranges **(Figure 3)**. In this illustration, there are sixteen cells; however, different engines may have fewer or more cells. Two of the cells are used only in idle, based on TPS and Park/Neutral switch inputs. Two cells are used for deceleration, based on TPS, rpm, and vehicle speed. The other twelve cells are based on rpm and load inputs.

As the engine enters a cell, the PCM monitors the short-term adaptive value. The goal of long-term adaptive memory is to keep the short-term adaptive value at zero. For example, short-term adaptive memory will correct for low oxygen content based on $O_2S$ inputs. After the short-term adaptive value reaches a predetermined threshold, long-term adaptive memory will start to correct in the same direction to bring the short-term adaptive value back to zero **(Figure 4)**.

The long-term adaptive value is retained in memory even after the engine is turned off. However, the long-term value does not get updated until the engine reaches about 170°F (76°C). Additionally, long-term adaptive memory correction values are used in open-loop operation. The long-term adaptive memory can alter pulse width an additional ±25 percent to ±33 percent to the correction being made by short-term adaptive, depending on the PCM model.

To describe this operation first you need to remember the pulse width calculation discussed in Chapter 18. The values that the inputs provide are assigned a multiplicative value. This is also true for the adaptives. If the short term adaptive value is 0 then its multiplicative is 1. If the value is +25%, then its multiplicative is 1.25, and if the value is −25% the multiplicative is 0.75. Assume that an engine is idling with a pulse width of 1.8 ms. This is the correct amount of fuel to maintain the stoichiometric ratio for the operating conditions. If the $O_2S$ indicates a high oxygen content that is suficient enough to require a full+33% correction from short-term adaptive, this would increase the injector pulse width to 2.39 ms (1.8 × 1.33). If this correction is not sufficient to return the engine to the stoichiometric ratio then long-term adaptive memory will start to make corrections. Let's say that it corrects an additional +33%. This would bring the total injector pulse width to 3.18 ms (2.39 × 1.33). Total correction is determined by reverting the percentages back to the multiplicative values. In this case we started with a pulse width of 1.8 ms and ended with one of 3.18 ms. Dividing 3.18 by 1.8 equals 1.7689, which is the multiplicative for the approximately 77 percent addition to the pulse width.

| | Open Throttle | | | | | | Idle | Decel |
|---|---|---|---|---|---|---|---|---|
| Above X rpm | 1 | 3 | 5 | 7 | 9 | 11 | 13 "D" | 15 |
| Below X rpm | 0 | 2 | 4 | 6 | 8 | 10 | 12 "N" | 14 |
| MAP (inches) | 20 | 17 | 13 | 9 | 5 | 0 | | |

**Figure 3** *Long-term adaptive memory cells.*

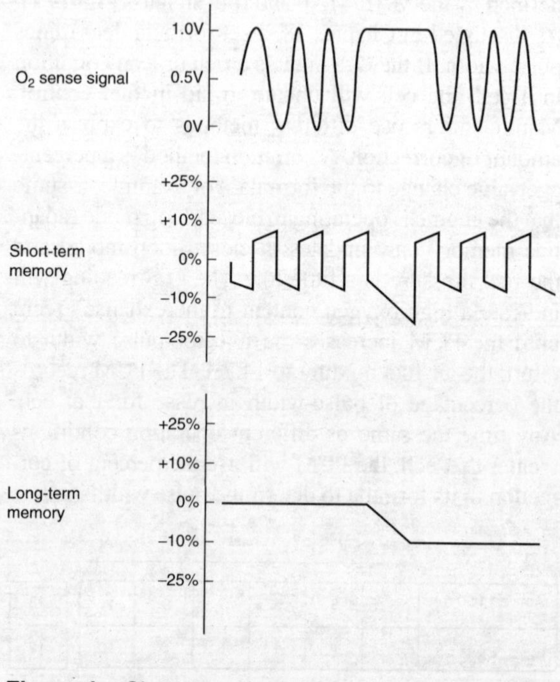


**Figure 4** *Short-term fuel trim and long-term adaptive memory activity based on $O_2S$ inputs. Long-term adaptive memory attempts to keep the short-term adaptive at zero.*

## General Motors's Integrator and Block Learn

Pre-OBD-II General Motors learning abilities are called **integrator** and **block learn**. Integrator, the first correction, represents a short-term correction to the pulse width. As the $O_2S$ senses a rich or lean condition, the integrator will attempt to correct for it by adjusting the pulse width. If integrator is being observed on a scan tool, the value is displayed by a number. The number displayed by the scan tool is a base 10 conversion of a binary code. If there is no correction, then the number is 128 (halfway between 0 and 255). A number greater than 128 represents increased pulse width over base. A number lower than 128 means pulse width is being reduced.

Block learn represents pulse-width adjustments that have become a trend over an extended period of time. Once the engine has been running in closed-loop mode long enough, the PCM will store the corrective factor (as a number) in memory. That number will be stored until additional corrections need to be made.

Block learn numbers are stored in sixteen different cells **(Figure 5)**. Each cell represents a different combination of engine speed and load. Integrator and block learn work together to correct pulse width. For example, if block learn is above 128 but the integrator is at 128, the system has a lean tendency that is being corrected by block learn, but the integrator is not having to make any additional changes. If block learn is at 128 but integrator is above 128, the system is attempting to run lean but it has not been doing so long enough for block learn to make any corrections.

**A** System operating at design specifications

| LOAD \ rpm | low | | | high |
|---|---|---|---|---|
| heavy | 128 | 128 | 128 | 128 |
| | 128 | 128 | 128 | 128 |
| | 128 | 128 | 128 | 128 |
| light | 128 | 128 | 128 | 128 |

**B** Result of slightly lean tendency

| LOAD \ rpm | low | | | high |
|---|---|---|---|---|
| heavy | 128 | 130 | 128 | 129 |
| | 129 | 130 | 130 | 131 |
| | 133 | 136 | 129 | 128 |
| light | 135 | 132 | 129 | 130 |

**C** Result of slightly rich tendency

| LOAD \ rpm | low | | | high |
|---|---|---|---|---|
| heavy | 121 | 124 | 128 | 128 |
| | 119 | 120 | 119 | 121 |
| | 115 | 117 | 116 | 121 |
| light | 122 | 119 | 126 | 120 |

**Figure 5** *Block learn information.*

Adaptive memory strategies for a General Motors OBD-II-certified vehicle are similar to those for the Chrysler system discussed earlier.

> **You Should Know**
>
> *On GM vehicles with an OBD-II, integrator is called short-term fuel trim and block learn is called long-term fuel trim.*

## Ford's Adaptive Strategies

Inside the PCM's ROM chip is a table that represents cells based on engine speed and load **(Figure 6)**. Each cell is assigned a base value of 1. Based on rpm and engine load, the PCM will determine which cell it is operating in. The value in that cell is then used to correct pulse width. The value is the multiplier used in pulse-width calculations. A value of 1 will not change pulse width over base.

| MAP | | | | | | |
|---|---|---|---|---|---|---|
| High | 1 | 1 | 1 | 1 | 1 | 1 |
| | 1 | 1 | 1 | 1 | 1 | 1 |
| | 1 | 1 | 1 | 1 | 1 | 1 |
| | 1 | 1 | 1 | 1 | 1 | 1 |
| Low | 1 | 1 | 1 | 1 | 1 | 1 |

**Figure 6** *Adaptive fuel-control table base values.*

| MAP | | | | | | |
|---|---|---|---|---|---|---|
| High | 1 | 1.2 | 1.1 | 0.95 | 0.9 | 1.0 |
| | 0.975 | 1.0 | 1.0 | 1.0 | 1.1 | 1.1 |
| | 0.95 | 1.0 | 1.0 | 1.3 | 1.2 | 1.2 |
| | 0.95 | 1.0 | 1.0 | 1.1 | 1.0 | 1.0 |
| Low | 0.9 | 1.1 | 1.1 | 1.1 | 1.0 | 0.9 |

**Figure 7** *Adaptive fuel-control table with adapted cell values.*

Notice this table is located in the computer's ROM chip. Since the values in ROM cannot be changed by the microprocessor, there is a second copy of the table in volatile RAM. The computer uses this second copy for corrections **(Figure 7)**. The adaptive control will add or subtract from 1. A number greater than 1 indicates an increase in pulse width over base; a number less than 1 represents a decrease in pulse width.

## Import Adaptive Memory Strategy

Many import manufacturers will use a variation of the strategies discussed; however, some do not report the adaptive value to the scan tool for the technician to review. The following is an example of the Mitsubishi adaptive memory.

The PCM recognizes the air-fuel mixture as being rich or lean based on which side of the $O_2S$ switch point (usually around 0.5 volt) the input voltage is. If the voltage is above the switching point, the system is running rich (low oxygen). A lean mixture (high oxygen) is recognized when the voltage is below the switching point. Once the system is in closed-loop mode, the upstream oxygen sensor(s) is used to update the short-term fuel trim. If the $O_2S$ stops switching, the PCM either lengthens or shortens the injector pulse width until the sensor begins switching again.

The only time emissions requirements are achieved is when the $O_2S$ is switching back and forth. The short-term fuel trim is used to make instantaneous corrections to the pulse width and is only updated in closed-loop mode. This correction factor is not stored in memory and is reset each time the ignition key is cycled to Off.

The short-term fuel trim, in turn, moves the correction factor to a long-term fuel cell, where it becomes part of the pulse-width equation. This memory location represents the amount of correction required by the vehicle under specific load to maintain $O_2S$ switching. The information stored in a long-term memory is nonvolatile so it is not lost during a key cycle.

The short-term fuel trim range correction can add 25 percent to the base pulse width for lean conditions. For rich conditions, the short-term trim value will have a maximum authority of −30.5 percent. Long-term fuel trim uses two cells with a range of authority from −17 percent to 17 percent.

## Review Questions

1. Describe the difference between short-term and long-term adaptive memory.
2. Technician A says that short-term fuel trim corrects fuel delivery in direct proportion to the voltage signals from the $O_2S$. Technician B says that on the General Motors integrator systems, a number greater than 128 represents an increase of pulse width over base. Who is correct?
   A. Technician A only
   B. Technician B only
   C. Both Technician A and Technician B
   D. Neither Technician A nor Technician B
3. Technician A says that block learn is an adjustment that represents a short-term correction to the pulse width. Technician B says that integrator represents pulse-width adjustments that have become a trend over an extended period of time. Who is correct?
   A. Technician A only
   B. Technician B only
   C. Both Technician A and Technician B
   D. Neither Technician A nor Technician B
4. Technician A says that in the Mitsubishi adaptive memory system, short-term fuel trim range correction can add 25 percent to the base pulse width for lean conditions. Technician B says that for rich conditions, the short-term trim value will have a maximum authority of −17 percent. Who is correct?
   A. Technician A only
   B. Technician B only
   C. Both Technician A and Technician B
   D. Neither Technician A nor Technician B
5. Short-term adaptive memory is directly driven by the ______________ activity.
6. Technician A says that long-term fuel trim updates only in closed-loop mode and when engine temperature is higher than the programmed values. Technician B says that long-term fuel trim is used only in open-loop strategy. Who is correct?
   A. Technician A only
   B. Technician B only
   C. Both Technician A and Technician B
   D. Neither Technician A nor Technician B
7. Explain Ford's adaptive strategy.

CHAPTER

# 20 Fuel-Injection Systems

## Introduction

Some European manufacturers have been using fuel injection for many years, but it was not until the 1980s that U.S. automotive engineers started to change engines from carburetors or computer-controlled carburetors to **electronic fuel-injection (EFI)** systems. This chapter discusses EFI systems. EFI refers to any fuel-injection system that is electronically controlled, including throttle body injection (TBI), multipoint fuel-injection (MFI), and sequential fuel-injection (SFI) systems. The automotive manufacturers made the change to EFI to improve fuel economy, performance, and emissions levels. They had to meet increasingly stringent **corporate average fuel economy (CAFE)** regulations and comply with stricter emissions standards at the same time. CAFE standards are federally imposed regulations requiring vehicle manufacturers to meet a specific average fleet fuel mileage for all their vehicles.

> **You Should Know**
>
> *The term electronic fuel injection was used by some manufacturers to describe their fuel systems. Today, this term is no longer used; it has been replaced with more specific descriptions. In this book, the term EFI is applied to any electronic or computer-controlled fuel-injection system.*

Many of the EFI systems of the 1980s were **throttle body injection (TBI)** systems in which the fuel was injected above the throttles. In a TBI system, the injector(s) are located in a throttle body assembly and above the throttle plates. Over time, the TBI systems have been changed to port fuel-injection (PFI) systems with the injectors located in the intake ports. On TBI engines, some intake manifold heating was required to prevent fuel condensation on the intake manifold passages. When the injectors are positioned in the intake ports, intake manifold heating is not required. The PFI system provided engineers with increased intake manifold design flexibility. Intake manifolds could now be designed with longer, curved air passages, which increased airflow and improved torque and horsepower.

Since intake manifolds no longer require heating, they can now be made from plastic materials such as fiberglass-reinforced nylon resin. This material is considerably lighter than cast iron or even aluminum. Saving weight means an improvement in fuel economy. The plastic-type intake manifold does not transfer heat to the air and fuel vapor in the intake passages; this improves fuel economy and hot-start performance.

## Injector Internal Design and Electrical Connections

The fuel injector is simply a fast-acting solenoid. Whenever the injector is opened, it sprays a constant amount of fuel for a given period of time. Because the pressure drop across the injector is fixed, the fuel flow through the injector is constant. With the injector connected to a pressurized fuel supply, atomized fuel sprays from the injector nozzle behind the intake valve.

Inside the injector, the plunger and valve seat are held downward by a spring. In this position, the seat closes the metering orifices in the end of the injector **(Figure 1)**. Openings in the sides of the injector allow fuel to enter the cavity surrounding the injector tip. A mesh screen filter inside the injector openings prevents dirt particles in the fuel from entering the metering orifices. In some injectors, a diaphragm is located between the valve seat and the housing. The tip of the injector may contain one to six metering orifices. Injector design varies depending on the manufacturer.

Each injector contains two terminals with an internal coil connected across them. The plunger is

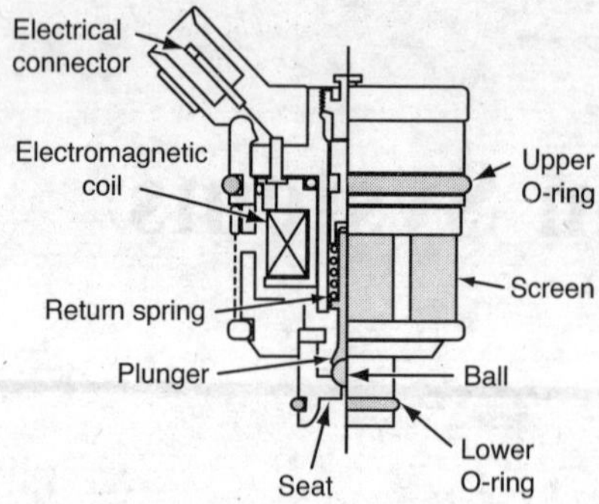


**Figure 1** *Throttle body injector design.*

positioned in the center of the coil and the lower end of the plunger has a tapered valve seat. When the ignition switch is turned on, 12 volts are supplied to one of the injector terminals. The other injector terminal is connected to the PCM that controls ground. When the PCM grounds this terminal, current flows through the injector coil to ground. When this action occurs, the injector coil magnetism moves the plunger and valve seat upward. The pressurized fuel sprays out the injector orifices into the air stream.

### Injector Deposits

Some EFI systems have had problems with deposits on injector tips as a result of small quantities of gum present in gasoline. Injector deposits usually occur when this gum bakes onto the injector tips after a hot engine is shut off. Most oil companies have added detergents to their gasoline to help prevent these deposits. Vehicle manufacturers and auto parts stores sell detergent additives to mix in the fuel tank to clean injector tips. Also, some vehicle manufacturers and parts suppliers have designed deposit-resistant injectors. These injectors have several different pintle tip and orifice designs to help prevent deposits. On one type of deposit-resistant injector, the pintle seat opens outward away from the injector body, providing more clearance between the pintle and the body. Another type of deposit-resistant injector has four orifices in a metering plate rather than a single orifice. Some deposit-resistant injectors come in distinct colors. For example, conventional injectors supplied by Ford Motor Company are painted black, whereas its deposit-resistant injectors have tan or yellow bodies.

## Throttle Body Injection Systems

TBI systems spray the fuel directly above the throttle plates. The **throttle body** assembly is mounted on top of the intake manifold. It contains a pressure regulator, injector(s), TPS, idle speed control motor, and throttle shaft and linkage assembly. Four-cylinder engines usually have a single throttle body assembly with one throttle plate, whereas V6 and V8 engines may be equipped with dual throttle bodies with two throttle plates on a common throttle shaft **(Figure 2)**. When the engine is cranking or running, fuel is supplied from the fuel pump through the lines and filter to the throttle body assembly. A fuel-return line connected from the throttle body to the fuel tank returns excess fuel to the tank.

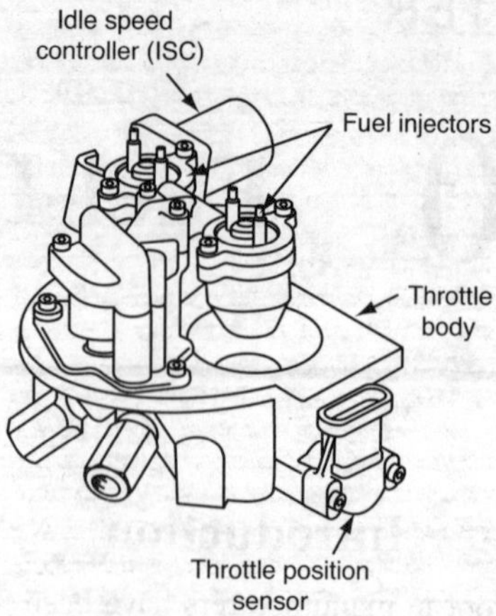


**Figure 2** *Dual throttle body assembly from a V8 engine.*

When fuel enters the throttle body fuel inlet, it surrounds the injector(s) at all times. Each injector is sealed into the throttle body with O-ring seals, which prevent fuel leakage around the injector at the top or bottom. Fuel is supplied from the injector through a passage to the pressure regulator. A diaphragm and valve assembly is mounted in this regulator and a diaphragm spring holds the valve closed. At a specific fuel pressure, the regulator diaphragm is forced upward to open the valve and some excess fuel is returned to the fuel tank.

**You Should Know**

*In Chrysler TBI assemblies, fuel pressure is regulated at 14.5 psi (100 kPa) with a temperature sensor. These systems, referred to as low-pressure TBI (LPTBI), were used from 1986 through 1990. A black plastic rivet is located on top of the pressure regulator in an LPTBI system. In 1991, Chrysler installed a pressure regulator on its TBI systems that controls the fuel pressure at 39.2 psi (270 kPa). These systems are called high-pressure TBI (HPTBI). A white plastic rivet is located on top of these pressure regulators and the TBI temperature sensor is no longer required with the higher fuel pressure.*

In most TBI systems, the fuel pressure regulator controls fuel pressure at 10 to 25 psi (70 to 172 kPa). The fuel pressure must be high enough to prevent fuel boiling in the TBI assembly. When the pressure on a liquid is increased, the boiling point is raised proportionately. If fuel boiling occurs in the TBI assembly, vapor and fuel are discharged from the injectors.

In the four-cylinder TBI system, there is a single injector event for each individual cylinder event. This means the injector pulses twice for each engine revolution. When the engine is being started, double injector pulses occur to enrich the mixture. Double pulsing at cranking occurs only for a programmed time interval to avoid flooding. Pulse width during cranking is based on engine coolant temperature.

**You Should Know**

*In many TBI systems, the computer grounds an injector each time a signal is received from the distributor pickup. This provides an injector firing for each intake valve opening. This type of TBI system is referred to as a synchronized system because the injector pulses are synchronized with the pickup signals. In a dual-injector throttle body assembly, the computer grounds the injectors alternately under most operating conditions.*

Once the engine is started, the PCM provides the pulse-width calculation based on coolant temperature and MAP inputs. Once the engine warms, allowing the mode to change to closed loop, the PCM uses the feedback formula. Some systems will pulse the injector every 12.5 ms, regardless of reference pulses, when the TPS signals it is near full throttle.

**You Should Know**

*Some General Motors TBI systems have a semi-closed-loop mode (highway mode). During extended cruise conditions at highway speeds, the PCM sets the air-fuel ratio as lean as 16.5: 1. This is done to increase fuel economy.*

## Port Fuel-Injection Systems

There are two basic types of **port fuel-injection (PFI)** systems: **multipoint injection (MPI)**, and **sequential fuel injection (SFI)**. *Port fuel injection* is a term that may be applied to any fuel-injection system with the injectors located in the intake ports. The MPI system has one injector per cylinder. The injectors are fired in pairs or in groups of three or four. Usually half of the fuel is delivered on each crankshaft revolution. In an SFI system each injector is fired individually in ignition firing order prior to the intake valve opening.

**You Should Know**

*Multipoint fuel injection is also called multiport fuel injection.*

The MPI and SFI systems supplied by the various domestic and import vehicle manufacturers have many similarities. For example, both MPI and SFI systems have injectors installed in the intake ports near the intake valve and many of these systems share similar inputs and outputs. One of the major differences between MPI and SFI systems is the method of connecting the injectors to the computer. In SFI systems, each injector is connected individually to the computer, so the computer energizes one injector at a time. In MPI systems, the injectors are grouped together in pairs or groups. Each group of injectors shares a common wire to the computer so they are energized together. For example, on some four-cylinder engines, the injectors are connected in pairs and each pair has a common connection to the computer. On some V6 engines, each group of three injectors has a common ground wire connected to the computer. Groups of four injectors share a common ground connection to the computer on some V8 engines.

The computer is programmed to ground the injectors well ahead of the actual intake valve opening so the intake ports are already filled with fuel vapor **(Figure 3)**. In both MPI and SFI systems, the computer supplies the correct injector pulse width to provide the stoichiometric air-fuel ratio. The computer increases the injector pulse width to provide air-fuel ratio enrichment while starting a cold engine. A clear flood mode is also available in the computer in MPI and SFI systems. On many such systems, the computer decreases injector pulse width while the engine is decelerating to improve emissions levels and fuel economy. On some of these systems, the computer stops operating the injectors while the engine is decelerating in a certain rpm range.

## Sequential Fuel-Injection Systems

It would be impossible to cover all SFI systems; however, most operate similarly. The main differences are the inputs used between speed-density and air-density systems. Manufacturers use a similar process to deliver fuel to the combustion chamber. The following is offered as an example of an SFI system.

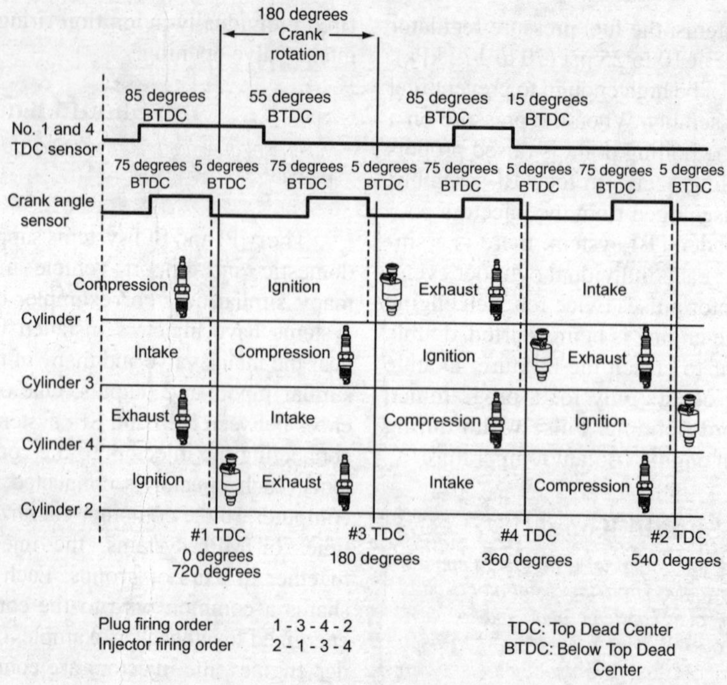


**Figure 3** *Fuel-injection timing strategy.*

The Chrysler SFI system used on a 3.5L engine is wired so each injector has a separate ground wire connected to the PCM **(Figure 4)**. When the ignition switch is in the Start position and an rpm signal is received, the automatic shutdown (ASD) relay contacts supply voltage to the injectors. If the rpm signal is lost, the ASD relay is turned off, shutting off the

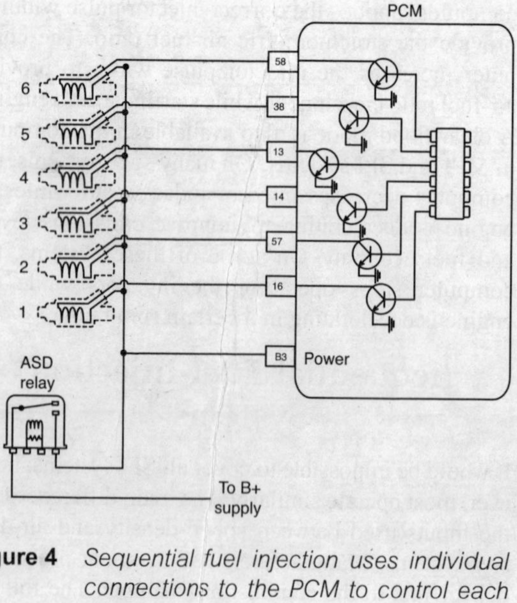


**Figure 4** *Sequential fuel injection uses individual connections to the PCM to control each injector.*

injectors. This engine is equipped with an electronic ignition (EI) system; the crankshaft and camshaft position sensors are inputs for this system.

The crankshaft position sensor informs the PCM of engine speed and that two pistons are approaching top dead center (TDC). The camshaft position sensor informs the PCM which piston is approaching TDC on the compression stroke. The two sensors provide alternating high- and low-voltage signals to provide input for pulse width and ignition strategies **(Figure 5).**

The different slots on the crankshaft position sensor inform the PCM of needed functions as follows:

- 69 degrees before top dead center (BTDC)

  MAP reading is updated and a value is stored in RAM.

  Ignition period is determined by the time between two consecutive 69-degree edges.

  Detonation is mainly controlled on the 69-degree edge.

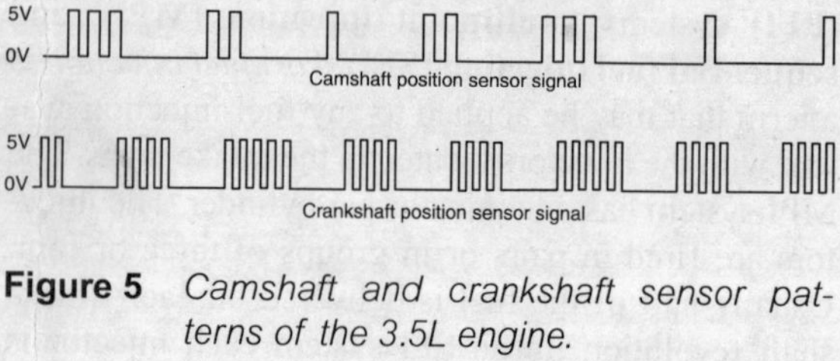


**Figure 5** *Camshaft and crankshaft sensor patterns of the 3.5L engine.*

Synchronization is mainly conducted on this edge along with camshaft interrupt.
Ignition coil firing for the current cylinder as well as for the next cylinder is generated.
Injection firing may occur based on operating conditions.

- 49 degrees BTDC
  Ignition coil firing adjustment may occur if the engine speed is below a programmed value.
  Injector firing may occur based on operating conditions.
- 29 degrees BTDC
  Ignition coil firing adjustment may occur if the engine speed is below a programmed value.
  Injector firing may occur based on operating conditions.
  Adaptive knock is updated.
- 9 degrees BTDC
  If the engine is being started, the coil firing is done on this edge.
  Cylinder indication is updated.
  Adaptive dwell is updated.
  MAP is read. The value is averaged between the reading at 69 degrees BTDC and this reading.

The PCM determines idle mode based on inputs from the TPS. When the ignition switch is turned to the Run position during engine starting, the PCM is programmed to monitor the TPS input. Once the engine is started, the PCM assumes that the lowest value it receives from the TPS must be where the throttle plate is fully closed. Normally this voltage is between 0.5 and 1.0 volt. At the low-voltage position, the PCM records the signal as idle or **minimum TPS**. Minimum TPS is the lowest voltage value the PCM received from the TPS during that key cycle. If the sensed voltage from the TPS goes lower than this value, the PCM will update its memory. However, the PCM will not make any corrections for an increase in voltage values. The PCM uses minimum TPS to determine other modes of operation.

If the throttle is opened and the TPS moves from its minimum TPS value by approximately 0.06 volt, the PCM enters off-idle strategies. At this time, spark advance is no longer used to control idle speed. Also, the idle air-control motor is positioned to act as a dashpot if the throttle is released suddenly.

If the PCM receives a rapidly increasing voltage signal from the TPS, it will enter acceleration mode. During this time, the injector pulse width is increased based on the rate of TPS voltage increase. The PCM may activate three injectors at the same time: the injector for the cylinder whose intake valve just closed, the injector for the cylinder whose intake valve is being opened, and the injector for the cylinder whose intake valve will be opened next.

Wide-open throttle (WOT) is determined by an increase in TPS volts of 2.608 volts above the recorded minimum TPS. The PCM enters open-loop mode and increases the injector pulse width. If a rapid deceleration is sensed, the PCM will make the air-fuel ratio leaner. In some instances, the injectors are turned off. This action reduces the amount of emissions emitted.

If the throttle is at WOT during engine cranking, the injectors are turned off and this provides a clear flood mode. This function only occurs during cranking and when the TPS voltage exceeds 2.608 volts above minimum TPS. The function is not dependent upon engine coolant temperature.

## Typical Import Sequential Fuel-Injection System

The Nissan electronic concentrated engine control system (ECCS) is an SFI system that has many of the same inputs and outputs as the other systems already discussed **(Figure 6)**. Like most systems, the ECCS has evolved over the years.

The Nissan system incorporates both sequential and multipoint (simultaneous) injection operation. The sequential mode injects fuel once every complete engine cycle (two crankshaft revolutions) during the corresponding cylinder's intake stroke. The simultaneous injection mode sprays all the injectors at once, twice per complete engine cycle. The simultaneous injection mode is used when the engine is starting, if the engine is in fail-safe mode, or if the camshaft position sensor signal is lost and the engine is running solely on the crankshaft position sensor signal.

Early systems used a vane-type mass airflow sensor. The throttle valve switch is mounted in the throttle chamber and the switch contacts are closed when the throttle is in the idle position **(Figure 7)**. When the throttle is opened from the idle position, the switch contacts open.

Later systems use a hot-film MAF sensor located in the intake air duct between the air filter and the throttle. The sensor will indicate a signal voltage of less than 1 volt with the ignition in the Run position and the engine off. At idle, the voltage will change to between 1.0 and 1.7 volts. At higher engine speeds and loads, the signal voltage can rise to as high as 4.0 volts.

Some systems use a barometric pressure sensor located in the PCM. This sensor sends a signal to the

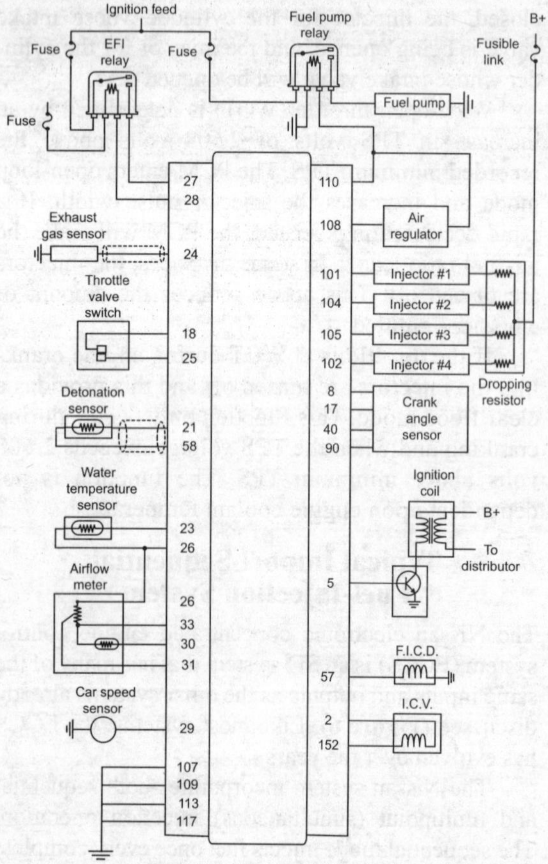


**Figure 6** *Inputs and outputs for the Nissan ECCS system.*

PCM in relation to barometric pressure as it varies with altitude and climatic conditions. Newer systems use an absolute pressure sensor. The Nissan absolute pressure sensor connects to the MAP/BARO switch solenoid through a hose. It detects both ambient pressure and intake manifold pressure. As the pressure rises, the voltage signal rises.

Early systems located the crank angle sensor in the distributor. Later systems have two crankshaft sensors. One of these, referred to as the REF (reference) sensor, is located next to the crankshaft accessory belt pulley. The REF signals the computer when the TDC position for individual cylinders occurs. This sensor is a coil-and-magnet inductive signal generator that produces a signal that varies in voltage with engine speed. The second crankshaft position sensor is referred to as the POS (position) sensor. It is located at the rear of the engine block and reads pulses from a trigger system on the flywheel or flexplate. The POS signals a 1-degree change in crankshaft position.

Later systems also use a camshaft position sensor, called a (PHASE sensor), located in the front engine cover facing the camshaft sprocket. The PHASE sensor provides information that distinguishes which cylinder is at the firing position. The camshaft position sensor is a pulse generator consisting of a permanent magnet wound in an electrical coil. When the gap between the camshaft sprocket and the magnet changes, the magnetic field changes and generates a voltage signal in the sensor output circuit. The computer uses the camshaft position signal to sequence the fuel injectors. At the beginning of the start-up cranking period, the camshaft position signal has not yet occurred and all the injectors are simultaneously pulsed for engine start-up.

The dropping resistor assembly contains a resistor connected in series with each injector. These

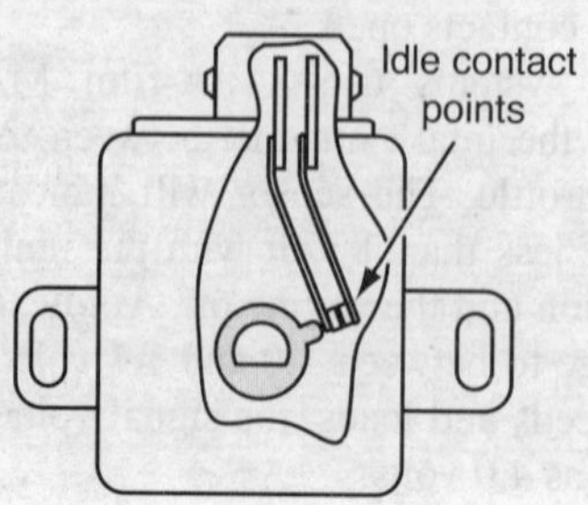


**Figure 7** *Typical idle switch.*

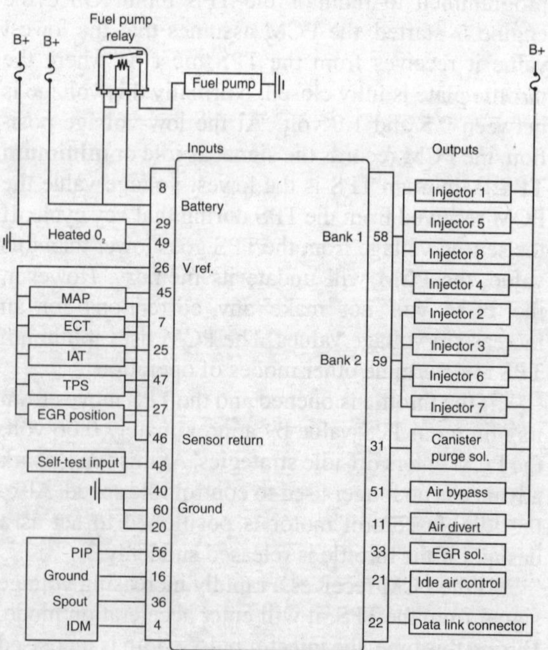


**Figure 8** *Inputs and outputs for the 5.8L multipoint injection system.*

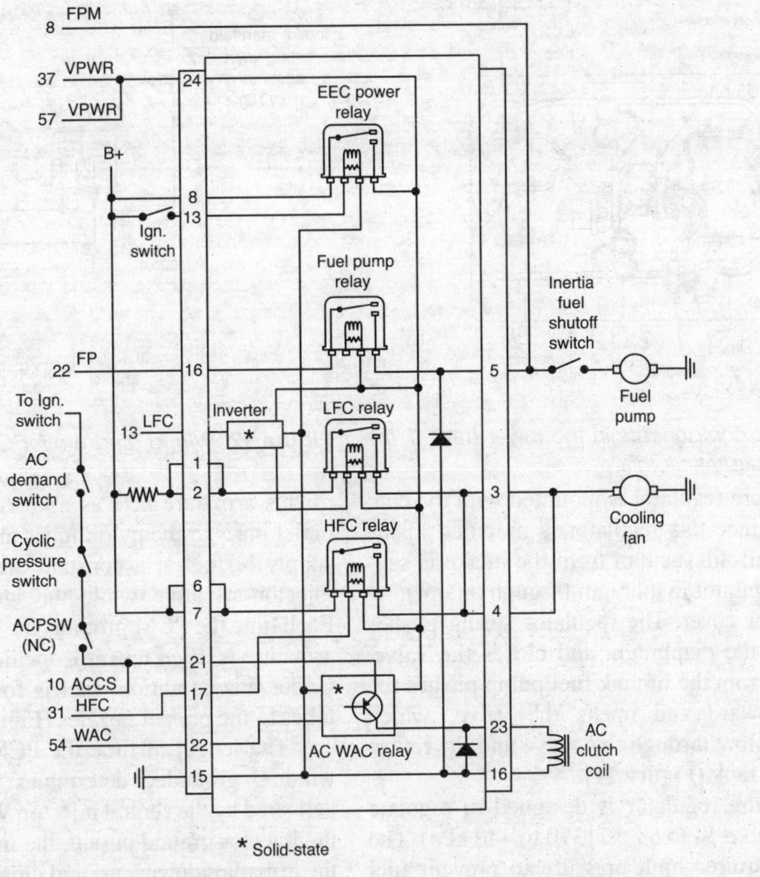


**Figure 9** *Wiring diagram for a constant control relay module.*

resistors protect the injectors from sudden voltage changes and provide constant injector operation.

## Multipoint Injection Systems

The following is offered as an example of a multipoint injection system. Like SFI systems, MPI systems are very similar among manufacturers.

In many Ford MPI systems on V8 engines, the injectors are connected in groups of four on the ground side. Each group shares a common ground wire to the power-train control module. On a Ford 5.8L V8 engine, injectors 1, 4, 5, and 8 are connected to PCM terminal 58 with a common ground wire. The ground sides of injectors 2, 3, 6, and 7 are connected to PCM terminal 59 **(Figure 8)**. When the injectors are connected in groups of four on the ground side, a group of injectors is grounded by the computer every crankshaft revolution.

Some Ford products are equipped with a **constant control relay module (Figure 9)** that contains several relays within its housing. The relays perform the same function whether they are mounted separately or located in the constant control relay module. When the ignition switch is turned on, voltage is supplied to the power relay winding. This action closes the power relay points, which supply voltage to the fuel pump relay winding, PCM terminals 37 and 57, and the electric drive fan (EDF) relay winding.

## Central Port Injection Systems

Another type of fuel injection system is central port injection. This system uses a central port injector assembly mounted in the lower half of the intake manifold. Fuel inlet and return lines are connected from the rear of the intake to the CPI assembly with a retaining clip. Small poppet nozzles are positioned in each intake port in the lower half of the intake. Nylon fuel lines connect these nozzles to the CPI assembly **(Figure 10)**.

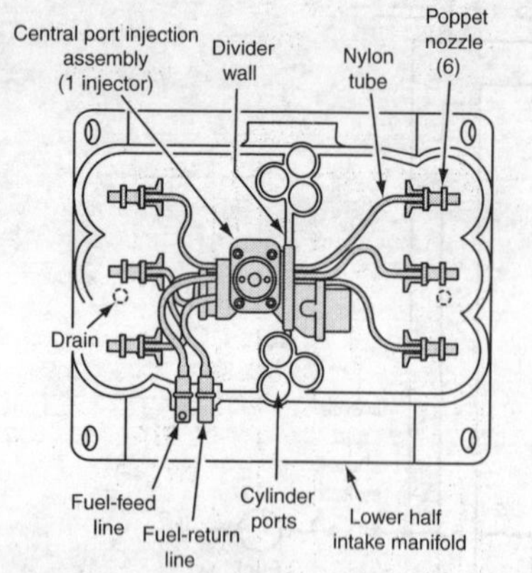


**Figure 10** *CPI components in the lower half of the intake manifold.*

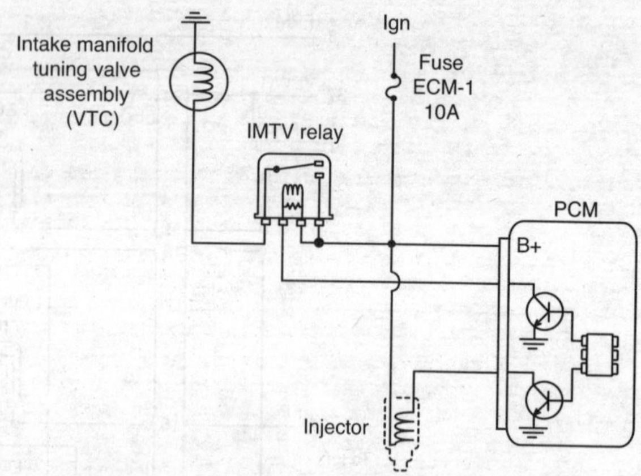


**Figure 12** *Wiring diagram for CPI system.*

The pressure regulator is mounted with the central injector. Since this regulator is mounted inside the intake manifold, vacuum from the intake is supplied to the regulator diaphragm through an opening in the regulator cover. The regulator spring pushes downward on the diaphragm and closes the valve. Fuel pressure from the in-tank fuel pump pushes the diaphragm upward and opens the valve, which allows fuel to flow through this valve and the return line to the fuel tank **(Figure 11)**.

The pressure regulator is designed to regulate fuel pressure from 54 to 64 psi (370 to 440 kPa). The CPI system requires high pressure to prevent fuel vaporization from the extra heat encountered with the CPI assembly, poppet nozzles, and lines mounted inside the intake manifold.

The central port injector contains a winding with two terminals extending from the ends of the winding through the top of the injector. When the ignition switch is on, voltage is supplied from the ignition switch to one of the injector terminals. The other injector terminal is connected to the PCM to control ground.

A pivoted armature is mounted under the injector winding in the central port injector. The lower side of this armature acts as a valve that covers the six outlet ports to the nylon tubes and poppet nozzles. A supply of fuel at a constant pressure surrounds the injector armature while the ignition switch is on. Each time the PCM grounds the injector winding, the armature is lifted upward, opening the injector ports. Under this condition, fuel is forced from the nylon tubes to the poppet nozzles **(Figure 12)**.

The length of time the PCM keeps the injector winding grounded determines the amount of fuel delivered by the central injector. When the PCM opens the injector ground circuit, the injector spring pushes the armature downward and closes the injector ports. The injector winding has a resistance of 1.5 ohms and the PCM operates the injector with a peak-and-hold current. When the PCM grounds the injector winding, the current flow in this circuit increases rapidly to 4 amperes. When the current flow reaches this value, a current-limiting circuit in the PCM limits the current

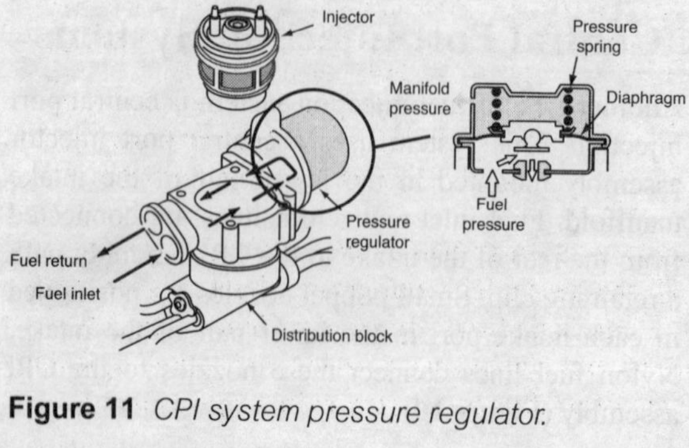


**Figure 11** *CPI system pressure regulator.*

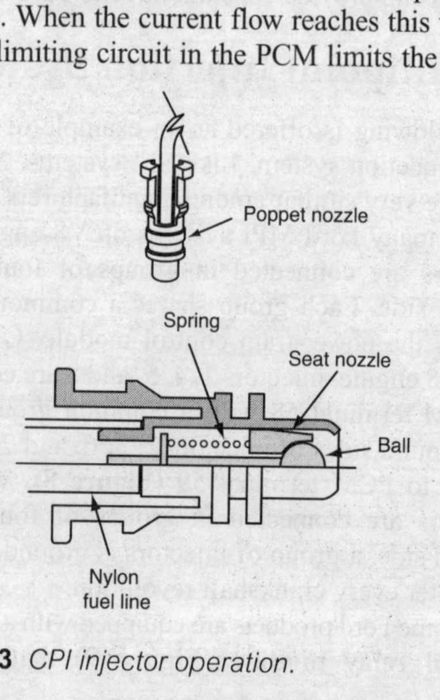


**Figure 13** *CPI injector operation.*

flow to 1 ampere for the remainder of the injector pulse width. The peak-and-hold function provides faster injector armature opening and closing.

The poppet nozzles are snapped into openings in the lower half of the intake manifold and the tip of each nozzle directs fuel into an intake port. Each poppet nozzle contains a valve with a check ball seat in the tip of the nozzle **(Figure 13)**. A spring holds the valve and check ball seat in the closed position. When fuel pressure is applied from the central injector through the nylon lines to the poppet nozzles, this pressure forces the valve and check ball seat open against spring pressure. The poppet nozzles open when the fuel pressure exceeds 37 to 43 psi (254 to 296 kPa) to spray fuel from these nozzles into the intake ports.

When the fuel pressure drops, the poppet nozzles close. Under this condition, approximately 40 psi (276 kPa) fuel pressure remains in the nylon lines and poppet nozzles. This pressure prevents fuel vaporization in the nylon lines and nozzles during hot engine operation or hot soak periods. If a leak occurs in a nylon line or any other CPI component, fuel drains from the bottom of the intake manifold through two drain holes to the center cylinder intake ports. The in-tank fuel pump, fuel filter, lines, and fuel pump circuit used with the CPI system are similar to those used with MFI and SFI systems.

## Gasoline Direct-Injection Systems

Diesel engines have used direct fuel injection for many years. However, this system has not been used in gasoline engines until recently. Manufacturers are designing engines that use a **gasoline direct-injection (GDI)** system. GDI engines do not inject fuel above an intake valve like other fuel-injection systems. Instead, the high-pressure fuel is injected directly into the combustion chamber. GDI has the potential to make the internal combustion engine up to 20 percent more fuel efficient.

The premise behind the use of direct injection is that under most circumstances, an air-fuel ratio can be much leaner than the stoichiometric ratio (14.7:1) without loss of engine performance. The main hindrance to a leaner air-fuel mixture is that it can be hard to ignite. In addition, lean air-fuel mixtures can lead to increased $NO_x$ emissions. However, computer technology can overcome these problems.

Advantages of GDI systems include the following:

- Fuel enters the cylinder as a mist of tiny droplets rather than a vapor. As a result, the cylinder cools as the fuel absorbs the heat needed to turn it into a vapor. This cooling reduces the engine's octane requirement. This allows for the use of increased compression ratios that mean better fuel efficiency and power output. Some GDI engines have compression ratios as high as 12.5:1.
- GDI systems produce a very fast burn of the air-fuel mixture. This makes the engine very tolerant of exhaust gas recirculation.
- Since the fuel is injected directly into the combustion chamber just prior to the spark plug firings, precise control of charge stratification is achieved. This is vital to ignite ultralean air-fuel mixtures.
- Direct injection also eliminates the need for a throttle. Efficiency is increased, since pumping loss that normally occurs when drawing air around a throttle plate is eliminated.

The shapes of the piston and the combustion chamber, the energy and direction of the injected pulses, the movement of the piston, and the heat of the engine all influence the position of each droplet of fuel and air. GDI engines use upright, straight intake ports and concave-section piston surfaces **(Figure 14)**. A swirl airflow is created during the compression stroke. When the fuel is injected directly into the combustion chamber, the swirl mixes the air with the fuel.

The GDI system uses injector pressures of about 100 bar (100 times atmospheric pressure) to force fuel into the combustion chamber. The shape of the injector spray pattern is critical for proper operation. The spray must be directed at the correct angle to the spark plug. Since the GDI engine can idle with an air-fuel mixture as lean as 40:1 (55:1 with EGR), igniting the mixture is accomplished by using two modes of injection: **stratified mode** and **homogeneous mode**.

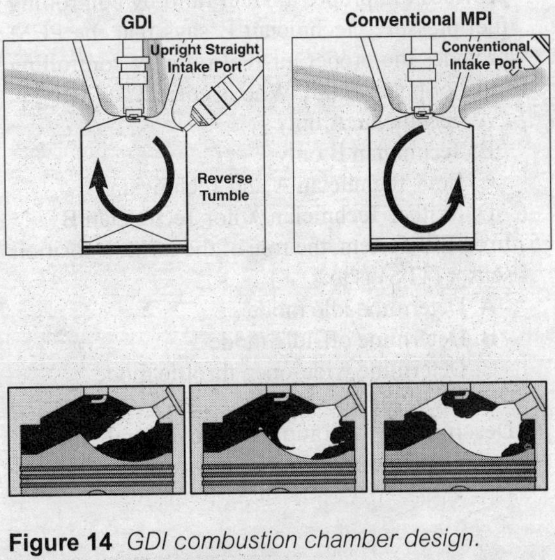


**Figure 14** *GDI combustion chamber design.*

The stratified mode is used when the engine is operating at low and medium loads. Fuel is injected into the combustion chamber just before the spark plug is ignited. This leaves very little time for the fuel to mix with air in the combustion chamber and results in a rich air-fuel mixture next to the spark plug. In addition, the concave-section piston concentrates more fuel around the spark plug. This mixture is rich enough to start the flame kernel. As a result of the injector spray pattern, the air-fuel mixture varies within the chamber. It is rich near the spark plug and leaner farther away from it. However, the overall composition of the mixture is lean.

When the spark plug ignites the rich air-fuel mixture next to it, a flame kernel is started. Because the size of the flame kernel is so much larger than a spark gap, this beginning of combustion can then spread to leaner areas of the combustion chamber.

When the GDI engine loads are increased, the system changes to a homogeneous mode to provide more power. In this mode, fuel is injected into the engine earlier during the compression stroke. This provides extra time for the fuel and air to mix uniformly in the chamber resulting in a rich, homogeneous mixture for maximum power.

To overcome the problem of $NO_x$ emissions with the leaner mixture, GDI vehicles may use a special catalytic converter. The $NO_x$ that is produced is trapped in the converter. When the converter becomes saturated, a special $NO_x$ sensor will inform the PCM. The PCM then supplies a rich mixture into the cylinders. The rich exhaust reacts with the $NO_x$ to create carbon dioxide ($CO_2$). Once the sensor indicates that the converter is flushed, the PCM returns to normal strategy. In addition to the special converter, the engine may use a 30 percent EGR rate.

**You Should Know**

*At present, GDI vehicles are sold primarily in European and Asian markets. This is because the sulfur content of U.S. gasoline is too high to allow effective use of the special catalytic converter and $NO_x$ sensor required on these vehicles.*

## Review Questions

1. Describe the difference between MPI and SFI systems.
2. Explain the stratified mode used in the GDI engine.
3. In an SFI system, each injector has an individual ____________________ connected to the computer.
4. TBI, MPI, and SFI systems are being discussed. Technician A says that in EFI systems the PCM provides the proper air-fuel ratio by controlling fuel pressure. Technician B says that the PCM provides the proper air-fuel ratio by controlling injector pulse width. Who is correct?
   A. Technician A only
   B. Technician B only
   C. Both Technician A and Technician B
   D. Neither Technician A nor Technician B
5. In an EFI system, the role of the throttle position sensor (TPS) is to:
   A. Determine idle mode
   B. Determine off-idle mode
   C. Determine wide-open throttle mode
   D. All of the above
6. Describe the operation of the central injector and poppet nozzles in a CPI system.
7. Explain the homogeneous mode used in the GDI engine.
8. In a CPI system, the air-fuel ratio is determined by the pulse width on the ____________________.
9. Technician A says that in a central port injection (CPI) system the poppet nozzles are opened by the computer during engine starting. Technician B says that the poppet nozzles are opened by fuel pressure when the engine is running. Who is correct?
   A. Technician A only
   B. Technician B only
   C. Both Technician A and Technician B
   D. Neither Technician A nor Technician B
   E. None of the above
10. Technician A says that piston design is a factor in the operation of the GDI system. Technician B says that the GDI system uses high pressure to inject fuel into the intake port. Who is correct?
    A. Technician A only
    B. Technician B only
    C. Both Technician A and Technician B
    D. Neither Technician A nor Technician B

CHAPTER 21

# Electronic Fuel-Injection Input Sensor Diagnosis and Service

## Introduction

When customers with fuel-injected vehicles complain about engine performance, emissions, or fuel economy, many technicians tend to think the problem is in the engine's computer system. However, many other defects can affect the engine and fuel-injection system operation. For example, an intake manifold vacuum leak causes a rough idle condition and engine surging at low speed. If the engine has an MAF sensor, an intake manifold vacuum leak allows additional air into the intake, and this air does not flow through the MAF sensor. Therefore, the MAF sensor signal indicates to the PCM that less airflow is entering the engine in relation to the throttle opening. Under this condition, the PCM supplies less fuel through the injectors. This creates a lean air-fuel ratio, which results in poor idle quality, engine surging, and acceleration stumbles. A technician can avoid expensive hours of diagnostic time by checking for the following items before diagnosing the fuel-injection system:

1. Intake manifold vacuum leaks
2. Emissions devices such as the EGR valve and related controls
3. Ignition system condition
4. Engine compression
5. Battery fully charged
6. Engine at normal operating temperature
7. All accessories turned off

If no defects are found, then the fuel-injection system is suspect. This chapter details common diagnostic procedures used to test the inputs of typical fuel-injection systems. Since 1996, all passenger cars sold in the United States are required to be OBD-II compliant. This system provides additional information for the technician to aid in the diagnosis and repair of emissions-related failures. However, the technician will still need to perform basic diagnostics of the system to find the root cause of the problem. Section 7 of this book covers the OBD-II system.

## Service Precautions

The following precautions must be observed when diagnosing and servicing TBI, MFI, and SFI systems:

1. Always relieve the fuel pressure before disconnecting any component in the fuel system.
2. Never turn the ignition switch to the Run position when any fuel system component is disconnected.
3. Use only the test equipment recommended by the vehicle manufacturer.
4. Always turn off the ignition switch before connecting or disconnecting any system component or test equipment.
5. Never allow electrical system voltage to exceed 16 volts. Disconnecting the circuit between the generator and the battery with the engine running can cause this condition.
6. Avoid static electric discharges when handling computers, modules, and computer chips.

## Preliminary Inspection

A complete visual inspection will often reveal the cause of a driveability problem. Connections must be tight and clean. In addition, all wires must be checked for signs of damage. While inspecting the wires, also look for signs of tampering and modifications.

Along with the inspection of wires, check the vacuum hoses. Use the underhood sticker to ensure the vacuum hose routing is correct. Check the hoses for cracks, loose connections, and restrictions.

Include in the inspection all emissions control devices. If the engine is equipped with an air pump, check the tension and condition of the belt. Also, check the condition of the diverter valve and all hoses connected to the system. Make sure the EGR valve and the canister purge system are functioning properly.

# Input Sensor Diagnosis and Service

Most input voltage values can be checked with a scan tool. Often this is the easiest way to get this information since it does not require access to the components or the PCM. However, remember the scan tool only provides an interpretation of these values. Sometimes the scan tool may display a value that is within specifications, but actual DVOM testing of the sensor may prove the component is defective. If a scan tool is not available or is suspected of erroneous display, there are other methods of input sensor testing. The following sections provide examples.

## Manifold Absolute Pressure Sensor Diagnosis

> **You Should Know**
>
> *Manifold absolute pressure sensors have a different calibration on turbocharged engines than on nonturbocharged engines. Be sure to use the proper specifications for the sensor being tested.*

A defective manifold absolute pressure (MAP) sensor can cause a rich or lean air-fuel ratio, excessive fuel consumption, and engine surging. In some systems, the engine will not run if the MAP sensor is defective. Diagnosis of MAP sensors differs among types. If the MAP sensor signal voltage is within the programmed value range of the PCM, a diagnostic trouble code (DTC) may not set. For example, an open signal circuit will set a DTC, but resistance in the circuit that causes the voltage to be off by a few millivolts may not set a fault code. Also, a vacuum leak will cause the voltage values from the signal circuit to be different from specifications.

To test a Wheatstone bridge or a silicon-type MAP sensor that produces an analog signal, place the ignition switch in the Run position (engine off), backprobe the MAP sensor signal wire, and connect a voltmeter from this wire to ground **(Figure 1)**. The voltage reading indicates the barometric pressure signal from the MAP sensor to the computer. Usually this is about 4 volts, depending on altitude. Many MAP sensors send a barometric pressure signal to the computer every time the ignition switch is turned on and when the throttle is in the wide-open position (engine running). If the voltage supplied by the barometric pressure signal in the MAP sensor does not match the vehicle manufacturer's specifications, replace the MAP sensor.

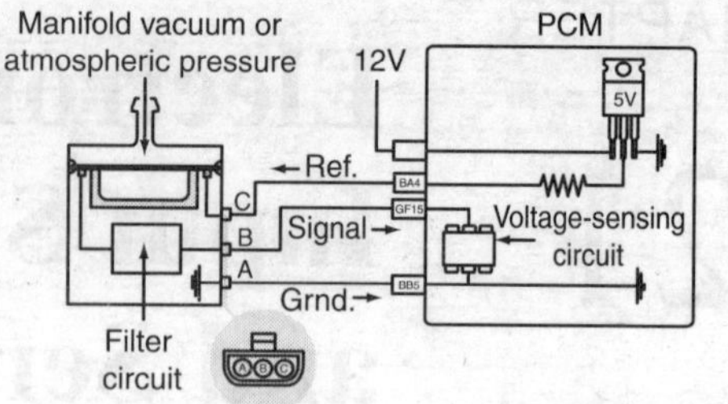


**Figure 1** *MAP sensor circuit.*

The barometric pressure voltage signal varies, depending on altitude and atmospheric conditions. Follow this calculation to obtain an accurate barometric pressure reading:

1. Determine present barometric pressure; for example, 29.85 in.
2. Multiply the altitude by 0.001; for example, 600 ft × 0.001 = 0.6.
3. Subtract the altitude correction from the present barometric pressure; e.g., 29.85 – 0.6 = 29.25.
4. Check the vehicle manufacturer's specifications to obtain the proper barometric pressure voltage signal in relation to the present barometric pressure **(Figure 2)**.

If the voltage is acceptable with the ignition in the Run position and the engine off, then start the engine and observe the signal voltage. The voltage should have changed to a low value. Allow the engine to warm to normal operating temperature while watching the voltage. Compare the voltage reading with the specifications. If the voltage is within specifications, the MAP sensor is not faulty. Therefore, other components will need to be tested.

If the voltage reading obtained in the above tests was 0 volt, then perform these additional steps: Unplug the MAP sensor connector and measure the voltage on each terminal with the ignition in the Run position. On a three-wire connector, the voltage will usually be 5 volts on the supply circuit, 5 volts on the

| Absolute BARO reading | Lowest allowable voltage at –40°F | Lowest allowable voltage at 257°F | Lowest allowable voltage at 77°F | Designed output voltage | Highest allowable voltage at 77°F | Highest allowable voltage at 257°F | Highest allowable voltage at –40°F |
|---|---|---|---|---|---|---|---|
| 31.0" | 4.548V | 4.632V | 4.716V | 4.800V | 4.884V | 4.968V | 5.052V |
| 30.9" | 4.531V | 4.615V | 4.699V | 4.783V | 4.867V | 4.951V | 5.035V |
| 30.8" | 4.514V | 4.598V | 4.682V | 4.766V | 4.850V | 4.934V | 5.018V |
| 30.7" | 4.497V | 4.581V | 4.665V | 4.749V | 4.833V | 4.917V | 5.001V |
| 30.6" | 4.480V | 4.564V | 4.648V | 4.732V | 4.816V | 4.900V | 4.984V |
| 30.5" | 4.463V | 4.547V | 4.631V | 4.715V | 4.799V | 4.883V | 4.967V |
| 30.4" | 4.446V | 4.530V | 4.614V | 4.698V | 4.782V | 4.866V | 4.950V |
| 30.3" | 4.430V | 4.514V | 4.598V | 4.682V | 4.766V | 4.850V | 4.934V |
| 30.2" | 4.413V | 4,497V | 4.581V | 4.665V | 4.749V | 4.833V | 4.917V |
| 30.1" | 4.396V | 4.480V | 4.564V | 4.648V | 4.732V | 4.816V | 4.900V |
| 30.0" | 4.379V | 4.463V | 4.547V | 4.631V | 4.715V | 4.799V | 4.883V |

**Figure 2** *Barometric pressure voltage values at various barometric pressures.*

signal circuit, and 0 volt on the ground circuit. However, some MAP sensor signal circuits may have no voltage when the connector is separated. Always confirm the expected voltage values with the proper service information. If there is no voltage on the supply or signal circuits, then check if the fault is an open circuit or the PCM. With the ignition in the Off position, test for continuity between the PCM connector and the MAP connector on the suspect circuit. If there is continuity in the circuit, confirm PCM powers and grounds. If these are good, replace the PCM.

If the voltage reading obtained when testing the signal circuit was 5 volts, the MAP sensor return (ground) circuit is suspect. With the ignition in the Off position, test continuity between the MAP sensor connector and the PCM connector.

### You Should Know

*If the sensor return circuit is faulty within the PCM, the engine will not start and all sensors that use this circuit of the MAP sensor indicates an open circuit, check the return circuit from another input component such as the TPS. If this circuit shows good continuity, then the problem is between a splice or inline connector in the return circuit and the MAP sensor connector. Use the wiring diagrams to locate these.*

If the voltage recorded in the first step is inaccurate, backprobe the 5-volt reference wire (connector plugged to the MAP sensor) with a voltmeter with the ignition in the Run position. If the reference wire is not supplying the specified voltage, check the voltage on this wire at the computer. If the voltage is within specifications at the computer but low at the sensor, repair the 5-volt reference wire for high resistance. If this voltage is low at the computer, disconnect the PCM connector and use an ohmmeter to test for a short to ground. If there is no short to ground indicated, check the voltage supply and ground circuits for the PCM. If these circuits are satisfactory, replace the PCM.

If the voltage supply circuit is good, then test the sensor return (ground) circuit. With the ignition switch in the Run position, backprobe the ground circuit terminal at the connector (with the connector plugged into the MAP sensor). Connect the voltmeter from the sensor ground wire to the battery ground terminal. If the voltage drop across this circuit exceeds specifications, test the voltage drop on the PCM's ground circuits. If this is good, repair the ground wire from the MAP sensor to the computer.

If the MAP sensor has a vacuum hose connected to it, the sensor can be tested for functionality using a hand vacuum pump. With the ignition switch in the Run position and the voltmeter connected to the MAP sensor signal wire, apply 5 inches of vacuum to the sensor. Check the MAP sensor voltage at 5-inch intervals from 0 to 20 inches. On some MAP sensors, the sensor voltage signal should change 0.7 to 1.0 volt for every 5 inches of vacuum change applied to the sensor. Always use the vehicle manufacturer's specifications. When applying vacuum to the sensor, normal test results should be as follows (if the MAP sensor voltage is not within specifications at any of the vacuum settings, replace the sensor):

1. If the barometric pressure voltage signal is 4.5 volts with 5 inches of vacuum applied to the MAP sensor, the voltage signal should be 3.8 to 3.5 volts.
2. With 10 inches of vacuum applied to the sensor, the voltage signal should be 3.1 to 2.5 volts.
3. With 15 inches of vacuum applied to the sensor, the voltage signal should be 2.4 to 1.5 volts.
4. With 20 inches of vacuum applied to the sensor, the voltage signal should be 1.7 to 0.5 volts.
5. With 25 inches of vacuum applied to the sensor, the voltage signal should be 0.7 to 0 volt.

If the MAP sensor produces a digital voltage signal of varying frequency, a voltmeter can be used to check the 5-volt reference and the ground wires. Follow the same procedure as that used on other MAP sensors. Testing the signal circuit of this type of sensor requires the use of a MAP sensor tester that changes the MAP sensor's varying frequency to a voltage reading **(Figure 3)**. To test the frequency-generating MAP sensor, follow these steps:

1. Remove the MAP sensor connector and vacuum hose.
2. Connect the MAP sensor tester leads to the MAP sensor.

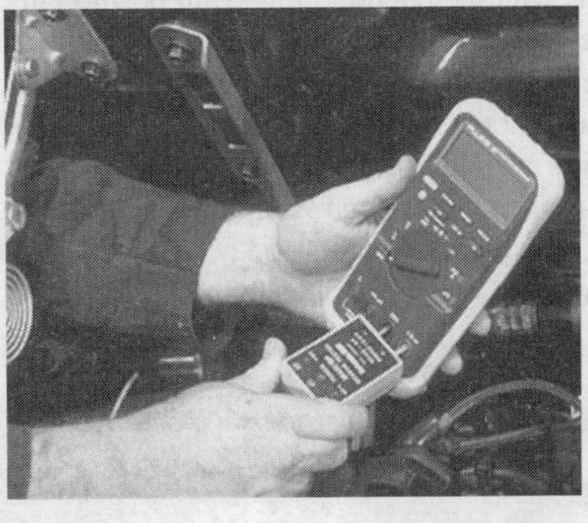

**Figure 3** *A special MAP sensor tester connects to the DVOM.*

| Vacuum Applied | Output Frequency |
|---|---|
| 0 in. Hg | 152–155 Hz |
| 5 in. Hg | 138–140 Hz |
| 10 in. Hg | 124–127 Hz |
| 15 in. Hg | 111–114 Hz |
| 20 in. Hg | 93–98 Hz |

**Figure 4** *Example of MAP sensor frequency with various vacuums.*

3. Connect the MAP sensor connector to the harness connector.
4. Connect the MAP sensor tester to the DVOM.
5. With the DVOM set to DC voltage, observe the MAP sensor BARO reading on the voltmeter and compare this reading with the specifications.
6. Connect a hand vacuum pump to the MAP sensor and supply 5 in. Hg. Observe the MAP sensor reading and compare it with the specifications.
7. Increase the vacuum applied to the MAP sensor to 10 in. Hg and observe the reading. Compare this reading with the specifications.
8. Apply 15 in. Hg to the MAP sensor and compare the reading with the specifications.
9. Increase the vacuum 20 in. Hg to the MAP sensor and compare the reading with the specifications. If any of the MAP sensor readings do not meet specifications, replace the MAP sensor.

If a MAP sensor tester is not available, the sensor can be checked with a DVOM that has a frequency measurement function. Connect the positive meter lead to the MAP sensor signal wire and the negative meter lead to ground. Set the meter to read hertz with a "plus" trigger. With the ignition in the Run position, measure the frequency on the sensor signal wire with no vacuum applied. Using a hand vacuum pump, increase the vacuum against the sensor in 5 in. Hg increments. Record the hertz reading at each step. A good MAP sensor will decrease hertz as vacuum is applied **(Figure 4)**. Normally, a frequency of about 155 hertz is expected at sea level with no vacuum applied to the MAP sensor.

### You Should Know

*Never connect any type of voltmeter directly to the voltage signal wire on a Ford MAP sensor. This action causes MAP sensor damage! Some Ford MAP sensor testers read the sensor hertz directly and other Ford MAP sensor testers convert the hertz to a voltage signal.*

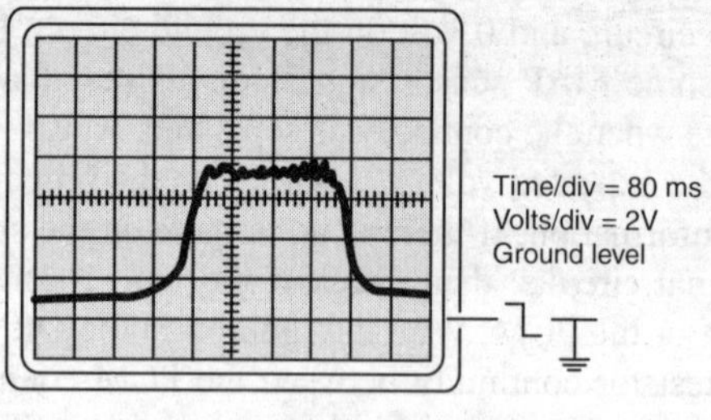


**Figure 5** *Normal MAP sensor waveform.*

**Lab Scope Testing of the MAP Sensor.** Another very good method of testing the MAP sensor is to use a lab scope. When the engine is accelerated and returned to idle, the MAP sensor voltage signal should increase and then decrease **(Figure 5)**. If the engine is accelerated and the MAP sensor voltage signal does not rise and fall (or the signal is erratic), the sensor or connecting wires may be defective. Remember that the MAP sensor output voltage changes as pressure in the intake manifold changes. Internal engine problems will cause improper MAP sensor voltage output.

A frequency generating MAP sensor produces a frequency that should change as the engine speed increases. The upper voltage value of the trace should be 5 volts, while the lower voltage should be close to 0 **(Figure 6)**. If the frequency is erratic or does not change, or the voltage values are incorrect, the sensor is defective.

## Mass Airflow Sensor Diagnosis

The mass airflow MAF sensor is used to determine the volume of air entering the combustion chamber. Based on this input, the PCM calculates how much fuel to inject to obtain the correct air-fuel mixture. Failure of this sensor will result in severe driveability problems. The PCM monitors this sensor for defects. A circuit sensor signal that is not within the expected range may set a DTC. Based on the DTC, a troubleshooting tree can be followed to isolate the cause. Diagnosis of the MAF sensor depends on the type of sensor used. Typical types of sensors include: vane-type, heated resistor, hot wire, and ultrasonic. Most provide for input signal monitoring through the use of a scan tool. The following are typical procedures that may be used if a scan tool is not available. Always refer to the proper service information for the vehicle.

**Vane-type MAF Sensors.** The following procedure will test the signal circuit of the MAF sensor. If the voltage on the signal circuit is not within specifications, test the voltage supply and ground circuits to the

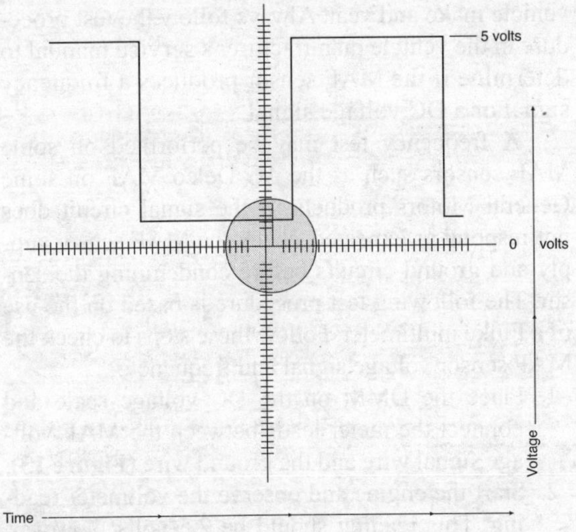


Figure 6 *Example of a Ford MAP sensor signal.*

MAF module before condemning it. This test is based on the use of a Fluke digital multimeter (DMM):

1. Set the DMM on the DC voltage scale. Connect the red meter lead to the MAF signal wire with a special piercing probe and connect the black meter lead to ground **(Figure 7)**.
2. Turn the ignition switch to the Run position and press the min/max button to activate the min/max feature.

**You Should Know**

*While pushing the MAF sensor vane open and closed, be careful not to mark or damage the vane or the sensor housing.*

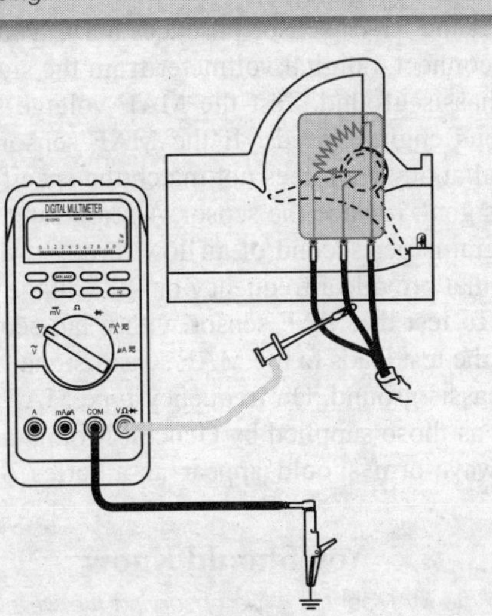

Figure 7 *Voltmeter connected to measure MAF sensor voltage signal.*

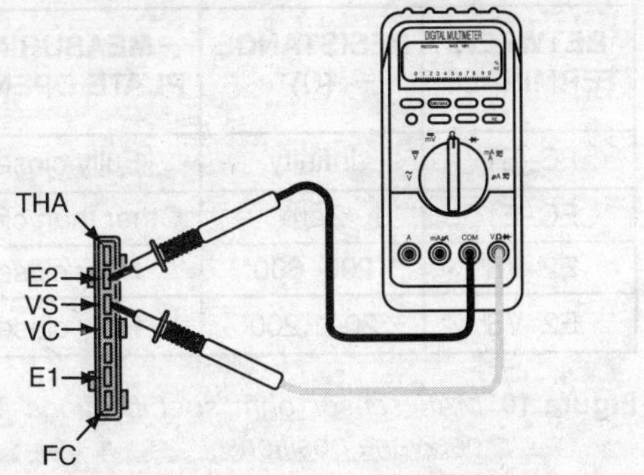


Figure 8 *Ohmmeter connections to the MAF sensor.*

3. Slowly push the MAF sensor vane from the closed to the wide-open position and allow the vane to return slowly to the closed position.
4. Touch the min/max button once to read the maximum voltage signal recorded. Press this button again to read the minimum voltage signal. If the minimum voltage signal is zero, there may be an open circuit in the MAF sensor's variable resistor. If the voltage signal is not within the manufacturer's specifications, replace the sensor.

A vehicle manufacturer that does not allow for taking voltage readings as just described will usually specify ohmmeter tests. The following example is for a typical Toyota MAF sensor. With the MAF sensor removed, connect the ohmmeter to the E2 and VS MAF sensor terminals **(Figure 8)**. The resistance at these terminals should be 200 to 600 ohms **(Figure 9)**.

Connect the ohmmeter leads to the other recommended MAF sensor terminals as shown in Figure 9 and record the resistance readings. Since the THA and E2 sensor terminals are connected to a thermistor, resistance is affected by temperature. If the specified resistance is not present in any of the test connections, replace the MAF sensor.

| BETWEEN TERMINALS | RESISTANCE (Ω) | TEMPERATURE °F (°C) |
|---|---|---|
| E2–VS | 200–600 | – |
| VC–E2 | 200–400 | – |
| E2–THA | 10,000–20,000 | –4 (–20) |
| | 4,000–7,000 | 32 (0) |
| FC–E1 | Infinity | – |

Figure 9 *MAF sensor resistance specifications between terminals.*

| BETWEEN TERMINALS | RESISTANCE (Ω) | MEASURING PLATE OPENING |
|---|---|---|
| FC–E1 | Infinity | Fully closed |
| FC–E1 | Zero | Other than closed |
| E2–VS | 200–600 | Fully closed |
| E2–VS | 20–1,200 | Fully closed |

**Figure 10** *MAF sensor ohm specifications at various vane positions.*

Connect the ohmmeter leads to the specified MAF sensor terminals **(Figure 10)** and move the vane from the fully closed to the fully open position. With each specified meter connection and vane position, the ohmmeter should indicate the specified resistance. When the ohmmeter leads are connected to the E2 and VS sensor terminals, the ohmmeter reading should increase smoothly as the sensor vane is opened and closed.

If testing the vane-type MAF sensor with a lab scope, the analog voltage value should transfer smoothly **(Figure 11)**. If there are any glitches in the trace **(Figure 12)**, the sensor is faulty.

**Heated Resistor or Hot-Wire MAF Sensor Frequency Test.** The test procedure for heated resistor and hot-wire MAF sensors varies depending on the vehicle make and year. Always follow the test procedure in the vehicle manufacturer's service manual to determine if the MAF sensor produces a frequency signal or a DC voltage signal.

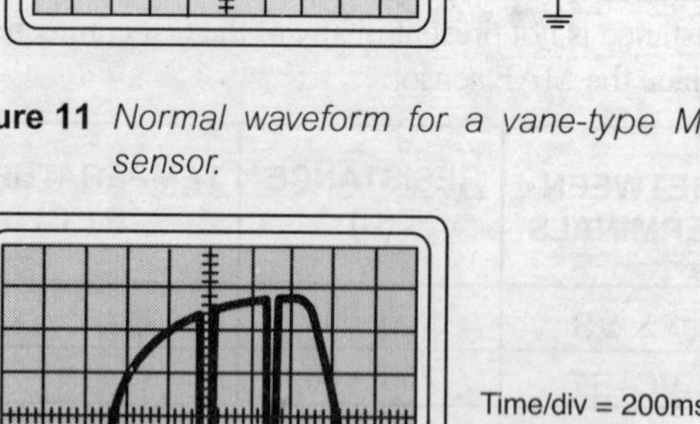


**Figure 11** *Normal waveform for a vane-type MAF sensor.*

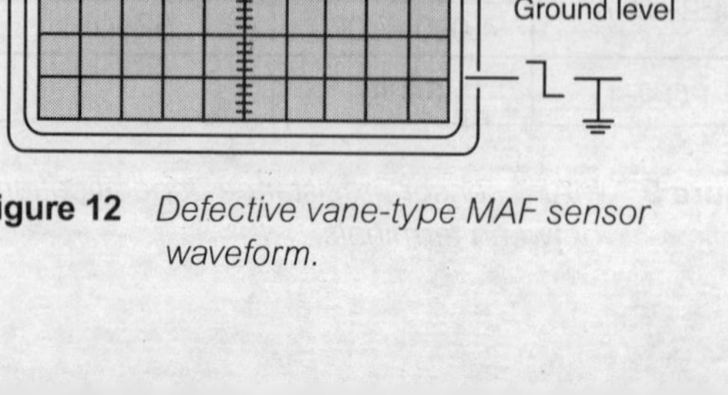


**Figure 12** *Defective vane-type MAF sensor waveform.*

A frequency test may be performed on some MAF sensors such as the AC Delco MAF on some General Motors products. If the signal circuit does not respond as expected, check the MAF sensor supply and ground circuits before condemning the sensor. The following test procedure is based on the use of a Fluke multimeter. Follow these steps to check the MAF sensor voltage signal and frequency:

1. Place the DMM on the DC voltage scale and connect the meter leads between the MAF voltage signal wire and the ground wire **(Figure 13)**.
2. Start the engine and observe the voltmeter reading. This reading should be 2.5 volts; however, always refer to the manufacturer's specifications.
3. Lightly tap the MAF sensor housing with a screwdriver handle and watch the voltmeter's analog scale. If the meter indicates a fluctuating voltage or the engine misfires, replace the MAF sensor.
4. With the meter dial set on DC volts, press the rpm button three times until the meter displays voltage frequency. The meter should indicate about 30 hertz with the engine idling.
5. Increase the engine speed and record the meter reading at various speeds.
6. Graph the frequency readings. The MAF sensor frequency should increase smoothly and gradually in relation to engine speed. If the MAF sensor frequency reading is erratic, replace the sensor **(Figure 14)**.

If the MAF sensor produces a DC voltage signal, connect a digital voltmeter from the signal wire to chassis ground. Test the MAF voltage signal at various engine speeds. If the MAF sensor voltage signal at any rpm does not match the specified voltage signal, replace the sensor. A scan tester displays the grams per second of airflow through MAF sensors that provide a frequency or DC voltage signal.

To test the MAF sensor with a lab scope, connect the test leads to the MAF sensor signal wire and to chassis ground. On frequency-type MAF sensors, such as those supplied by General Motors and Ford, the waveform should appear as a series of digital

**You Should Know**

*Tap the MAF sensor while observing the waveform. If this action causes an erratic waveform, the sensor or the connecting wires are defective.*

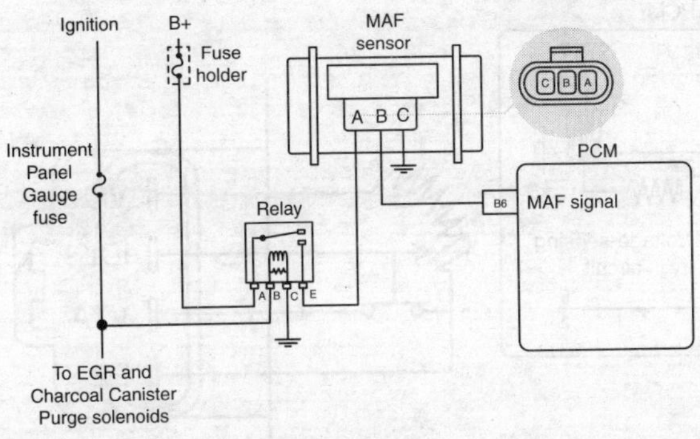


**Figure 13** *MAF sensor circuit.*

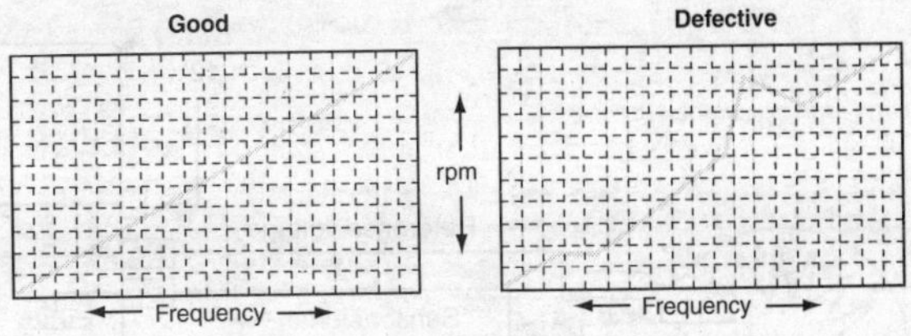


**Figure 14** *Satisfactory and unsatisfactory MAF sensor frequency readings.*

signals **(Figure 15)**. When engine speed and intake airflow are increased, the frequency of the MAF sensor signal increases proportioately. If the MAF or connecting wires are defective, there is an intermittent change in signal frequency **(Figure 16)**.

## Throttle Position Sensor Diagnosis

> **You Should Know**
>
> *The order of diagnostic testing is not set in stone. The presentation given is an example of testing that can be done on the TPS circuit to locate the fault. All the tests will not need to be done to isolate the problem.*

A defective throttle position sensor (TPS) may cause acceleration stumbles, engine stalling, improper idle speed, or a no-start. This sensor is used to modify the pulse-width calculation based on the position of the throttle plate and its rate of change. The TPS may use three or four wires. To test the three-wire TPS, disconnect the TPS connector from the sensor and measure the voltage on the three terminals with the ignition in the Run position. Compare the results with specifications since some systems will have 5 volts, 5 volts, 0 volt on the terminals, respectively, while others will have 5 volts, 0 volt, 0 volt. If these voltages are within specifications, reconnect the connector and backprobe the signal wire from the sensor.

With the ignition switch in the Run position, connect a voltmeter between the signal wire and ground **(Figure 17)**. Typical TPS voltage readings are 0.5 to 1.0 volt with the throttle in the idle position. Always refer to the vehicle manufacturer's specifications. As the throttle blade is opened slowly, observe the voltmeter reading. It should climb smoothly to the maximum specified voltage. Typical maximum voltage is between 3.5 to 4.5 volts at wide-open throttle. If the TPS does not have the specified voltage or if the voltage signal is erratic, replace the sensor.

> **You Should Know**
>
> *When testing the TPS voltage signal, use an analog voltmeter because the gradual voltage increase on this wire is quite visible on the meter pointer. If the sensor voltage increase is erratic, the voltmeter pointer will fluctuate.*

> **You Should Know**
>
> *When the throttle is opened gradually to check the TPS voltage signal, tap the sensor lightly and watch for fluctuations on the voltmeter pointer. Fluctuations indicate a defective sensor.*

To test the sensor ground (return) circuit, perform a voltage drop test. With the ignition switch in the Run position, connect the voltmeter from the sensor ground wire to the battery's ground terminal. If the voltage drop across this circuit exceeds specifications (and the computer ground circuit is good), repair the ground wire from the sensor to the computer.

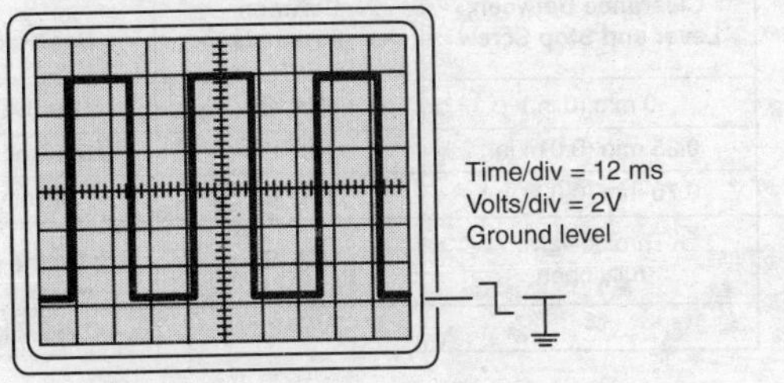


**Figure 15** *Normal waveform of a frequency-type MAF sensor.*

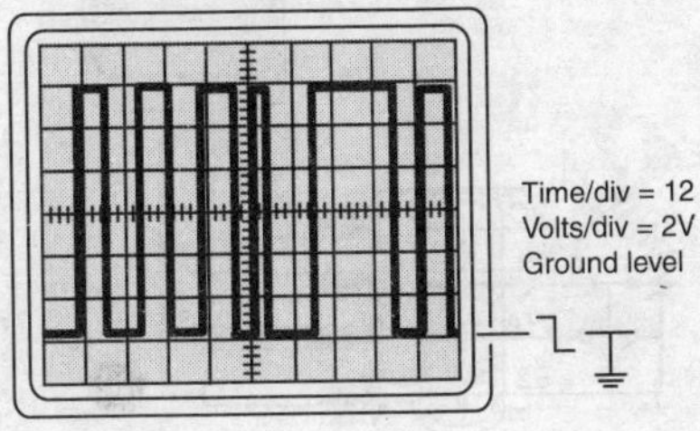


**Figure 16** *Defective waveform of a frequency-type MAF sensor.*

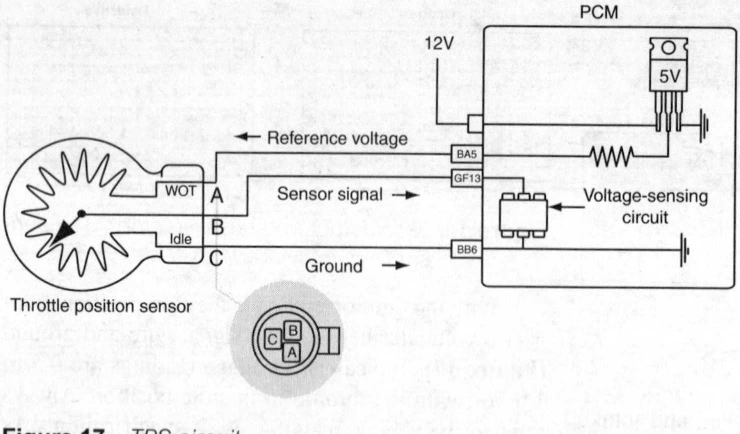


**Figure 17** *TPS circuit.*

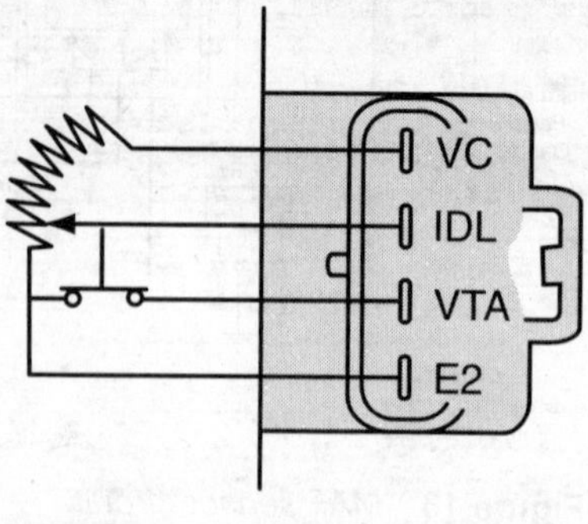


**Figure 18** *Four-wire TPS with idle switch.*

Some EFI systems use a fourth circuit in the TPS that contains an **idle switch**. These sensors have the same wires as a three-wire TPS and an extra wire for the idle switch **(Figure 18)**. The idle switch is used to identify a closed-throttle condition. This information is used to enable the computer's target idle control strategy. This switch input is also used for deceleration fuel shutoff control.

Test the four-wire TPS with an ohmmeter connected from the sensor ground terminal to each of the other terminals **(Figure 19)**. For some of the ohmmeter tests, a feeler gauge blade of a specified thickness is inserted between the throttle lever and the stop. When the ohmmeter is connected from the ground terminal to the valve throttle angle (VTA) terminal, the throttle must be held in the wide-open position **(Figure 20)**.

**Lab Scope Testing of the TPS.** The lab scope is an excellent tool for testing the TPS since it displays every voltage value it detects. This means there is less chance of missing something than there is with a voltmeter or an ohmmeter. Each time the throttle is opened, the TPS should provide a smooth analog voltage signal **(Figure 21)**. If the TPS is defective, glitches will appear in the sensor signal as the throttle is opened **(Figure 22)**. Depending on the TPS circuit design, the glitch for an open can go either upward or downward. A glitch for a short will always go downward.

Remember that the scope trace shows the voltage over time. Evaluate the trace to determine if the results could occur with a good sensor. For example, when the TPS is opened and closed, there should still be a ramping of the voltages involved. As the voltage moves from low to high and back again there should also be a "rounded corner" indicated on the trace. If the trace spikes straight up or down, this would indicate a problem with the sensor or circuit.

**TPS Adjustment.** A TPS adjustment may be performed on some vehicles, but not on many newer applications. Check the vehicle manufacturer's service manual for the TPS adjustment procedure. An improper TPS adjustment can cause inaccurate idle speed, engine stalling, and acceleration stumbles. If the TPS is adjustable, use a scan tool to monitor the TPS signal voltage. If a scan tool is not available,

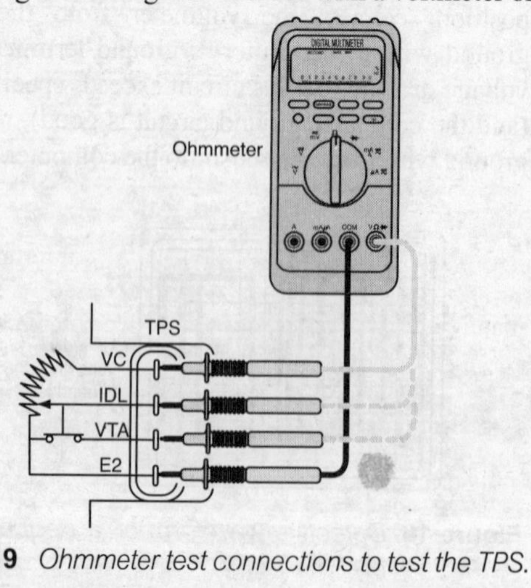


**Figure 19** *Ohmmeter test connections to test the TPS.*

| Clearance Between Lever and Stop Screw | Between Terminals | Resistance |
|---|---|---|
| 0 mm (0 in.) | VTA–E2 | 0.28–6.4 kΩ |
| 0.35 mm (0.014 in.) | IDL–E2 | 0.5 kΩ or less |
| 0.70 mm (0.028 in.) | IDL–E2 | Infinity |
| Throttle valve fully open | VTA–E2 | 2.0–11.6 kΩ |
| – | VC–E2 | 2.7–7.7 kΩ |

**Figure 20** *Example of specifications for TPS ohmmeter tests.*

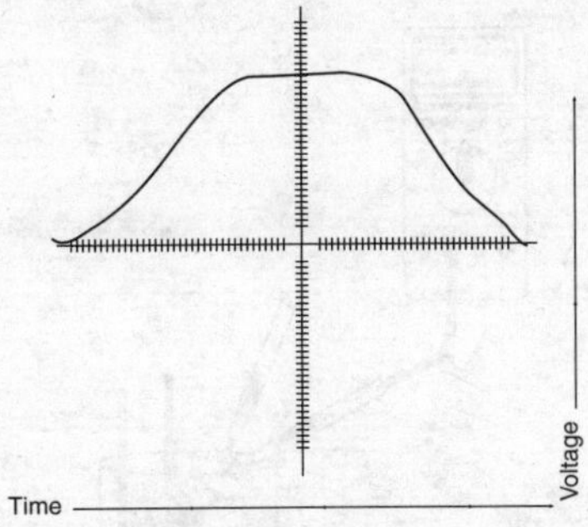


**Figure 21** *Normal TPS scope pattern as it is opened and closed.*

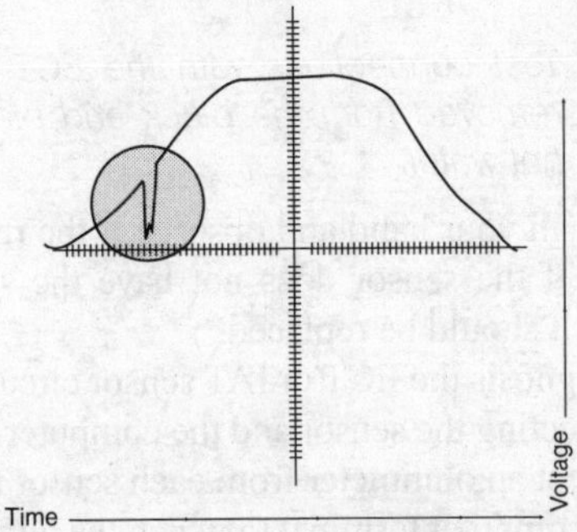


**Figure 22** *Defective TPS waveform.*

backprobe the TPS signal wire and connect a voltmeter from this wire to chassis ground. Turn the ignition switch to the Run position and observe the scan tool or voltmeter reading with the throttle in the idle position. If the TPS does not provide the specified signal voltage, loosen the TPS mounting bolts and rotate the sensor housing until the specified voltage is indicated **(Figure 23)**. While holding the sensor at the correct voltage reading, tighten the mounting bolts to the specified torque.

## Engine Coolant and Idle Air Temperature Sensor Diagnosis

Both the engine coolant temperature (ECT) and the idle air temperature (IAT) sensors are thermistor circuits that operate the same way. Most are negative coefficient thermistors. Since they operate the same, diagnosis is similar *except* an IAT sensor is not placed in liquid.

A defective ECT sensor can cause some of the following problems:

1. Hard engine starting
2. Rich or lean air-fuel ratio
3. Improper operation of emissions devices
4. Reduced fuel economy

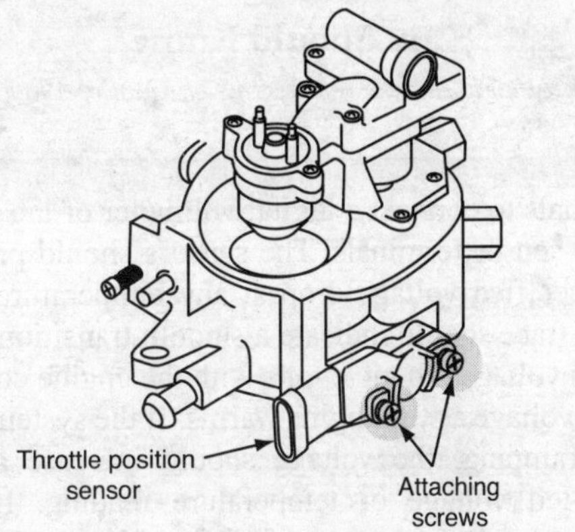


**Figure 23** *Some throttle position sensors have elongated slots to allow adjustments.*

5. Improper converter clutch lockup
6. Hesitation on acceleration
7. Engine stalling

The results of a defective idle IAT sensor may vary depending on the vehicle make and year. A defective IAT sensor may cause the following problems:

1. Rich or lean air-fuel ratio
2. Hard engine starting
3. Engine stalling or surging
4. Acceleration stumbles
5. Excessive fuel consumption

**You Should Know**

*The idle air temperature sensor is also referred to as an intake temperature sensor.*

Some computers use **dual ramping** in the ECT and IAT sensor circuits. The dual-ramping ECT sensor circuit provides a more precise method of determining the engine temperature. This circuit starts with a high voltage value when cold. As the temperature increases, the voltage value will decrease. When the voltage reaches a predetermined value (1.25 volts, for example), the PCM will turn on a transistor that puts a 1,000-ohm resistor in parallel to the 10,000-ohm resistor. This will cause the voltage reading to go high again and provide a second set of inputs.

The computer switches these resistors at approximately 120°F (49°C). This resistance change inside the computer causes a significant change in voltage drop across the sensor as indicated in the specifications **(Figure 24)**. This is a normal condition on any computer with this feature. This change in voltage drop is always detailed in the vehicle manufacturer's specifications.

The ECT and IAT sensors can be tested while they are installed in the engine by backprobing the

> **You Should Know**
>
> *Dual ramping is also referred to as dual resolution or dual range.*

terminals to connect a digital voltmeter or lab scope to the sensor terminals. The sensors should provide the specified voltage drop at any temperature. The scope trace should indicate a smooth transition from a high voltage (about 4 volts with the engine cold) to a low voltage as the engine warms. If the system uses dual ramping, the voltage should increase at the specified voltage or temperature reading. If this change does not occur, the fault is within the PCM.

The ECT sensor may be removed and placed in a container of water with an ohmmeter connected across the sensor terminals **(Figure 25)**. A thermometer is also placed in the water. When the water is heated, the sensor should have the specified resistance at any temperature **(Figure 26)**. Always use the vehicle manufacturer's specifications. If the sensor does not have the specified resistance, replace the sensor.

> **You Should Know**
>
> *Before disconnecting any computer system component, be sure the ignition switch is turned off. Disconnecting components with the switch on may induce high voltages and damage the computer.*

> **You Should Know**
>
> *Never apply an open flame to ECT or IAT sensors for test purposes. This action will damage the sensors.*

The IAT sensor can be removed from the engine and tested with an ohmmeter. With the ohmmeter leads connected to the sensor terminals, check the resistance value at room temperature. Next, hold the

| Cold: 10K-ohm resistor | Hot: 909-ohm resistor |
|---|---|
| – 20°F (– 29°C) 4.7V | 110°F (43°C) 4.2V |
| 0°F (–18°C) 4.4V | 130°F (54°C) 3.7V |
| 20°F (–7°C) 4.1V | 150°F (65°C) 3.4V |
| 40°F (4°C) 3.6V | 170°F (70°C) 3.0V |
| 60°F (15°C) 3.0V | 180°F (82°C) 2.8V |
| 80°F (27°C) 2.4V | 200°F (93°C) 2.4V |
| 100°F (38°C) 1.8V | 220°F (104°C) 2.0V |
| 120°F (49°C) 1.25V | 240°F (116°C) 1.62V |

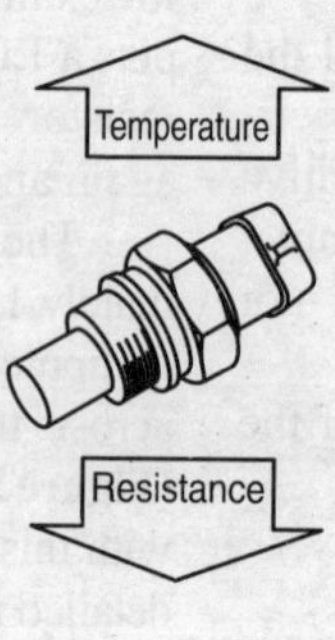


**Figure 24** *Example of voltage specifications for the ECT sensor.*

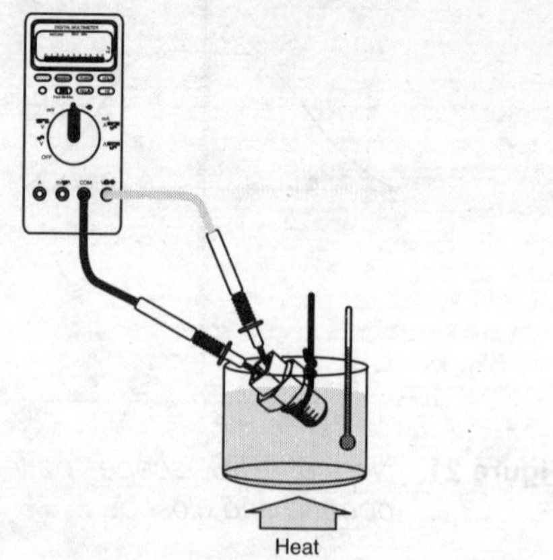


**Figure 25** *Test connections with the ECT sensor removed from the block and placed in hot water.*

IAT sensor in your hand and observe if the resistance decreases. If the sensor does not have the specified resistance, it should be replaced.

To diagnosis the ECT or IAT sensor circuit, begin by disconnecting the sensor and the computer connectors. Connect an ohmmeter from each sensor terminal and to the computer terminal to which the wire is connected. Both of the sensor wires should indicate less resistance than specified by the vehicle manufacturer. If the wires have higher resistance than specified, the wires or wiring connectors must be repaired.

## Oxygen Sensor Diagnosis

A faulty oxygen sensor will result in poor engine performance, increased pollutants, and loss of fuel economy. Generally, a faulty $O_2S$ will not cause an engine no-start. In fact, since the $O_2S$ is not used when the engine is first started, the engine may run very well until the PCM enters closed-loop mode. It is also important to understand cause-and-effect relationships. For example, if a spark plug fails to ignite

| Thermistor temperature to resistance valves | | |
|---|---|---|
| °F | °C | Ohms |
| 210 | 100 | 185 |
| 160 | 70 | 450 |
| 100 | 38 | 1,800 |
| 40 | 4 | 7,500 |
| 0 | –18 | 25,000 |
| –40 | –40 | 100,700 |

**Figure 26** *An example of ECT sensor wiring and resistance specifications.*

the air-fuel charge, the unburned hydrocarbons will pass through the exhaust and into the atmosphere. Since there was no oxygen consumed in the combustion process (since there was no combustion in that cylinder) the $O_2S$ will indicate a high oxygen condition (low $O_2S$ signal voltage). If the technician used only the $O_2S$ signal as the bases of their diagnosis, the constantly low-voltage reading may mistakenly lead to the determination that the sensor is faulty. Actually, the sensor may be working just fine. Other things to consider before condemning the $O_2S$ include vacuum leaks, intake air restrictions, exhaust restrictions, fuel pressure, condition of injectors, EGR operation, PCV function, and any other component or system that may have an effect on the air-fuel ratio.

To test the operation of the $O_2S$, the engine must be at normal operating temperature and the fuel system must be in closed-loop mode. Always follow the test procedure in the vehicle manufacturer's service manual and use the manufacturer's supplied specifications.

**You Should Know**

*An $O_2S$ must be tested with a 10M impedance digital voltmeter. Using an analog meter for this purpose may damage the sensor or computer driver circuit.*

**You Should Know**

*A contaminated $O_2S$ may provide a continually high-voltage reading because the oxygen in the exhaust stream does not contact the sensor.*

A digital voltmeter connected between the $O_2S$ signal wire and chassis ground can be used to test this sensor. Use an electric probe to backprobe the sensor signal wire at the $O_2S$ connector. Connect the positive lead of the voltmeter to the probe. Operate the engine at normal temperature and at idle speed. With the engine in closed-loop mode, the $O_2S$ voltage should be cycling from low to high. A typical $O_2S$ cycles from 0.3 to 0.8 volts.

**You Should Know**

*If possible, avoid probing the insulation to connect a meter to a wire. Over time, the hole left from the probe may allow condensation and/or battery acid to enter the wire and cause problems. Although this will take some to time to happen, the customer will inevitably return for a repair as the result of the hole.*

If the voltage is continually high, the air-fuel ratio may be rich or the sensor may be contaminated. The $O_2S$ may be contaminated from room temperature vulcanizing (RTV) sealant, antifreeze, or additives in the gasoline that contain lead. A rich mixture can be caused by a faulty input sensor (such as an ECT sensor or TPS), a dirty air cleaner element, a leaking injector, and so forth.

When the $O_2S$ voltage is continually low, the air-fuel ratio may be lean due to a vacuum leak, a stuck or plugged injector, an open injector winding, or an open injector circuit. Also, the sensor may be tricked lean by a leaking or cracked exhaust manifold. Other reasons include a defective sensor or high resistance in the wire between the sensor and the computer. A low-voltage $O_2S$ output can also be the result of a faulty spark plug or spark plug cable. Since none of the oxygen is used during the combustion process, a high oxygen content passes the $O_2S$. Remember, the $O_2S$ does not measure hydrocarbons, only oxygen content.

If the $O_2S$ voltage signal remains in a midrange position, the computer may be in open-loop mode, the sensor may be defective, or the sensor circuit is open.

The $O_2S$ can also be tested once it is removed from the engine. Connect a digital voltmeter between the signal wire and the sensor case. Heat the sensor element with a propane torch while observing the voltmeter. The propane flame keeps the oxygen in the air away from the sensor element while heating it, causing the sensor to produce voltage. While the sensor element is in the flame, the voltage should be nearly 1 volt. Once the sensor is removed from the flame, the voltage should immediately drop to zero. If the sensor does not produce the specified voltage, it should be replaced.

If a defect in the $O_2S$ signal wire is suspected, backprobe the sensor signal wire at the computer and connect a digital voltmeter from the signal wire to ground with the engine idling. The difference between the voltage readings at the sensor and at the computer should not exceed the vehicle manufacturer's specifications (usually 0.2 volt). Also, the circuit can be checked with an ohmmeter by disconnecting the connector at the PCM and at the $O_2S$ and measuring the resistance of the wire between these two points.

Some oxygen sensors are grounded through the threaded connection to the exhaust manifold or pipe. Most new model vehicles have a designated ground circuit for the $O_2S$. A voltage drop test of the ground circuit is the best way to determine any circuit problems. With the engine idling, connect a digital voltmeter from the sensor case to the sensor ground wire on the computer. A typical maximum voltage drop reading across the sensor ground circuit is 0.2 volt. Always use the vehicle manufacturer's specifications. If the voltage drop across the sensor ground exceeds those specifications, repair the ground wire or the sensor ground in the exhaust manifold.

**Oxygen Sensor Testing with a Lab Scope.** The more accurate way to diagnose automotive computer systems is with a lab scope (DSO).The DSO has a much faster signal-sampling speed than most other types of test equipment. Before diagnosing an $O_2S$, the engine must be at normal operating temperature and the computer must be in closed-loop mode. When diagnosing input sensor waveforms, adjust the horizontal time base on the lab scope so three waveforms appear on the screen **(Figure 27)**. When testing any input sensor, connect the scope leads from the sensor signal wire to ground. The $O_2S$ voltage should be cycling continually in the 0- to 1-volt range. Observe the number of **cross counts** that occur **(Figure 28)**. $O_2S$ voltage signal cross counts are the number of times the $O_2S$ voltage signal changes above or below 0.45 volt per second. At 2,000 rpm, there should be two to three cross counts every second (use the record function of the scope). If the $O_2S$'s cross counts are below specifications, it may be defective or partially contaminated. If the voltage remains about 0.5 volt with very little fluctuation, the signal wire or sensor may have an open circuit.

### You Should Know

*When using a DSO, always follow the DSO manufacturer's recommended operating procedures. Improper DSO use may result in tester damage.*

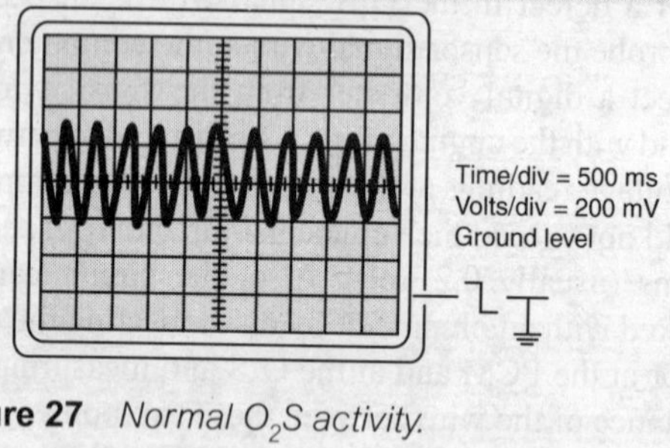


**Figure 27** *Normal $O_2S$ activity.*

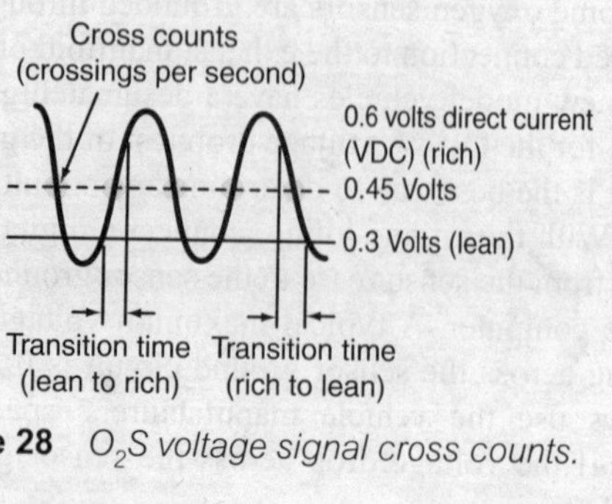


**Figure 28** *$O_2S$ voltage signal cross counts.*

If the sensor's voltage toggles between 0 volt and 500 millivolts, it is biased low or lean. As a result, the computer will be constantly adding fuel to try to reach the upper limit of the sensor. Something is causing the sensor to be biased lean, such as low fuel pump pressure, a plugged injector, an open spark plug wire, a faulty spark plug, a vacuum leak, and so on. If the toggling only occurs at the higher limits of the voltage range, the sensor is biased rich. In either case, the computer does not have true control of the air-fuel mixture because of the faulty $O_2S$ signals.

### You Should Know

*Always maintain the propane cylinder in an upright position and position the cylinder and the hose away from rotating components.*

Propane enrichment may be used to force the fuel system into a very rich condition and to verify the $O_2S$ operation. To verify rich-to-lean $O_2S$ switching, connect the propane hose to a large intake manifold port such as the brake booster hose or simply insert it into the nozzle of the air-cleaner assembly. Connect the lab scope to the sensor and a good ground. Set the scope to display the trace at 200 ms per division and 500 ms per division. With the engine running, inject some propane while observing the $O_2S$ signal's trace. This will drive the mixture rich. Most good oxygen sensors will produce almost 1 volt when driven full rich. The typical specification is at least 800 mV. Under this condition, the $O_2S$ signal should increase from 300 to 600 mV in 100 ms **(Figure 29)**. Quickly disconnect the propane hose and expose the brake booster hose to the atmosphere.

### You Should Know

*If equipped with an air pump, it should be disabled during the propane enrichment test since unwanted air may bias the results.*

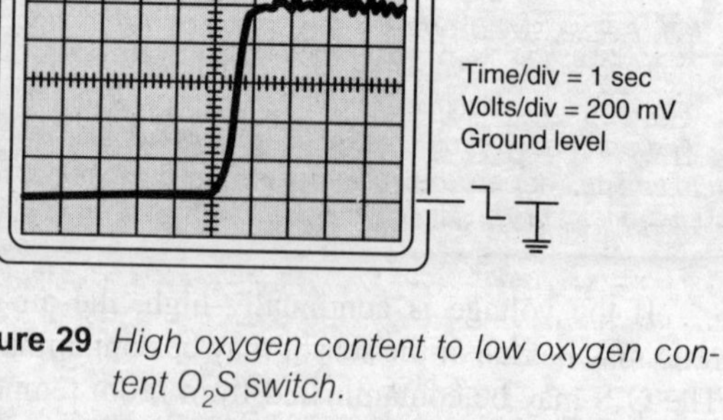


**Figure 29** *High oxygen content to low oxygen content $O_2S$ switch.*

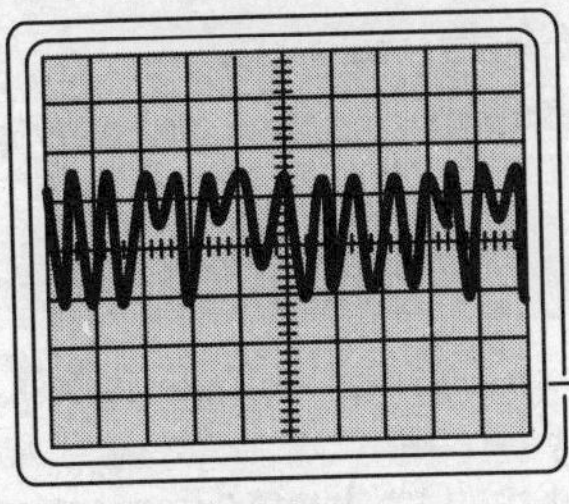


**Figure 30** *$O_2S$ waveform indication of a shorted spark plug.*

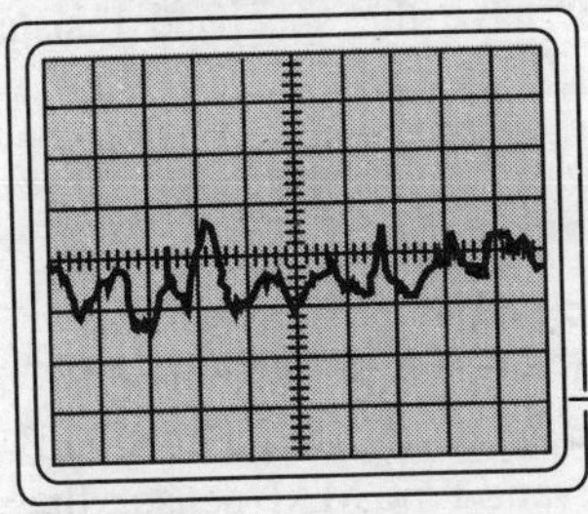


**Figure 31** *$O_2S$ waveform indication of an open spark plug cable.*

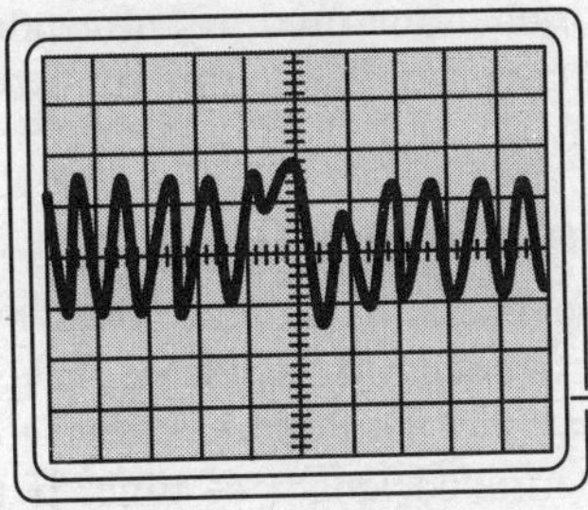


**Figure 32** *$O_2S$ waveform indication of a leaking injector.*

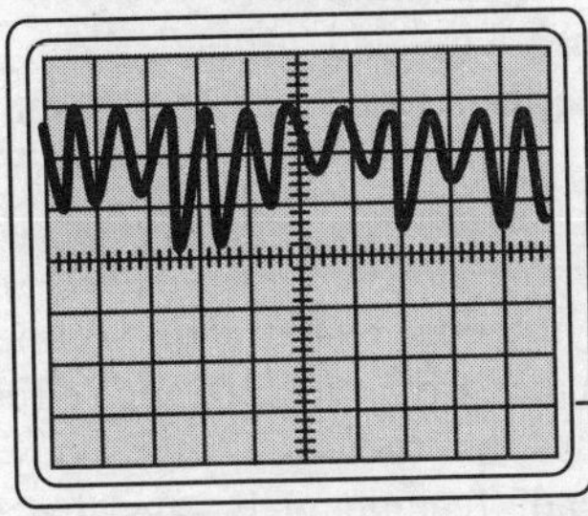


**Figure 33** *$O_2S$ waveform indication of a defective injector.*

The $O_2S$ signal should drop from 600 to 300 mV in less than 125 ms and it should go below 175 mV. If the voltage fails to meet any of the specifications, the $O_2S$ is bad and should be replaced.

With practice and experience, it is possible to diagnose other engine performance problems by observing the trace of an $O_2S$. **Figure 30** and **Figure 31** show how ignition problems affect the signal from the $O_2S$. Keep in mind that during complete combustion, nearly all of the oxygen in the combustion chamber is combined with the fuel. This means there will be little oxygen in the exhaust of a very efficient engine. As combustion becomes more incomplete, the levels of oxygen increase. Ignition problems cause incomplete combustion and there is much oxygen in the exhaust. This is also true of lean mixtures, over-advanced ignition timing, or anything else that causes incomplete combustion.

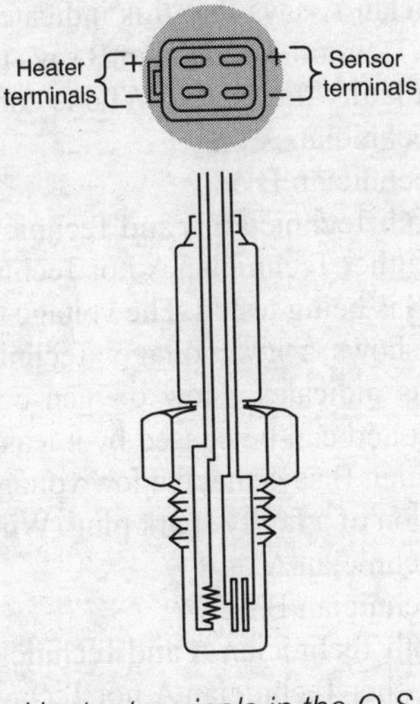


**Figure 34** *Heater terminals in the $O_2S$.*

When the mixture is rich, combustion has a better chance of being complete. Therefore, the oxygen levels in the exhaust decrease. The $O_2S$ output will respond to the low oxygen with a high voltage signal **(Figure 32** and **Figure 33)**. Remember that the PCM will always drive the air-fuel mixture to try to do the opposite direction of what it receives from the $O_2S$. When the $O_2S$ shows lean, the PCM commands the air-fuel ratio to go rich, and vice versa. When a lean exhaust signal is not caused by an air-fuel problem, the PCM does not know the

> **You Should Know**
>
> *On some vehicles, the heaters are supplied voltage from a relay such as the fuel pump relay or the ASD relay. If all other circuits that receive voltage from these relays function properly, the relays are not at fault. However, for these relays to stay energized, the engine must be running. Voltage will be present at the heater terminal at the $O_2S$ only momentarily if the sensor is checked with the ignition switch in the Run position.*

true cause and will richen the mixture in response to the signal. This may make the engine run worse than it did before.

## Oxygen Sensor Heater Diagnosis

If the $O_2S$ heater is not operating properly, the sensor warm-up time is extended. Most PCMs have a closed-loop timer that puts the PCM to closed-loop mode after a predetermined length of time. If the heater fails, the computer supplies a richer air-fuel ratio. To test the heater, disconnect the $O_2S$ connector and connect a digital voltmeter between the heater voltage supply wire and ground. With the ignition switch in the Run position, 12 volts should be supplied on this wire. If the voltage is less than that, repair the supply circuit.

The resistance of the $O_2S$ heater element can be checked and compared with the specifications. With the $O_2S$ disconnected from the engine harness, connect an ohmmeter across its heater terminals **(Figure 34).** If the heater does not have the specified resistance, replace the sensor.

> **You Should Know**
>
> *Be sure to measure the resistance of the heater element and not the $O_2S$. Ohmmeter testing of the sensor will damage it.*

# Review Questions

1. A MAP sensor is being tested. The voltmeter reads zero when measuring the reference circuit. Technician A says that the circuit wire may be shorted to ground. Technician B says that the circuit wire has an open. Who is correct?
   A. Technician A only
   B. Technician B only
   C. Both Technician A and Technician B
   D. Neither Technician A nor Technician B
2. The TPS is suspected of being faulty. Voltmeter testing indicates 5 volts on the signal circuit when backprobed at the sensor, regardless of the throttle blade position. Technician A says that this can be caused by an open sensor return circuit. Technician B says that this can be caused by a short to ground in the voltage reference circuit. Who is correct?
   A. Technician A only
   B. Technician B only
   C. Both Technician A and Technician B
   D. Neither Technician A nor Technician B
3. Technician A says that a four-wire TPS contains an idle switch. Technician B says that all throttle position sensors are adjusted at wide-open throttle. Who is correct?
   A. Technician A only
   B. Technician B only
   C. Both Technician A and Technician B
   D. Neither Technician A nor Technician B
4. Technician A says that higher-than-normal idle speed may be caused by low electrical system voltage. Technician B says that higher-than-normal idle speed may be caused by a defective ECT sensor. Who is correct?
   A. Technician A only
   B. Technician B only
   C. Both Technician A and Technician B
   D. Neither Technician A nor Technician B
5. When testing a vane-type MAF sensor, the vane is moved from the closed to wide-open position. The voltmeter reading on the signal circuit drops to zero. Technician A says that this is the reset function of the sensor and it is normal. Technician B says that this indicates an open in the variable resistor. Who is correct?
   A. Technician A only
   B. Technician B only
   C. Both Technician A and Technician B
   D. Neither Technician A nor Technician B
6. The ECT sensor is being tested. While observing the lab scope trace, the voltage steadily decreases as the engine warms. At about 1.25 volts, the voltage value suddenly jumps to about 4.0 volts. Technician A says that this indicates an open in the ECT circuit. Technician B says that this indicates a faulty thermistor. Who is correct?
   A. Technician A only
   B. Technician B only
   C. Both Technician A and Technician B
   D. Neither Technician A nor Technician B
7. An $O_2S$ is being tested. The voltage trace of a lab scope shows a low voltage. Technician A says that this indicates a low oxygen content in the exhaust and can be caused by a leaking injector. Technician B says that the low voltage may be an indication of a faulty spark plug. Who is correct?
   A. Technician A only
   B. Technician B only
   C. Both Technician A and Technician B
   D. Neither Technician A nor Technician B

CHAPTER

# 22 Injector Testing and Service

## Introduction

Injectors can fail due to electrical continuity problems, worn internal parts, or from deposits forming on them. In particular, port fuel-injection (PFI) injectors, because of their location, are subjected to high amounts of heat and are prone to forming tip deposits. Regardless of the type of failure, the common symptoms of injector failures are as follows:

1. Lean surge at low speeds
2. Acceleration stumbles
3. Hard starting
4. Acceleration sag, cold engine
5. Engine misfiring
6. Rough engine idle
7. Lack of engine power
8. Slow starting when cold
9. Stalling after a cold start

This chapter covers typical procedures used to diagnose and service fuel injectors.

## Injector Sound Test

The **injector sound test** is a method of quickly checking the operation of the pintle on engines where the injectors are accessible. The injector sound test is simply listening to the injector as it is running. A port fuel injector that is not functioning may cause a cylinder misfire at low engine speeds. With the engine idling, a stethoscope pickup may be placed on the side of the injector body **(Figure 1)**. Each injector should produce the same rhythmic clicking noise. If the injector clicking noise is erratic, the injector plunger may be sticking. If an injector does not produce any clicking noise, the injector, the connecting wires, or the PCM may be defective. Checking the injector resistance and performing the noid light test should help to locate the cause of the problem.

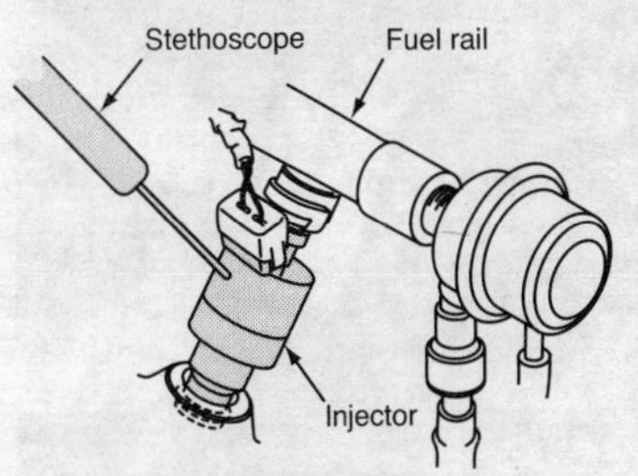


**Figure 1** *Performing an injector sound test.*

> **You Should Know**
>
> *If the connectors to the injectors are difficult to access, the sound test may be a better option than the injector balance test.*

## Injector Ohmmeter Test

By disconnecting the injector wire harness from the injector, an ohmmeter may be connected across the injector terminals to check the injector winding resistance. If the ohmmeter reading is infinite, the injector winding is open. An ohmmeter reading below the manufacturer's specified value indicates the injector winding is shorted and the injector should be replaced. If the injector winding has the correct amount of resistance, the problem is likely in the wire harness or the PCM.

> **You Should Know**
>
> *The injector part number may change in mid–model year. Be sure to replace the injector with the correct unit. Usually the connector end of the injector is color coded by the manufacturer.*

## Noid Light Test

If the injector is accessible, a noid light can be used to check the circuit operation. Connect the noid light to the harness side of the connector. When the engine is cranked, the noid light flashes as the computer cycles the injector on and off. If the light is not flashing, the computer or the connecting wires are defective.

## Lab Scope Testing of the Injector

To use the lab scope, connect the leads to the switched side of the injector and to ground. When the PCM energizes the injector, the voltage should decrease **(Figure 2)**. The voltage should remain low while the PCM keeps the injector grounded. When the PCM opens the injector ground and the injector shuts off, a voltage spike should occur from the induced voltage in the injector winding. If the voltage spike for one injector is lower than the spikes for the other injectors, the injector with the low spike probably has a shorted winding or resistance in the circuit.

Resistance in the ground (control) side of the injector will also be indicated by a ramping up in voltage while the injector is energized **(Figure 3)**. In addition, the induced voltage will not be consistent.

If the PCM operates the injectors with a **peak-and-hold current**, the waveform will have a voltage spike when the PCM reduces the injector current and a second spike when the injector shuts off **(Figure 4)**. In peak-and-hold current injector systems, the PCM grounds the injector winding and the current flow increases rapidly to 4 amperes. When the current flow reaches this value, a current-limiting circuit in the PCM limits the current flow to 1 ampere for the remainder of the injector pulse width. The peak-and-hold function provides faster injector armature opening and closing. When the waveform is aligned properly with the horizontal grids on the screen, the injector pulse width can be measured **(Figure 5)**. If

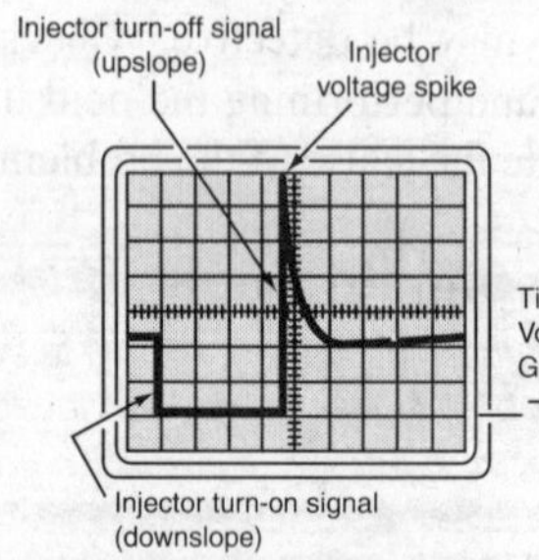


**Figure 2** *Normal injector waveform.*

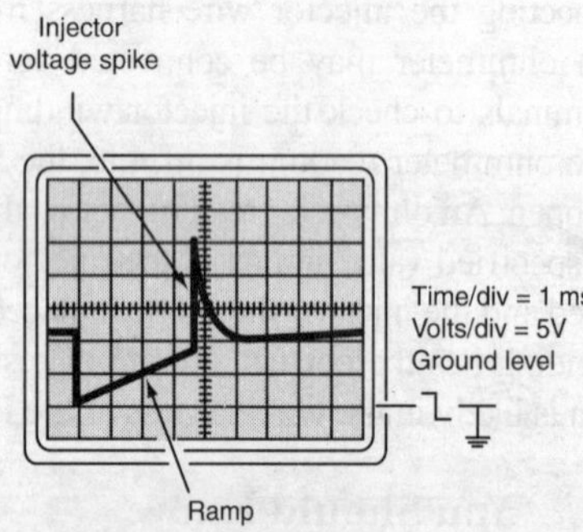


**Figure 3** *High resistance in the control circuit of the injector.*

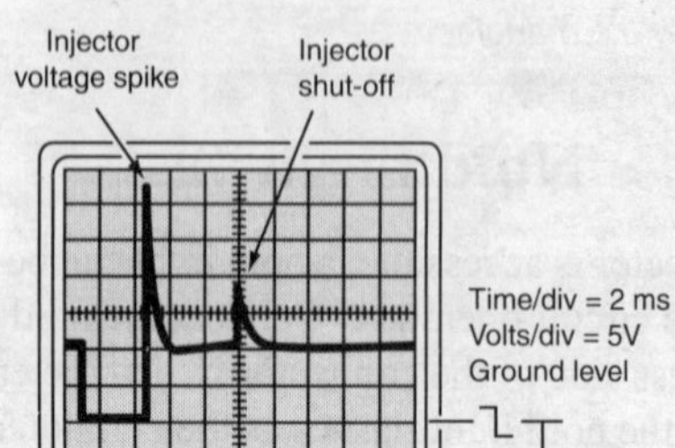


**Figure 4** *Normal peak-and-hold injector waveform.*

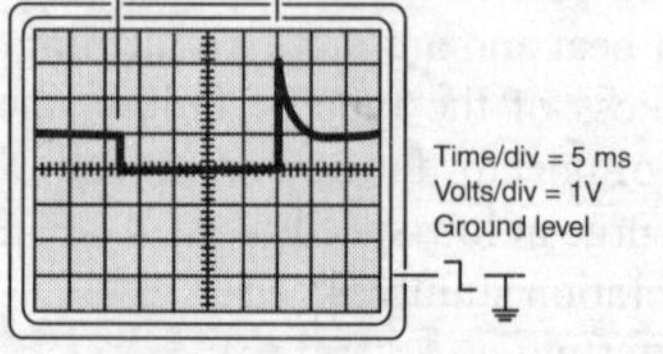


**Figure 5** *Measuring injector pulse width with a lab scope.*

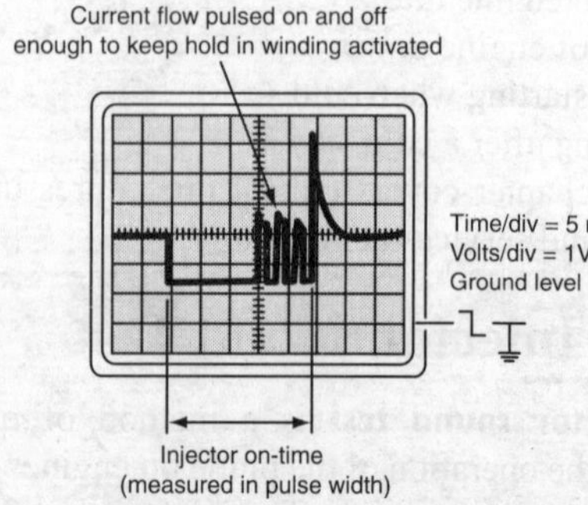


**Figure 6** *Pulse-width modulated signal used to reduce injector current.*

**Figure 7** *Special injector pulse tester.*

the injector waveform is ragged, the injector driver in the PCM may be degraded. In some fuel systems, the PCM grounds the injector winding and then uses a pulse-width modulated (PWM) signal to reduce injector current **(Figure 6)**.

## Injector Balance Test

An **injector balance test** may be performed to diagnose restricted injectors. The test determines if each cylinder is receiving the same amount of fuel through the injectors. Restricted injector orifices cause acceleration stumbles, engine stalling, and erratic idle operation. Perform this test if all the electrical tests fail to find the problem. A fuel pressure gauge and an **injector pulse tester** are required for this test **(Figure 7)**. That injector pulse tester contains a timer circuit that energizes each injector for an exact time period when the timer button is pressed.

Check the fuel pressure before performing the injector balance test. Begin by installing the tester leads to the injector terminals and to the battery terminals. Cycle the ignition switch on and off until the specified fuel pressure appears on the fuel gauge. Leave the ignition switch in the Off position. Push the timer button on the tester and record the gauge reading. When the timer energizes the injector, fuel is discharged from the injector into the intake port and the fuel pressure drops in the fuel rail. Repeat this procedure for all the injectors and record the fuel pressure after each injector is energized by the timer. Compare the gauge readings on each injector. When the injectors are in satisfactory condition, the fuel pressure will be the same after each injector is energized by the timer. If an injector orifice or tip is restricted, the fuel pressure will not drop as much when the injector is energized by the timer. If an injector's fuel pressure is 1.4 psi (10 kPa) above or below the average pressure when tested, the injector is defective. The injector plunger is sticking open if excessive pressure drop occurs when the injector is energized. Sticking injector plungers may result in a rich air-fuel mixture.

In addition to this method, many scan tools provide for an **injector kill test** for sequential fuel-injection (SFI) systems. For this test, the scan tool will shut off the injectors one at a time so the technician can observe rpm drop. With the engine running and the rpms stabilized, the scan tool will request the PCM to kill one injector at a time. Engine rpm should drop equally for each injector killed. With this test, it is important to remember that factors other than the injector may cause a cylinder to have a low rpm drop. Low compression, faulty spark plugs, a worn spark plug cable, or an open injector circuit will also cause a low rpm drop.

> **You Should Know**
>
> *In some engines, it is very difficult to gain access to the injector terminals to hook up the injector balance test. Some scan tools provide an activation test that can be substituted for the injector balance test.*

## Injector Flow Test

Some vehicle manufacturers recommend an **injector flow test** rather than the balance test. The injector flow test measures the amount of fuel that flows through an injector during a specified length of time. This test requires removing the fuel rail and injectors from the engine. Leave all the injectors attached to the fuel rail. Place the tip of each injector into a calibrated container. Connect a jumper wire between the B+ and FP terminals in the data link connector (DLC) in the same fashion as the fuel pump pressure test. Turn the ignition switch to the Run position and connect a special jumper wire from the terminals of the injector being tested to the battery terminals **(Figure 8)**. Energize the injector for 15 seconds and

> **You Should Know**
>
> *Due to the high cost of injectors, attempt to clean the injectors first. If cleaning fails to restore injector operation, replace the injector.*

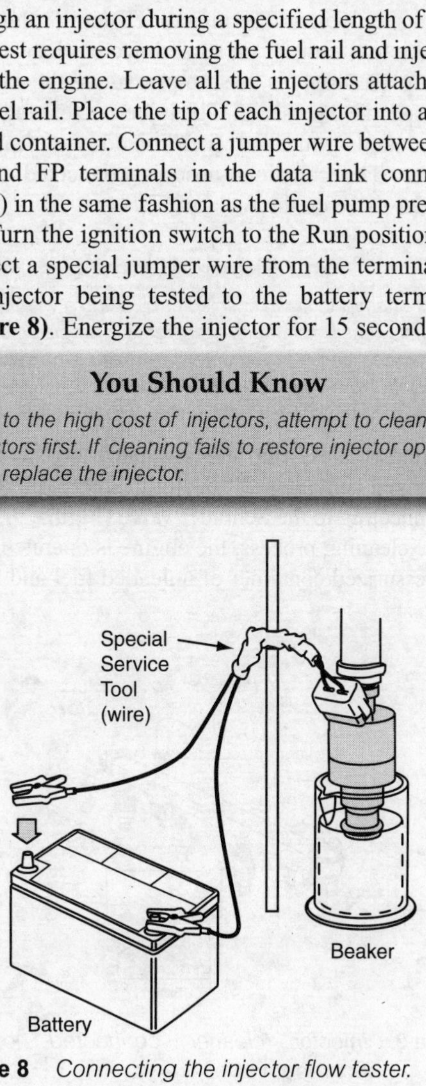


**Figure 8** *Connecting the injector flow tester.*

then disconnect the jumper wire from the negative battery cable. Record the amount of fuel in the calibrated container.

Repeat this procedure for each injector. If the volume of fuel discharged from any injector varies more than 0.3 cu. in. (5 cc) from the specifications, the injector should be replaced.

## Injector Service

If the injector tests indicate that some of the injectors are restricted, the injectors will need to be cleaned or replaced. The following are typical procedures for performing this service.

### Injector Cleaning

Tool manufacturers market a variety of injector cleaning equipment. The injector cleaning solution is poured into a canister on some injector cleaners and the shop air supply is used to pressurize the canister to the specified pressure. The injector cleaning solution contains unleaded fuel mixed with injector cleaner. The container hose is connected to the Schrader valve on the fuel rail.

> **You Should Know**
>
> *Never attempt to clean an injector by using a wire brush or other utensil. The metering holes are precision sized and any scraping may cause the orifices to become distorted, resulting in inaccurate operation.*

Automotive parts stores usually sell a sealed pressurized container of injector cleaner with a hose for connecting to the Schrader valve **(Figure 9)**. During the cleaning process, the engine is operated from the pressurized container of unleaded fuel and injector cleaner. The fuel pump operation must be stopped to prevent the pump from forcing fuel up the fuel rail and the fuel-return line must be plugged to prevent the solution in the cleaning container from flowing through the return line into the fuel tank. Follow these steps for the injector cleaning procedure:

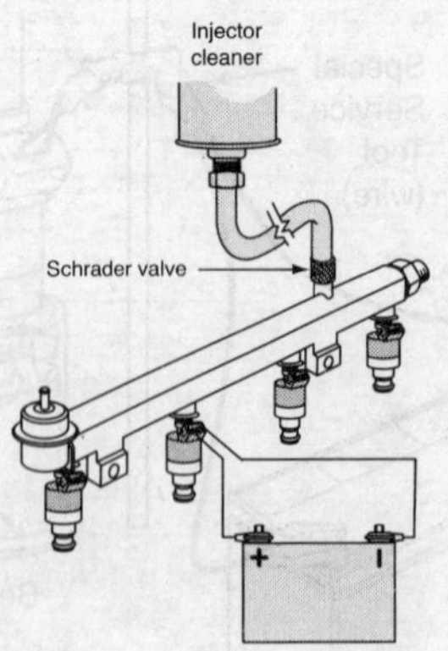


**Figure 9** *Injector cleaner connected to the Schrader valve.*

1. Disconnect the wires from the in-tank fuel pump or the fuel pump relay to disable the fuel pump. To disconnect the fuel pump relay on General Motors products, the oil pressure switch in the fuel pump circuit must also be disconnected to prevent current flow through this switch to the fuel pump.
2. Plug the fuel-return line from the fuel rail to the tank.
3. Connect a can of injector cleaner to the Schrader valve on the fuel rail.
4. Open the cleaner's control valve one-half turn and start the engine.
5. Adjust the pressure from the cleaner to 5 psi (34.47 kPa) below the fuel system operating pressure.
6. Run the engine at 1,000 rpm for 10 to 15 minutes on the injector solution.

> **You Should Know**
>
> *After cleaning the injectors, you might suggest that the customer use a fuel additive to help keep the injectors clean.*

### Injector Replacement

Whenever it becomes necessary to replace an injector, always consult the proper service information for the procedure. Some systems are straightforward and replacement is fairly simple. On the other hand, some require the removal of several components just to access the injectors. Be sure to install the O-rings in their proper locations. Lubricate the O-rings to aid in installation.

The following is a typical procedure for replacing injectors in a port fuel-injection system:

1. Disconnect the negative battery cable.
2. Disconnect the vacuum hose to the pressure regulator.
3. Disconnect the wiring harness connectors to the injectors.
4. Remove all hold-down fasteners to the fuel rail. The clips will keep the injectors attached to the fuel rail. Clean the area around the injectors.
5. Pull up on the fuel rail to lift the injectors out of the intake manifold. Be sure to retrieve all O-rings.

**You Should Know**

*Be sure to relieve system pressure before disconnecting any components.*

6. Remove the old injector and install a new O-ring in the new injector. Lubricate the O-ring.
7. Align the injectors to their holes and set the fuel rail assembly in place.
8. Torque the fuel rail hold-down fasteners to the specified value.
9. Reinstall all components and turn the ignition switch to the Run position. Check fuel pressure and check for any leaks.

## Resetting Adaptive Memory

After the injectors are cleaned or replaced, the engine may still run rough. This happens because the adaptive memory in the computer has been correcting for the restricted injectors. Since the plugged injectors were supplying a lean air-fuel ratio, the computer increased the pulse width to try to bring the air-fuel ratio back to stoichiometric. With the cleaned or replaced injectors, the adaptive memory is still supplying the increased pulse width. This action makes the air-fuel ratio too rich since the restricted injector problem no longer exists. With the engine at normal operating temperature, drive the vehicle for at least 5 minutes to allow the adaptive memory to relearn. After this time, the computer should supply the correct injector pulse width and the engine should run smoothly. This same problem may occur when any defective computer system component is replaced.

# Review Questions

1. Technician A says that a defective injector may cause cylinder misfire at idle speed. Technician B says that restricted injector tips may result in acceleration stumbles. Who is correct?
   A. Technician A only
   B. Technician B only
   C. Both Technician A and Technician B
   D. Neither Technician A nor Technician B
2. Explain the purpose of the injector sound test.
3. Technician A says that a noid light can be used to test the injector circuit. Technician B says that if the noid light does not flash while the injector is being activated by the PCM, the injector must be replaced. Who is correct?
   A. Technician A only
   B. Technician B only
   C. Both Technician A and Technician B
   D. Neither Technician A nor Technician B
4. Explain the purpose of the injector balance test.
5. Explain why adaptive memory should be reset after repairing a fuel-injection system problem.
6. Describe how to test an injector using an ohmmeter.
7. Technician A says that if the lab scope trace of an injector firing shows a voltage spike that is 2 volts higher than the source voltage, the circuit is good. Technician B says that if the voltage on the trace does not go to zero while the injector is energized, the PCM is faulty. Who is correct?
   A. Technician A only
   B. Technician B only
   C. Both Technician A and Technician B
   D. Neither Technician A nor Technician B
8. Technician A says that the injector kill test is a function of some scan tools. The tool will shut off the injectors one at a time so the technician can observe rpm drop. Technician B says that the injector flow test measures the amount of fuel that flows through an injector during a specified length of time. Who is correct?
   A. Technician A only
   B. Technician B only
   C. Both Technician A and Technician B
   D. Neither Technician A nor Technician B

CHAPTER

# 23 Scan Tester Diagnostics

## Introduction

The scan tester (also referred to as a scanner) provides an easy method for the technician to retrieve diagnostic trouble codes (DTCs), evaluate input and output data, and perform activation tests. Some scan testers offer additional functions and can be used as DMMs, lab scopes, and digital pressure gauges. In addition, many scan testers have the ability to record the data stream information for the technician. This chapter discusses typical uses of the scan tester.

## Scan Tester Functions

Several makes of scan testers are available to read fault codes and perform other diagnostic functions. The exact tester buttons and test procedures vary on these testers, but many of the same basic diagnostic functions can be performed regardless of the tester make. When test procedures are performed with a scan tester, the following precautions must be observed:

1. Always follow the directions in the manual supplied by the scan tester manufacturer.
2. Do not connect or disconnect any connectors or components with the ignition switch in the Run position. This includes the scan tester power wires and the connection from the tester to the vehicle diagnostic connector.
3. Never short across or ground any terminals in the electronic system except those recommended by the vehicle manufacturer.
4. If the computer terminals must be removed, disconnect the scan tester diagnostic connector first.
5. Observe the service precautions listed previously in this section.

A typical scan tester keypad contains these buttons **(Figure 1)**:

1. *Numbered keys* are digits 0 through 9.
2. *Up/Down arrow keys* allow the technician to move back and forward through test modes and menus.
3. *Enter key* enters information into the tester.
4. *Mode key* allows the technician to interrupt the current procedure and go back to the previous modes.

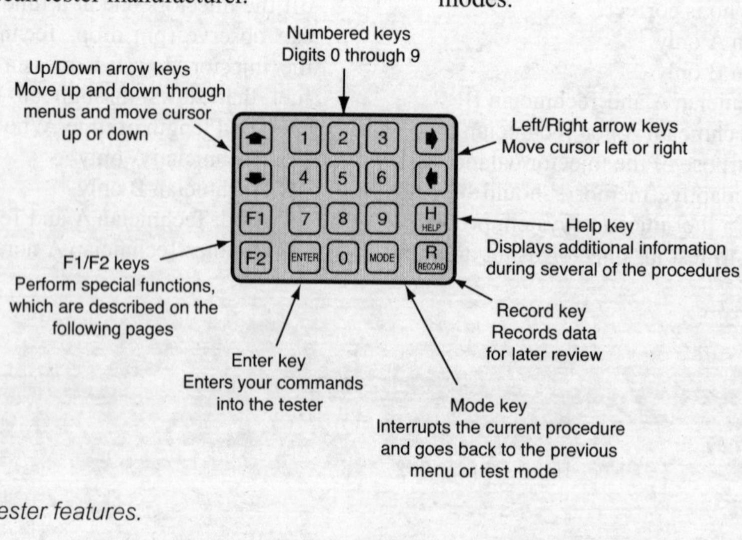


**Figure 1** *Scan tester features.*

5. *F1 and F2 keys* allow the technician to perform special functions described in the scan tester manufacturer's manual.

When the technician has programmed the scan tester by performing the initial entries, some entry options will appear on the screen. These entry options vary depending on the scan tester and the vehicle being tested. The technician presses the number beside the desired selection to proceed with the test procedure. In the first four selections, the tester asks the technician to select the computer system to be tested. If data line is selected, the scan tester provides a voltage reading from each input sensor in the system.

When the technician makes a selection from the initial test selection menu, the scan tester moves on to the actual test selections. These selections vary depending on the scan tester and the vehicle being tested. The following list includes many of the possible test selections and a brief explanation:

1. *Faults codes* display fault codes on the scan tester display.
2. *Switch tests* allow the technician to operate switch inputs such as the brake switch to the PCM. Each switch input should change the reading on the tester.
3. *Actuator test mode (ATM) tests* force the PCM to cycle all the solenoids and relays in the system for 5 minutes or until the ignition switch is turned off.
4. *Sensor tests* provide a voltage reading from each sensor.
5. *Automatic idle speed (AIS) motor test* forces the PCM to operate the AIS motor when the Up and Down arrows are pressed. The engine speed should increase 100 rpm each time the Up arrow is pressed. Rotations per minute are limited to 1,500 or 2,000 rpm.
6. *Solenoid state tests* or *output state tests* display the on or off status of each solenoid in the system.
7. *Emission maintenance reminder (EMR) tests* allow the technician to reset the EMR module. The EMR light reminds the driver when emissions maintenance is required.
8. *Wiggle test* allows the technician to wiggle solenoid and relay connections. The scan tester emits an audible beep if a loose connection is present.
9. *Key on, engine off (KOEO) test* allows the technician to perform this test with the scan tester on Ford products.
10. *Computed timing check* forces the PCM to move the spark advance 20 degrees ahead of the initial timing setting on Ford products.
11. *Key on, engine running (KOER) test* allows the technician to perform this test on Ford products.
12. *Clear memory* or *erase codes* quickly erase fault codes in the PCM's memory.
13. *Code library* reviews fault codes.
14. *Basic test* allows the technician to perform a faster test procedure without prompts.
15. *Cruise control test* allows the technician to test the cruise control switch inputs if the cruise module is in the PCM.

## Trouble Code Retrieval

With a typical scan tool, DTCs are retrieved by first installing the proper module into the scan tool **(Figure 2)**. With the ignition key in the Off position, connect the scan tester power cables to the vehicle's battery terminals, observing polarity. Select the proper scan tool data cable for the vehicle being tested and connect the scan tester cable to the DLC. Once the scan tool is powered up, enter the model year and the vehicle identification number (VIN)

**Figure 2** *Some scan testers require modules for different vehicle manufacturers.*

**Figure 3** *The scan tester will provide screen prompts.*

code in the scan tester (if required). Finally, follow the screen prompts to retrieve the DTCs **(Figure 3)**.

## Input Circuit Diagnostics

Since the scan tool provides the voltage value that the PCM is sensing from an input circuit, it is easier to diagnose circuit fault. For example, if the technician is monitoring the input voltage of a two-wire input sensor such as the ECT and the scan tool indicates 5 volts on the circuit, this reveals that the circuit has an open. The open could be anywhere in the circuit between the PCM and ground (including the sensor). If the monitored voltage showed zero, this usually indicates a short to ground. The short could be in the sensor or in the circuit between the PCM and the sensor. A reading of 0 volt may also be caused by a faulty PCM not sending the reference voltage to the circuit.

## Snapshot

Many scan testers have **snapshot** capabilities for some vehicles that allow the technician to operate the vehicle under the exact condition when a certain problem occurs and freeze the sensor voltage readings into the tester memory. This feature is very useful for locating intermittent problems. After driving the vehicle back to the shop, the technician can play back the recorded readings **(Figure 4)**. During the playback, the technician watches closely for a momentary change in any sensor or actuator readings that indicates a defective sensor, actuator, or circuit. This action is similar to taking a series of "snapshots" and then reviewing the pictures later. The snapshot test procedure can be performed on most vehicles equipped with a data line from the computer to the DLC.

## Block Learn and Integrator

While using a scan tester on General Motors vehicles, one of the test modes will display block learn and integrator. A scale of 0 to 255 is used for both displays; a midrange reading of 128 is preferred. These numbers represent the binary code used by the PCM for fuel correction. The number 128 represents the middle of this range and is then considered as no correction. The $O_2S$ signal is sent to the integrator chip and then to the pulse-width calculation chip in the PCM; the block learn chip is connected in parallel to the integrator chip. If the $O_2S$ voltage changes once, the integrator chip and the pulse-width calculation chip change the injector pulse width. If the $O_2S$ provides four continually high- or low-voltage signals then the block learn chip makes another injector pulse-width change. When the integrator or block learn numbers are considerably greater than 128, the PCM is continually attempting to increase fuel; therefore, the $O_2S$ voltage signal must be continually low (lean). If the integrator or block learn numbers are considerably below 128, the PCM is

91VR3301 E-Z Event Plus
Chrysler 1991 3.3/3.8 MFI VIN
Event 1 Tag – Frame -30
Fault Codes: No

| | -15 | -14 | -13 | -12 | -11 | -10 | -9 | -8 | -7 | -6 | -5 | -4 | -3 | -2 | -1 | 0 |
|---|---|---|---|---|---|---|---|---|---|---|---|---|---|---|---|---|
| Coolant Temp | 120 | 120 | 120 | 120 | 120 | 120 | 120 | 120 | 120 | 120 | 120 | 120 | 120 | 120 | 120 | 120 |
| RPM | 0000 | 0000 | 0000 | 0000 | 0000 | 0000 | 0000 | 0000 | 0000 | 0000 | 0000 | 0000 | 0000 | 0000 | 0000 | 0000 |
| Inj Pulse Width | 14.7 | 14.7 | 14.7 | 14.7 | 14.7 | 14.7 | 14.7 | 14.7 | 14.7 | 14.7 | 14.7 | 14.7 | 14.7 | 14.7 | 14.7 | 14.7 |
| MAP | 29.4 | 29.4 | 29.4 | 29.4 | 29.4 | 29.4 | 29.4 | 29.4 | 29.4 | 29.4 | 29.4 | 29.4 | 29.4 | 29.4 | 29.4 | 29.4 |
| O2 Volts | 0.50 | 0.50 | 0.50 | 0.50 | 0.50 | 0.50 | 0.50 | 0.50 | 0.50 | 0.50 | 0.50 | 0.50 | 0.50 | 0.50 | 0.50 | 0.50 |
| Throt Pos Sensor | 0.66 | 0.66 | 0.66 | 0.66 | 0.66 | 0.66 | 0.66 | 0.66 | 0.66 | 0.66 | 0.66 | 0.66 | 0.66 | 0.66 | 0.66 | 0.66 |
| Adaptive Fuel Fac | -03 | -03 | -03 | -03 | -03 | -03 | -03 | -03 | -03 | -03 | -03 | -03 | -03 | -03 | -03 | -03 |
| Baro Pressure | 29.4 | 29.4 | 29.4 | 29.4 | 29.4 | 29.4 | 29.4 | 29.4 | 29.4 | 29.4 | 29.4 | 29.4 | 29.4 | 29.4 | 29.4 | 29.4 |
| Baro Pressure Vol | 4.76 | 4.76 | 4.76 | 4.76 | 4.76 | 4.76 | 4.76 | 4.76 | 4.76 | 4.76 | 4.76 | 4.76 | 4.76 | 4.76 | 4.76 | 4.76 |
| Battery Temp | 065 | 065 | 065 | 065 | 065 | 065 | 065 | 065 | 065 | 065 | 065 | 065 | 065 | 065 | 065 | 065 |
| Battery Volts | 12.2 | 12.2 | 12.2 | 12.2 | 12.2 | 12.2 | 12.2 | 12.2 | 12.2 | 12.2 | 12.2 | 12.2 | 12.2 | 12.2 | 12.2 | 12.2 |
| Charge Temp | 105 | 105 | 105 | 105 | 105 | 105 | 105 | 105 | 105 | 105 | 105 | 105 | 105 | 105 | 105 | 105 |
| Charg Sys Targ Vol | 13.9 | 13.9 | 13.9 | 13.9 | 13.9 | 13.9 | 13.9 | 13.9 | 13.9 | 13.9 | 13.9 | 13.9 | 13.9 | 13.9 | 13.9 | 13.9 |
| Cruise Control Cut | KEY | KEY | KEY | KEY | KEY | KEY | KEY | KEY | KEY | KEY | KEY | KEY | KEY | KEY | KEY | KEY |
| Cruise Control | OFF | OFF | OFF | OFF | OFF | OFF | OFF | OFF | OFF | OFF | OFF | OFF | OFF | OFF | OFF | OFF |
| Cruise Control Sta | CRSW | CRSW | CRSW | CRSW | CRSW | CRSW | CRSW | CRSW | CRSW | CRSW | CRSW | CRSW | CRSW | CRSW | CRSW | CRSW |
| Cruise Set Speed | 000 | 000 | 000 | 000 | 000 | 000 | 000 | 000 | 000 | 000 | 000 | 000 | 000 | 000 | 000 | 000 |
| Dis Cam Status | NO | NO | NO | NO | NO | NO | NO | NO | NO | NO | NO | NO | NO | NO | NO | NO |
| Dis Crank Status | NO | NO | NO | NO | NO | NO | NO | NO | NO | NO | NO | NO | NO | NO | NO | NO |
| Electronic Adv/Re | 016 | 016 | 016 | 016 | 016 | 016 | 016 | 016 | 016 | 016 | 016 | 016 | 016 | 016 | 016 | 016 |
| Idle Air Control | 041 | 041 | 041 | 041 | 041 | 041 | 041 | 041 | 041 | 041 | 041 | 041 | 041 | 041 | 041 | 041 |
| Idle RPM x 10 | 078 | 078 | 078 | 078 | 078 | 078 | 078 | 078 | 078 | 078 | 078 | 078 | 078 | 078 | 078 | 078 |
| Knock Active | NO | NO | NO | NO | NO | NO | NO | NO | NO | NO | NO | NO | NO | NO | NO | NO |
| Knock Volts | 0.00 | 0.00 | 0.00 | 0.00 | 0.00 | 0.00 | 0.00 | 0.00 | 0.00 | 0.00 | 0.00 | 0.00 | 0.00 | 0.00 | 0.00 | 0.00 |
| Map Volts | 4.56 | 4.56 | 4.56 | 4.56 | 4.56 | 4.56 | 4.56 | 4.56 | 4.56 | 4.56 | 4.56 | 4.56 | 4.56 | 4.56 | 4.56 | 4.56 |
| TPS-Minimum | 0.84 | 0.84 | 0.84 | 0.84 | 0.84 | 0.84 | 0.84 | 0.84 | 0.84 | 0.84 | 0.84 | 0.84 | 0.84 | 0.84 | 0.84 | 0.84 |
| Vacuum | 00 I | 00 I | 00 I | 00 I | 00 I | 00 I | 00 I | 00 I | 00 I | 00 I | 00 I | 00 I | 00 I | 00 I | 00 I | 00 I |
| Vehicle Speed Sen | 000 | 000 | 000 | 000 | 000 | 000 | 000 | 000 | 000 | 000 | 000 | 000 | 000 | 000 | 000 | 000 |

**Figure 4** *Example of a snapshot.*

continually decreasing fuel. This indicates that the $O_2S$ voltage must be always high (rich).

## Cylinder Output Test

A **cylinder output test** can be performed on electronic engine control IV (EEC IV) SFI engines at the end of the KOER test. The cylinder output test is available on many SFI Ford products regardless of whether the KOER test was completed with a scan tester or with the flash code method. In the cylinder output test, the PCM stops grounding each injector (one at a time) for about 20 seconds, causing each cylinder to misfire. While each cylinder is misfiring, the PCM looks at the amount of rpm that the engine slows down.

When the KOER test is completed, momentarily push the throttle wide open to start the cylinder output test. After this throttle action, the PCM may require up to 2 minutes to enter the cylinder output test.

If the engine rpm does not drop while a cylinder is misfiring, there is a problem in the injector, in the ignition system, or in the engine compression, or there is a vacuum leak. If the engine does not slow down as much on one cylinder as compared with all the other cylinders, a fault code is set in the PCM memory; for example, Code 50 indicates a problem in the Number 5 cylinder. The correct DTC list must be used for each model year. Repeat the test three times or until a Code 90 (pass code) is received.

# Ford Breakout Box Testing

In Chrysler and General Motors products, a data line is connected from the computer to the DLC. The sensor voltage signals are transmitted on this data line and read on the scan tester. If these sensor voltage signals are normal on the scan tester, the technician knows that these signals are reaching the computer.

In 1989, Ford introduced data links on some Lincoln Continental models. Since that time, Ford has gradually installed data links on its engine computers to comply with OBD-II requirements. If a Ford vehicle has data links, there are two extra pairs of wires in the DLC. Ford has introduced a new generation star (NGS) tester that has the capability to read data on some Ford engine computers. This technology is available on other scan testers as well.

Since many Ford products prior to 1996 do not have a data line from the computer to the DLC, the sensor voltage signals cannot be displayed on the scan tester. Therefore, some other method must be used to prove that the computer is receiving the sensor voltage signals. A breakout box is available from Ford and some other suppliers **(Figure 5)**.

**Figure 5** *A breakout box can be connected in series with the PCM terminals on Ford vehicles to allow meter test connections.*

Two large wiring connectors on the breakout box allow it to be connected in series with the PCM wiring. Once the breakout box is connected, each PCM terminal is connected to a correspondingly numbered breakout box terminal. These terminals match the terminals on Ford PCM wiring diagrams. For example, in many Ford EEC IV systems, PCM terminals 37 and 57 are 12-volt supply terminals from the power relay to the PCM. Therefore, with the ignition switch in the Run position and the power relay closed, a digital voltmeter may be connected from breakout box terminals 37 and 57 to chassis ground and battery voltage should be available at these terminals. Use the wiring diagram for the model year and system being tested to identify the breakout box terminals.

# Fuel-Injection Data Recording Interpretation

As stated earlier, many scan tools have a snapshot feature. In addition, some have the ability to take a **data recording (Figure 6)** that can be played back in a dynamic data display. The data recording function helps to diagnose a problem that cannot be duplicated in the shop by recording PCM management functions in an actual driving mode. This is ideal for diagnosing intermittent problems. In addition, some systems come with a smaller version of the scan tool that can be placed in the vehicle and operated by the customer. When a problem is encountered, the customer pushes a button and the tool will record the event. The customer then returns to the shop and the technician retrieves the recording.

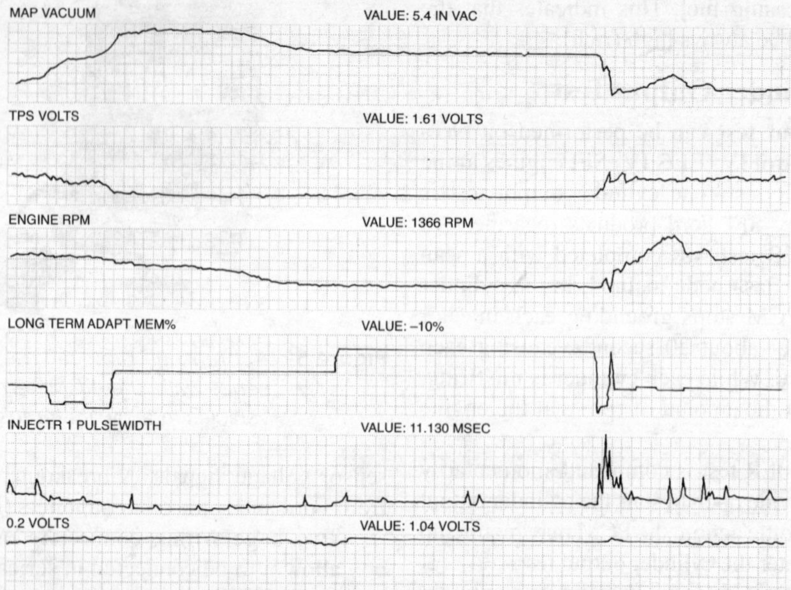


**Figure 6** *A data recording provides a method to capture the data stream information.*

Although setup and operation differ among scan tools from different manufacturers, proper interpretation of the recordings is critical to diagnosing the problem. Usually the data recording is about 45 seconds in length, starting 35 seconds before the trigger and running for 10 seconds after that. With many scanners, the default time after trigger can be changed. Once the recording is displayed, there are usually means to manipulate the graphs. The operator's manual for the scan tool will indicate how to record, retrieve, and manipulate the recordings.

## Method of Approach to Interpreting Data Recordings

There are several things to consider when looking at a data recording. These include:

1. Can the condition shown really happen? For example, is it possible for the ECT reading to go from 102°F (39°C) to 1°F (−17°C) in 5 seconds? **(Figure 7)**
2. If there is a sudden change in input voltages or output activation, find two additional supporting indicators. For example, if the TPS is reading 3.8 volts, the engine should be operating at wide-open throttle. If this is really the case, then MAP and engine rpm should support this.
3. Study the line characteristics. Since the graph line is a representation of values over time, vertical changes indicate changes that *may* be too fast to be physically possible. In addition, notice the pattern the line makes when it changes direction. Also note if the line pattern is consistent or erratic.

Examine the data recording shown in **Figure 8**. Notice the MAP vacuum drops suddenly and the MAP voltage increases suddenly. Is it possible for the MAP vacuum to drop that much under normal conditions? If so, what will support this conclusion? Obviously MAP vacuum cannot drop that fast with a wide-open throttle condition. The engine rpm graph does not indicate any increase in engine speed. In fact, it shows a decrease due to the increased injector pulse width. Injector pulse width is increased because the MAP indicates the engine is under load. Notice the TPS value does not support a low-vacuum condition either. This graph indicates an open in the MAP circuit. Notice that the second time the MAP volts go to 5 volts, the MAP vacuum does not drop as much. This is because the PCM has entered limp-in mode and the vacuum is now a calculated value. This is why it is important to look at sensor voltage more so than the interpreted value.

Notice the engine rpm, MAP vacuum, and MAP volts graphs in **Figure 9**. These lines are sawtooth shaped with constant amplitudes. Since MAP voltage is acceptable, MAP vacuum can be used to

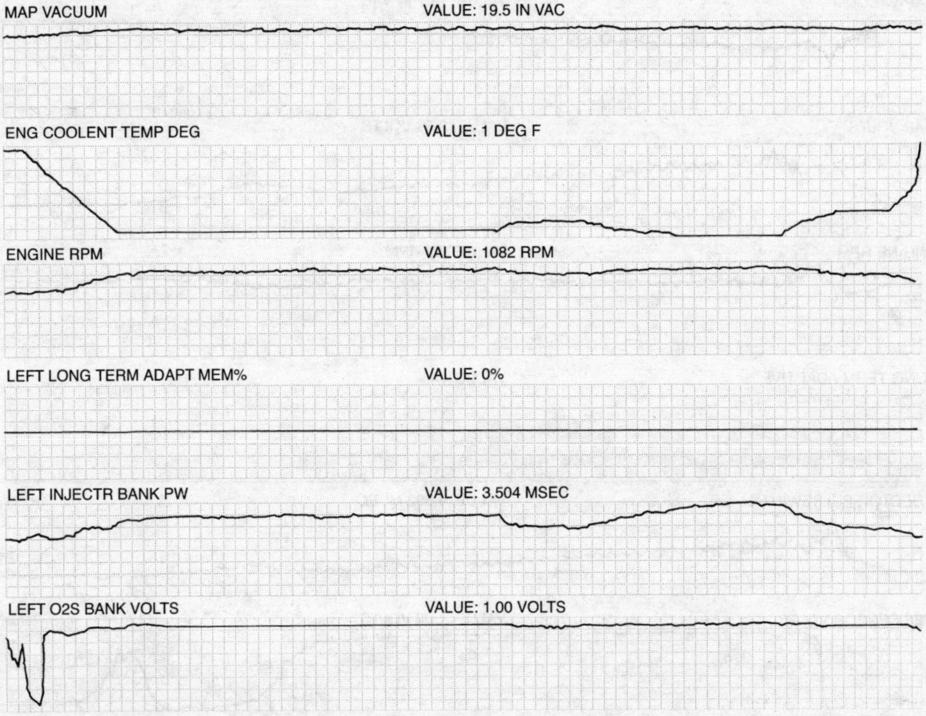


**Figure 7** *Is it really possible for the engine temperature to drop so quickly?*

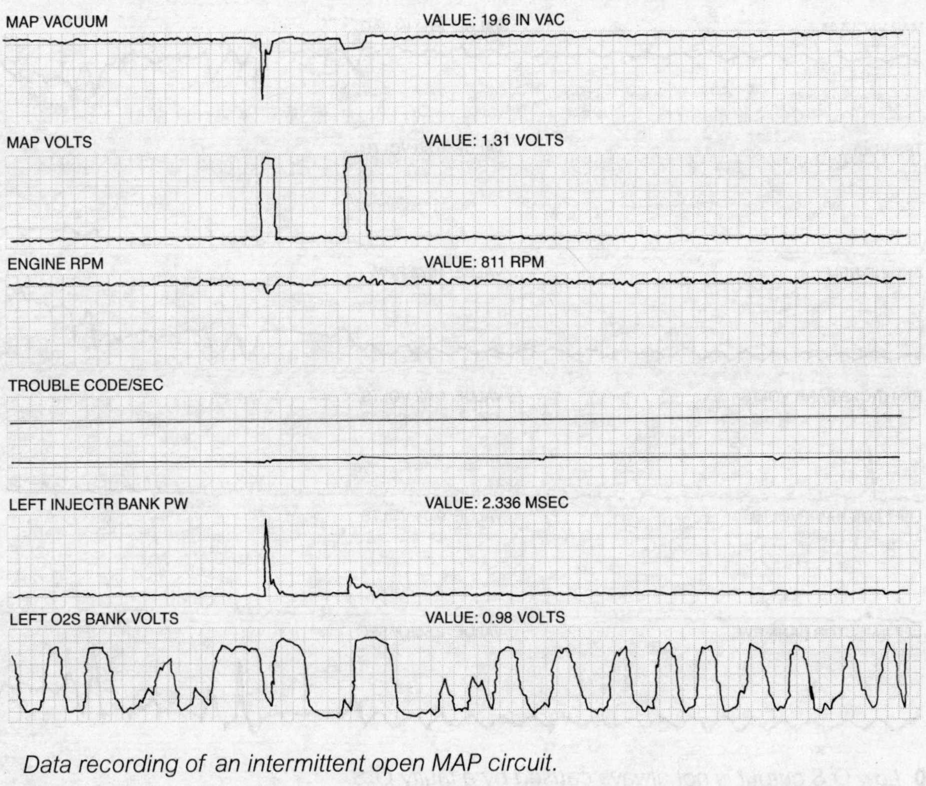


**Figure 8** *Data recording of an intermittent open MAP circuit.*

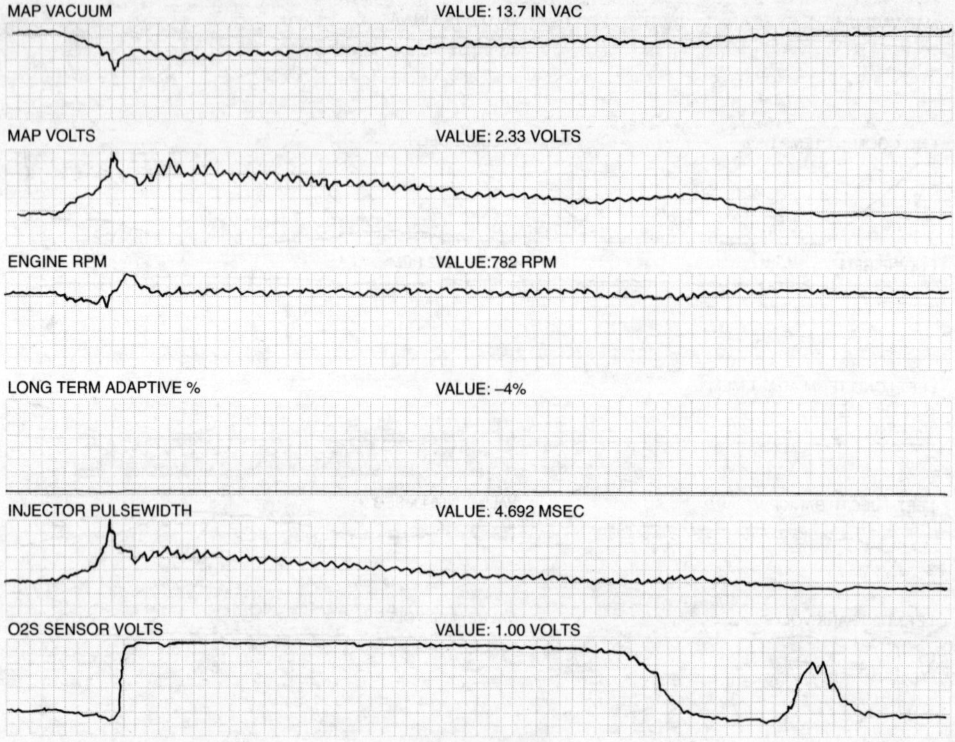


**Figure 9** *Use the MAP reading in a similar fashion to a vacuum gauge.*

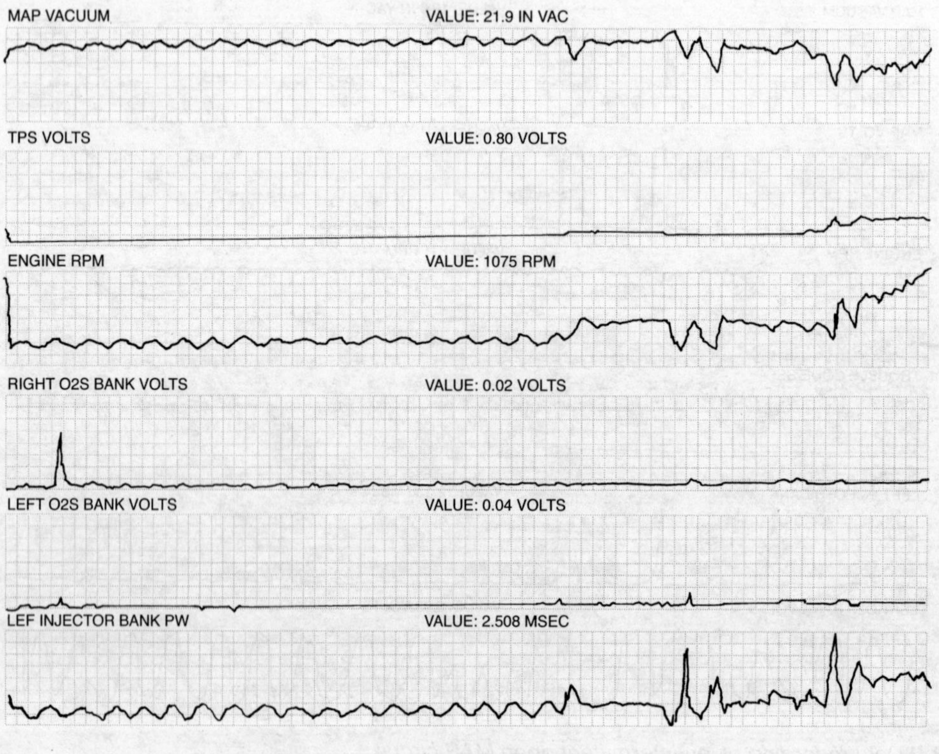


**Figure 10** *Low $O_2S$ output is not always caused by a faulty $O_2S$.*

determine the cause of this condition. Think of the vacuum reading on the graph as a vacuum gauge. The sawtooth pattern indicates a gauge needle that would be fluctuating. This usually indicates a leaking valve.

**Figure 10** is a data recording graph of a vehicle proceeding after a stop sign. Notice the graphs indicate that the engine is stumbling. Both left and right oxygen sensors indicate a low voltage. It is unlikely that two sensors would fail at the same time. Also, the vehicle drives fine at speeds above 30 mph. In this case, the $O_2S$ heaters are not working. At higher engine speeds, the exhaust gases are hot enough to heat the sensors, but at idle they cool and do not produce a valid signal. This is not a problem with the oxygen sensors. Rather, it is due to a common splice to the 12-volt supply.

### You Should Know

*Diagnosis is an extremely important part of a technician's job on today's high-tech vehicles. Always take time to diagnose a customer's vehicle accurately. Fast, inaccurate diagnosis of automotive problems leads to unnecessary and expensive repairs. It also leads to unhappy customers who may take their business to another shop. Accurate diagnosis may take more time but in the long term it will improve customer relations and keep them coming back.*

Interpreting data recordings requires an understanding of how the system is supposed to work. Based on that understanding, scan through the graphs looking for something out of the ordinary. As in reading fault codes, an incorrect graph line does not necessarily mean that a component is faulty.

## Review Questions

1. Technician A says that the scan tester can be used to erase fault codes on many EFI systems. Technician B says that many scan testers will store sensor readings during a road test and then play back the results in a snapshot test mode. Who is correct?
   A. Technician A only
   B. Technician B only
   C. Both Technician A and Technician B
   D. Neither Technician A nor Technician B
2. Explain how a scan tester can make input circuit testing easier.
3. Technician A says that when interpreting a data recording, the technician should consider if the condition shown can really happen. Technician B says that if there is a sudden change in input voltage or output activation, two additional supporting indicators should be found. Who is correct?
   A. Technician A only
   B. Technician B only
   C. Both Technician A and Technician B
   D. Neither Technician A nor Technician B
4. While viewing a scan tool, the technician observes that the integrator number is 180 and the block learn number is 185. Technician A says that these numbers indicate a normal condition. Technician B says that these numbers indicate that the PCM is trying to increase fuel delivery; therefore, the $O_2S$ signal must be continually lean. Who is correct?
   A. Technician A only
   B. Technician B only
   C. Both Technician A and Technician B
   D. Neither Technician A nor Technician B
5. Why do some vehicles require the use of a breakout box?

SECTION 6

# EMISSIONS CONTROL SYSTEMS

## Section Objectives

After you have read, studied, and practiced the contents of this section, you should be able to:

- Explain the emissions laws resulting from the Clean Air Act and their impact on the automotive industry.
- Describe the combustion process and resulting emissions.
- Explain how hydrocarbon exhaust emissions are produced and what affects their levels.
- Explain how carbon monoxide is produced and what affects its levels.
- Explain the production of nitrogen oxide and what affects its levels.
- Describe the nonpollutant exhaust gases and how they are used to determine combustion efficiency.
- Detail the importance of the air-fuel ratio in controlling exhaust emissions.
- Explain typical types of precombustion controls such as engine design, computerized engine controls, ignition control, evaporative controls, exhaust gas recirculation (ERG), and crankcase ventilation.
- Explain how evaporative emissions are contained.
- Describe how ERG is used to control nitrogen oxide emissions as well as to control the operation of different types of EGR valves .
- Explain the function of the PCV system.
- Describe the components used for postcombustion control, including air injection and catalytic converters.
- Zero, calibrate, purge, and flush an exhaust analyzer in preparation of testing the vehicle's exhaust.
- Properly precondition the engine for an exhaust analysis.
- Diagnose and service evaporative (EVAP) systems.
- Diagnosis the canister and the canister purge valve.
- Test the PCV system for proper operation.
- Diagnose engine performance problems caused by improper EGR operation.
- Diagnose and service port EGR valves.
- Diagnose and service negative-back-pressure EGR valves.
- Diagnose and service positive-back-pressure EGR valves.
- Diagnose and service digital EGR valves.
- Diagnose and service linear EGR valves.
- Diagnose EGR back-pressure transducers.
- Diagnose and service air-injection systems.
- Check the efficiency of a catalytic converter.
- Diagnose catalytic converter malfunctions.

### Interesting Fact

*The first emissions control device required by law was one that prevented crankcase vapors from escaping to the atmosphere. Prior to the introduction of positive crankcase ventilation (PCV) systems, crankcase blowby gases were dumped into the atmosphere through a "road draft tube." Fresh air entered the crankcase through an open breather cap on the oil filler tube. The fresh air then circulated through the engine and exited through the road draft tube along with moisture and blowby gases from the crankcase. It was not a very efficient way to vent the crankcase nor did it help the growing air pollution problem. In fact, prior to PCV systems, nearly 20 percent of the pollutants vehicles dumped into the atmosphere came from open crankcases.*

CHAPTER

# 24 Emissions and the Automotive Industry

## Introduction

There are three types of harmful automotive emissions. The first is **evaporative emissions** from the fuel system. Evaporative emissions are raw fuel vapors that drift into the atmosphere. This evaporation is a direct result of the property of gasoline to vaporize at low temperatures; it is not the result of the combustion process. The second type is **crankcase emissions** resulting from blowby around the piston rings and engine oil vapors. The third type is **exhaust emissions**, a product of combustion. Exhaust emissions are produced when the engine is running as a by-product of combustion or incomplete combustion. The exhaust by-products this chapter is concerned with are hydrocarbon (HC), carbon monoxide (CO), and nitrogen oxide ($NO_x$).

## Emissions Laws and the Automotive Industry

In the beginning of the automotive era, emissions were not of much concern. The old engines would chug and belch out smoke, but to most people the pollution produced from these "newfangled contraptions" was more of a nuisance. However, as the number of automobiles being driven increased, it soon became clear that pollution had to be controlled. According to the EPA, pollutant emissions estimates for the period of 1940 to 1990 showed that air pollution in the United States included 18.7 million metric tons of volatile organic compounds (mainly hydrocarbon), 60.1 million metric tons of carbon monoxide, and 19.6 million metric tons of nitrogen oxide. Automobiles are responsible for 17.8 percent of the total hydrocarbon emissions, 30.9 percent of the total carbon monoxide emissions, and 11.1 percent of the total nitrogen oxide emissions. As a result, the United States has passed and amended legislation several times to limit the pollutant output of the automobile.

Under Title II of the Clean Air Act, manufacturers are required to control motor vehicle emissions of the criteria pollutants: hydrocarbon, carbon monoxide, nitrogen oxide, and particulate matter (PM). The EPA and the California Air Resources Board (CARB) have established standards and measurement procedures for exhaust, evaporative, and refueling emissions of these criteria pollutants. Manufacturers are responsible for developing and certifying vehicles with emissions systems and emissions-related components that are durable for the full useful life of the vehicles. Regulations require emissions certification of vehicles prior to their commercial introduction. Regulations also require in-use compliance to ensure that standards are met during the useful life of the vehicle.

## History of Emissions Laws

In 1959, California established the first standards for automotive emissions. Over the years, the federal government has passed legislation in an effort to maintain air quality standards **(Table 1)**. The original Clean Air Act became law in 1963. The Clean Air Act Amendments of 1970 established the EPA and gave the agency broad authority to regulate vehicle pollution. Specific responsibilities to reduce emissions were set for government and private industry. Since then, the agency's pollution and automotive emissions standards have become increasingly more stringent. And over the years the EPA has passed legislation regulating various automotive systems.

The 1970 legislation required a 90 percent reduction in tailpipe emissions for new 1975 and 1976 automobiles. This presented the automakers with major technical and economic challenges. Nevertheless, the EPA successfully forced the adoption of two major control technologies: the catalytic converter in 1975 and the three-way catalyst in 1981.

| Year | Legislation |
|---|---|
| 1963 | Original Clean Air Act passed into law |
| 1970 | Clean Air Act amendments to current policies amended |
| 1970 | Environmental Protection Agency formed |
| 1971 | Evaporative emissions standards enacted |
| 1972 | First inspection and maintenance (I/M) program introduced |
| 1973 | $NO_x$ exhaust standards enacted |
| 1975 | First catalytic converter introduced |
| 1989 | Gasoline vo latility standards enacted |
| 1990 | New Clean Air Act amended to current policies |
| 1995 | I/M 240 testing for gasoline vehicles required in nonattainment zones |
| 1996 | OBD-II vehicle compliance required |

**Table 1** *EPA legislation history.*

The Clean Air Act Amendments of 1990 added more clout to many parts of the earlier law. Some features of these amendments:

- Stricter tailpipe emissions standards for cars, trucks, and buses
- Expansion of inspection and maintenance (I/M) programs with more stringent testing
- Attention to fuel technology (the development of alternative fuels)
- Study of nonroad engines (that is, boats, farm equipment, home equipment, construction equipment, lawn mowers, and so on)
- Mandatory alternative transportation programs (carpooling) in heavily polluted cities

California created the CARB in 1967, prompted by the original Clean Air Act. CARB passed legislation to regulate onboard diagnostics (OBD) for vehicles sold in California in 1988. This regulation required that all vehicles sold in California have onboard diagnostics by 1991. The first phase, OBD-I, required monitoring of the fuel-metering system, the EGR system, and additional emissions-related electrical components. A malfunction indicator lamp (MIL) was required to alert the driver of a malfunction. Along with the MIL, OBD-I required the setting of diagnostic trouble codes (DTCs) to identify the area at fault.

After the passage of the federal Clean Air Act Amendments in 1990, CARB developed regulations for the second generation of onboard diagnostics (OBD-II). These amendments also prompted the EPA to develop somewhat different OBD requirements. All passenger cars, light-duty trucks, and medium-duty vehicles sold in the United States were required to comply with OBD-II standards by 1996.

In spite of the tremendous progress that has been made in new-car emissions levels, the total U.S. vehicle population continues to be a significant source of air pollution. The reason is the number of vehicle-miles traveled on American roads has doubled in the last 20 years to 2 trillion miles per year. The total volume of emissions offsets a great deal of the technological progress made over that same period. As the number of vehicle-miles continues to increase, ongoing efforts are being made to reduce emissions levels in individual vehicles.

The Clean Air Act amendments of 1990 further tightened emissions standards by enactment of the Tier 1 standard. This new standard set the $NO_x$ standard at 0.6 grams per mile (gpm) for passenger cars by the 1994 model year. For trucks and SUVs, the Tier 1 standard ranges from 0.6 to 1.53 gpm, depending on the weight of the vehicle. The tightened emissions standards for Tier 1 required a 40 percent reduction from the 1981 standard. The Clean Air Act amendments of 1990 provided the following approaches to reduce the emissions levels in individual vechicles:

- Cleaner new cars

  TLEV—transitional low-emissions vehicle
  LEV—low-emissions vehicle
  ULEV—ultra low-emissions vehicle
  ZEV—zero-emissions vehicle

| | HC | CO | $NO_x$ |
|---|---|---|---|
| 1990—U.S. | 0.41 | 3.4 | 1.0 |
| 1994—U.S. | 0.25* | 3.4 | 0.4 |
| 1993—Calif. | 0.25 | 3.4 | 0.40 |
| 1994—TLEV | 0.125 | 3.4 | 0.40 |
| 1997—LEV | 0.075 | 3.4 | 0.20 |
| 2000—ULEV | 0.040 | 1.7 | 0.20 |

*nonmethane HC

**Table 2** *Emissions standards, 1990 to 2000.*

- Cleaner fuels (compressed natural gas, methanol)
- Improved I/M testing
- Second-generation onboard diagnostics (OBD-II)

The first two points represent an evolution of a process that has been going on for more than two decades. Modern technology allows manufacturers to build cleaner-burning engines without having to go through the same learning curve experienced by manufacturers in the early years of pollution control. Cleaner vehicle and fuel technology has already helped to reduce tailpipe exhaust emissions and gasoline evaporation. Some parts of the country use oxygenated fuels, particularly during the winter months.

During the 1990s, emissions standards in the United States became increasingly stringent **(Table 2)**. In 1994, California began an ambitious emissions program. This program specified emissions standards for TLEVs. In 1997, the California emissions program specified a further reduction in emissions levels for LEVs, and in the year 2000, emissions levels were specified for ULEVs.

In 1998, California emissions regulations required 2 percent of the cars sold in the state to be zero-emission vehicles. The regulations also called for a higher percentage of ZEVs in the years 2003 and 2005. With present technology, only electric cars meet the ZEV standards. California repealed these regulations in 2002.

California then put partial-zero-emission vehicle (PZEV) certification in place in 2001 since a ZEV was not obtainable. This certification requires that vehicles emit zero evaporative emissions and near-zero tailpipe exhaust emissions. To meet the zero evaporative emissions standards, new styles of fuel tanks are being designed. Some manufacturers have returned to steel tanks while others have improved the porosity of plastic tanks. In addition, fuel lines are multilined to prevent vapors from escaping. PZEV standards also require a 15-year/150,000-mile emissions system warranty.

In addition to the emissions programs discussed, the National Low-Emission Vehicle (NLEV) program introduced California low-emissions cars and light-duty trucks into the **Northeast Trading Region (NTR)**. The NTR includes Massachusetts, New York, Maine, and Vermont. Section 177 of the Clean Air Act allows states with nonattainment areas to adopt standards identical to California standards in lieu of federal standards. Standards must be adopted at least 2 model years prior to implementation and the state action cannot result in a different motor vehicle than those certified in California.

This program began with the 1999 model year vehicles in the NTR and the rest of the country started with the 2001 model year. The NLEV program incorporates the California LEV program vehicle categories and standards: TLEV, LEV, and ULEV.

Present government goals call for reductions of more than 98 percent of HC, 96 percent of CO, and 95 percent of $NO_x$, compared with pre-1970 engine emissions. These standards are referred to as Tier 2.

Most of the publicity surrounding the Clean Air Act Amendments of 1990 centers around improved I/M testing of vehicles on the road. Increased I/M testing requirements are important because despite all the improvements in new-car emissions and fuel quality, it will be many years before these cleaner cars dominate the overall population of vehicles in the United States.

In many cities, vehicles contribute 35 to 70 percent of HC emissions and at least 90 percent of CO emissions. Concentrations of one or both of these pollutants exceed air quality standards in almost every major city in the United States.

Under this law, more cities began I/M programs. With the Clean Air Act Amendments of 1990, 154 areas require either basic or enhanced I/M testing.

## Canadian Requirements

Through the 1997 model year, Canada voluntarily adopted the Tier I standards (measured in grams per kilometer), including alternative vehicles, with a Memorandum of Understanding. The Tier I standards were regulated beginning in the 1998 model year. The NLEV requirements were adopted with a Memorandum of Understanding for 2001–2003 model year vehicles. Canadian OBD and evaporative requirements also parallel U.S. federal requirements. Tier 2 standards were enacted beginning with the 2004 model year.

Canadian fuel economy requirements are part of a voluntary program that parallel U.S. fuel economy protocol and standards (measured in liters per 100 kilometers). A voluntary standard "EnerGuide" fuel efficiency label was implemented for 2000 model year vehicles.

In British Columbia, Canadian/U.S. federal emissions requirements apply through the 2000 model year. Beginning in 2001, light-duty vehicles must be California-certified. Medium- and heavy-duty vehicles continue to meet the Canadian/U.S. federal emissions standards.

The province mandates that vehicles in the Vancouver area pass the "AirCare" emission inspection and maintenance requirements. Beginning with the 1992 model year, vehicles are dynamometer tested on the transient I/M 240 test in government-operated facilities. OBD are monitored in 1998 and newer vehicles.

The province of Ontario mandates that vehicles in southern Ontario pass the "Drive Clean" emission inspection and maintenance requirements. Vehicles 20 years and newer are dynamometer tested with a steady state ASM2525 test at independent facilities. OBD is monitored in 1998 and newer vehicles.

## The Combustion Process and Exhaust Gases

A gasoline engine derives its power from the controlled combustion of the air-fuel mixture in the combustion chamber. Three elements are required for a burn to occur: fuel, oxygen, and heat. If any one of these three elements is missing, combustion will not occur.

Perfect combustion **(Figure 1)** results in the formation of three elements: heat, carbon dioxide ($CO_2$), and water ($H_2O$) vapor. Carbon atoms from the hydrocarbon fuel combine with oxygen ($O_2$) from the atmosphere to form $CO_2$. Hydrogen atoms from the hydrocarbon fuel combine with oxygen from the atmosphere to form $H_2O$. In a perfect combustion reaction, there is no excess fuel or oxygen. In other words, there is exactly the right amount of oxygen to burn the existing fuel and exactly the right amount of fuel to consume the existing oxygen.

If it were possible to obtain perfect combustion all the time, then all the emissions controls that have been developed over the last three decades would not have been necessary. However, perfect combustion rarely occurs in a gasoline engine **(Figure 2)**. Although the air-fuel ratio is controlled more precisely now with the use of computer controls, it is not always perfect. Compounding the problem is the

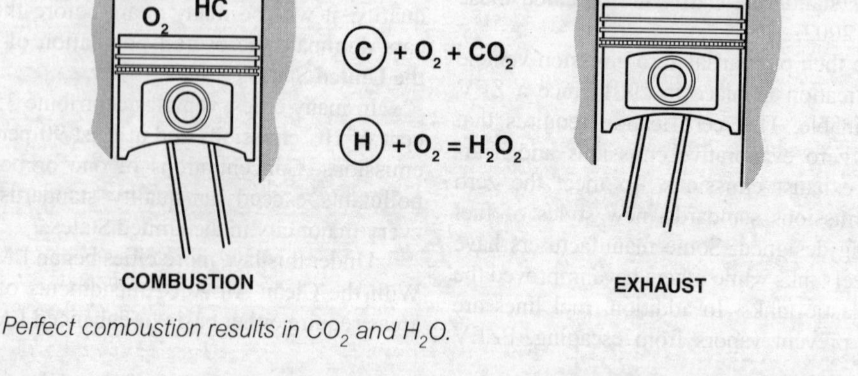


**Figure 1** *Perfect combustion results in $CO_2$ and $H_2O$.*

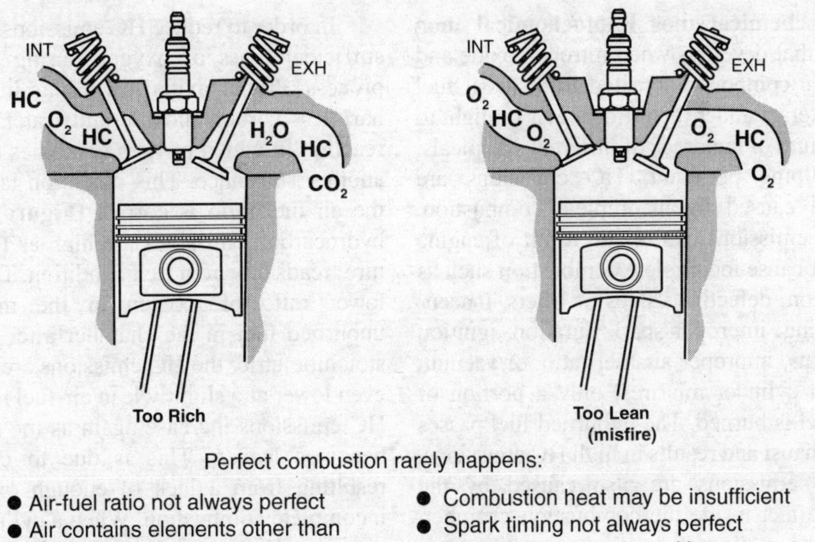


**Figure 2** *There are many factors that prevent perfect combustion.*

fact that even the highest-quality gasoline has impurities that travel along with the hydrocarbon. In addition, the air that is drawn into the combustion chamber contains more than just oxygen. Air contains about 21 percent oxygen and 78 percent nitrogen. Nitrogen under certain conditions can produce harmful nitrogen oxide.

Other factors impede perfect combustion (even when the proper air-fuel ratio exists). For instance, the combustion heat may be insufficient and the timing of the spark may be less than ideal. These conditions lead to incomplete combustion and the production of carbon monoxide and hydrocarbon in the exhaust.

## The Exhaust Gases

There are generally five exhaust gases being monitored in today's automobile:

- Carbon monoxide
- Hydrocarbon
- Nitrogen oxide
- Carbon dioxide
- Oxygen

Of these five exhaust gases, CO, HC, and $NO_x$ are pollutants. $CO_2$ and $O_2$ are not pollutants but they are informational gases that are observed for diagnostic purposes.

Studies indicate that the automobile accounts for about 50 percent of the hydrocarbon, more than 75 percent of the carbon monoxide, and nearly 50 percent of the nitrogen oxide that pollute our atmosphere **(Figure 3)**. Legislation places limits on how much of these pollutants can be produced by vehicles sold in the United States and Canada.

**Hydrocarbon.** Hydrocarbon consists of hydrogen and carbon atoms in various combinations. Gasoline is a hydrocarbon fuel and is the source of automotive hydrocarbon emissions. Hydrocarbon, in combination with other compounds, can produce ground-level

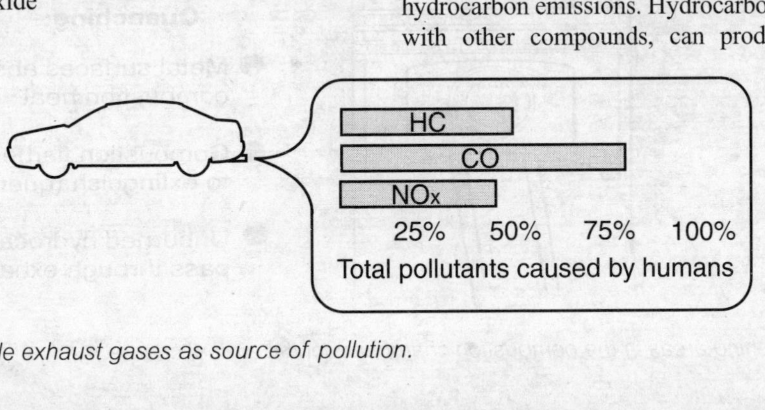


**Figure 3** *Vehicle exhaust gases as source of pollution.*

ozone or **photochemical smog**. **Photochemical smog** is a condition that develops when nitrogen oxide and volatile organic compounds created from fossil fuel combustion interact under the influence of sunlight to produce a mixture of hundreds of different chemicals.

As a tailpipe pollutant, HC emissions are unburned fuel caused by incomplete combustion. Excessive HC emissions can be the result of engine conditions that cause incomplete combustion such as low compression, defective valves or lifters, inaccurate spark timing, improper spark duration, ignition system problems, improper air-fuel ratio, or vacuum leaks. When a cylinder misfires, only a portion or none of the fuel is burned. The unburned fuel passes through the exhaust and results in high HC emissions.

High HC emissions are also caused by the "**quenching**" effect inside the combustion chamber. Quenching is the cooling of gases as a result of compressing them into a thin area. Temperatures are not consistent throughout the combustion chamber. The areas near the cylinder head and engine block are cooler than the other areas since the metal surfaces absorb much of the heat of combustion **(Figure 4)**. In these cooler areas, the combustion flame tends to extinguish or quench. Since the heat is not sufficient to complete the burn of the air-fuel mixture, unburned hydrocarbon passes through the exhaust. Engine designers reduce the effects of quenching through combustion chamber design and piston ring location.

In order to reduce HC emissions, there must be a sufficient mass of oxygen during the combustion process. Proper **oxidation** results in the creation of harmless carbon dioxide and water. Oxidation is a reaction in which oxygen combines chemically with another substance. This oxidation takes place when the air-fuel ratio is correct **(Figure 5)**. Notice that hydrocarbon emissions are high as the air-fuel mixture treads toward a rich condition. This is due to the lower ratio of oxygen in the mixture, leaving unburned fuel in the chamber after combustion. At stoichiometric, the HC emissions are low but will go even lower at a slightly lean air-fuel mixture. Finally, HC emissions increase again as the air-fuel mixture becomes leaner. This is due to cylinder misfire resulting from a lack of enough oxygen and thus incomplete combustion. When a cylinder has incomplete combustion, unburned fuel is pushed into the exhaust.

**Carbon Monoxide.** Carbon monoxide (CO) exists as a molecule containing one carbon (C) atom and one oxygen atom. It is a colorless, odorless, and very poisonous gas resulting from incomplete combustion of the air-fuel mixture. With perfect combustion, one carbon atom from the fuel will combine with atmospheric oxygen. This will form harmless carbon dioxide ($CO_2$). $CO_2$ is used by plants to manufacture oxygen.

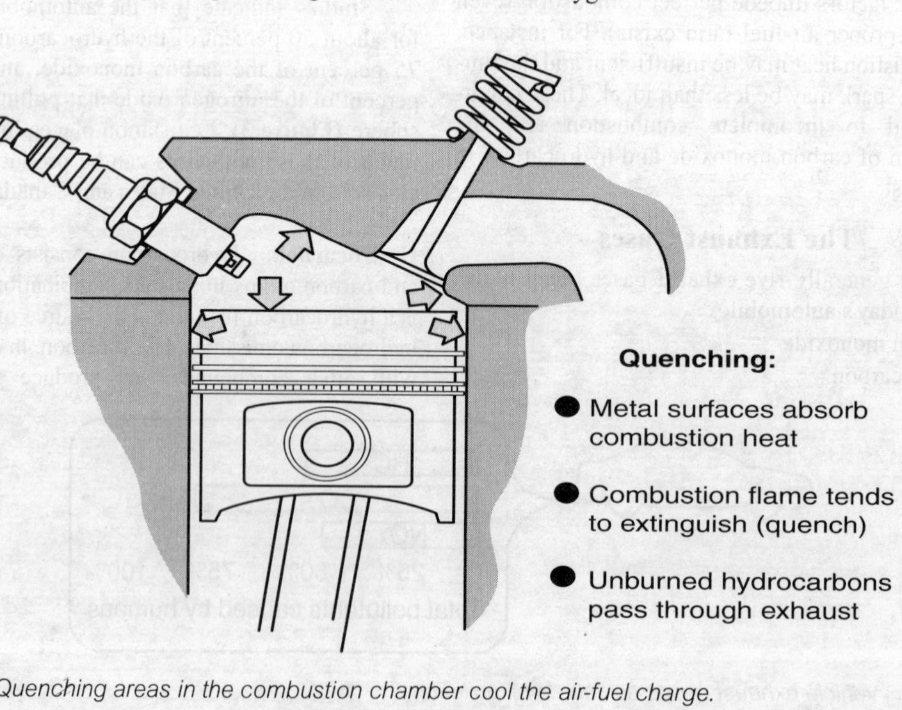


**Figure 4** *Quenching areas in the combustion chamber cool the air-fuel charge.*

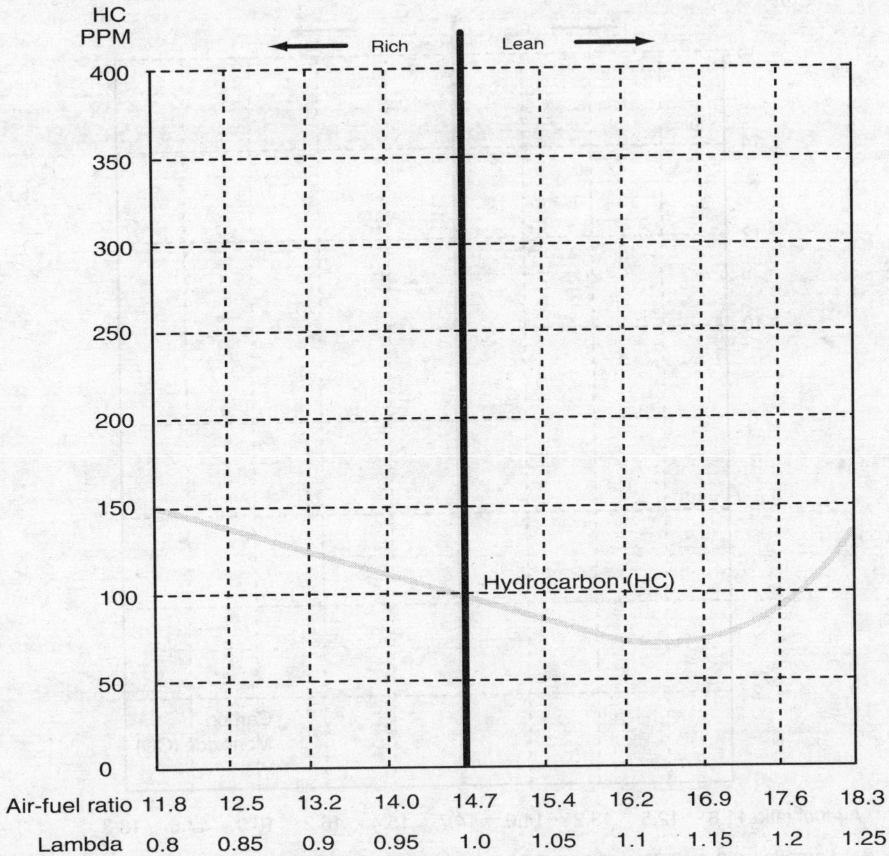


**Figure 5** *Hydrocarbon emissions are most controllable at stoichiometric.*

If the air-fuel mixture is too rich, there is a shortage of oxygen in the combustion chamber. Without sufficient oxygen, the burning of the mixture stops prematurely. In this case, most of the oxygen will be used up, so the burn stops. This results in one carbon atom combining with one oxygen atom to form harmful carbon monoxide.

Anything that restricts airflow to the combustion chamber, such as a blockage in the intake manifold or cylinder heads, can cause high CO emissions levels. Carbon monoxide emissions are a direct indication of the air-fuel mixture **(Figure 6)**. Generally, the lower the CO reads on an exhaust-gas analyzer, the leaner the air-fuel mixture. This means that high CO levels indicate a rich mixture. Keep in mind that the rich mixture may be caused by leaking injectors and higher-than-normal fuel pressures.

Since combustion is required to create CO, cylinder misfire does not increase CO emissions. In fact, the CO emissions level may decrease slightly while a misfire is occurring.

**Nitrogen Oxide.** Nitrogen oxide is gas composed of molecules that contain one nitrogen atom combined with a varying number of oxygen atoms. Combining a nitrogen atom with one oxygen atom results in the formation of nitrous oxide (NO). The combining of one nitrogen atom with two oxygen atoms forms nitrous dioxide ($NO_2$). The variable number of oxygen atoms results in the formula "$NO_x$" with "x" being the unknown number of oxygen atoms. Nitrogen oxide is the result of the combustion process but it is formed differently than CO and HC.

$NO_x$ is an environmental hazard because ultraviolet radiation from the sun acts upon a combination of $NO_x$ and HC in the atmosphere to produce photochemical smog. This smog appears as a brownish haze that irritates the eyes and respiratory system. It is also a contributor to acid rain.

The formation of $NO_x$ is dependent on the combustion temperature. Nitrogen is normally inert and does not tend to combine readily with other elements. However, when the combustion temperature

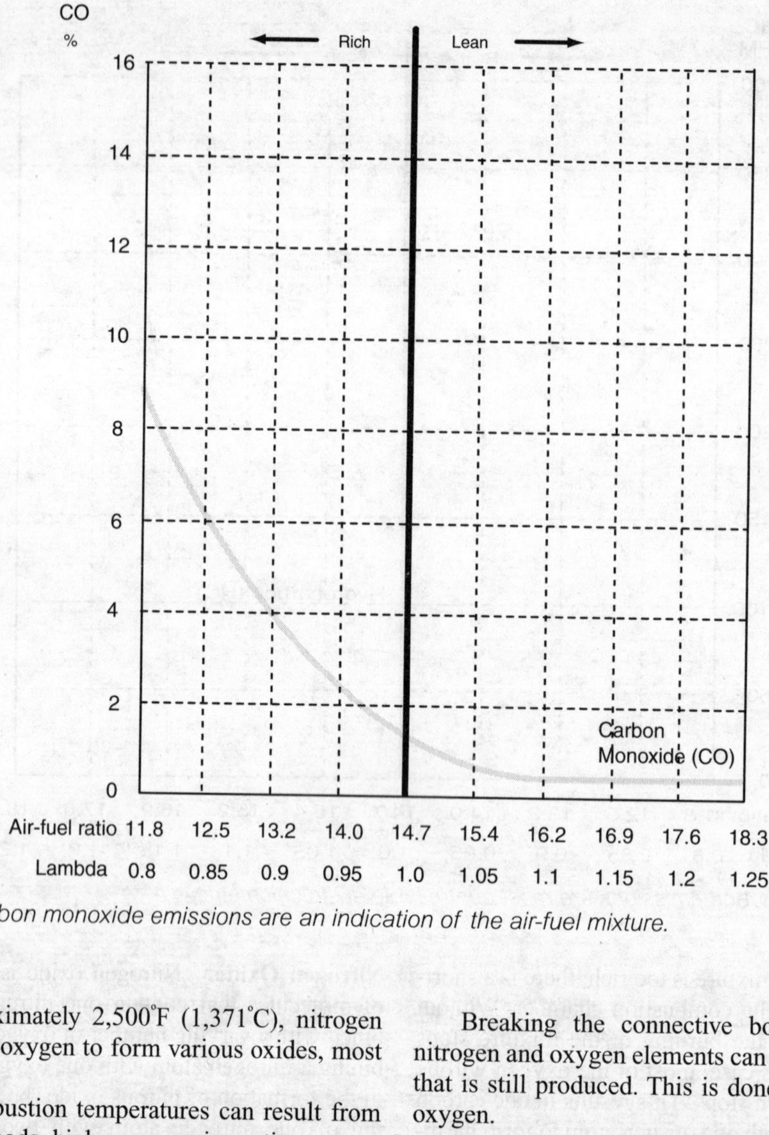


**Figure 6** *Carbon monoxide emissions are an indication of the air-fuel mixture.*

exceeds approximately 2,500°F (1,371°C), nitrogen combines with oxygen to form various oxides, most notably $N_2O$.

High combustion temperatures can result from heavy engine loads, high compression ratios, excessively lean air-fuel mixtures **(Figure 7)**, advanced ignition timing, engine overheating, and vacuum leaks.

$NO_x$ emissions are controlled by lowering the combustion temperature. This can be achieved through lower compression ratios, camshaft timing adjustments, air-fuel mixture control, and EGR systems.

### You Should Know

*Prior to these controls for $NO_x$, combustion temperatures in excess of 4,000°F (2,204°C) were not uncommon.*

Breaking the connective bonds between the nitrogen and oxygen elements can minimize the $NO_x$ that is still produced. This is done by removing the oxygen.

### You Should Know

*The process of removing the oxygen is called a reduction reaction and takes place inside a three-way catalytic converter.*

## Nonpollutant Exhaust Gases

As mentioned earlier, there are two major exhaust gases that are not considered to be pollutants; carbon dioxide and oxygen. Technicians should be aware of these two gases since their presence helps to diagnose certain engine conditions by determining combustion efficiency.

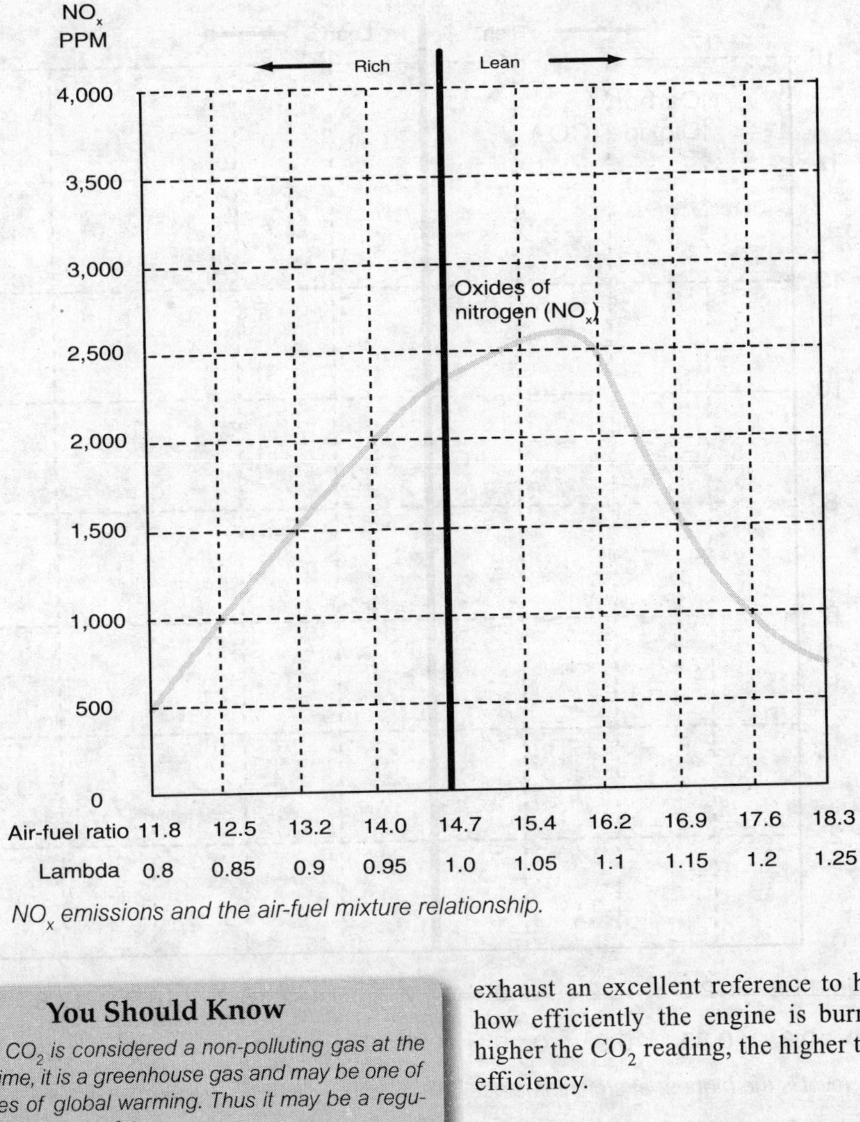


**Figure 7** *$NO_x$ emissions and the air-fuel mixture relationship.*

> **You Should Know**
>
> *Although $CO_2$ is considered a non-polluting gas at the present time, it is a greenhouse gas and may be one of the causes of global warming. Thus it may be a regulated emission in the future.*

**Carbon Dioxide.** Carbon dioxide is a by-product of perfect combustion. It is formed when one atom of carbon bonds with two atoms of oxygen from the air during combustion. An essentially harmless gas, it is present at levels of 14 to 15 percent in the exhaust of a properly running engine. Carbon dioxide is also produced when carbon monoxide is oxidized in the catalytic converter.

The amount of $CO_2$ in the exhaust is directly related to the air-fuel ratio **(Figure 8)**. As the fuel mixture approaches stoichiometric, the level of $CO_2$ peaks. It decreases when the mixture becomes richer or leaner. This fact makes $CO_2$ in the exhaust an excellent reference to help determine how efficiently the engine is burning fuel. The higher the $CO_2$ reading, the higher the combustion efficiency.

**Oxygen.** The air we breathe is made up of about 21 percent oxygen. Oxygen is mixed with gasoline for combustion and is consumed during the combustion process **(Figure 9)**. As long as there is sufficient oxygen, the combustion burn continues. If the combustion chamber is delivered too much fuel for the amount of oxygen, all the available oxygen is used up. This results in low $O_2$ content in the exhaust.

As the air-fuel mixture moves toward lean, the amount of oxygen in the exhaust steadily increases. This is because the fuel burns faster than the oxygen is used so oxygen is left over. Therefore, a higher level of $O_2$ in the exhaust is a direct indication of a leaner air-fuel ratio.

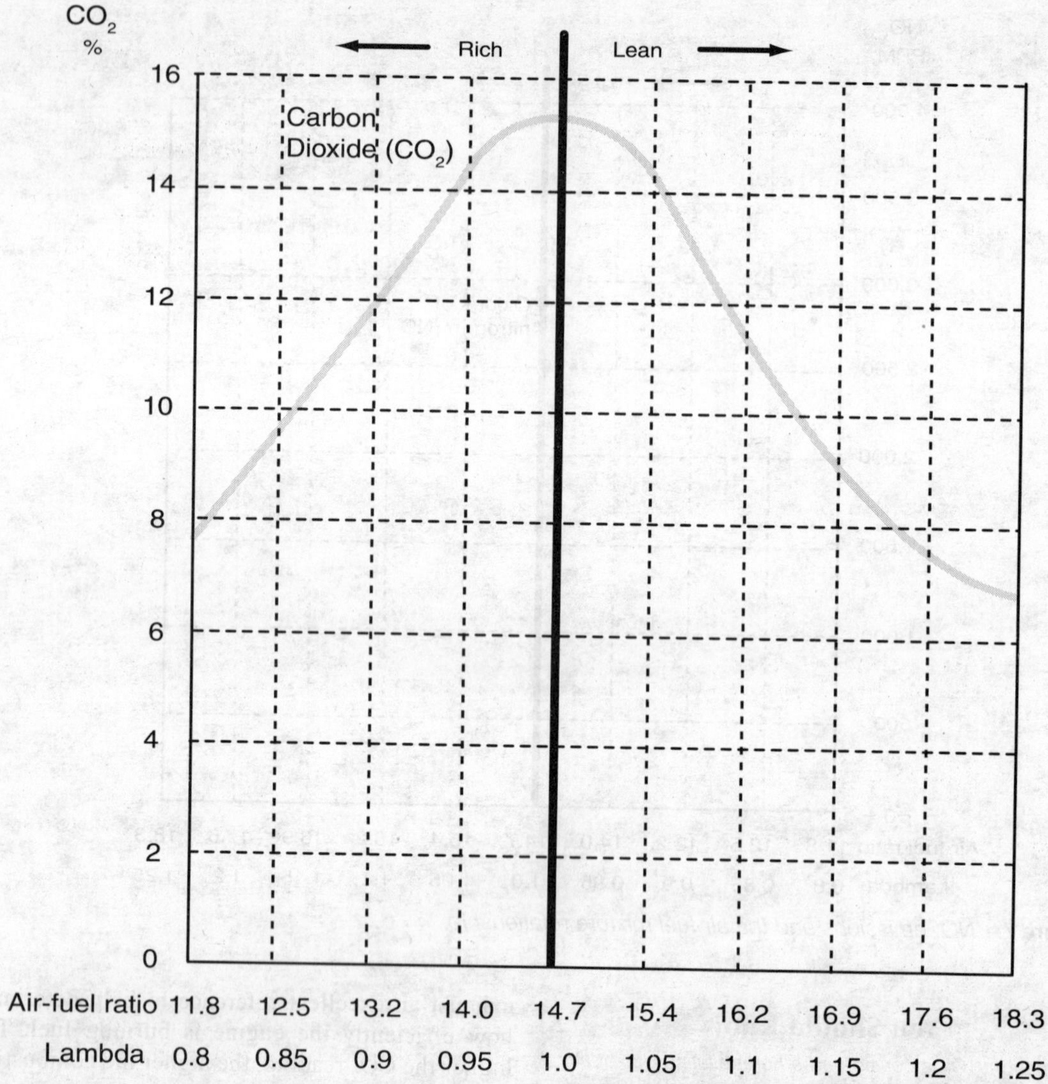


**Figure 8** *$CO_2$ level is the highest at stoichiometric.*

When the air-fuel mixture is either rich or lean, the levels of oxygen and carbon monoxide will be opposite (when $O_2$ is high, CO is low and vice versa). At the stoichiometric air-fuel ratio, the levels of $O_2$ and CO in the exhaust are approximately equal. At stoichiometric, $O_2$ content is about 1 to 2 percent.

## The Importance of the Air-Fuel Ratio

As just discussed, the air-fuel ratio is the common element in the formation of each of the three main pollutant gases (HC, CO, and $NO_x$). Each type of emissions can be caused by an improper ratio of air to fuel in the combustion process. Therefore, maintaining the correct air-fuel ratio has become one of the main goals of emissions-control system design. As seen in Chapter 18, the control of the air-fuel ratio has become a main function of the PCM.

As related to exhaust emissions, an air-fuel ratio that reduces the level of one pollutant can increase the level of another. Consider the following explanation.

Depending on other circumstances, a lean mixture can be responsible for excess levels of hydrocarbon and

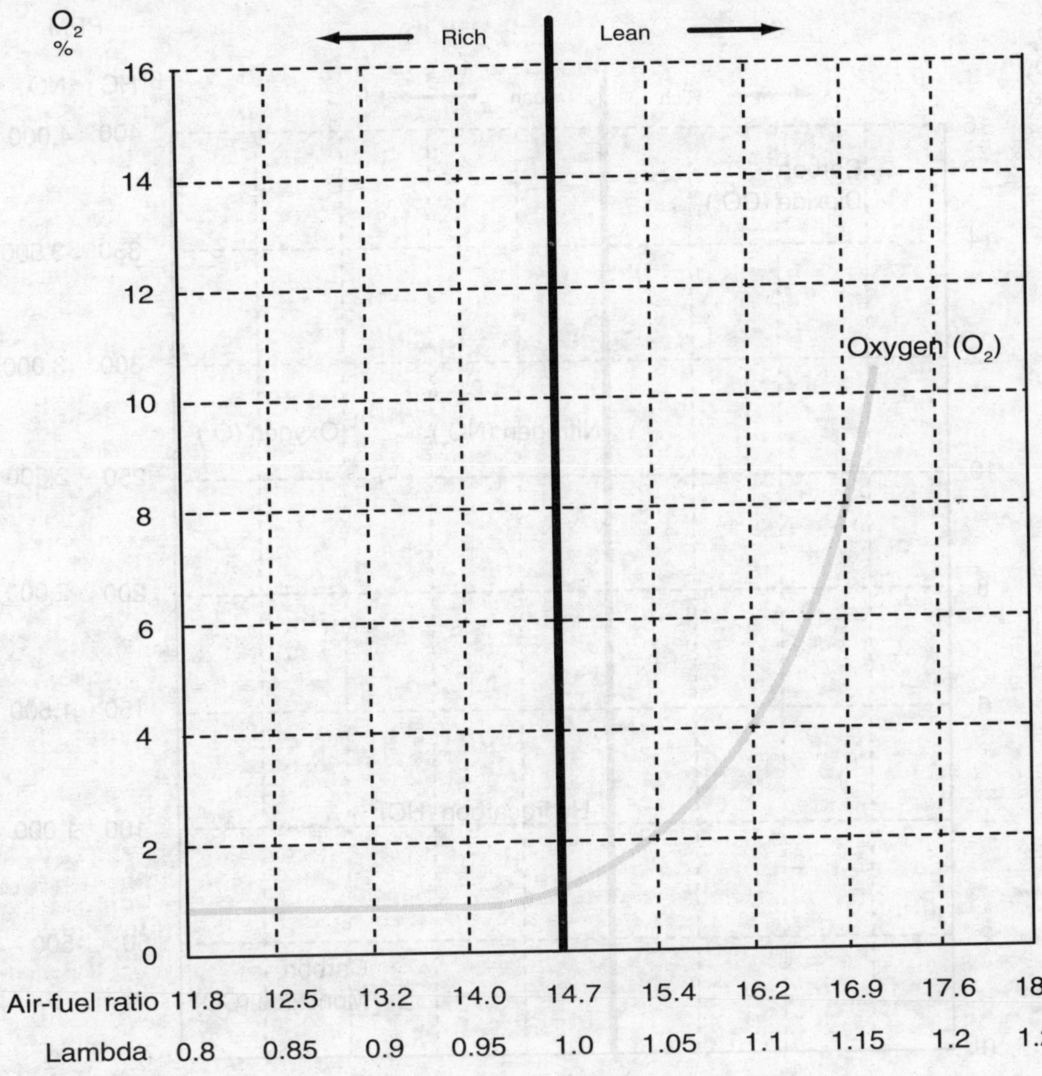


**Figure 9** *$O_2$ emissions increase with a leaner air-fuel mixture.*

nitrogen oxide. An excessively lean mixture can cause a lean-burn misfire resulting when a given volume of fuel mixture does not contain enough gasoline to ignite. What gasoline there is passes through the exhaust system unburned, resulting in high HC levels.

When the fuel mixture is slightly lean, the excess oxygen produces a very hot flame. This results in an increase in combustion temperature and causes the nitrogen in the intake air to combine with oxygen to form harmful nitrogen oxides $NO_x$.

On the other hand, a lean mixture results in low to zero levels of CO since all the fuel is burned. This is because there is sufficient oxygen to burn the fuel.

A rich mixture can lead to increased levels of HC and CO, but $NO_x$ emissions will generally be low. A high CO level is a good indicator of a rich fuel mixture, especially if a high HC level is also present. In a rich mixture, there is insufficient oxygen to combine with the carbon atoms from the gasoline to form harmless $CO_2$. Instead, each carbon atom combines with one oxygen atom to form poisonous CO. $NO_x$ levels will usually be low in a rich-mixture situation because the shortage of oxygen results in lower combustion temperatures.

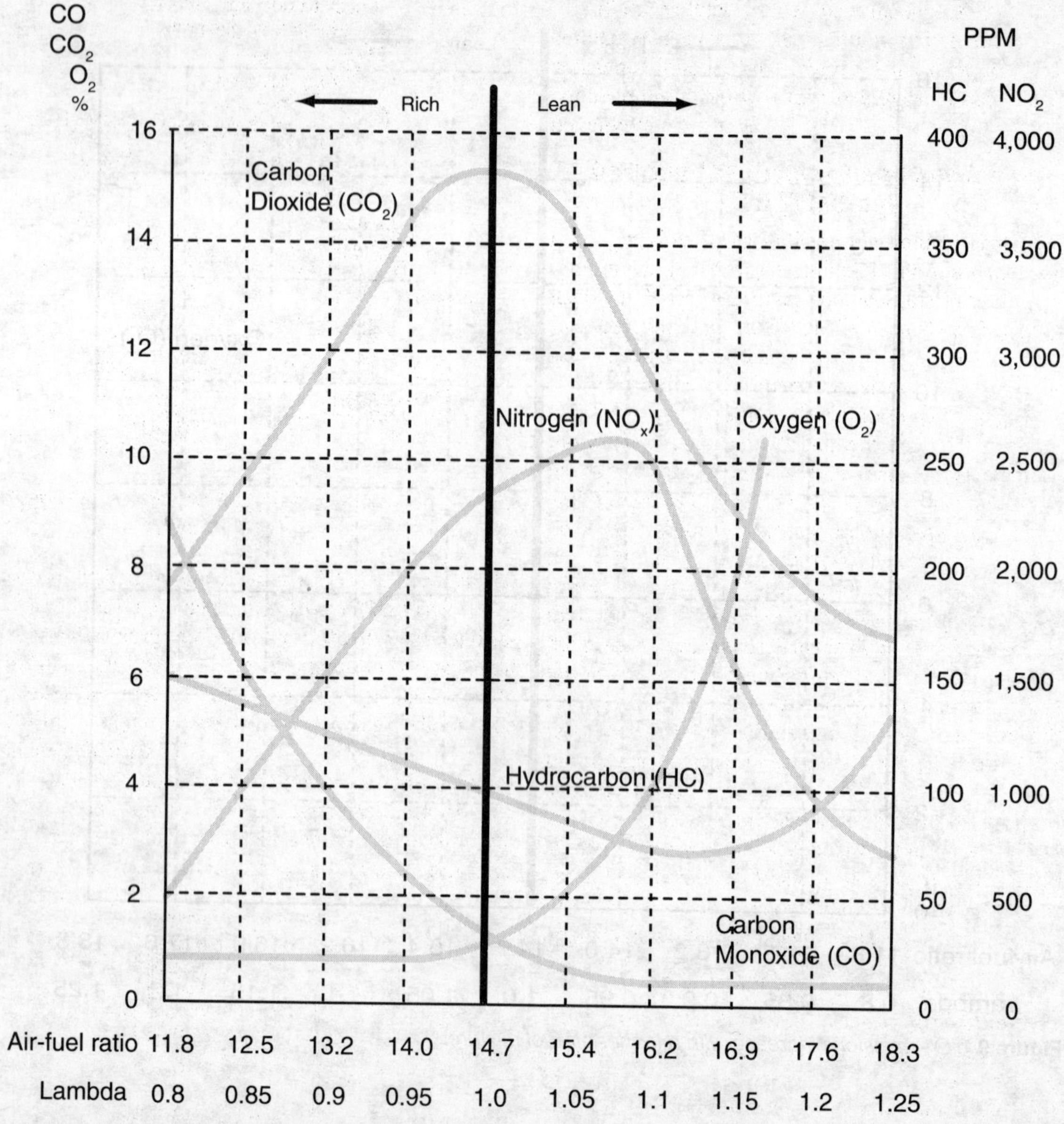


**Figure 10** *Illustration of the total effect of the air-fuel ratio on exhaust gases.*

However, HC emissions can be caused both by an excessively lean or an excessively rich mixture. In the case of a lean mixture, there is not enough gasoline to ignite and it passes unburned through the exhaust system. In a very rich mixture, there may not be enough oxygen present to support combustion. The end result is the same unburned HC goes through the exhaust system.

**Figure 10** illustrates the relationship between the air-fuel ratio and the emissions gases. As the air-fuel mixture moves away from stoichiometric (either too lean or too rich), a number of conditions could occur, including emissions test failures and engine problems. **Table 3** lists some of these relationships.

| Air-Fuel Mixture | Effect on Vehicle |
|---|---|
| Too lean | Increased $NO_x$ emissions<br>Poor engine power<br>Misfiring at cruising speeds<br>Burned valves<br>Burned pistons<br>Scored cylinders<br>Spark knock or ping |
| Slightly lean | Low exhaust emissions<br>High gas mileage<br>Reduced engine power<br>Slight tendency to knock or ping |
| Stoichiometric | Best all-around performance and emissions levels |
| Slightly rich | Increased CO emissions<br>Increased HC emissions<br>Maximum engine power<br>Higher fuel consumption<br>Less tendency to knock or ping |
| Too rich | Increased CO emissions<br>Increased HC emissions<br>Poor fuel mileage<br>Misfiring<br>Oil contamination<br>Black exhaust |

**Table 3** *Air-fuel mixture and emissions relationships.*

## Review Questions

1. What emissions are manufacturers required to control under Title II of the Clean Air Act?
2. How are HC exhaust emissions produced?
3. How are $NO_x$ emissions produced?
4. As a tailpipe pollutant, _________ emissions are caused by partially burned fuel.
5. The formation of _________ is dependent on the combustion temperature.
6. _________ emissions can be caused by both an excessively lean or an excessively rich mixture.
7. Technician A says that HC emissions are high as the air-fuel mixture moves toward a rich condition. Technician B says that HC is a result of the combustion process. Who is correct?
   A. Technician A only
   B. Technician B only
   C. Both Technician A and Technician B
   D. Neither Technician A nor Technician B
8. Technician A says that depending on other circumstances, a lean mixture can be responsible for excess levels of HC and $NO_x$. Technician B says that a high CO level is a good indicator of a rich fuel mixture. Who is correct?
   A. Technician A only
   B. Technician B only
   C. Both Technician A and Technician B
   D. Neither Technician A nor Technician B
9. Perfect combustion would result in the formation of what elements?
10. How are CO emissions produced?
11. Of five exhaust gases, _________ and _________ are not pollutants but are informational gases that are observed for diagnostic purposes.
12. _________ is a direct indication of the air-fuel mixture.

13. A ___________ air-fuel mixture can lead to increased levels of HC and CO, but $NO_x$ emissions will generally be low.
14. Under Title II of the Clean Air Act, manufacturers are required to control which of the following pollutants?
    A. Hydrocarbon
    B. Carbon monoxide
    C. Nitrogen oxides
    D. Particulate matter
    E. All of the above
15. Technician A says that high levels of $O_2$ indicate a rich air-fuel mixture. Technician B says that the formation of $NO_x$ is dependent on the combustion temperature. Who is correct?
    A. Technician A only
    B. Technician B only
    C. Both Technician A and Technician B
    D. Neither Technician A nor Technician B
16. Technician A says that HC emissions can be caused by both an excessively lean or an excessively rich mixture. Technician B says that high $NO_x$ emissions result from rich air-fuel mixtures. Who is correct?
    A. Technician A only
    B. Technician B only
    C. Both Technician A and Technician B
    D. Neither Technician A nor Technician B

CHAPTER

# 25 Precombustion Controls

## Introduction

All regulated automotive emissions result from the properties and combustion of fuel. Three types of emissions have been identified and are regulated by federal and state governments: evaporative emissions, crankcase emissions, and exhaust emissions.

All three types produce HC. Crankcase and exhaust emissions also include CO and $NO_x$. Various designs have evolved to control emissions. All of the systems described in this chapter can be classified as the group of systems that prevents the formation of harmful emissions. This group is known as **precombustion control** systems. This is the preferred approach to pollution control because it is more effective to prevent the formation of pollutants than it is to get rid of them after they have been produced. **Table 1** identifies the major precombustion emissions control systems, and lists the pollutants they control.

## Precombustion Controls

Precombustion controls include fuel formulation, engine design, engine operation, fuel control, ignition control, exhaust gas recirculation (EGR), evaporative controls, and crankcase ventilation. Fuel formulation was discussed in Chapter 9. The other precombustion controls are discussed here.

### Engine Design and Operation

Since it is better to prevent emissions than it is to treat them after they are produced, good engine design is the best approach to automotive pollution control. A well-designed engine reduces the need for add-on devices and posttreatment methods.

In the 1970s, engineers attempted to meet federal and state emissions regulations with a host of devices and techniques. Unfortunately, the result was reduced performance along with decreased emissions. In the 1980s, the trend shifted toward designing new engines from the ground up with low emissions levels and good performance. Advances in the areas of manifold design, combustion chamber design, fuel injection, and computerization have enabled new engines to provide both better performance and lower emissions. For example, CO can be controlled by a fuel system design that utilizes less fuel. This reduces the chance for partial combustion. In addition, a major consideration for CO control is equal fuel distribution. This is accomplished through the intake manifold design and the use of fuel injection. Leaner air-fuel mixtures ensure enough air for complete oxidation to $CO_2$ during normal driving conditions. The heated air inlet causes better vaporization of fuel in a cold engine. Advanced ignition timing allows more time for complete burning. However, this can raise HC as well as combustion temperatures, which leads to the formation of $NO_x$.

| System | Classification | Pollutant Controlled |
|---|---|---|
| Engine design/operation | Pre-combustion | HC, CO, $NO_x$ |
| Computerized engine control | Pre-combustion | HC, CO, $NO_x$ |
| Spark control | Pre-combustion | HC, CO, $NO_x$ |
| Exhaust gas recirculation | Pre-combustion | $NO_x$ |
| Evaporative controls | Pre-combustion | HC |

**Table 1** *Classification of emissions control types.*

$NO_x$ forms at high combustion temperatures. These occur with lean mixtures and advanced timing (which also produce high power output and good fuel economy). Cooler combustion is achieved through fuel control and moderate ignition timing advanced. It is also achieved through the use of EGR to dilute the air-fuel mixture with exhaust gases.

One way to control emissions through engine design is by the use of a **heated air inlet**. Heated air inlet systems, also called thermostatic air systems, direct warm air into the air-cleaner assembly during cold engine operation to help vaporize the fuel. The main component in most of these systems is the thermostatic air cleaner **(Figure 1)**. Traditionally, a cold engine required a rich mixture because the fuel did not vaporize well when it was mixed with cold air. A thermostatic air cleaner takes warm air from around the exhaust manifold and directs it through a hot air duct to the air cleaner. This achieves the following results:

- Heated inlet air provides better fuel vaporization.
- Better vaporization means leaner mixtures can be used.
- Leaner mixtures result in reduced levels of HC and CO.

**You Should Know**

*Earlier six- and eight-cylinder vehicles used a heated crossover in the manifold to assist in fuel atomization during cold engine operation.*

A bimetal sensor reacts to the air temperature. The sensor is calibrated to provide a specific output vacuum to the air control valve based on air temperature. This allows the amount of heated air to be regulated **(Figure 2)**.

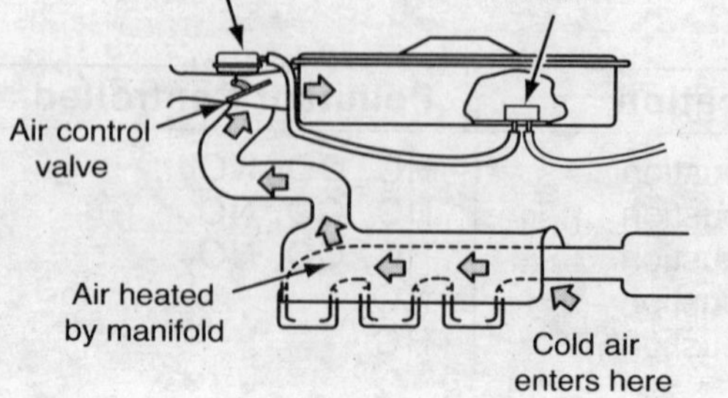


**Figure 1** *Typical heated air inlet system.*

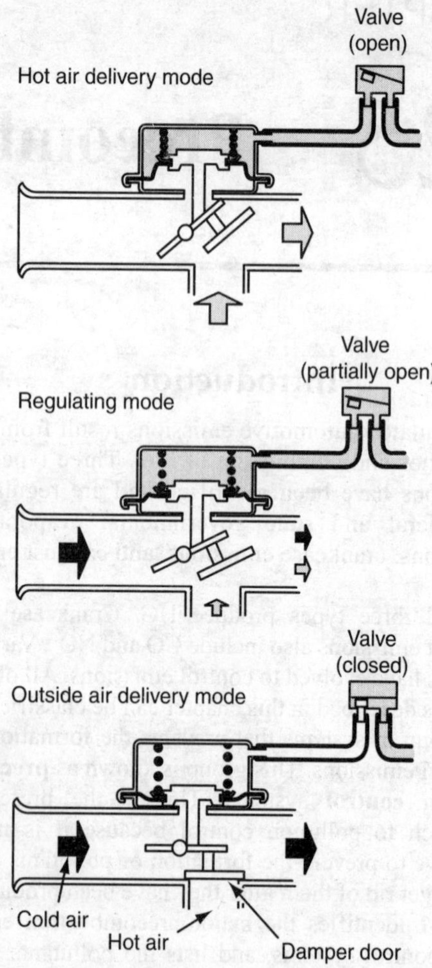


**Figure 2** *The bimetal sensor operates the air control valve to direct heated air to the throttle body.*

## Computerized Engine Control

The power-train control module (PCM) is considered a precombustion device since its function is to maintain optimum fuel and ignition and to control the operation of other emissions-control systems. As regulations tightened, manufacturers added more and more pollution-control devices to vehicles. Precomputerized systems responded to mechanical feedback such as vacuum signals and there was a great deal of conflict as to which devices and conditions should take precedence. The PCM brought a degree of precision and control to the emissions-control system. The PCM constantly reads electronic signals from various sensors throughout the power train and responds by sending control signals to various devices as needed.

Technicians should be careful not to assume that the PCM is at fault in all repair situations. Items not monitored by the computer that can affect driveability and emissions include:

- Engine mechanical condition
- Ignition system condition
- Exhaust system condition
- Basic timing
- Routing of hoses and tubing

As discussed in Chapter 7, the oxygen sensor ($O_2S$) is located in the exhaust manifold and is used to detect the amount of oxygen in the exhaust stream. This information tells the computer whether the fuel mixture is too rich or too lean. The computer can then have the fuel system adjust the fuel mixture as necessary.

With most computerized engines, the heated intake system is no longer needed. A thermistor in the intake air system provides the PCM with the temperature of the air entering the combustion chamber. The PCM can then alter the air-fuel mixture and/or the ignition timing to provide good performance and lower emissions.

With an ability to analyze several inputs and make corrective decisions in real time, the engine computer has made possible the design of smaller, high-performance, low-polluting automotive engines.

## Ignition Control

As discussed in Chapter 24, the ignition-control system has a tremendous effect on overall emissions output. Ignition timing determines the precise moment at which the electrical impulse is sent to the spark plug to ignite the air-fuel mixture. Controlling this function as precisely as possible before combustion prevents harmful emissions. Improper ignition timing can affect the levels of all three exhaust emissions. Excessive advance can result in partially burned fuel and increased CO. The unburned fuel also causes HC to increase. Finally, high $NO_x$ is created due to excessive combustion temperatures. On the other hand, if the ignition timing is retarded excessively, CO will also increase as a result of partially burned fuel and high HC will result from the unburned fuel.

To maintain optimum ignition control, the engine computer responds to various sensors and it is programmed with instructions that tell it how to change the ignition timing for different situations.

## Evaporative Control

Evaporative emissions result from the direct release of fuel vapors into the atmosphere. This can occur during refueling, while the vehicle is parked, and during engine operation. Gasoline is a volatile substance and vaporizes easily. Prior to the use of evaporative controls, the fuel system was vented directly to the atmosphere. This was necessary to prevent unsafe pressure levels inside the system. This type of atmospheric venting led to HC emissions and the production of photochemical smog. Regulations no longer allow atmospheric venting of the fuel vapors. Today's automobiles have an evaporative emissions-control system to perform the following functions:

- Contain liquid during a vehicle rollover
- Trap fuel vapors
- Store fuel vapors
- Deliver vapors to the engine for burning

**Figure 3** illustrates a typical evaporative emissions-control system. Most systems include the following components **(Figure 4)**:

- Fuel tanks designed to limit the amount of fuel that can be put into the tank. This allows for expansion of the fuel as it warms.
- Pressure/vacuum relief fuel caps
- A purge solenoid
- An evaporative canister

All the evaporative controls trap fuel vapors and reduce the tendency for fuel to evaporate to the atmosphere. The fuel tank is sealed with relief valve filler caps to raise the fuel boiling point. The pressure/vacuum filler cap releases vapors to the atmosphere at 44 inches of water column and draws air in at a system vacuum of 4 inches of water column **(Figure 5)**. Hoses route internal vapors into an **evaporative canister (Figure 6)** for storage. This canister is a charcoal-filled container used to store evaporative emissions temporarily. The charcoal absorbs the fuel vapors until a control valve "purges" the vapors from the charcoal. The vapors are then burned in the engine instead of evaporating into the atmosphere. Early vehicles typically located the canister in the engine compartment. In current vehicles, the canister is located near the fuel tank.

**You Should Know**

*The evaporative canister is also called a charcoal canister or carbon canister.*

The canister contains a liquid fuel trap that prevents liquid fuel from saturating the canister. As the fuel tank level lowers or the temperature of the fuel decreases, a vacuum is created that draws any trapped fuel back into the tank.

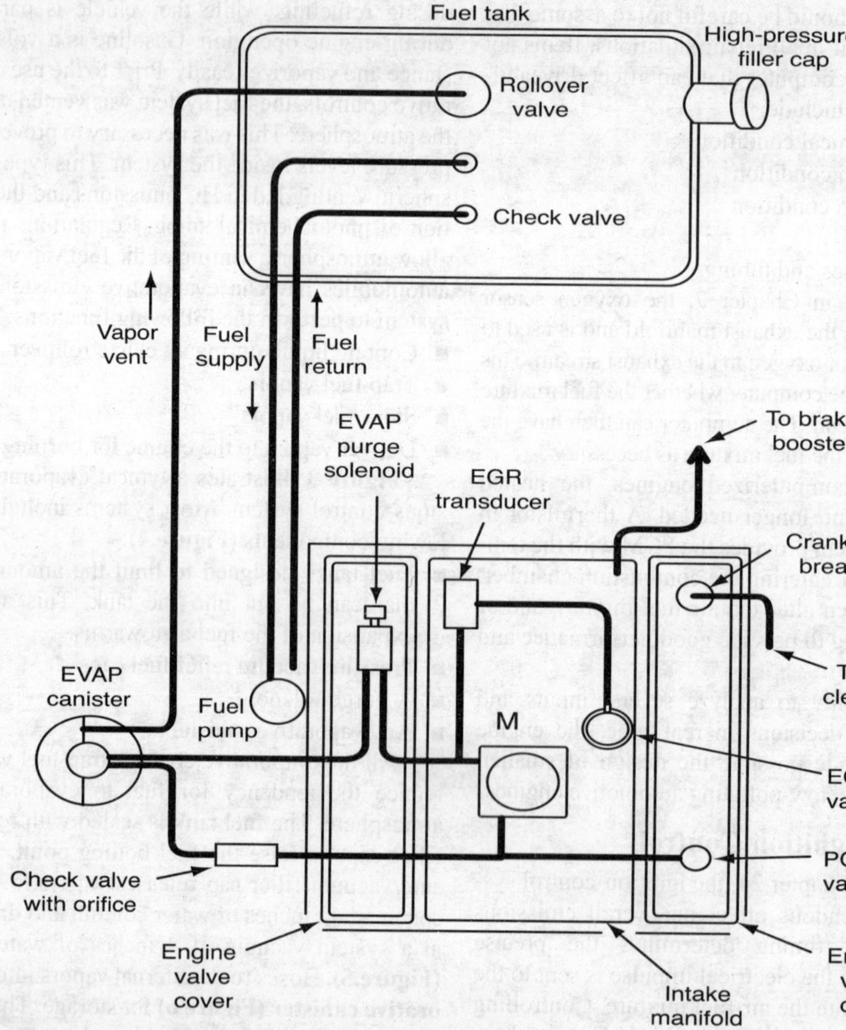


**Figure 3** *Typical evaporative emissions-control system.*

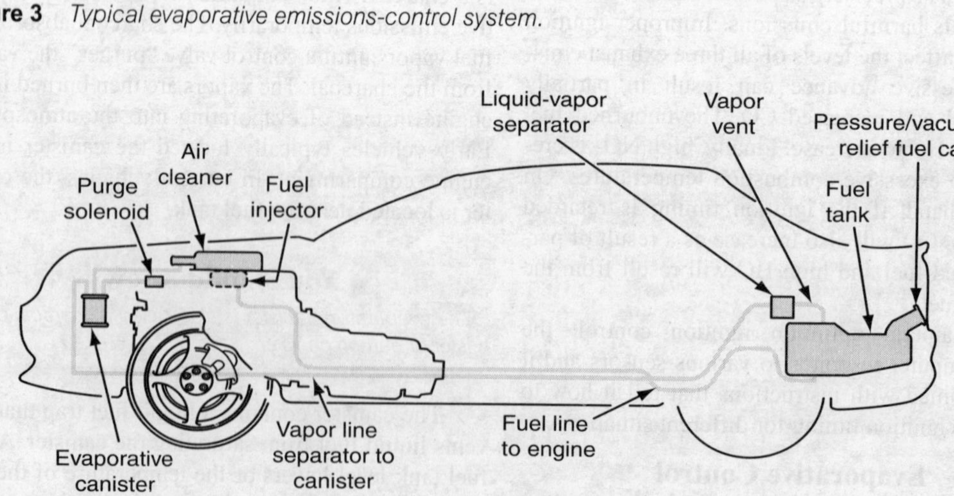


**Figure 4** *Typical evaporative emissions-control system components.*

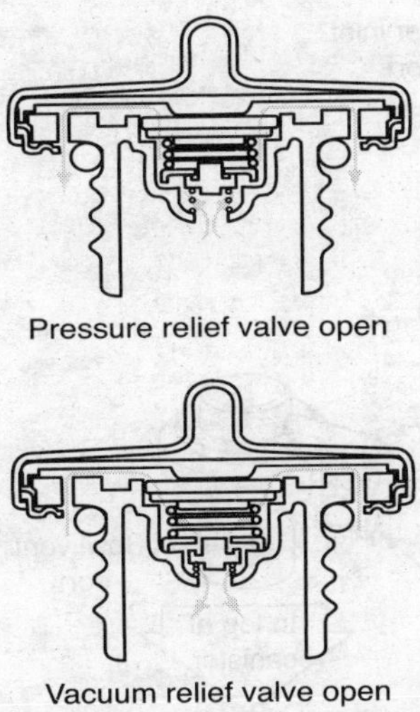


**Figure 5** *Sealed fuel tank cap operation.*

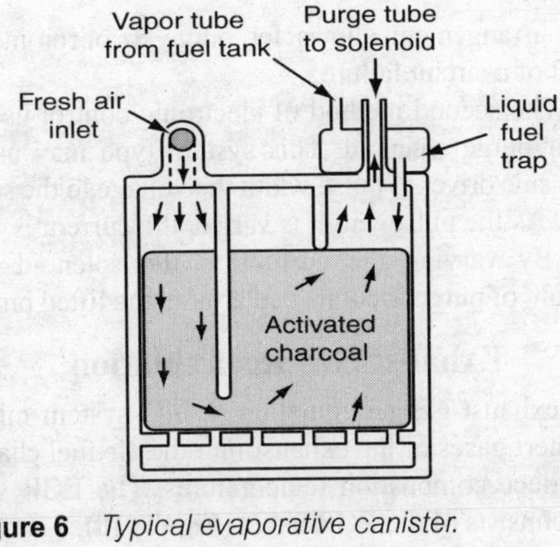


**Figure 6** *Typical evaporative canister.*

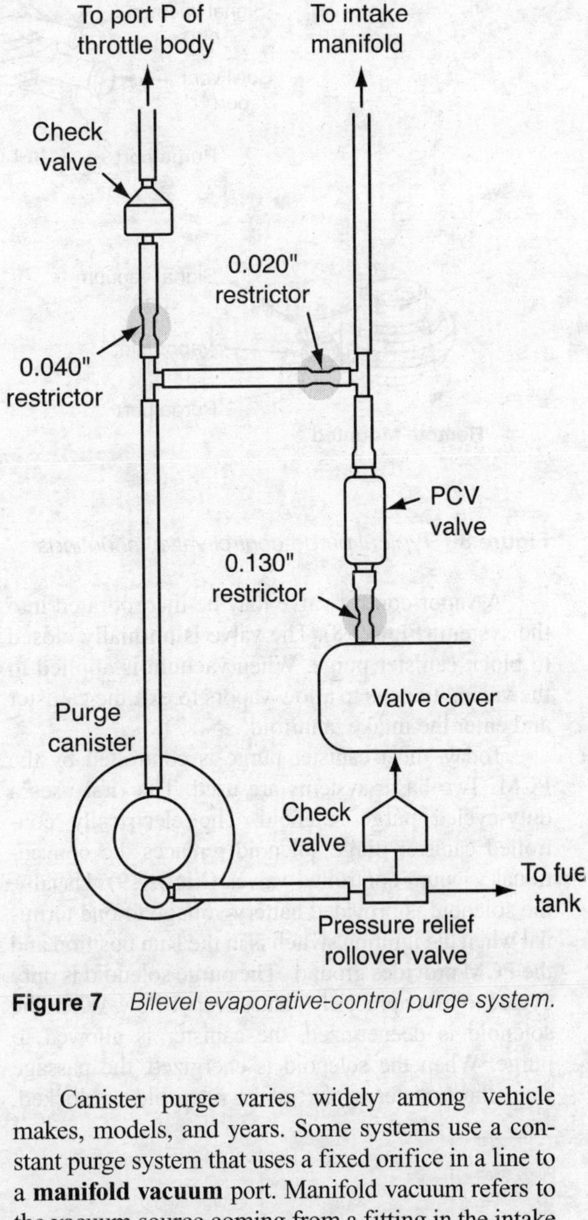


**Figure 7** *Bilevel evaporative-control purge system.*

As fuel is drawn from the tank to be used by the engine, a slight negative pressure (vacuum) is created in the tank. To prevent the tank from collapsing, it needs to be vented to allow air to enter it and replace the fuel. Evaporative controls allow this ventilation to take place within a closed circuit so that HC stays in the fuel system. The sealed fuel system is vented into the charcoal-filled canister, which retains the vapors until they can be drawn into the intake manifold and burned.

An evaporative line runs from the fuel tank to the canister. A purge valve is normally used to control the flow of the vapors from the canister into the intake system. On carbureted vehicles, a line also runs from the carburetor fuel bowl to the canister to contain fuel vapors from the float bowl.

Canister purge varies widely among vehicle makes, models, and years. Some systems use a constant purge system that uses a fixed orifice in a line to a **manifold vacuum** port. Manifold vacuum refers to the vacuum source coming from a fitting in the intake manifold, downstream of the throttle blade. Other systems connect the purge line to ported vacuum. **Ported vacuum** refers to a vacuum source above the throttle plates so the throttle needs to open before vacuum is present. This allows purge to only occur when the throttle is open. Some systems use bilevel purging **(Figure 7)** in which different-sized orifices control purge into an intake manifold port or a port above the throttle blades. Still other systems use a thermal delay valve to prevent the canister from purging until the engine reaches a specified temperature.

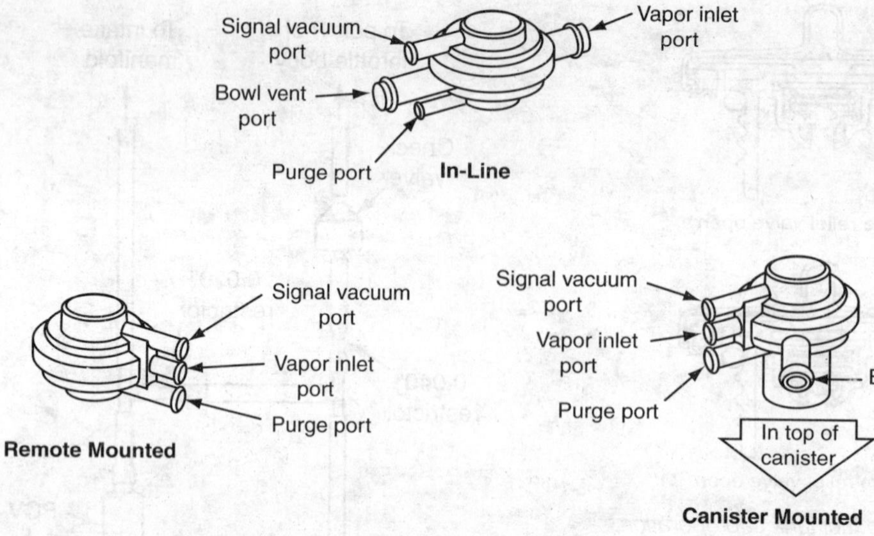


**Figure 8** *Typical purge-control valve mountings.*

A vapor-control valve may be incorporated into the system **(Figure 8)**. The valve is normally closed to block canister purge. When vacuum is applied to the valve, it opens to allow vapors to exit the canister and enter the intake manifold.

Today, most canister purge is controlled by the PCM. Two basic systems are used. The first uses a duty-cycled purge solenoid. The electrically controlled canister purge solenoid replaces the conventional vacuum-controlled valves **(Figure 9)**. Usually, the solenoid is provided battery voltage to one terminal when the ignition switch is in the Run position and the PCM provides ground . The purge solenoid is duty cycled to more precisely control vapor flow. When the solenoid is deenergized, the canister is allowed to purge. When the solenoid is energized, the passage from the canister to the intake manifold is blocked.

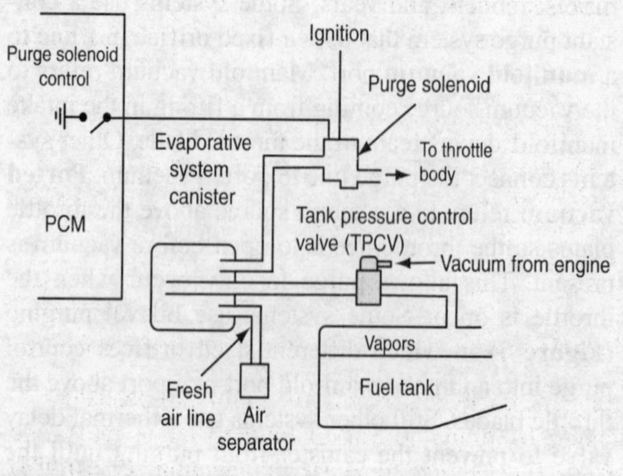


**Figure 9** *Computer-controlled EVAP system.*

This arrangement allows for purge to occur in the event of a circuit failure.

The second method of electronic control uses a linear purge solenoid. This system type may use a high-side driver to pulse width the voltage to the solenoid. As the pulse width is varied, the current is varied. By varying the current to the solenoid, the amount of purge is controlled around the lifted pintle.

## Exhaust Gas Recirculation

The exhaust gas recirculation (EGR) system mixes the inert gases of the exhaust into the air-fuel charge to reduce combustion temperatures. The EGR system consists of the EGR valve **(Figure 10)**, a control mechanism, and related hoses and passages. Knowledge of this system is particularly important because EGR directly affects engine performance.

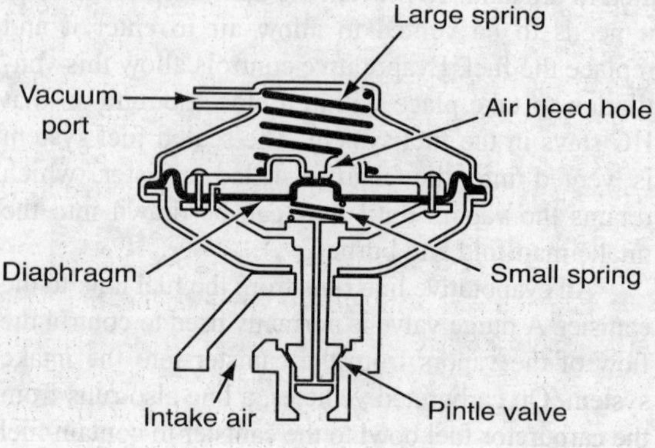


**Figure 10** *Typical EGR valve design.*

| Don't Want EGR @ | Because |
|---|---|
| Idle | Causes roughness |
| WOT | Reduces maximum power |
| Cold Start / Light Throttle | Engine not hot enough to produce much $NO_X$; causes hard starting, roughness, hesitation, stalls |

**Table 2** *Effects that improper EGR has on performance.*

Since $NO_x$ forms in high-temperature combustion, the solution is to prevent $NO_x$ from forming by keeping combustion temperatures low enough to prevent the bonding of nitrogen and oxygen atoms. However, lean mixtures reduce HC and CO but increase $NO_x$. In addition, advanced ignition timing, which improves fuel economy, also increases $NO_x$. The EGR valve is used to prevent excessive combustion temperatures and reduce the formation of $NO_x$ by reducing the volume in the combustion chamber. When operating, the EGR system mixes a metered amount of exhaust gases with the combustion mixture in the intake manifold. With the chamber's reduced volume, there is less room for the air-fuel mixture and the combustion chamber temperature is reduced. A side benefit of EGR is a slight increase in fuel economy due to less fuel being burned.

Burned exhaust gases are basically inert and will not react chemically or continue to burn. A percentage of exhaust gases mixed with the air-fuel mixture "dilutes" the charge strength. The result is less volume of air-fuel mixture. In turn, this reduces combustion temperature and pressure, resulting in less $NO_x$.

Since the decrease in the volume of the air-fuel mixture by EGR interferes with combustion, the EGR system must be designed to add just enough EGR to control $NO_x$, without interfering with engine performance. If EGR operation is allowed to occur when it should not, driveability problems may result **(Table 2)**. For best driveability, the EGR valve opens in proportion to throttle opening (but not at wide-open throttle). Also, EGR is not allowed during cold engine operation.

Most EGR valves are vacuum-operated, poppet-type valves **(Figure 11)**. The valve fastens to the intake manifold delivery port. The mounting block has two passages: one for exhaust gas into the valve, and one for metered exhaust gas flow out of the valve. The stem operates the poppet valve, which seats in the mounting block to close off EGR flow.

The upper part of the valve contains a vacuum diaphragm that is attached to the stem. The diaphragm and stem are spring-loaded to close the poppet valve against the seat. The valve can move from closed to open or to metering positions in between, depending on the amount of vacuum applied to the diaphragm. In precomputerized systems, a temperature vacuum switch (TVS) was located between the vacuum source and the valve to prevent operation of the EGR valve when the engine was cold **(Figure 12)**.

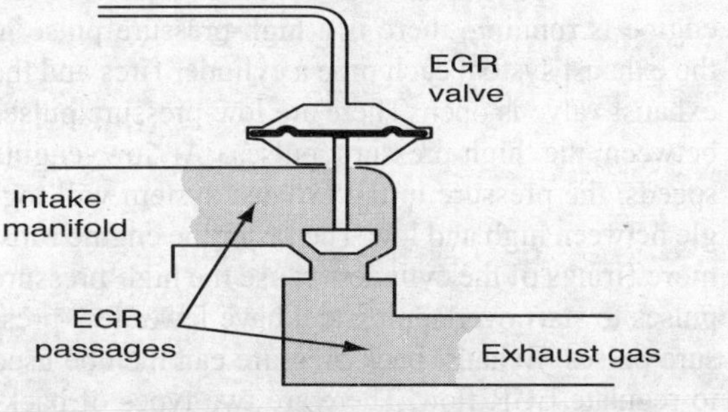


**Figure 11** *Most EGR valves use vacuum controls to lift the valve.*

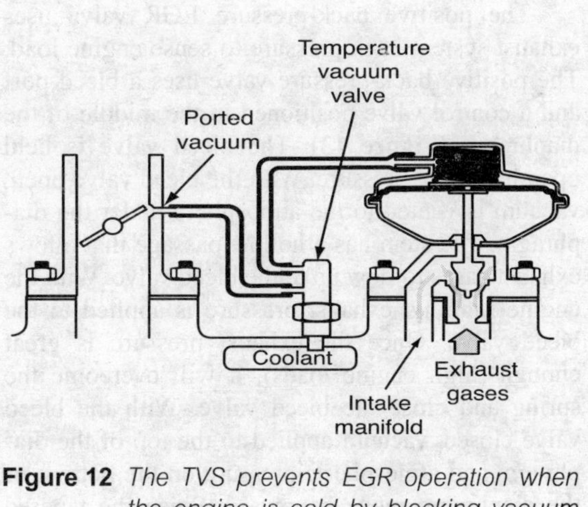


**Figure 12** *The TVS prevents EGR operation when the engine is cold by blocking vacuum to the valve.*

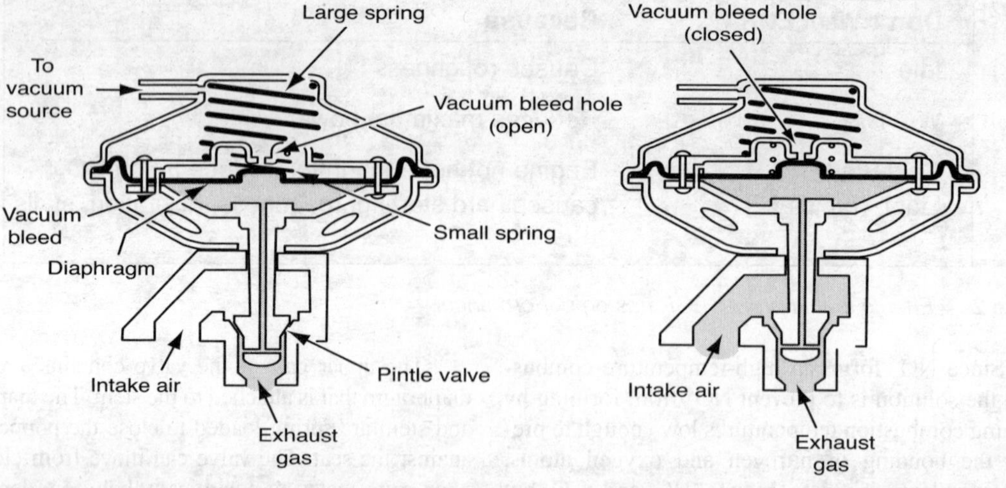


**Figure 13** *Positive back-pressure EGR valve operation.*

Remembering that $NO_x$ is formed under high heat conditions, increases in engine load require more EGR flow. A good indication of engine load is exhaust back pressure. **Back pressure** is an exhaust system's resistance to the flow of gases. As the engine is running, there is a high-pressure pulse in the exhaust system each time a cylinder fires and the exhaust valve is open. There are low-pressure pulses between the high-pressure pulses. At low engine speeds, the pressure in the exhaust system will toggle between high and low. The faster the engine runs, more firings of the cylinders cause the high-pressure pulses to start overlapping and have fewer low-pressure pulses. Exhaust back pressure can then be used to regulate EGR flow. There are two types of back-pressure EGR valves: **positive back-pressure EGR** valve, and **negative back-pressure EGR** valve.

The positive back-pressure EGR valve uses exhaust system back pressure to sense engine load. The positive back-pressure valve uses a bleed port and a control valve positioned in the middle of the diaphragm **(Figure 13)**. The bleed valve is held open by spring pressure. With the bleed valve open, vacuum is vented to the atmosphere under the diaphragm. The stem has a hollow passage that allows exhaust gases to flow up to the bleed valve. With the engine running, exhaust pressure is applied to the bleed valve. Once the exhaust pressure is great enough (high engine loads), it will overcome the spring and close the bleed valve. With the bleed valve closed, vacuum applied to the top of the diaphragm and atmospheric pressure on the bottom of the diaphragm move it upward. Since the tapered stem is attached to the diaphragm, the stem is lifted and it opens the passage to allow EGR flow into the intake manifold. Once the stem opens to allow flow, back pressure is reduced slightly. As a result of reduced back pressure, the valve will start to close again. With the valve closed, back pressure increases and opens the valve. This oscillation of the valve with changing exhaust back pressure maintains a balanced flow of EGR. EGR will not flow at idle since there is insufficient back pressure.

The negative back-pressure EGR opens the bleed valve when exhaust back pressure decreases to stop EGR flow. The negative back-pressure EGR valve operates similarly to the positive back-pressure EGR valve, but it uses decreasing back pressure. The bleed valve is normally closed and the hollow stem directs exhaust pressure to the valve **(Figure 14)**. At low engine speed, the negative pressure pulses from cylinder firing are more predominant. The negative pulses will hold the bleed valve open and prevent EGR flow.

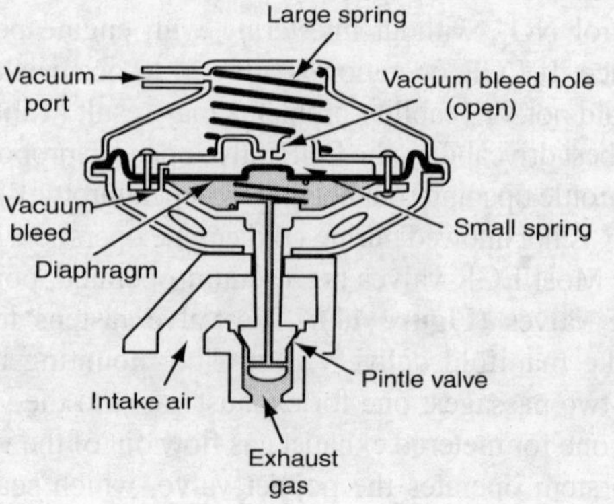


**Figure 14** *Negative back-pressure EGR valve.*

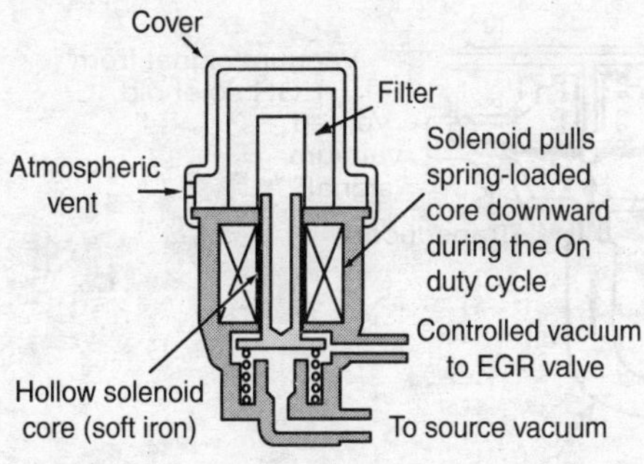


**Figure 15** *EGR vacuum regulator (EVR) operation.*

At higher speeds, negative pulses decrease and close the bleed valve. With the bleed valve closed, the stem lifts and allows EGR flow.

Computer-controlled EGR systems can use an EGR vacuum regulator (EVR), an electronic EGR transducer, or linear solenoids. Some systems may also use a pressure feedback electronic (PFE) sensor.

An EVR uses the PCM to control the normally closed EVR solenoid **(Figure 15)**. The EVR supplies vacuum to the EGR valve. If the EVR solenoid is not energized, the solenoid plunger tip is seated in the vacuum passage and shuts off vacuum to the EGR valve. In this position, any vacuum in the EGR valve and hose is vented through the EVR solenoid to prevent vacuum from being locked in the system and holding open the EGR valve.

When the PCM inputs indicate the EGR valve should be open, the PCM provides a ground for the EVR solenoid winding. This action moves the solenoid plunger and opens the vacuum passage through the solenoid to the EGR valve. In some systems, the PCM pulses the EVR on and off to supply the precise vacuum and EGR valve opening required by the engine.

The PCM uses inputs such as engine temperature, throttle position, and vehicle speed to operate the EGR valve. The PCM will not energize the EVR solenoid if the engine coolant temperature is below a preset value. The PCM opens the EGR valve when the vehicle is operating at normal temperature in the cruising speed range. When the vehicle is operating at low speed or near wide-open throttle, the PCM does not open the EGR valve. If the EGR valve is open while the engine is idling or operating at low rpm, engine operation is erratic.

The electronic EGR transducer system contains an electrically controlled solenoid and a back-pressure transducer. The EGR transducer is mounted in-line above the EGR valve **(Figure 16)**. The transducer controls the operation of the EGR valve when the PCM allows EGR operation. The EGR transducer consists of a diaphragm and three ports; one port receives an exhaust back-pressure signal from a vent in the EGR valve housing, one port receives a vacuum signal from the intake manifold, and one port routes controlled vacuum out to the EGR valve diaphragm.

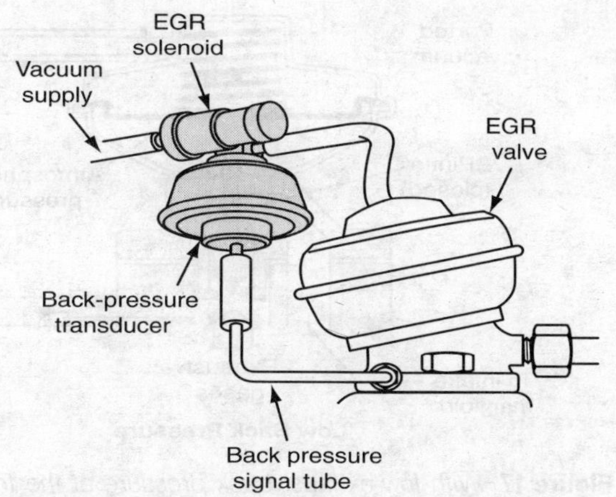


**Figure 16** *Electrically controlled back-pressure transducer.*

The EGR solenoid is normally open, allowing the free passage of vacuum to the EGR valve. When the PCM energizes the solenoid, it restricts vacuum to the EGR transducer.

The PCM controls when EGR can occur by turning the EGR solenoid on and off. However, the actual amount of EGR valve movement depends on the amount of back pressure. If there is low-exhaust back pressure reaching the transducer, the transducer diaphragm is not lifted and the vacuum supplied through the solenoid is bled off through the vent **(Figure 17)**. As back pressure increases, the diaphragm is pushed up and the vacuum bleed closes. With the bleed closed, vacuum is allowed to reach the EGR diaphragm **(Figure 18)**. When the vacuum reaches the EGR valve, the diaphragm is pulled up, causing the EGR valve pintle to move off its seat and allow exhaust gases to enter the intake manifold and then the combustion chamber. The more back pressure, the more EGR flow. The PCM energizes the EGR solenoid during cold engine operation or wide-open throttle. Under both conditions EGR could reduce vehicle performance but under either condition EGR is not needed due to the richer fuel mixtures.

The linear solenoid EGR valve consists of three major components:

- Pintle, valve seat, and housing that contain and regulate gas flow

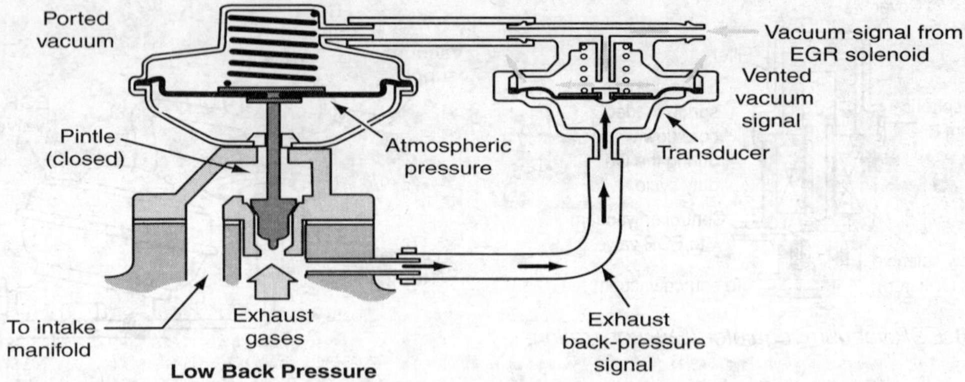


**Figure 17** *With low-exhaust back pressure at the transducer, the vacuum supplied through the solenoid is bled off.*

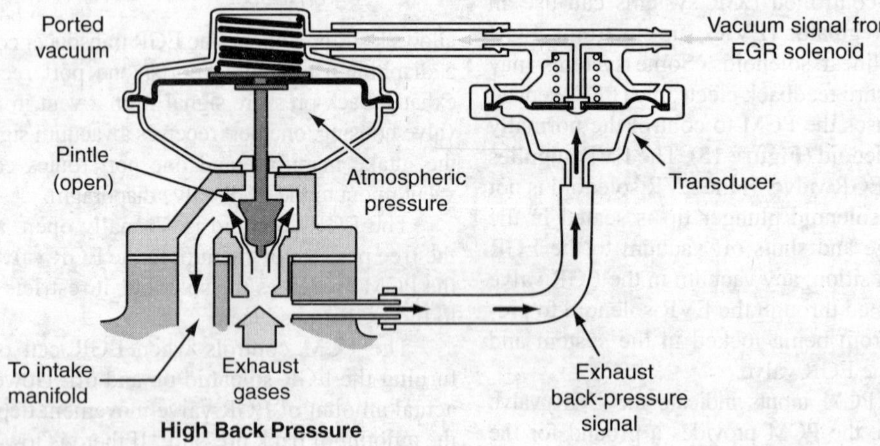


**Figure 18** *EGR valve opens when enough back pressure closes the transducer vent.*

- Armature, return spring, and solenoid coil to provide the operating force to regulate exhaust gas flow by changing the pintle position
- EGR pintle position sensor to return information concerning the position of the pintle to the PCM

The PCM determines the EGR flow. For a given set of conditions, the PCM knows the ideal EGR flow to optimize $NO_x$ and fuel economy as a function of the pintle position. When the solenoid is energized, the valve is lifted to allow exhaust gases to flow into the intake manifold **(Figure 19)**. Pintle position is obtained from a linear potentiometer that is integral to the valve. The PCM adjusts the duty cycle of 128 Hz power supplied to the solenoid coil to obtain the correct position.

Some manufacturers use a digital EGR valve that regulates EGR flow according to computer control. The digital EGR valve has three metering orifices that are opened and closed by solenoids **(Figure 20)**. By opening various combinations of these three solenoids, different flow rates can be achieved to match EGR to the engine's requirements. The solenoids are normally closed and opened only when the computer completes the ground to each.

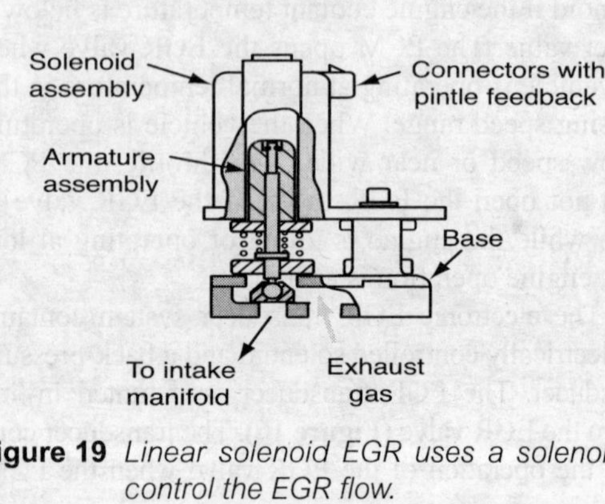


**Figure 19** *Linear solenoid EGR uses a solenoid to control the EGR flow.*

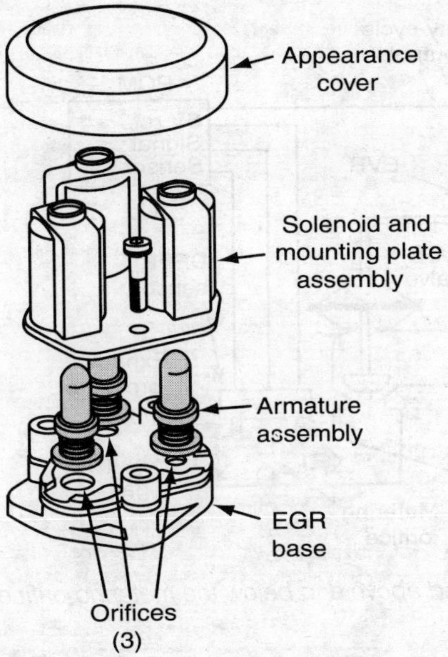


**Figure 20** *Digital EGR valve with three solenoids.*

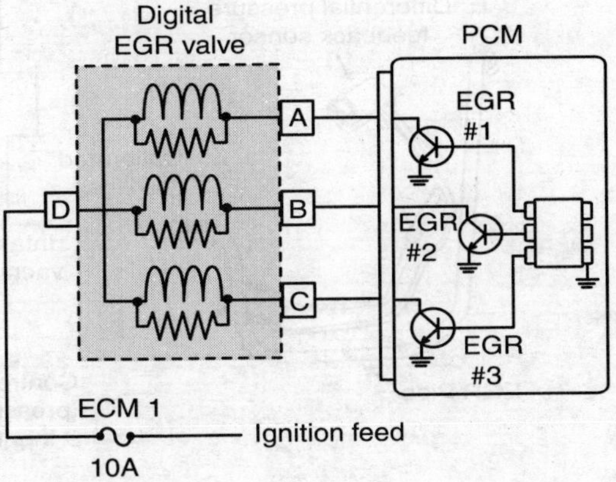


Some EGR systems have a PFE sensor. These systems have an orifice located in the exhaust passage below the EGR valve that attaches a small pipe to the PFE sensor **(Figure 21)**.

The PFE sensor operates in the same fashion as a MAP sensor and changes the exhaust pressure signal to a voltage signal. The voltage signal is sent to the PCM. The exhaust pressure in the orifice chamber is proportional to the EGR valve flow. The PFE signal informs the PCM regarding the amount of EGR flow and the PCM compares this signal with the EGR flow requested by the input signal. If there is some difference between the actual EGR flow indicated by the PFE signal and the requested EGR flow, the PCM makes the necessary correction to the EVR output signal.

In some EGR systems, two pipes supply exhaust pressure to the differential PFE (DPFE) sensor **(Figure 22)** from above and below the orifice under the EGR valve.

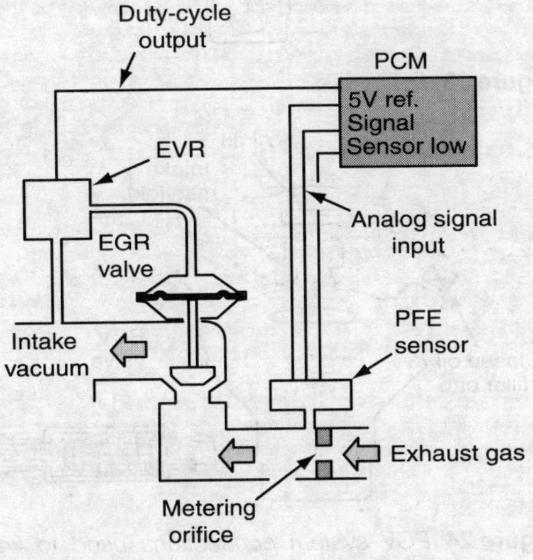


**Figure 21** PFE *sensor measures back pressure.*

### You Should Know

*Despite this discussion on EGR systems, some manufacturers control $NO_x$ without the use of an EGR valve. Instead, they use camshaft profile and valve overlap to provide EGR. The camshaft profile holds the exhaust valve open a little longer as the intake valve is opening. The downward piston movement of the intake stroke then "sucks" exhaust gases around the two valves and into the combustion chamber.*

## Crankcase Ventilation

The leakage of fuel and combustion gases into the engine crankcase is called **blow-by** and occurs toward the end of the engine's power stroke. Some unburned fuel and other products of combustion such as water vapor leak past the engine's piston rings into the crankcase. Blow-by gases must be removed from the crankcase before they condense and react with the oil to form sludge. Sludge, if

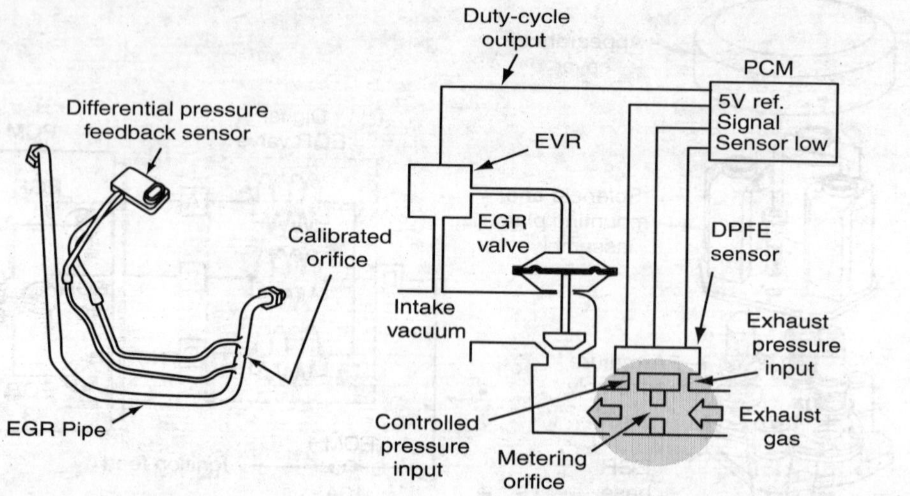


**Figure 22** *Differential PFE with dual pressure pipes connected above and below the metering orifice.*

allowed to circulate with engine oil, corrodes and accelerates the wear of pistons, piston rings, valves, bearings, and other internal working parts of the engine. Blow-by gases must also be removed from the crankcase to prevent seal and gasket failures. Since blow-by gases pass the piston rings as a result of the pressure formed during combustion, they will pressurize the crankcase. The gases exert pressure on the oil pan gasket and crankshaft seals. If the pressure is not relieved, oil will eventually be forced out of these seals.

In addition, blow-by also carries some unburned fuel into the crankcase. If the unburned fuel is not removed, it dilutes the crankcase oil. When oil is diluted with gasoline it does not lubricate the engine properly, causing excessive wear.

Combustion gases that enter the crankcase are removed by a **positive crankcase ventilation (PCV)** system that uses engine vacuum and a PCV valve **(Figure 23)** to draw fresh air through the crankcase. The PCV system picks up crankcase gases and sends them to the intake manifold. The PCV valve is usually mounted in a rubber grommet in one of the valve covers or in the intake manifold. A hose connects the PCV valve to the valve cover and intake manifold **(Figure 24)**. A fresh air intake hose is connected from the air cleaner to the valve cover. On V-type engines, the PCV valve is usually connected to one valve cover and the fresh air intake is connected to the other. Some manufacturers connect the fresh air intake to the oil fill spout. Regardless of the configuration, the fresh air intake is filtered.

When the engine is running, intake manifold vacuum is supplied to the PCV valve. This vacuum moves air through the fresh air intake hose and into the crankcase where it mixes with blow-by gases. The mixture of blow-by gases and air flows up through cylinder head openings to the rocker arm cover and the PCV valve. Intake manifold vacuum

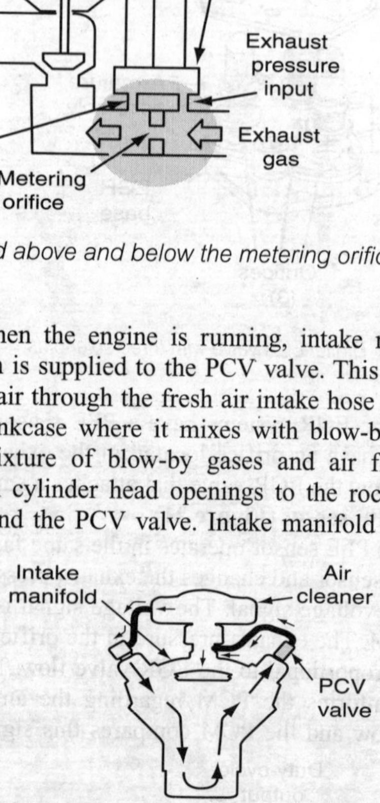


**Figure 23** *PCV valve.*

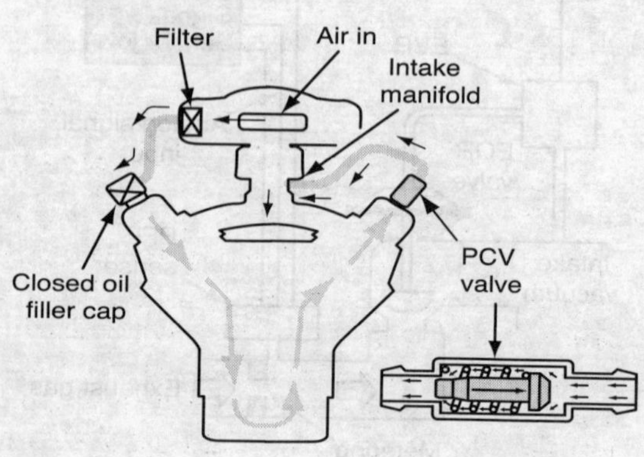


**Figure 24** *PCV system connections used to keep crankcase emissions from escaping.*

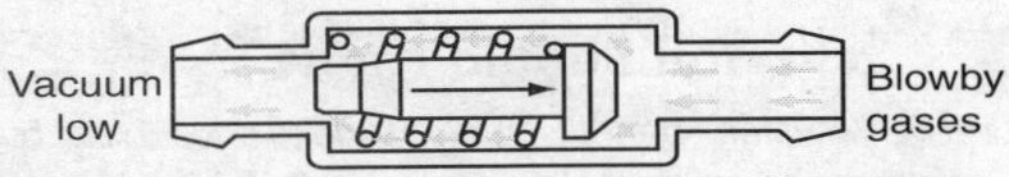


**Figure 25** *Fresh air is drawn into the crankcase as blow-by gases flow into the intake manifold.*

moves the blow-by gas mixture through the PCV valve into the intake manifold **(Figure 25)**. The blow-by gases are then moved through the intake valves into the combustion chamber where they are burned.

Because the vacuum supply for the PCV system is from the engine's intake manifold, the airflow through this system must be controlled so it varies in proportion to the regular air-fuel ratio being delivered. If this were not controlled, the additional air that is drawn into the system would cause the air-fuel mixture to become too lean for efficient engine operation. Therefore, the PCV valve is placed in the flow just before the intake manifold to regulate the flow according to vacuum.

On many engines, the PCV system delivers blow-by gases to one location in the intake manifold. This type of system may not deliver these gases equally to all the cylinders. This action may result in an air-fuel ratio variation among the cylinders, which results in rougher idle operation. Some engines such as the Ford 4.6L V8 and Dodge 5.7L Hemi have passages from the PCV valve through the intake manifold to supply blow-by gases equally to each cylinder, resulting in smoother idle operation.

The PCV valve contains a tapered valve surrounded by a spring. When the engine is not running, the spring seats the valve against the valve housing **(Figure 26)**. During idle or deceleration, the high intake manifold vacuum moves the tapered valve against the spring tension. With the valve moved this far, the taper of the valve creates a small opening **(Figure 27)**. Since the engine is not under heavy load during idle or deceleration, blow-by gases are minimal and the small PCV valve opening is adequate to move the blow-by gases out of the crankcase.

Compared with idle vacuum, intake manifold vacuum is lower during part-throttle operation. Under this condition, the spring pressure overcomes the vacuum against the tapered valve and moves the valve downward. This creates a larger opening between this valve and the PCV valve housing **(Figure 28)**. Since engine load is higher at part-throttle operation than at idle operation, blow-by gases are increased. The larger opening between the tapered valve and the PCV valve housing allows all the blow-by gases to be drawn into the intake manifold.

When the engine is operating under heavy load conditions with a wide throttle opening, the decrease in intake manifold vacuum allows the spring to move the tapered valve farther downward in the PCV valve. This action provides a larger opening between the tapered valve and the PCV valve housing. Since higher engine

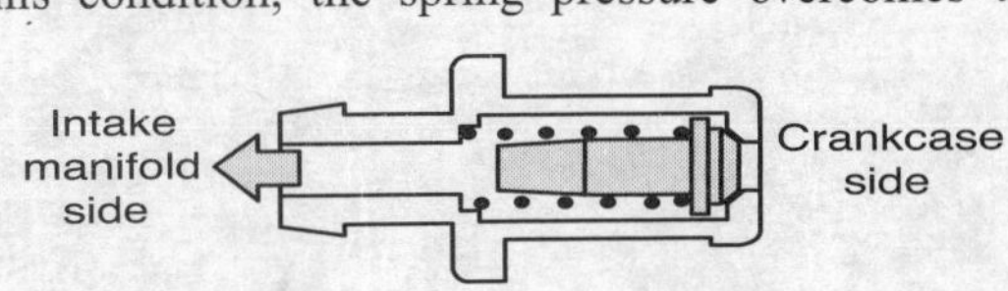


**Figure 26** *PCV valve position with engine off or during a backfire occurrence.*

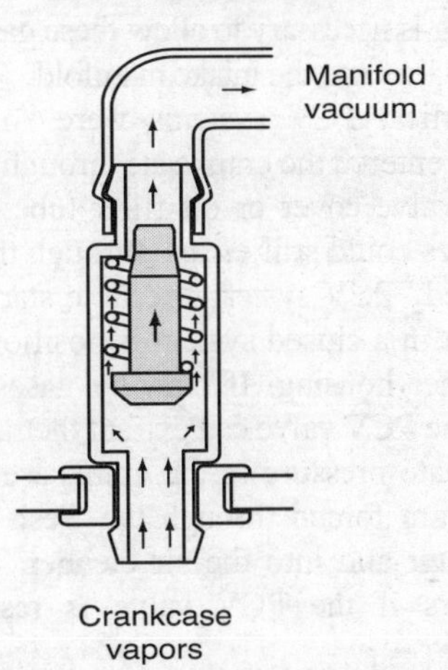


**Figure 27** *PCV valve position at idle with limited flow.*

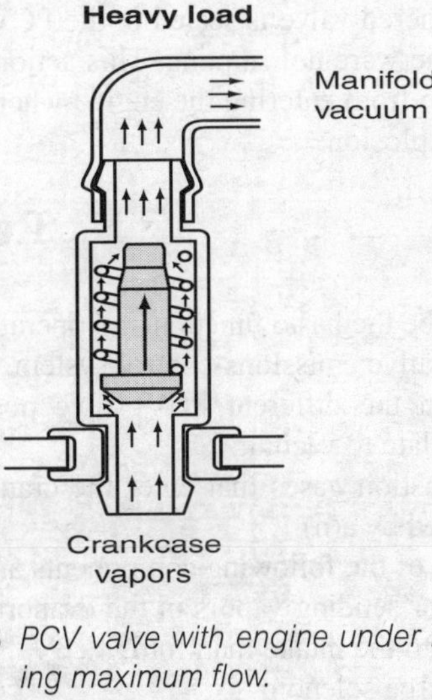


**Figure 28** *PCV valve with engine under load allowing maximum flow.*

load results in more blow-by gases, the larger PCV valve opening is necessary to allow these gases to flow through the valve into the intake manifold.

The earliest PCV systems were "open"-style systems. Air entered the crankcase through a breather cap on the valve cover or oil filler tube, but some blow-by gases could still escape through the vent. In 1968, "closed" PCV systems became standard. The air inlet vent in a closed system is positioned inside the air cleaner housing. If blow-by gases build up faster than the PCV valve can extract them, the blow-by gases create pressure in the crankcase. Some of these gases are forced through the fresh air intake hose and filter and into the air cleaner. This same action occurs if the PCV valve is restricted or plugged.

If the PCV valve sticks in the wide-open position, excessive airflow through the valve causes a rough idle. If a backfire occurs in the intake manifold, the tapered valve is seated in the PCV valve as if the engine were not running. This action prevents the backfire from entering the engine where it could cause an explosion.

**You Should Know**

*The PCV system benefits the vehicle's driveability by eliminating harmful crankcase gases, reducing air pollution, and promoting fuel economy. An inoperative PCV system could shorten engine life by allowing harmful blow-by gases to remain in the engine, causing corrosion and accelerating wear. Also, the excessive pressures that are built up can cause seals and gaskets to fail.*

Some engines do not use a PCV valve. Two examples include the Ford Escort and the General Motors Quad 4. Instead, these engines use a calibrated orifice in a breather box to siphon blow-by vapors back into the intake manifold. The basic system works the same as if it had a valve, except that the system is regulated only by the vacuum on the orifice. The size of the orifice limits the amount of blow-by flow into the intake. The engine's air-fuel system is calibrated for this air leak. Since the action of the PCV allows unmetered air into the intake, the air-fuel system must be set for this amount of extra air.

## Review Questions

1. Describe the basic function and operation of the evaporative emissions-control system.
2. Explain the different PCV valve positions as they relate to vacuum.
3. Combustion gases that enter the crankcase are removed by a(n) ________.
4. Which of the following components are responsible for sending vapors in the evaporative canister into the intake manifold?
   A. Purge solenoid
   B. PCV valve
   C. Switchover valve
   D. EGR transducer
5. Technician A says that the positive back-pressure EGR valve contains a normally closed bleed valve in the center of the diaphragm. Technician B says that the negative back-pressure EGR valve contains a normally open bleed valve in the center of the diaphragm. Who is correct?
   A. Technician A only
   B. Technician B only
   C. Both Technician A and Technician B
   D. Neither Technician A nor Technician B
6. Describe how EGR is used to control $NO_x$ emissions.
7. A ________ back-pressure EGR valve has a normally open bleed valve in the center of the diaphragm. This bleed valve is closed by exhaust pressure at a specific throttle opening.
8. Of the following items, which would be considered precombustion control?
   A. Oxidation catalytic converter
   B. Air injection
   C. Three-way catalytic converter
   D. EGR
9. Which pollutant is the EGR valve used to control?
   A. HC
   B. $NO_x$
   C. CO
   D. $CO_2$

CHAPTER

# 26 Postcombustion Controls

## Introduction

**Postcombustion control** systems reduce emissions levels after combustion has taken place. The two postcombustion controls commonly found in vehicles today are air injection and catalytic converters. This chapter details the operation and function of these two emissions-control components.

## Air Injection

**Air injection** was one of the earliest postcombustion devices for controlling HC and CO emissions. The air-injection system injects air into the exhaust system to heat the catalysts faster and to supply additional oxygen to the center section of some converters. First appearing in 1968, most air-injection systems use an engine-driven air pump to force atmospheric air into the exhaust stream **(Figure 1)**. This additional oxygen causes unburned fuel to continue burning in the exhaust manifold. This oxidizes the HC emissions as well as heats the catalytic converter so it will be more efficient.

> **You Should Know**
> *Air-injection systems are also called air-injection reaction (AIR) systems.*

Secondary air systems are also used to supply oxygen to the second stage of a dual-bed catalytic converter. The oxygen supplied by the air pump enhances the reaction by which HC and CO emissions are converted to harmless carbon dioxide and water. There are four basic types of secondary air systems; they differ by the method in which they supply air to the exhaust system:

- Air pump systems **(Figure 2)** employ a belt-driven air pump to inject air into the exhaust system.
- Aspirated systems **(Figure 3)** utilize negative pressure from the exhaust system to "pull" air into the system.
- Pulse air feeder systems **(Figure 4)** use engine crankcase pressures acting on a diaphragm to draw air from the air cleaner.
- Electronic control air pump systems are similar to regular air pump systems except they use PCM control of an electric air pump to inject air into the exhaust system.

> **You Should Know**
> *As an add-on system, air injection does an especially effective job of reducing pollutants without sacrificing much in the way of performance. Its only burden on the engine is a small amount of drag from the air pump. Even though air injection is one of the older emissions-control technologies, its effectiveness is such that it is still in use in many vehicles today.*

The following are typical components of an air pump system:

- A belt-driven vane pump
- A vacuum-operated diverter valve
- A pressure relief valve
- A one-way check valve
- The plumbing and nozzles necessary to distribute and inject the air

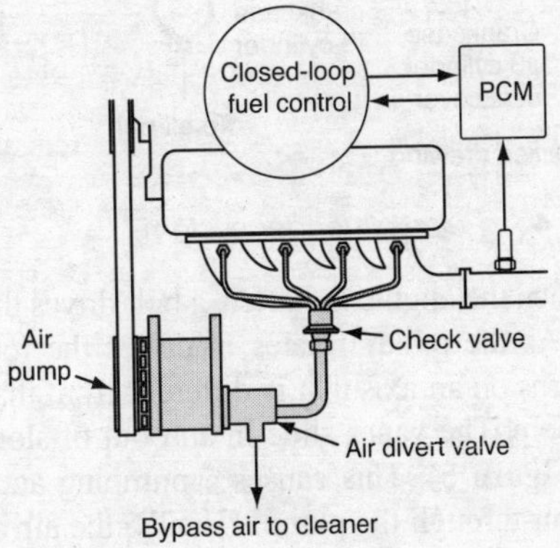


**Figure 1** *Typical air-injection system components.*

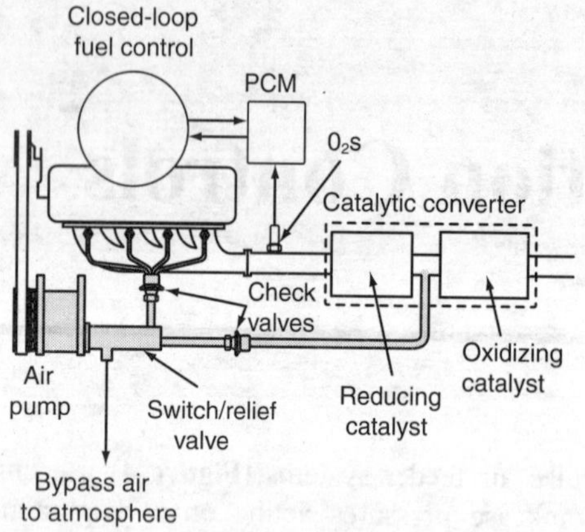


**Figure 2** *Air pump system.*

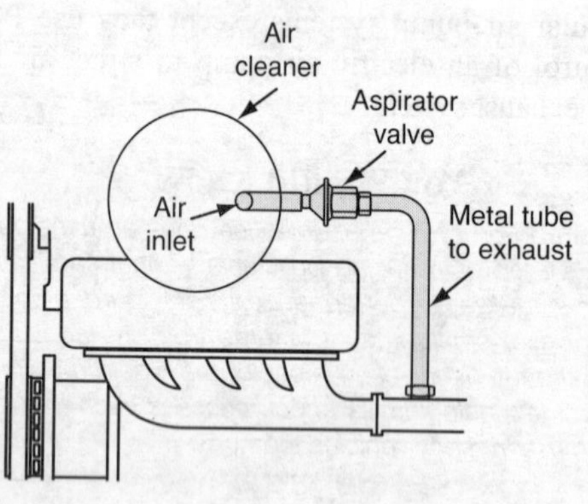


**Figure 3** *Aspirated air-injection system.*

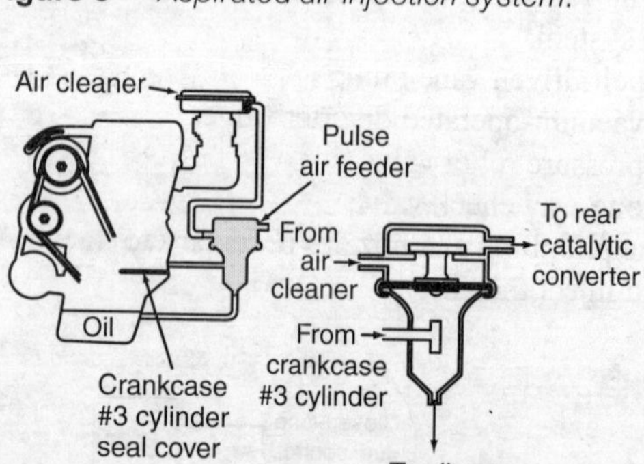


**Figure 4** *Pulse air feeder system.*

When the engine is started, a belt drives the pump pulley. As the pulley rotates, it drives the rotor. The rotor turns on an axis that is different from that of the pump bore. The vanes slide in and out of slots in the rotor **(Figure 5)**. This causes a pumping action that moves air through the pump. Usually, the air intake is through a centrifugal filter mounted behind the drive pulley. Some applications use a separate intake filter.

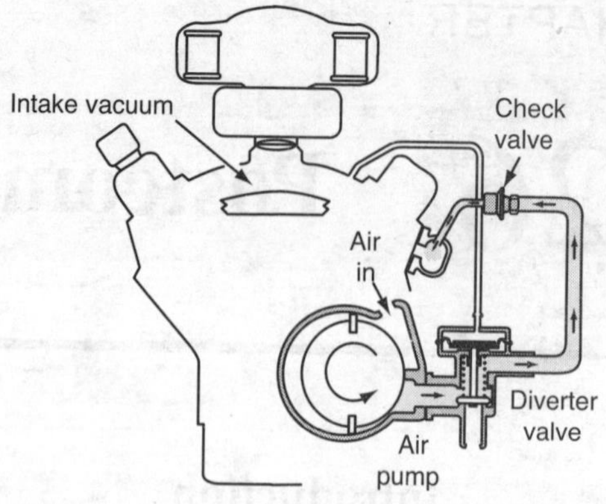


**Figure 5** *Air pump operation.*

Most systems use a relief valve that allows excess pump pressure to escape. The relief valve may be located internal to the pump or it may be incorporated into the diverter valve. The pressurized air exits into a large-diameter hose that routes it to this valve.

During all modes of engine operation except deceleration, air from the pump flows into the hose to the check valve. The check valve is a one-way device that allows air to enter the air-injection manifold but keeps exhaust from backing up into the pump if a belt should break or if the pump stops working. From the check valve, the air flows into the air-injection manifold, which directs the air into each exhaust port.

The **diverter valve** vents pump output to the atmosphere during deceleration so the combination of a rich mixture and extra oxygen does not cause backfiring **(Figure 6)**. The diverter valve receives a manifold vacuum signal to operate the valve during deceleration. The valve has a vacuum chamber, diaphragm, and spring arrangement. This valve moves the stopper from one seat to the other, thereby controlling the switching operation. During closed-throttle deceleration, engine vacuum is high enough for the diaphragm in the valve's chamber to overcome the force of the spring. This opens the passage to dump the airflow to a small muffler, which is usually mounted on the pump.

Another antibackfire device is the gulp valve **(Figure 7)**. The gulp valve has a diaphragm chamber, a spring, and a normally closed valve inside. During deceleration, the vacuum signal pulls the diaphragm against spring pressure. This action opens the valve. It also allows some of the air pump's output to flow into the intake manifold to dilute the rich mixture present in this mode.

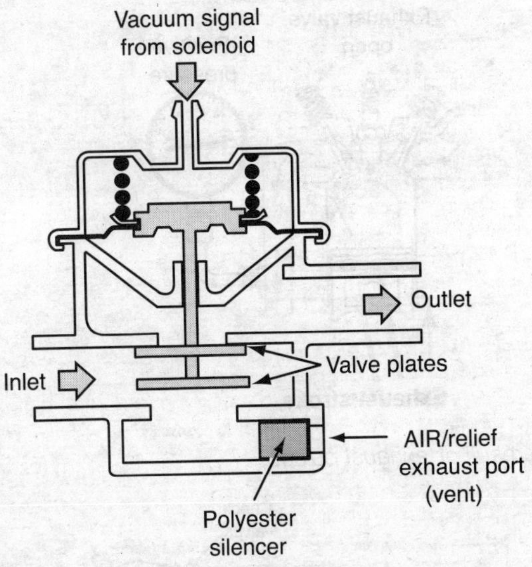


**Figure 6** *Typical diverter valve design.*

In air pump systems used with three-way converters (discussed next), an air switching/relief valve directs the air pump output upstream (into the exhaust manifold) while the engine is cold and downstream to the catalyst after the engine has warmed up. This switching activity is controlled by a coolant vacuum switch, cold open (CVSCO); an air switching solenoid; or the PCM. Manifold vacuum is applied to the air switching/relief valve at start-up and when the vehicle is cold. Very little $NO_x$ is produced during this time so the efficiency of the reduction process is not important. The vacuum supply is discontinued when the vehicle is warm, causing the air pump output to be directed downstream to the middle bed of the catalytic converter. This boosts the performance of the oxidation catalyst.

Some manufacturers use the PCM to switch air-injection flow from the exhaust ports to the catalytic converter. The PCM makes decisions about where to send the air pump output on the basis of input from various engine sensors, notably the coolant temperature sensor. It works in conjunction with a pair of solenoid valves: the bypass solenoid, which directs air pump flow to the atmosphere when energized; and the diverter solenoid, which switches airflow to either the exhaust ports or the catalytic converter.

Some air-injection systems do not use an air pump. Rather, these systems use the vacuum that is present in the exhaust manifold momentarily after each cylinder's exhaust stroke **(Figure 8)**. When the exhaust valve is open and the piston is moving upward on its exhaust stroke, the volume of exhaust

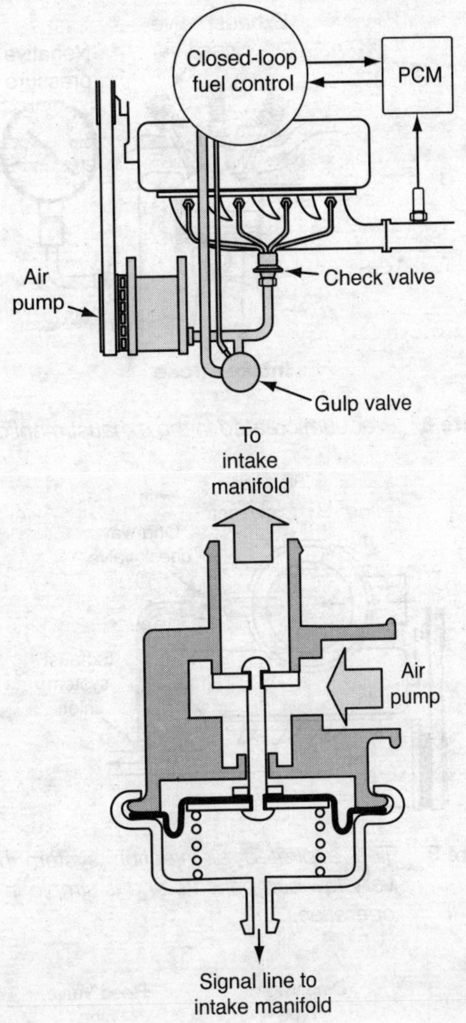


**Figure 7** *Gulp valve system.*

gases rushes into the exhaust manifold. This flow of gases results in a positive pressure. When the exhaust valve closes, inertia keeps the column of spent gases speeding through the exhaust system and creates a negative pressure (vacuum). This phenomenon is the basis of **aspirated** air-injection systems. A one-way valve allows fresh air from the air cleaner to enter the exhaust stream whenever a vacuum opens it **(Figure 9)**. The valve stops the hot gases from escaping by closing when it encounters positive pressure **(Figure 10)**. Depending on the application, there may be only one aspirator valve or one for each cylinder **(Figure 11)**. When the air is injected into the exhaust ports by the pulsed secondary air-injection system, most of the unburned HC coming out of the exhaust ports is ignited and

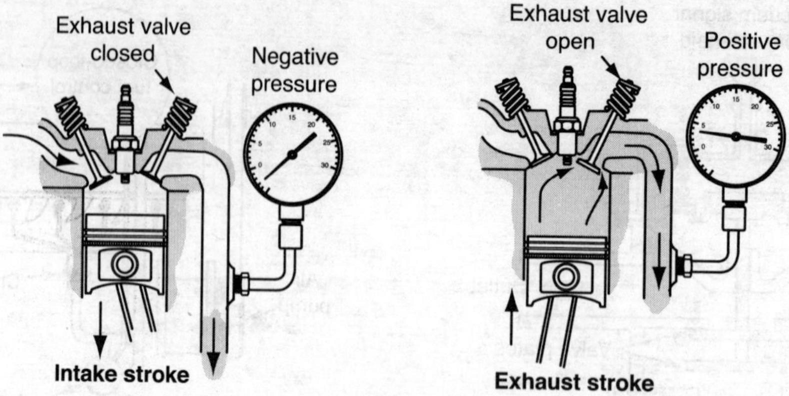


**Figure 8** *Vacuum created in the exhaust manifold as a result of exhaust pulses.*

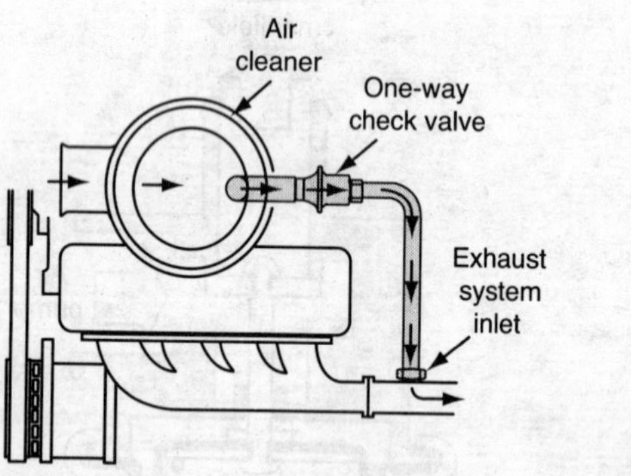


**Figure 9** *The aspirated air-injection system has very few components and is simple in its operation.*

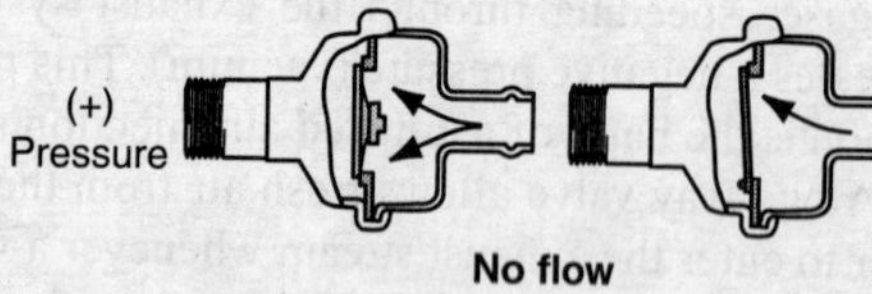


**Figure 10** *Pulse air check valve operation.*

burned in the exhaust manifold. Since the duration of low-pressure pulses in the exhaust ports decreases with engine speed, this system is more effective in reducing HC emissions at lower engine speeds.

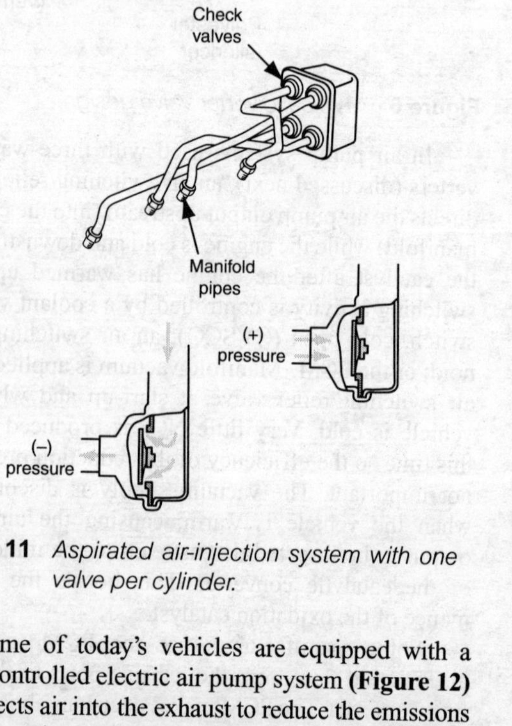


**Figure 11** *Aspirated air-injection system with one valve per cylinder.*

Some of today's vehicles are equipped with a PCM-controlled electric air pump system **(Figure 12)** that injects air into the exhaust to reduce the emissions during engine warm-up. The air injected into the exhaust will cause the catalytic converters to heat up more quickly. This will improve the emissions levels during a cold start. The following example of the Mercedes system illustrates typical operation.

The system consists of the following components **(Figure 13)**:

- An electric air pump
- Two air pump switchover valves
- An air pump switchover solenoid
- An air pump relay
- A vacuum check valve
- A PCM

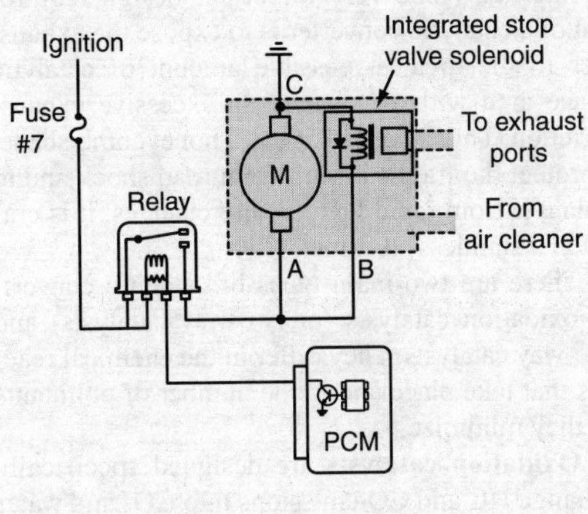


**Figure 12** *Electric air pump circuit.*

The air pump draws in air past a filter and pumps it to the two **air pump switchover valves**. These valves prevent exhaust gases from flowing back into the air pump. The air pump switchover solenoid is supplied with vacuum from the intake manifold through a check valve. When the air pump switchover solenoid is activated, it allows engine vacuum to be applied to the air pump switchover valves. The air from the pump is forced through the valves into the cylinder head openings to the exhaust. The injected air reacts with the hot exhaust gases in the outlet port. An oxidation of CO and HC takes place and results in an additional increase in the exhaust temperature.

Air is allowed to enter the exhaust when the PCM activates the air pump after engine start-up for up to 2½ minutes. The following conditions must also be met in order for the system to become active:

- Coolant temperature greater than 50°F (10°C) but less than 140°F (60°C)
- Engine speed less than 3,000 rpm
- Throttle valve not wide open

After activation, the air-injection system will remain deactivated until the coolant temperature drops from 140°F (60°C) to 104°F (40°C).

## Catalytic Converter

The most critical element of the emissions-control system is the **catalytic converter**. As the name implies, the catalytic converter is simply a catalyst that converts emissions into harmless gases. A catalyst triggers a chemical reaction that changes the exhaust gases that pass through it without itself being changed by the reaction.

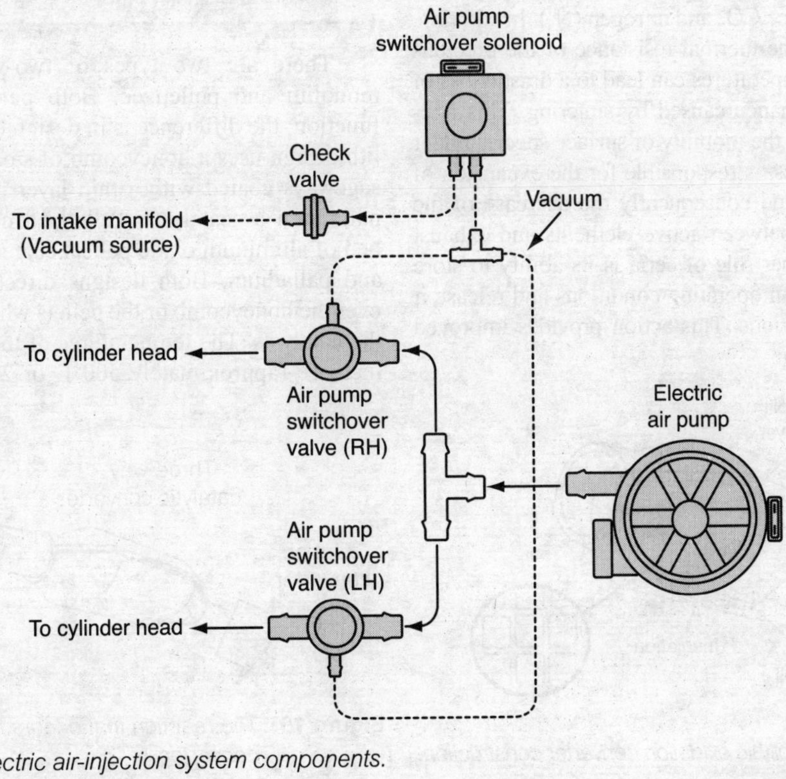


**Figure 13** *Electric air-injection system components.*

Potentially harmful amounts of the three pollutant gases still enter the exhaust system. The catalytic converter is the "last chance" to render these gases harmless before they enter the environment. Automotive engines emit CO and unreacted HC that have to be oxidized, as well as $NO_x$ that has to be reduced.

The catalytic converter is a unique emissions-control device. It has no moving parts. It is not dependent on engine vacuum, mechanical devices, or any other special control technology. It is designed for the sole purpose of decreasing exhaust emissions levels.

The catalyst uses a **monolithic** construction that is thinly coated by certain "precious metals" such as platinum, palladium, and rhodium **(Figure 14)**. Monolithic is a type of construction in which a ceramic honeycomb is coated with a thin layer of the catalyst metals. The effectiveness of the catalytic converter is largely controlled by the performance of cerium oxide, which provides oxygen needed to oxidize CO and HC.

**You Should Know**

*The monolithic construction in the converter is commonly referred to as the biscuit.*

The role of ceria (cerium oxide, or $CeO_2$) is very important since it enhances the reaction between NO and CO to produce $CO_2$ and nitrogen ($N_2$). In addition, ceria improves the thermal resistance of the precious metals. High temperatures can lead to a drastic loss of catalytic performance caused by sintering. This phenomenon, due to the mobility of surface species under high temperatures, is responsible for the expansion of metal particles and consequently the decrease of the contact surface between active elements and exhaust pollutants. Another role of ceria is its ability to store oxygen under lean operating conditions and release it under rich conditions. This action provides improved oxidation reaction.

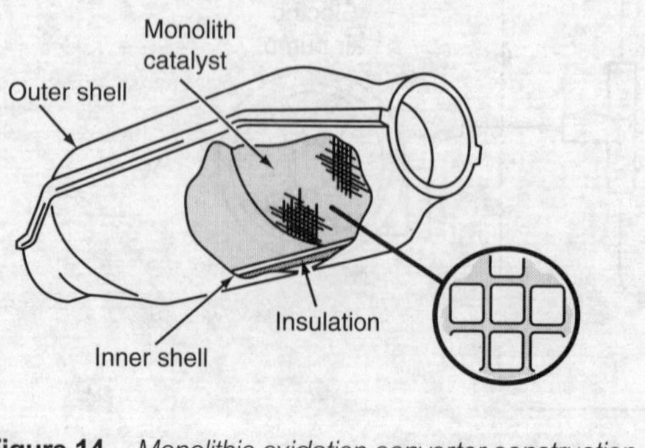


**Figure 14** *Monolithic oxidation converter construction.*

Internally, one very important design requirement of a catalytic converter is to expose the exhaust gases to the greatest possible amount of catalytic surface area without causing an excessive exhaust restriction. Thus, the biscuit has a honeycomb shape. To protect the fragile biscuit from road shock and to insulate it from rapid temperature changes, it is cradled in a stainless steel mat.

There are two main types of catalytic converters: oxidation catalysts (or two-way catalysts) and three-way catalysts. They differ in the chemical reactions that take place and in the number of pollutants that they minimize.

**Oxidation catalysts** are designed specifically to change HC and CO emissions into $CO_2$ and water vapor. By supplying a sufficient amount of oxygen to HC and CO in the presence of heat, a chemical reaction results **(Figure 15)**. This reaction breaks the HC and CO bonds and results in the formation of harmless $CO_2$ and water ($H_2O$). In some engine systems, the needed oxygen is supplied by an air pump or aspirator from the secondary air system.

**You Should Know**

*Oxidation converters are commonly called two-way converters. The name refers to the number of pollutants affected by the converter.*

There are two types of two-way converters: monolith and pelletized. Both perform the same function; the difference is in design only. The monolith design uses a honeycomb of small ceramic passageways coated with a thin layer of platinum and palladium as catalysts. Pelletized converters use a bed of aluminum oxide pellets coated with platinum and palladium. Both designs direct exhaust gases over the honeycomb or the pellets where they contact the catalysts. The temperatures of the exhaust gases increase (approximately 500°F, or 260°C) and they

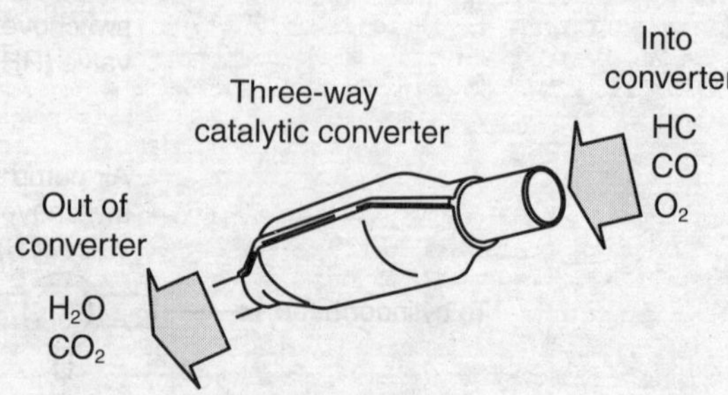


**Figure 15** *The reaction that occurs in an oxidation converter.*

continue to oxidize. This converts the HC and CO into $H_2O$ and $CO_2$ before the gases enter the muffler.

> **You Should Know**
>
> *Reaching the proper operating temperature for oxidation reaction is commonly known as lighting off.*

Two special types of oxidation catalysts are known as the mini-oxidation and the close-coupled converter **(Figure 16)**. These catalysts have essentially the same function and are named as such from their small size and the fact that they are installed close to the engine. The mini-oxidation and close-coupled converters are *fast-light-off* oxidation catalysts that help reduce start-up emissions and pretreat the exhaust before it reaches the main converter.

As government regulations became more stringent, automobile manufacturers needed a more effective means to reduce $N0_x$ emissions. To solve this problem, oxidation and oxidation-reduction catalysts **(Figure 17)** were introduced in the late 1970s. The purpose of the **three-way converter** is to help clean up the $NO_x$, HC, and CO emissions in the engine exhaust (hence the term *three-way*).Three-way converters use rhodium(Rh), which is an effective catalyst for $NO_x$ pollutants **(Figure 18).** The chemical reaction that occurs is quite different from the oxidizing process used to control HC and CO. In this process, the $NO_x$ is "reduced" to harmless nitrogen and oxygen. Instead of oxygen being added, it is taken away. Thus, this chemical reaction is just the opposite of the oxidation reaction just described. Excessive amounts of oxygen in the exhaust stream could actually inhibit the reduction process. To decrease all three pollutants, the three-way catalyst relies heavily on proper air-fuel control. It is the most common type of catalyst used today.

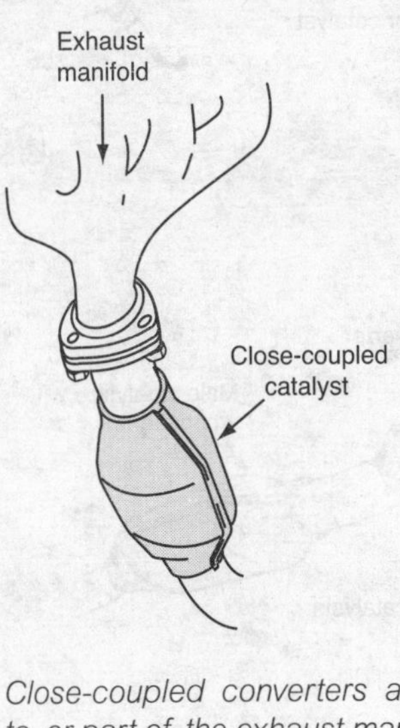


**Figure 16** *Close-coupled converters are attached to, or part of, the exhaust manifold.*

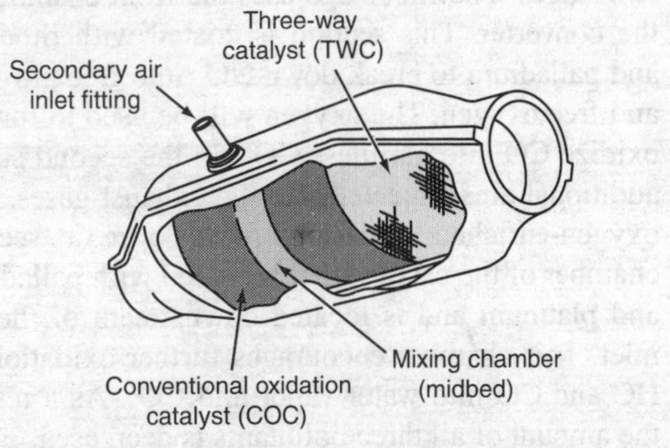


**Figure 17** *Typical three-way converter.*

There are two types of three-way converters: without air and with air. A three-way converter without air looks like a two-way converter inside, but the substrate is coated with rhodium and palladium.

Because of varying loads on the vehicle, additional air may sometimes be required to provide enough oxygen for the catalyst to work. To effectively reduce all three pollutants, a three-way converter with air may be used. This converter has two chambers that are separated by an inlet tube from the air-injection system or aspirator. The process of the air-injected three-way converter to reduce $NO_x$, HC, and CO emissions in the engine exhaust occurs in

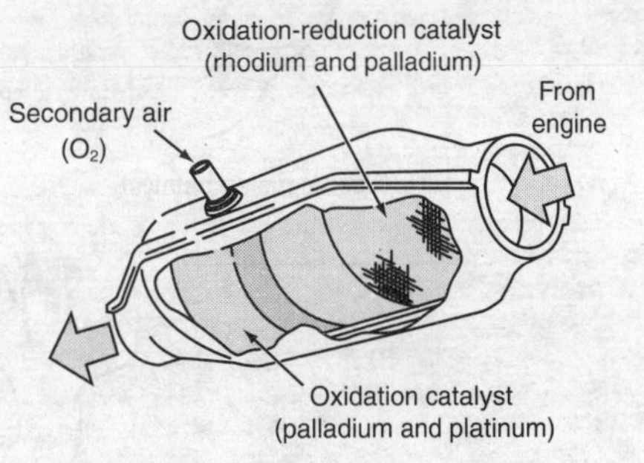


| | From engine | Between chamber | Exhaust |
|---|---|---|---|
| **CO** | 0.1–1.0% | 0.0–0.5% | 0% |
| **HC** | 50–200 ppm | 0–50 ppm | 0 ppm |
| **$CO_2$** | 13.6–14.3% | 14.7–15.5% | 9.6–12% |
| **$O_2$** | 0.3–0.7% | 0–0.5% | 2.5–5.5% |
| **NOx** | 1,200 ppm | 120 ppm | 120 ppm |
| | | | Water vapor |
| | | | Nitrogen |

**Figure 18** *Oxidation and reduction reaction of the three-way converter.*

two stages. The first stage uses the front chamber of the converter. This section is coated with rhodium and palladium to break down $NO_x$ into free nitrogen and free oxygen. The oxygen will be used to further oxidize CO into harmless $CO_2$. In the second stage, additional air is injected into the exhaust gases. The oxygen-enriched emissions pass over a second chamber of the catalyst that is coated with palladium and platinum and is located downstream of the air inlet. This chamber encourages further oxidation of HC and CO into water vapor and $CO_2$. As a result, the amount of all three pollutants is decreased.

Effective catalytic control of all three pollutants is possible only if the exhaust gas contains a very small amount of oxygen. It is necessary, therefore, to maintain precise control of the air-fuel mixture entering the engine and to keep it very close to the stoichiometric level. Like the engine, the three-way catalytic converter is sensitive to the air-fuel ratio, providing maximum overall (HC, CO, and $NO_x$) conversion efficiency when a stoichiometric air-fuel ratio is being supplied to the engine. A richer-than-stoichiometric ratio will reduce HC and CO converter efficiency more than it increases $NO_x$ converter efficiency. A leaner-than-stoichiometric ratio will reduce $NO_x$ converter efficiency more than it increases HC and CO converter efficiency. Accordingly, the $O_2$ feedback system will maintain a stoichiometric air-fuel ratio whenever engine and driving conditions permit.

The curve of efficiency, as a function of air-fuel ratio, indicates that when the air-fuel ratio is lean, the control of HC and CO is very good but control of $NO_x$ is poor. On the other hand, when the air-fuel ratio is rich, the control of $NO_x$ is very good but control of HC and CO is poor. At the chemically correct mixture, a narrow window exists where the control of all three pollutants is quite good. Maintaining the exhaust contents at this precise value at which the three-way catalyst is most effective is the purpose of $O_2$ feedback systems.

Catalyst types vary not only by their internal makeup but also by their design configurations **(Figure 19)**. Catalysts mounted near the engine are

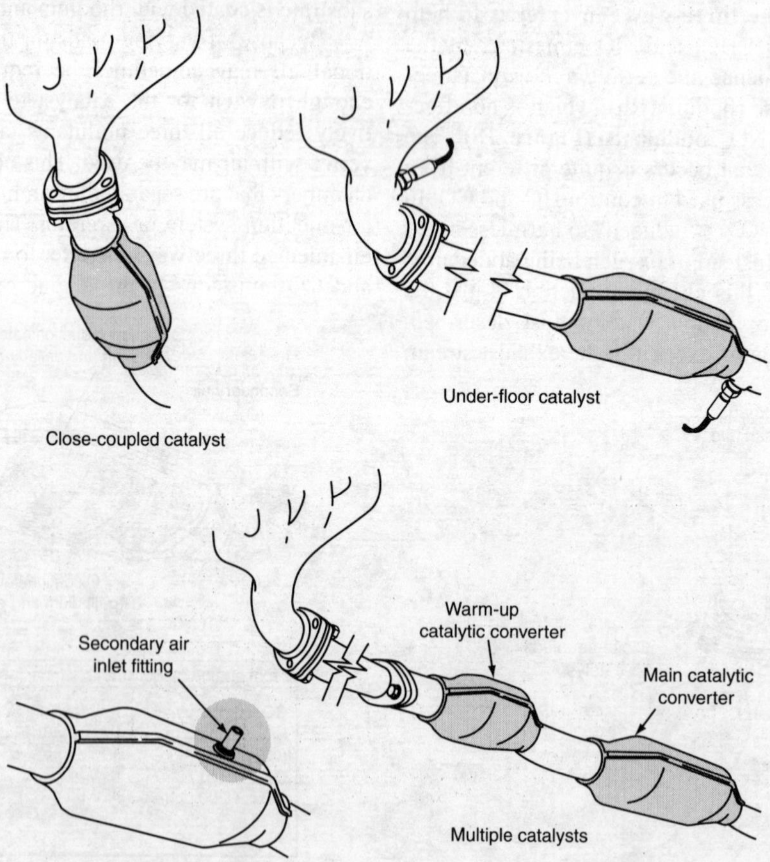


**Figure 19** *There are several different catalytic converter exhaust system configurations.*

referred to as close-coupled. Catalysts mounted farther from the engine are called under-floor catalysts. Some systems have only one catalyst; others have two or three. As discussed, some catalysts also require air injection. To help the catalyst become hotter faster, many are now mounted in the exhaust manifold.

Exhaust temperatures are extremely important to the function of catalytic converters. The desired chemical reactions most readily occur at high temperatures. In fact, the catalyst starts to operate when the exhaust temperatures reach about 500°F (260°C). Catalytic converters experience reductions in conversion efficiency through the normal course of operation. The three primary causes of these reductions are:

- High temperatures oxidize Rh, irreversibly affecting the conversion of HC, CO, and $NO_x$.
- Contamination or poisoning of catalyst materials is caused by low catalytic converter temperature. Contaminants collect on the catalyst surface, restricting exhaust gases to and from active catalyst sites where the metals are located. Hydrocarbon is primarily affected because it consists of larger molecules than CO and $NO_x$ molecules. shows the common source of contaminants and their effect on emissions.
- Inadequate air-fuel control can result in lean mixtures and high temperatures. A lean air-fuel ratio promotes irreversible reactions of Rh with other metal oxides when the converter bed exceeds temperatures of 932°F (500°C). Also, base metals are irreversibly converted to a different, less reactive form.

**Common Contaminants**

| Contaminant | Major Source | Gas Most Affected |
|---|---|---|
| Lead (Pb) | Fuel | HC |
| Phosphorous (P) | Oil | HC, CO |
| Silicone (Si) | Fuel | HC, CO, $NO_x$ |
| Sulfur (S) | Fuel | HC, CO, $NO_x$ (esp. rich) |

**Table 1** *Table of common contaminates and their effects.*

## Review Questions

1. Explain the purpose of the air-injection system.
2. _______ _______ are designed specifically to change HC and CO emissions into CO2 and water vapor.
3. List the four types of air-injection systems.
4. Explain the function of the aspirated air-injection system.
5. Explain the function of the three-way catalyst.
6. Technician A says that with air-injected exhaust systems, air is injected into the exhaust manifold to heat the catalysts faster. Technician B says that air is sent to the catalytic converter to oxidize HC and CO. Who is correct?
   A. Technician A only
   B. Technician B only
   C. Both Technician A and Technician B
   D. Neither Technician A nor Technician B
7. Technician A says that the diverter valve vents pump output to the atmosphere during deceleration to prevent backfiring. Technician B says that an air switching/relief valve directs the air pump output downstream to the catalyst when the engine is cold. Who is correct?
   A. Technician A only
   B. Technician B only
   C. Both Technician A and Technician B
   D. Neither Technician A nor Technician B
8. Technician A says that the three-way converter adds oxygen to break down NOx into free nitrogen and free oxygen. Technician B says that a leaner-than-stoichiometric ratio will reduce NOx converter efficiency more than it increases HC and CO converter efficiency. Who is correct?
   A. Technician A only
   B. Technician B only
   C. Both Technician A and Technician B
   D. Neither Technician A nor Technician B

CHAPTER

# 27 Exhaust Analyzer Operation

## Introduction

The first Clean Air Act prompted California to create the California Air Resources Board. CARB's purpose was to implement strict air standards. These became the standards for federal mandates. One of CARB's approaches to clean the air was to start periodic motor vehicle inspection (PMVI). The purpose for the PMVI, or **inspection and maintenance (I/M)**, is to inspect a vehicle's emissions controls once a year. This inspection includes a tailpipe emissions test and an underhood inspection. The tailpipe test certifies that the vehicle's exhaust emissions are within the limits set by law. The underhood inspection verifies that the pollution-control equipment has not been tampered with or disconnected.

The 1970 Clean Air Act made it illegal for professional technicians to remove, disconnect, or tamper with factory emissions controls. The law did not make it illegal for the owner of the vehicle to disable emissions-control devices. The Clean Air Act was revised in 1990 to make it illegal for *anyone* to remove, disconnect, or tamper with emissions controls on any vehicle used on public roads.

Many states have incorporated an emissions test with their annual vehicle registration procedures. Some states have or are planning to implement the I/M 240 program that was first implemented in California. The I/M 240 tests the emissions of a vehicle while it is operating under a variety of load conditions.

When diagnosing emissions problems, verifying base engine performance should be the first consideration. This is especially true when no trouble codes are being set. Many mechanical problems can lead to excess emissions and the following items should be checked as possible causes:

- Engine vacuum—must be above the minimum specified for the engine. Refer to the service manual for specifications.
- Engine valve timing—must be within specifications.
- Engine compression—must be within specifications.
- Engine exhaust system—must be free of any restrictions.
- Engine PCV system—must be the correct application.
- Torque converter stall speed—must be within specifications.
- Power brake booster—must be no internal vacuum leaks.
- Fuel—must be of proper quality and free of contamination.
- Fuel injectors—check for plugged, leaking, or restricted injectors, or a control wire not connected to the correct injector.

This chapter covers the testing of the emissions system and components. It provides the necessary information to successfully diagnosis state emissions testing failures and driveability problems associated with emissions controls.

## Exhaust Analyzers

As stated in Chapter 2, testing the quality of the exhaust is both a procedure for testing emissions levels *and* a diagnostic routine. One of the most valuable diagnostic aids is the exhaust analyzer **(Figure 1)**. The quality of an engine's exhaust indicates the effects of the combustion process. Any defect can cause a change in exhaust quality. The amount and type of change serves as the basis of diagnostic troubleshooting.

By using the exhaust analyzer to measure the efficiency of the engine's combustion system (by measuring the emissions levels of the gases at the tailpipe), the technician can determine which of the vehicle's components are defective or misadjusted based on which gas readings are too high or too low.

**Figure 1** *The exhaust gas analyzer is a very useful diagnostic tool.*

The defective or misadjusted components can then be repaired or brought back within specifications.

Modern exhaust analyzers are designed so they can examine the efficiency of an engine in spite of the effectiveness of the emissions controls. These types of analyzers measure five exhaust gases. In addition to measuring HC and CO levels, a five-gas exhaust analyzer monitors $CO_2$ and $O_2$ levels in the exhaust. These two gases are changed only slightly by the emissions controls and therefore can be used to check engine efficiency.

In addition, a five-gas analyzer can detect $NO_x$. The analyzer does not measure $NO_x$ directly. Instead, it measures nitric oxide (NO), one of the components of $NO_x$. The NO reading is not an accurate measurement of the amount of $NO_x$ but a high NO reading does indicate that there is a $NO_x$ problem.

For the technician, $NO_x$ levels are determined by assumption and can only be measured with the vehicle under load. Make sure the vehicle does not have any driveability problems, particularly anything that might cause a lean fuel condition (such as a vacuum leak). Also, check for anything that may cause the engine to run at higher-than-normal operating temperatures. This could include a faulty thermostat, a faulty water pump, a worn water pump drive belt, a damaged fan, a damaged or missing fan shroud, a faulty electric fan operation, and so forth. Then, if all possible conditions are confirmed as being acceptable, if HC and CO levels are acceptable, and if the vehicle's emissions-control systems are in good working order (especially the EGR system and three-way catalytic converter), $NO_x$ levels can be assumed to be acceptable also.

Prior to testing a vehicle's exhaust, the technician must ensure that the analyzer will display accurate information about the sample. This is done by checking the zeroing and the calibration of the analyzer. Procedures for performing these tasks differ among manufacturers; however, calibrating the analyzer is typically done by zeroing the analyzer with clean air and then calibrating it with known concentrations of HC, CO, and $CO_2$.

## Preparing the Vehicle for Exhaust Gas Testing

Visual inspection is an important part of emissions diagnosis. Technicians can often locate problems by simply doing a thorough visual inspection of the emissions-control system. Check for the following conditions and correct as necessary before testing:

- Dirty or restricted air filter
- Leaks in vacuum hoses
- Damaged or discolored catalytic converter
- Dirty throttle body
- Damaged distributor cap
- Damaged or restricted fuel filter and hoses
- Open or damaged sensor wires
- Malfunctioning PCV valve
- Defective spark plug wires
- Evidence of tampering to any emissions system component
- Malfunctioning PCM and systems
- Check to see if the Check Engine light is on
- Check for fault codes and correct as necessary before testing

After the visual inspection has been completed and any problems have been corrected, the vehicle is ready to be preconditioned for the exhaust test.

## Preconditioning the Engine

Preconditioning is an important step to take before performing the exhaust test. Omitting this procedure can lead to false failures. The purpose of preconditioning the engine is to bring it to normal operating temperature so the $O_2S$ and the catalytic converter are hot enough to function properly and all other components of the emissions-control system have stabilized.

### You Should Know

*To avoid a false failure, make the customer aware of the procedure necessary to precondition the engine prior to having a state emissions test.*

Exhaust analyzers have two primary diagnostic functions: engine performance and emissions compliance. The instruction manual for the analyzer may contain one preconditioning procedure for diagnosing engine performance and another for

| Model Year | HC less than | $CO_2$ greater than | CO less than | $O_2$ in the range of |
|---|---|---|---|---|
| 1971–1975 | 400 ppm | 8.0% | 3.0% | 0.1–7.0% |
| 1976–1979 | 200 ppm | 8.0% | 2.0% | 0.1–7.0% |
| 1980–present | 100 ppm | 8.0% | 0.5% | 0.1–7.0% |

**Table 1** *Typical exhaust emissions readings at idle speed on a vehicle equipped with a catalytic converter with the air-injection system disabled.*

emissions compliance. Follow the correct procedure for the function to be performed. When performing an emissions compliance test, do not use the preconditioning procedure intended for diagnosing engine performance. This procedure involves disabling the vehicle's air pump and catalytic converter so that the gases can be tested before they have been treated by the catalyst. If you are specifically analyzing $CO_2$ for engine performance, you should disable the air-injection system. Air injected into the catalytic converter combines with CO to form $CO_2$. This gives false $CO_2$ readings and is especially noticeable with rich fuel mixtures. When using the exhaust analyzer as a diagnostic tool, it is important to realize that the severity of the problem dictates how much higher than normal the readings will be.

For emissions compliance, test the final exhaust product *after* all treatment has taken place. Preconditioning for emissions compliance does not involve disabling any components. It just requires warming up the engine to normal operating temperature.

## Exhaust Testing

### You Should Know

*Inhaled CO enters the bloodstream and takes the place of oxygen in the red blood cells. Carbon monoxide "absorbs" oxygen ten times more readily than the red blood cells do. Oxygen is essentially blocked from the body, possibly resulting in headaches, reduced mental alertness, and death.*

Most emissions analyzers have a warning light to indicate if the exhaust flow through the analyzer is restricted due to a plugged filter, probe, or hose. Be sure the warning light is off before proceeding with the exhaust emissions analysis.

When performing an exhaust emissions analysis, always follow the recommended procedure in the vehicle manufacturer's service manual or the equipment operator's manual. Exhaust emissions of HC, CO, and $NO_x$ must conform to state and federal emissions standards for the model year being tested.

Typical exhaust emissions readings at idle speed on a catalytic converter–equipped vehicle with the air-injection system disabled are shown in **Table 1**. These emissions levels are not intended as accurate specifications on all vehicles, but rather as a general guide to help baseline a vehicle.

Since a catalytic converter produces some $CO_2$, these emissions levels are higher for vehicles with catalytic converters than for vehicles without catalytic converters.

Some exhaust analyzers also include a lambda reading. A lambda reading of 1 indicates the stoichiometric air-fuel ratio; a reading above 1 indicates a lean air-fuel ratio; and a reading below 1 represents a rich air-fuel ratio. Some exhaust analyzers also display the air-fuel ratio **(Figure 2)**.

As mentioned earlier, the exhaust analyzer can be used for state I/M programs and for general engine diagnosis. Basic I/M tests are covered here. Infrared analyzers used in state I/M testing must meet specific standards such as California BAR-80 or ETl-80 or BAR-90. Some analyzers may require a software change or the addition of a printer to comply with these programs. During the emissions test,

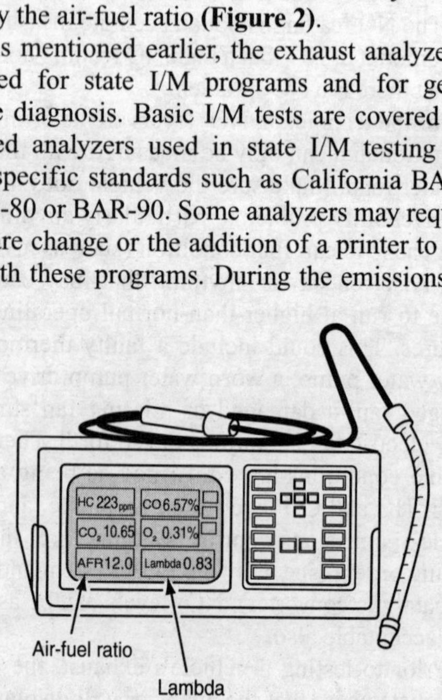


**Figure 2** *Four-gas analyzer display with lambda and air-fuel ratio readings.*

the vehicle must be operated under the conditions specified in the state emissions standards. Regardless of the purpose, the following is a guide for a typical procedure for five-gas emissions analysis:

1. Run the engine until it reaches normal operating temperatures.
2. Connect the emissions analyzer to the proper voltage supply.
3. Allow the analyzer to warm up and perform the calibration function **(Figure 3)**.
4. Connect the analyzer probe to the tailpipe.
5. With the engine held at 2,500 rpm, record the emissions readings
6. Allow the engine to return to idle and record the emissions readings **(Figure 4)**.
7. Compare the emissions readings with state or federal standards.

Basic I/M tests use the **two-speed idle test**. This test is performed by holding the engine speed at 2,500 rpm for 30 to 90 seconds while the emissions readings are taken. The engine is then allowed to return to idle (it must be below 1,100 rpm) for 30 to 90 seconds. During this time, a second set of readings is taken.

To pass, the emissions concentration must be below the standard set by the state for that particular test area. The pass/fail standards can vary with the age and technology of the vehicle, whether it is a car or a truck, and so on. Present federal and California state standards for basic I/M testing as of the date of this publication are shown in **Table 2**.

> **You Should Know**
>
> *Some test areas only take emissions readings during the idle portion of the test. The 2,500 rpm portion of the procedure is used to warm up (precondition) the vehicle for the idle test.*

> **You Should Know**
>
> *Leaks in the exhaust system cause inaccurate emissions analyzer readings. Always inspect the exhaust system for leaks prior to infrared exhaust analysis.*

> **You Should Know**
>
> *When performing any test with the engine running, be sure the parking brake is applied with the automatic transmission in Park or the manual transmission in Neutral. Ignoring these precautions may cause personal injury and/or property damage.*

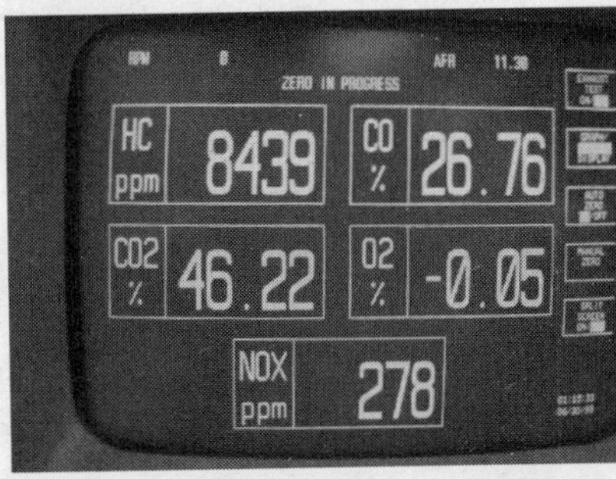


**Figure 3** *Allow the analyzer to warm up and perform the calibration function.*

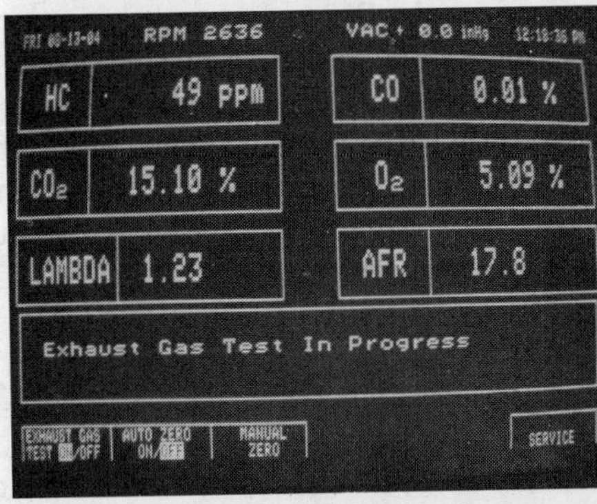


**Figure 4** *Allow the engine to return to idle and record the emissions readings.*

## Analyzer Results

Maximum limits for the measured gases are set by regulations according to the model year of the vehicle. All gases except HC and $NO_x$ are measured in percentages. The percentage reading for the gas is a comparison of that gas with the total exhaust gases. Hydrocarbon and $NO_x$ are measured in parts per million (ppm). It is always desirable to have low amounts of four of the five measured gases at all engine speeds. The only gas that is desired in high percentages is carbon dioxide. Normally the desired amount of $CO_2$ is 13.4 percent or greater. The other gases should be kept to the lowest levels possible.

If a measured exhaust gas is not at an acceptable level, consider what causes the formation of this gas. For example, if the engine is producing excessive HC emissions, any of the following conditions could be the cause:

- Ignition system misfiring
- Improper ignition timing
- Excessively lean or rich air-fuel ratio
- Low cylinder compression
- Defective valves, guides, or lifters

| Geographic Area | HC (ppm) | CO (%) |
|---|---|---|
| Federal | 220 | 1.2 |
| California | 100 | 1.0 |

**Table 2** *Present federal and California state standards for basic I/M testing.*

- Defective rings, pistons, or cylinders
- Vacuum leaks

It is the responsibility of the emissions technician to isolate the cause. Keep in mind that sensors and actuators used by the PCM need to be functioning properly to prevent the formation of emissions. And the emissions components designed to reduce the excessive gas also need to be functioning properly.

If the engine is producing excessive CO emissions, any of the following conditions may be the cause:

- Rich air-fuel mixture
- Dirty air filter
- Faulty injectors
- Higher-than-normal fuel pressure
- Defective system input sensor

If both HC and CO emissions are excessive, any of the following conditions may be the cause:

- Plugged PCV system
- Excessively rich air-fuel ratio
- Stuck open heat riser valve
- AIR pump inoperative or disconnected
- Engine oil diluted with gasoline

If the $O_2$ emissions are lower than normal, any of the following conditions may be the cause:

- Rich air-fuel mixture
- Dirty air filter
- Faulty injectors
- Higher-than-normal fuel pressure
- Defective system input sensor
- Restricted PCV system
- Charcoal canister purging at idle and low speeds

If the $O_2$ emissions are higher than normal, any of the following conditions may be the cause:

- Engine misfire
- Lean air-fuel mixture
- Vacuum leaks
- Lower-than-specified fuel pressure
- Defective fuel injectors
- Defective system input sensor

Lower-than-normal $CO_2$ emissions can be caused by any of the following conditions:

- Leaking exhaust system
- Loss of compression
- Rich air-fuel mixture
- Lean air-fuel mixture
- Misfire
- Anything that prevents complete combustion

Higher-than-normal $NO_x$ emissions can be caused by any of the following conditions:

- Overheated engine
- Lean air-fuel mixture
- Vacuum leaks
- Overadvanced ignition timing
- Defective EGR system

**Table 3** shows typical exhaust gas readings for various conditions for model year vehicles 1981 to present. The reading for vehicles equipped with a catalytic converter is with the air-injection system disabled.

An understanding of possible causes for excessive emissions can help isolate the problem to that of the engine (mechanical), fuel, spark, or emissions components. Look at the "whole picture," then focus on those items that could be the cause of the results.

## Locating Cylinder Misfiring

When a cylinder misfires there is a large increase in HC emissions. Since CO is a byproduct of combustion, it usually decreases slightly when a cylinder misfires. Using these facts, engine analyzers with a cylinder disable function can be used along with the exhaust analyzer to determine which cylinder is misfiring. When the engine analyzer prevents the spark from firing a cylinder, the rpm drop is recorded. If a cylinder is not contributing to engine power, there is very little rpm drop when the spark plug stops firing. When a cylinder is misfiring and not contributing to engine power, the HC emissions from that cylinder are high. Therefore, when the spark plug stops firing, there is not much change in HC emissions.

## Head Gasket and Combustion Chamber Leaks

Allow the engine to completely cool, then remove the radiator cap. Place the infrared analyzer probe

| Gases | Idle | | 1,500 rpm | | 2,500 rpm | | Condition/Possible Cause |
|---|---|---|---|---|---|---|---|
| | Conv. | Non conv. | Conv. | Non conv. | Conv. | Non conv. | |
| HC (ppm) | 0–150 | 75–250 | 0–135 | 50–200 | 0–75 | 25–150 | Normal gas readings—stable values |
| CO (%) | 0.1–1.5 | 0.5–3.0 | 0.0–1.1 | 8.0–10.0 | 0.5–2.0 | 0.1–1.5 | |
| $CO_2$ (%) | 10.0–12.0 | 10.0–12.0 | — | — | 11.0–13.0 | 11.0–13.0 | |
| $O_2$ (%) | 0.5–2.0 | 0.5–2.0 | 0.5–2.0 | 0.5–2.0 | 0.5–1.3 | 0.5–2.0 | |
| HC (ppm) | 0–150 | 75–250 | 0–135 | 50–200 | 0–75 | 0–100 | ***Rich mixture:*** Idle mixture too rich/choke set too rich or not opening fully/power valve leaking/float level too high/restricted air cleaner/PCV restricted/contaminated crankcase |
| CO (%) | 3.0+ | 4.0+ | 3.0+ | 3.5+ | 3.0+ | 3.0+ | |
| $CO_2$ (%) | 8.0–10.0 | 8.0–10.0 | — | — | 9.0–11.0 | 9.0–11.0 | |
| $O_2$ (%) | 0.0–0.5 | 0.0–0.5 | 0.0–0.5 | 0.0–0.5 | 0.0–0.5 | 0.0–0.5 | |
| HC (ppm) | 0–150 | 75–250 | 0–135 | 50–200 | 0–75 | 0–100 | ***Lean mixture:*** Low float level/idle mixture lean/cruise mixture lean/small air leaks/cracked or disconnected vacuum lines |
| CO (%) | 0.0–1.0 | 0.0–1.0 | 0.0–0.8 | 0.0–0.9 | 0.0–0.25 | 0.0–0.75 | |
| $CO_2$ (%) | 8.0–10.0 | 8.0 | 10.0 | 9.0 | 11.0 | 9.0–11.0 | |
| $O_2$ (%) | 1.5–3.0 | 1.5–3.0 | 1.0–2.5 | 1.0–2.5 | 1.0–2.0 | 1.0–2.0 | |
| HC (ppm) | 50–850 | 400–1,200 | 50–850 | 400–1,200 | 50–750 | 400–1,200 | ***Lean misfire:*** Severe air leak/bad spark plug or wire/stuck PCV/misadjusted or defective carb |
| CO (%) | 0.0–0.3 | 0.0–0.75 | 0.0–0.3 | 0.0–0.75 | 0.0–0.3 | 0.0–0.75 | |
| $CO_2$ (%) | 5.0–9.0 | 5.0–9.0 | — | — | 6.0–10.0 | 6.0–10.0 | |
| $O_2$ (%) | 4.0–9.0 | 4.0–9.0 | 4.0–9.0 | 2.0–7.0 | 2.0–7.0 | 2.0–7.0 | |
| HC (ppm) | 50–850 | 1,000+ | 50–850 | 1,000+ | 50–750 | 1,000+ | ***Misfire:*** Overadvanced timing/fouled plug/open or grounded plug wire/EGR stuck open |
| CO (%) | 0.1–1.5 | 0.5–3.0 | 0.0–1.1 | 0.5–2.0 | 0.0–0.8 | 0.1–1.5 | |
| $CO_2$ (%) | 6.0–8.0 | 6.0–8.0 | — | — | 8.0–10.0 | 8.0–10.0 | |
| $O_2$ (%) | 4.0–12.0 | 5.0–12.0 | 4.0–12.0 | 5.0–12.0 | 4.0–12.0 | 5.0–12.0 | |

**Table 3** *Typical exhaust gas readings for various conditions.*

*above* the radiator filler neck and run the engine. Do not immerse the probe in the coolant. Head gasket or combustion chamber leaks into the cooling system result in HC and CO readings **(Figure 5)**.

> **You Should Know**
>
> *Do not place the probe into the coolant. Coolant will be sucked into the analyzer and damage the unit.*

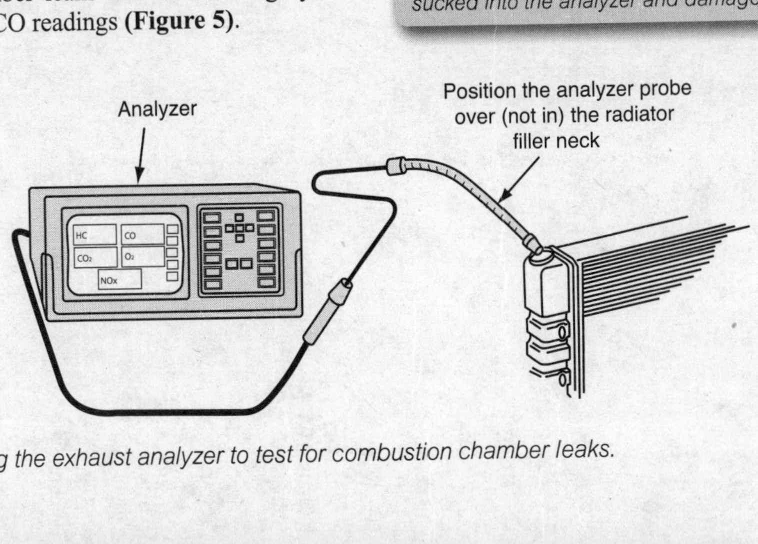


**Figure 5** *Using the exhaust analyzer to test for combustion chamber leaks.*

## Review Questions

1. Technician A says that a low level of HC indicates that a misfire is occurring. Technician B says that a low CO level indicates an excessively rich air-fuel mixture. Who is correct?
   A. Technician A only
   B. Technician B only
   C. Both Technician A and Technician B
   D. Neither Technician A nor Technician B
2. Technician A says that $CO_2$ is a good rich/lean indicator. Technician B says that HC is a good rich/lean indicator. Who is correct?
3. An engine is producing excessive $NO_x$ emissions. Technician A says that this can be caused by a stuck open thermostat. Technician B says that this can be caused by lean air-fuel mixture. Who is correct?
   A. Technician A only
   B. Technician B only
   C. Both Technician A and Technician B
   D. Neither Technician A nor Technician B
4. Two technicians are preparing to test a vehicle's exhaust. Technician A says that it is important to warm the engine to normal operating temperatures before emissions compliance testing. Technician B says that to disable the vehicle's air pump and catalytic converter before testing for emissions compliance. Who is correct?
   A. Technician A only
   B. Technician B only
   C. Both Technician A and Technician B
   D. Neither Technician A nor Technician B
   A. Technician A only
   B. Technician B only
   C. Both Technician A and Technician B
   D. Neither Technician A nor Technician B
5. Technician A says that excessive HC emissions can be the result of worn piston rings. Technician B says that excessive CO emissions can be the result of leaking injectors. Who is correct?
   A. Technician A only
   B. Technician B only
   C. Both Technician A and Technician B
   D. Neither Technician A nor Technician B

CHAPTER

# 28 Diagnosing the Emissions System

## Introduction

If a vehicle has excessive emissions or poor driveability or it has failed I/M testing or the MIL is illuminated due to an emissions failure, the components of the emissions system will need to be tested. Not all emissions failures are the result of the fuel or ignition systems; often the emissions component itself has failed. Sometimes the failure of an emissions-control device will cause other driveability problems.

## Evaporative Emissions-Control System Diagnosis and Service

**You Should Know**

*If a gasoline odor is present in or around a vehicle, check the EVAP system for cracked or disconnected hoses and check the fuel system for leaks. Gasoline leaks or escaping vapors may result in an explosion, causing personal injury and/or property damage. The cause of fuel or vapor leaks should be repaired immediately.*

Faults in the EVAP system **(Figure 1)** can include leaks or an improper purging function. A leaking or inoperative EVAP system allows gasoline vapors to escape to the atmosphere. These vapors contain HC emissions. When the EVAP system allows vapor purging at idle or very low vehicle speeds, rough engine idle operation may be experienced.

As with most other diagnostic routines, begin with a thorough visual inspection of the system. Check the hoses in the EVAP system for leaks, restrictions, and loose connections. Check the electrical connections in the EVAP system for looseness, corroded terminals, and worn insulation. When a defect occurs in the canister purge solenoid and related circuit, a DTC is usually set in the PCM memory. If a DTC related to the EVAP system is set in the PCM memory, always correct the cause of this code before further EVAP system diagnosis.

### EVAP Leak Testing—Vacuum Purge Systems

There are several methods that can be used to check for leaks in the EVAP system including use of the exhaust analyzer and flow testing.

To test for leaks using the exhaust analyzer, warm the engine to a normal operating temperature. Place the analyzer probe near any component in the EVAP system where a gasoline vapor leak is suspected. Follow the hoses and test over individual components while observing the HC reading. When the HC reading increases, the probe is near the gasoline vapor leak. All the hoses in the EVAP system should be checked for leaks, restrictions, and loose connections.

Another method to test for leaks is to use a flow meter. This test will verify that there is sufficient

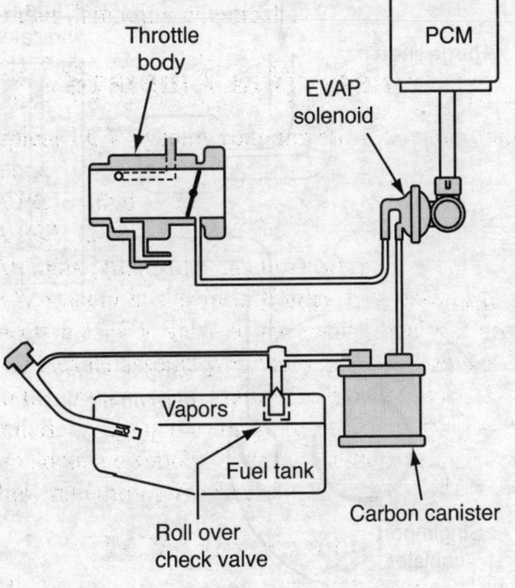


**Figure 1** *Typical EVAP system.*

vacuum to draw vapors from the canister. Warm the engine to a normal operating temperature and let it idle for at least 2 minutes. Remove the purge hose coming from the canister and install the purge flow indicator in series **(Figure 2)**. Observe the purge flow indicator. The ball should be at the top of the meter ("bouncing" due to the on/off action of the duty cycle purge solenoid). If the ball reaches about halfway up the meter, check the fuel tank vent line for cracks.

If the flow meter ball does not lift, verify vacuum supply to the canister. Inspect all vacuum hoses and ensure that they are properly connected and free of cracks. Remove the manifold vacuum source hose to the purge solenoid. With the engine running, manifold vacuum should be present (at least 12 in. Hg). If there is too little vacuum, the passage is blocked. If vacuum is present, verify the operation of the purge solenoid. This is done by utilizing a scan tool and reading the diagnostic fault codes. A code may be set if the duty cycle purge solenoid is not being energized at the correct time. If no DTCs are set, remove the electrical connector from the purge solenoid and connect a jumper wire to the positive battery terminal. Connect a second jumper wire to a good ground source. With the purge flow indicator installed and the engine running, touch the other ends of the jumper wires to the two open terminals on the purge solenoid (not the connector terminals). If flow is indicated now, check the electrical circuit to the solenoid.

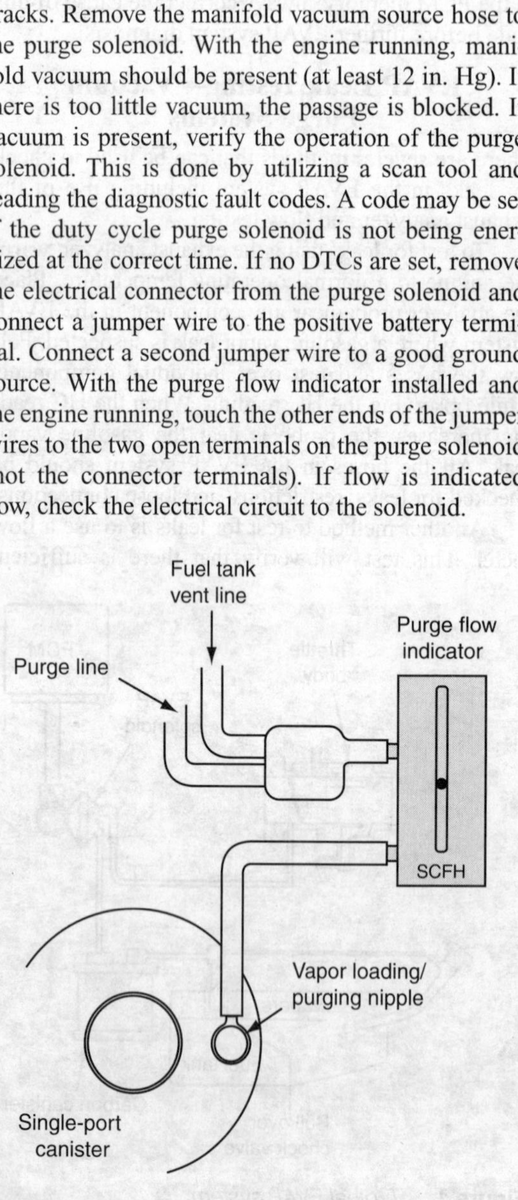


**Figure 2** *Flow testing the EVAP system.*

If no flow was indicated when the solenoid was jumped, disconnect the purge flow meter from the canister side but leave it connected to the solenoid side. Jump the solenoid again and observe the flow meter. If flow is now indicated, replace the canister. If there still is no flow indicated, replace the solenoid.

## EVAP Diagnostics—Pressure Purge Systems

The pressure purge system is unique in that it purges the canister by using pressure from the air pump to flow air through the canister. Purged fuel vapors are directed into the air cleaner.

The canister purge is activated when pressure from the downstream side of the air pump closes the canister vent (on top of the canister) to allow air pump pressure to purge the canister through a second "orificed" line. The pressure in this second hose serves two functions. First, it pushes fuel vapors from the canister to the air cleaner. Second, it directs vapors from the fuel tank through the tank vapor bypass valve to the air cleaner **(Figure 3)**.

If pressure from the downstream side of the air pump is passing through the canister to the air cleaner, the system will operate as designed. Thus, the strategy for repairing this system is to inspect all hoses and connectors to be sure that they are properly connected and they are not cracked or broken. Next, verify that sufficient pressure is reaching the air cleaner. If this pressure is not sent, all individual components (hose routings, valves, restrictors, and connectors) must be checked for either a leak or an obstruction. If no leak or obstruction is found, then the canister must be replaced. **Table 1** is a typical diagnostic routine for testing these systems.

## EVAP System Component Diagnosis

The evaporative canister should be inspected for cracks or other damage **(Figure 4)**. Some canisters will have a serviceable air filter (usually located on the bottom) that should be inspected and replaced if it is dirty. Also check the canister filter for indications of fuel saturation. A saturated evaporative canister can cause symptoms that can be mistaken for fuel system problems. Rough idle, flooding, and other conditions can indicate a canister saturation problem. A canister that is filled with liquid fuel causes back pressure in the fuel tank. It also causes richness and flooding symptoms during purge or start-up.

While the engine is running and noticeable driveability problems are occurring, test for saturation by unplugging the canister momentarily and observing

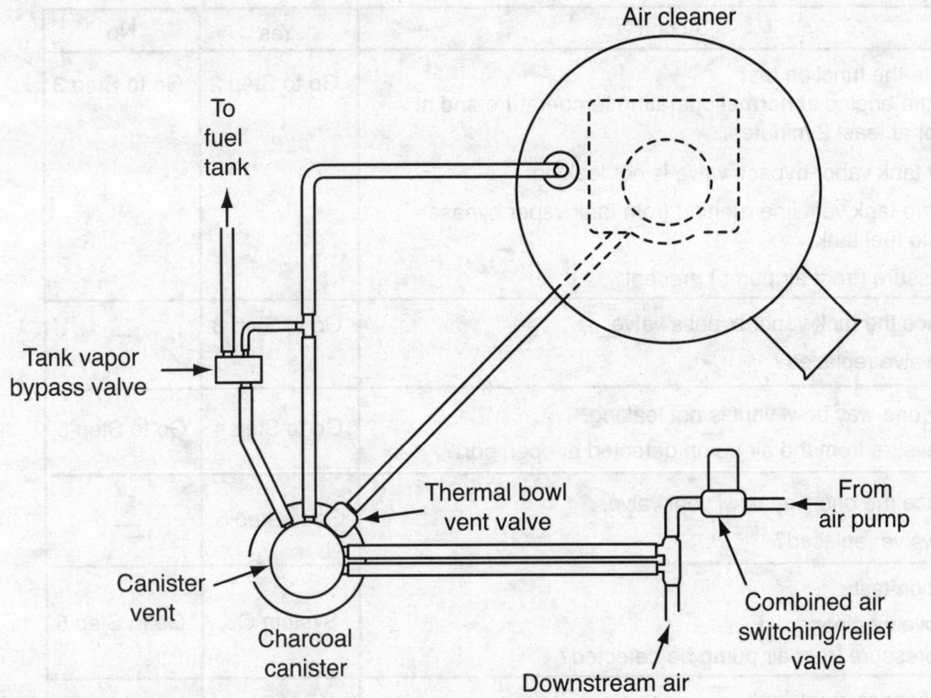


**Figure 3** *Example of a pressure purge EVAP system.*

the engine's operation. An exhaust analyzer can also be used to measure the amount of HC difference experienced when the canister is unplugged. If the canister is saturated, it must be replaced.

If the fuel tank has a pressure valve and a vacuum valve in the filler cap, check them for dirt and damage. The cap may be washed in cleaning solvent. If the valves are sticking or damaged, replace the cap.

The **purge solenoid**, an electrical control orifice in the EVAP system, controls the admission of fuel vapors into the combustion chamber. The canister purge solenoid winding **(Figure 5)** may be checked with an ohmmeter. Remove the tank pressure control valve and try to blow air through the valve with regulated shop pressure at 20 psi or less from the tank side of the valve. Some restriction to airflow should be noticed until the air pressure opens the valve. Connect a vacuum hand pump to the vacuum fitting on the valve and apply 10 in. Hg of vacuum to the valve. Attempt to blow air through the valve from the tank side. Under this condition, there should be no restriction to airflow. If the tank pressure control valve does not operate properly, replace it.

A scan tester may be used to diagnose the EVAP system. In the appropriate scan tester mode, the tester indicates whether the purge solenoid is on or off. Connect the scan tester to the DLC and start the engine. With the engine idling, the purge solenoid should be off. Leave the scan tester connected and road test the vehicle. Be sure all the conditions required to energize the purge solenoid are present and observe the solenoid status on the scan tester. The tester should indicate that the purge solenoid is on when all the conditions are present for canister purge operation. If the purge solenoid is not on under the necessary conditions, check the power supply wire to the solenoid, the solenoid winding, and the wire from the solenoid to the PCM. A defective input, such as an ECT sensor, may cause improper operation of the purge solenoid. Verify proper input sensor operation from the serial data on the scan tester.

## Exhaust Gas Recirculation (EGR) System Diagnosis

Manufacturers calibrate the amount of EGR flow for each engine. If there is too much or too little, it can cause performance problems by changing the engine "breathing" characteristics **(Figure 6)**. With too little EGR flow, the engine can overheat, detonate, and emit excessive amounts of $NO_x$. When any of these problems exists and it seems likely that the EGR system is at fault, check the system. Typical problems that may appear in ported EGR systems are:

| Step | Action | Yes | No |
|---|---|---|---|
| 1 | Prior to the function test:<br>With the engine at normal operating temperature and at idle for at least 2 minutes.<br>Verify tank vapor bypass valve is not leaking.<br>Remove tank vent line running from tank vapor bypass valve to fuel tank.<br>Is pressure (from air pump) present? | Go to Step 2 | Go to Step 3 |
| 2 | Replace the tank vapor bypass valve.<br>Was valve replaced? | Go to Step 3 | — |
| 3 | Verify one-way bowl vent is not leaking.<br>Is pressure from the air pump detected at open port? | Go to Step 4 | Go to Step 5 |
| 4 | Replace the one-way bowl vent valve.<br>Was valve replaced? | Go to Step 5 | — |
| 5 | Function test:<br>Remove air cleaner lid.<br>Can pressure from air pump be detected? | System OK | Go to Step 6 |
| 6 | Maintenance steps:<br>Verify air pump supply to the canister.<br>Is pressure from the air pump detected? | Go to Step 9 | Go to Step 7 |
| 7 | Inspect lines from canister to air pump.<br>Were any problems found? | Go to Step 8 | See secondary air system |
| 8 | Were repairs made? | Go to Step 5 | — |
| 9 | Verify canister operation:<br>Reconnect air pump lines. Disconnect purge line at canister.<br>Can pressure be detected at canister port? | Go to Step 11 | Go to Step 10 |
| 10 | Replace the canister.<br>Was canister replaced? | Go to Step 5 | — |
| 11 | Check routing from canister to air cleaner.<br>Reconnect air pump to canister hose. Disconnect tee at tank vapor bypass valve. Block open tee port.<br>Check for pressure at air cleaner (same as Step 5). | Go to Step 13 | Go to Step 12 |
| 12 | Inspect hose and tee from canister to air cleaner for cracks and blockage.<br>Was repair made? | Go to Step 5 | — |
| 13 | Replace the tank vapor bypass valve and reconnect all hoses.<br>Was valve replaced and hose reconnected? | Go to Step 5 | — |

**Table 1** *Typical diagnostic routine for testing pressure purge systems.*

**Figure 4** *The evaporative canister should be inspected for cracks and other damage.*

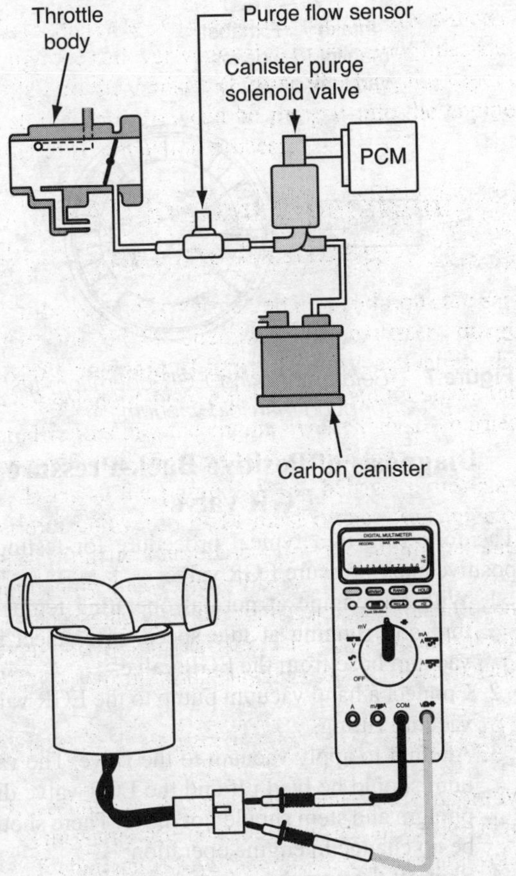


**Figure 5** *The canister purge solenoid controls vapor flow into the combustion chamber.*

- **Rough idle**. Possible causes are an EGR valve is stuck open, the portal vacuum switch (PVS) fails to open, there is dirt on the valve seat, or there are loose mounting bolts. Loose mounting causes a vacuum leak and a hissing noise.
- **Surges, stalls, or does not start**. Probable cause is the valve is stuck open.
- **Detonation (spark knock)**. Any condition that prevents proper EGR flow can cause detonation. This includes a valve that is stuck closed, a leaking valve diaphragm, restrictions in flow passages, an EGR that is disconnected, or a problem in the vacuum source.
- **Excessive NOx emissions**. Any condition that prevents the EGR from allowing the correct amount of exhaust gases into the cylinder can cause this problem. High combustion temperatures allow NOx to form; therefore, anything that allows combustion temperatures to rise can cause this problem as well.
- **Poor fuel economy**. This is an EGR condition only if it relates to detonation or other symptoms of restricted EGR flow.

Before attempting to test the EGR system, visually inspect the condition of all vacuum hoses for kinks, bends, cracks, and flexibility. Replace defective hoses as required. Check vacuum hose routing (see the underhood decal or the manufacturer's service manual for correct routing) and correct any misrouted hoses. If the emissions system is fitted with an EVP sensor, the wires routed to it should also be checked.

To test a ported EGR valve, begin by connecting a vacuum gauge to read the manifold vacuum. Connect a tachometer or use the scan tool to obtain engine speed information. Start the engine and gradually increase speed to 2,000 rpm with the transmission in neutral. The reading from the manifold vacuum gauge **(Figure 7)** should be above 16 in. Hg. If it is not, there could be a vacuum leak or an exhaust restriction. Before continuing to test the EGR, first correct the problem of low vacuum.

Remove the vacuum supply hose from the EGR valve and connect a hand vacuum pump to the EGR valve port. Apply 18 in. Hg of vacuum while observing the EGR diaphragm for movement. When the vacuum is applied, the diaphragm should move. If the valve diaphragm does not move or hold the vacuum, replace the EGR valve.

> **You Should Know**
>
> *In some applications, a mirror may be held under the EGR valve to see the diaphragm movement.*

If the valve passes this test, start the engine and warm it to normal operating temperature. Use a vacuum gauge and check the vacuum supply hose to make sure there is no vacuum to the EGR valve at

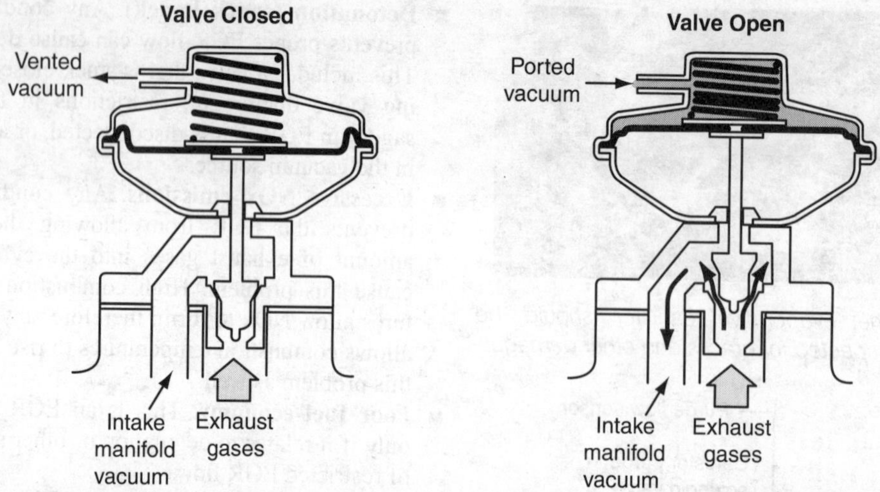


**Figure 6** *Operation of a simple ported EGR valve.*

idle. Then plug the hose. On many EFI engines (multipoint injection), it is necessary to disconnect the throttle air bypass valve solenoid so the PCM will not change engine speed. Observe the engine's idle speed. If necessary, adjust the idle speed to the manufacturer's specification. Slowly apply 5 to 10 in. Hg of vacuum to the EGR valve vacuum port using the hand vacuum pump. The idle speed should drop more than 100 rpm (the engine may stall) and then return to normal (±25 rpm) when the vacuum is removed.

If the idle speed does not drop as the EGR valve is opened, remove the valve and check for carbon in the passages under the valve. Clean the passages as required or replace the EGR valve. Carbon may be cleaned from the lower end of the EGR valve with a wire brush but do not immerse the valve in solvent.

## Diagnosis of Negative Back-Pressure EGR Valve

To diagnosis a negative back-pressure EGR valve, first warm the engine to normal operating temperature. With the ignition switch in the Off position, disconnect the vacuum hose from the EGR valve. Connect a hand vacuum pump to the vacuum fitting on the valve and supply 18 in. Hg of vacuum to the EGR valve while observing the valve operation and the vacuum gauge. The EGR valve should open and hold the vacuum for 20 seconds. If the valve does not operate properly, replace the valve.

While maintaining 18 in. Hg of vacuum to the EGR valve with the hand pump, start the engine. The vacuum should drop to zero and the valve should close. If the valve does not pass this test, replace the valve.

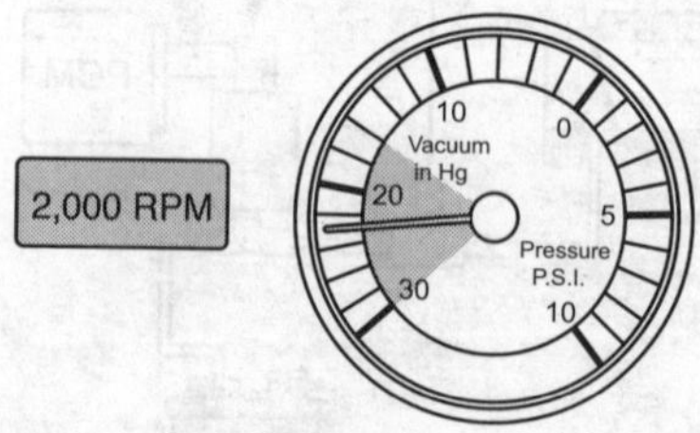


**Figure 7** *Before checking EGR operation, test for good manifold vacuum.*

## Diagnosis of Positive Back-Pressure EGR Valve

The following is a typical procedure for testing a positive back-pressure EGR valve:

1. With the engine at normal operating temperature and running at idle speed, disconnect the vacuum hose from the EGR valve.
2. Connect a hand vacuum pump to the EGR valve vacuum fitting.
3. Attempt to apply vacuum to the valve. The vacuum should be bled off and the EGR valve diaphragm and stem should not move. There should be no change in engine operation.
4. Shut off the engine.
5. Disconnect the EGR vacuum supply hose from the TBI unit or intake manifold.
6. Connect a long hose from this port directly to the EGR valve.
7. Use a T fitting to connect a vacuum gauge into the direct vacuum line to the EGR valve. The T fitting allows the vacuum to be measured while the EGR valve operates.
8. Start the engine and accelerate to 2,000 rpm while observing the vacuum gauge. Vacuum

should be supplied to the EGR valve and the valve should be open.

9. Shut off the engine, disconnect all test equipment, and reconnect the EGR.

## EGR Efficiency Tests

The EGR tests discussed thus far involve the valve's ability to open and close. However, the system needs to be tested further to determine if it is actually performing its function and if it is performing at the proper time. Checking the efficiency of the system may also uncover the cause of other problems. EGR efficiency testing is extremely important while diagnosing the cause of high $NO_x$ and detonation.

Just because the engine stalls or idles very rough when the EGR valve is opened does not mean the system is functioning properly. Actually, the results of this test indicate that the valve was closed and it will open. A good EGR valve does not just open and close; it must allow the correct amount of exhaust gas to enter the cylinders. EGR valves are normally closed at idle and open at approximately 2,000 rpm. It is at this engine speed at which the EGR system should be checked. Keep in mind that anything that increases combustion temperatures can cause an increase in $NO_x$. Common causes of high $NO_x$ are faulty cooling systems, incorrect ignition timing, lean mixtures, faulty air cleaner intake heat controls, and faulty EGR valves.

Proper operation of the EGR system can be tested using a five-gas exhaust analyzer. With the engine at normal operating temperature, increase the engine speed to about 2,000 rpm. Observe the $NO_x$ readings on the analyzer. In most cases, $NO_x$ should be below 1,000 ppm. If the $NO_x$ is above 1,000 ppm, the EGR system is not doing its job. The exhaust passage in the valve is probably clogged with carbon.

### You Should Know

*It is normal to have some temporary increases above 1,000 ppm; however, the reading should generally be under 1,000 ppm.*

If only a small amount of exhaust gas is entering the cylinder, $NO_x$ will still be formed. A restricted exhaust passage of only ⅛ in. will still cause the engine to run rough or stall at idle but it is not enough to control combustion chamber temperatures at higher engine speeds. Never assume the EGR passages are functioning properly just because the engine stalls at idle when the EGR is fully open.

## Electronic EGR Control Diagnosis

Apart from the electronic control, the electronic EGR system can experience many of the problems of any non- electronic EGR system. Some EGR valves are fully electrically controlled and do not use a vacuum signal to open and close. This type of EGR system can exhibit the same problems as basic EGR systems, with the exception of vacuum leaks and other vacuum-related problems. Sticking valves, obstructions, and loss of vacuum produces the same symptoms as with standard systems. If an electronic control component is not functioning, the condition is usually recognized by the PCM. The EGR vent (EGRV) and EGR control (EGRC) solenoids **(Figure 8)** or the electronic vacuum regulator (EVR) should normally cycle on and off very frequently when EGR flow is being controlled (warm engine and cruising-speed rpm). If they do not, a problem in the electronic control system or the solenoids is indicated. Generally, an electronic control failure results in low or zero EGR flow and might cause symptoms like overheating, detonation, and power loss.

In many EGR systems, the PCM uses inputs from the ECT, TPS, and MAP sensors to operate the EGR valve. A defect in one of these sensors may cause improper EGR operation. A scan tester can be connected to the DLC to check for an EGR DTC or a DTC from another sensor that may affect EGR operation. Correct the cause of any DTC before doing any further EGR diagnosis.

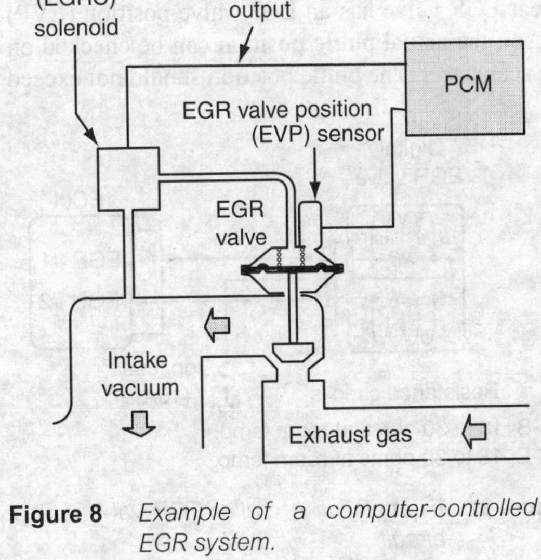


**Figure 8** *Example of a computer-controlled EGR system.*

## Digital EGR Valve Diagnosis

A **digital EGR valve** is operated electronically with no vacuum connections. It contains one to three solenoids that are open or closed. The digital EGR valve can be diagnosed with a scan tester. With the engine at normal operating temperature and the ignition switch in the Off position, connect the scan tester to the DLC. Start the engine and allow it to operate at idle speed. Select EGR control on the scan tester and then energize EGR solenoid #1 with the scan tester. When this action is taken, the engine rpm should decrease slightly. Repeat the test for the other EGR solenoids. The engine rpm should drop slightly as each EGR solenoid is energized with the scan tester. If the EGR valve does not operate properly, check the following items before replacing the EGR valve:

1. Check for 12 volts at the power supply wire on the EGR valve **(Figure 9)**.
2. Check the wires between the EGR valve and the PCM.
3. Remove the EGR valve and check for plugged passages under the valve.

## Linear EGR Valve Diagnosis

A **linear EGR valve** contains one solenoid and a valve pintle that is operated electronically. The PCM operates this type of valve with a pulse-width modulation principle. The linear EGR valve diagnostic procedure varies depending on the vehicle make and model year. Always follow the recommended procedure in the vehicle manufacturer's service manual. The scan tester can be used to diagnose a linear EGR valve. The engine should be at normal operating temperature prior to EGR valve diagnosis. Since the linear EGR valve has an EGR valve position (EVP) sensor, the actual pintle position can be checked on the scan tester. The pintle position should not exceed 3 percent at idle speed. The scan tester can be operated to command a specific pintle position, such as 75 percent, and this commanded position should be achieved within 2 seconds. With the engine idling, select various pintle positions and check against the actual pintle position. The pintle position should always be within 10 percent of the commanded position. If the linear EGR valve does not operate properly, do the following:

1. Check the fuse in the 12-volt supply wire to the EGR valve.
2. Check for open circuits, grounds, and shorts in the wires connected from the EGR valve winding to the PCM.
3. Use a DVOM to check for 5 volts on the reference wire to the EVP sensor.
4. Check for excessive resistance in the EVP sensor ground wire.
5. Leave the wiring harness connected to the valve and remove the valve. Connect a DVOM from the pintle position wire at the EGR valve to ground and manually push the pintle upward **(Figure 10)**. The voltmeter reading should change from approximately 1 volt to 4.5 volts.

If the EGR valve does not operate properly when tested with the scan tester and the above checks are satisfactory, replace the valve.

## EGR Vacuum Regulator (EVR) Tests

Connect the ohmmeter test leads across the EVR terminals to check the winding for open circuits and

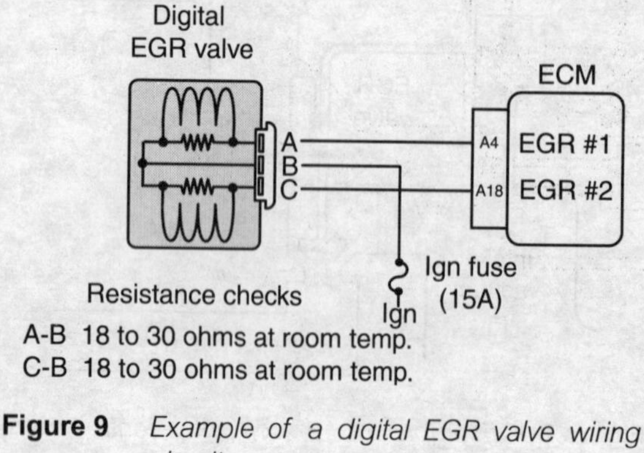


**Figure 9** *Example of a digital EGR valve wiring circuit.*

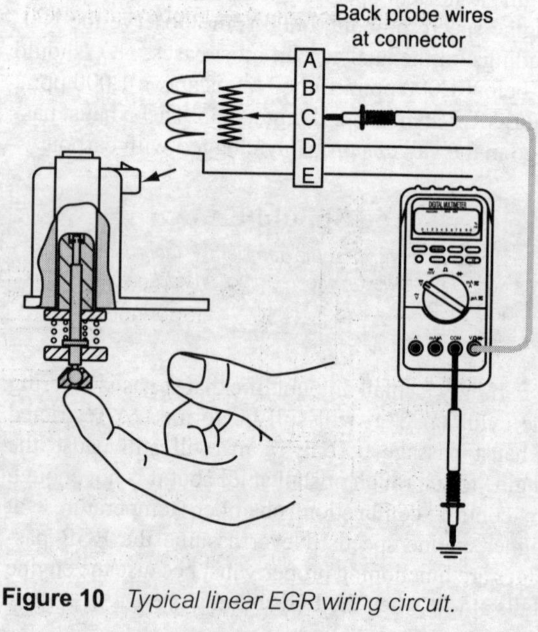


**Figure 10** *Typical linear EGR wiring circuit.*

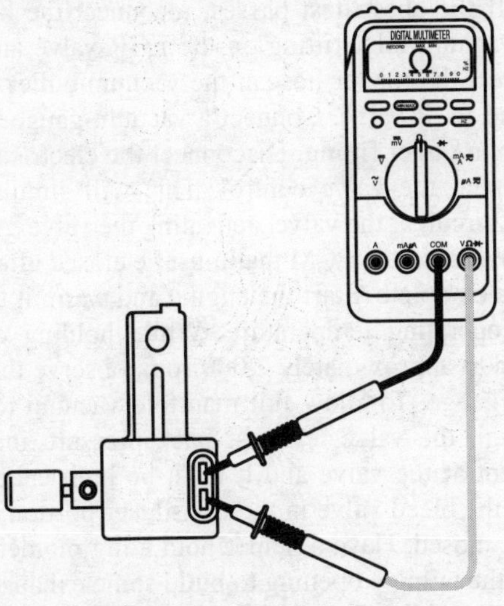


**Figure 11** *Testing the resistance of the EVR solenoid winding.*

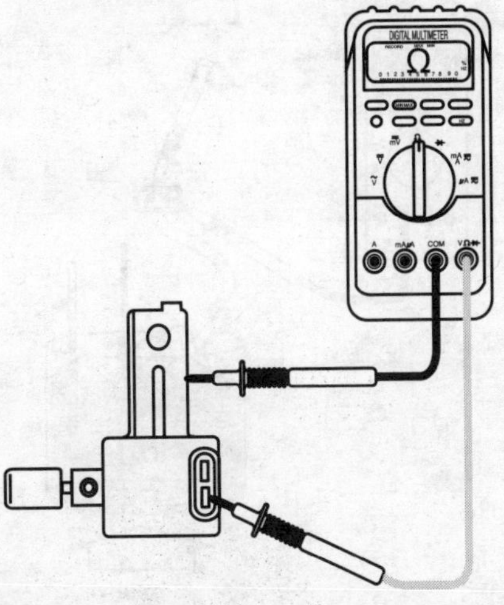


**Figure 12** *Testing for winding shorted to ground of an EVR solenoid.*

shorts **(Figure 11)**. An infinity reading indicates an open circuit whereas a lower-than-specified reading means the winding is shorted. Then, connect the ohmmeter leads from one of the EVR solenoid terminals to the solenoid case **(Figure 12)**. A low ohmmeter reading indicates that the winding is shorted to ground. If the winding is not shorted, an infinity reading will be observed.

> **You Should Know**
>
> *The same quad driver in a PCM may operate several outputs. For example, a quad driver may operate the EVR solenoid and the torque converter clutch solenoid. In General Motors computers, the quad drivers sense high current flow. If a solenoid winding is shorted and the quad driver senses high current flow, the quad driver shuts off all the outputs it controls rather than be damaged by the high current flow. When the PCM fails to operate an output or outputs, always check the resistance of the solenoid windings in the outputs before replacing the PCM. A lower-than-specified resistance in a solenoid winding indicates a shorted condition. This problem may explain why the PCM quad driver stops operating the outputs.*

A scan tester can be used to diagnose the EVR solenoid operation. In the appropriate mode, the scan tester displays the EVR solenoid status as on or off. With the engine idling, the EVR solenoid should remain off. Drive the vehicle until the conditions required to open the EVR solenoid are present. Once these conditions are present, the scan tester should indicate that the EVR solenoid is on.

> **You Should Know**
>
> *In some EGR systems, the PCM energizes the EVR solenoid at idle and low speeds. Under these conditions, the solenoid shuts off vacuum to the EGR valve. When the proper input signals are available, the PCM deenergizes the EVR solenoid and allows vacuum to the EGR valve.*

## Exhaust Gas Temperature Sensor Diagnosis

To test the exhaust gas temperature sensor, remove it from the engine and connect the ohmmeter test leads across its terminals. Place the sensor in a container of oil along with a thermometer. Heat the oil **(Figure 13)** while observing the temperature and the ohmmeter readings. The exhaust gas temperature sensor should have the specified resistance at various temperatures.

## EGR Back-Pressure Transducer Diagnosis

The **back-pressure transducer** directs vacuum from the solenoid to the EGR valve or vents it based on back-pressures against its diaphragm. Diagnosis of an EGR back-pressure transducer will vary depending on the vehicle make and model year. Always follow the manufacturer's instructions. The following procedure is typical.

The first part of this test determines if the transducer diaphragm at the back-pressure side of the valve is leaking. The second part of the test determines if engine vacuum is flowing from the inlet to the outlet side of the valve.

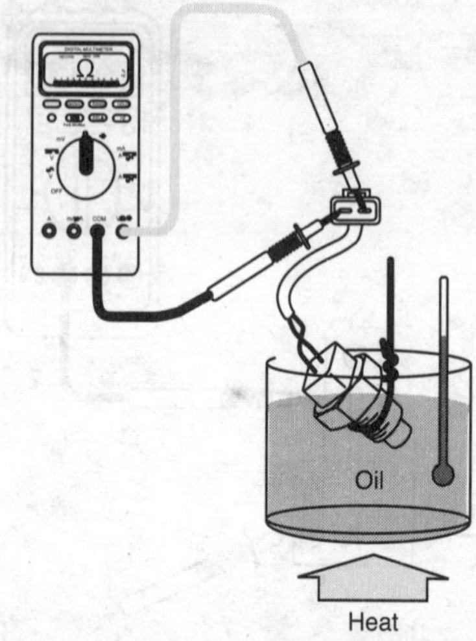


**Figure 13** *Testing an exhaust gas temperature sensor.*

Begin by disconnecting the back-pressure hose from the fitting at the bottom of EGR valve **(Figure 14)**. Connect a handheld vacuum pump to this fitting and apply 10 in. Hg of vacuum to this fitting. If vacuum falls off, the valve diaphragm is leaking.

If the diaphragm passes the above test, reconnect the hose to the EGR valve and remove the vacuum supply hose at the vacuum inlet fitting on the EGR solenoid. Connect a vacuum gauge to this disconnected vacuum line. Start the engine and allow it to warm to normal operating temperature. While holding engine speed at approximately 1,500 rpm, check for steady engine vacuum on the gauge. If engine vacuum is not present, check the vacuum line to the engine. If necessary, repair the vacuum line before proceeding.

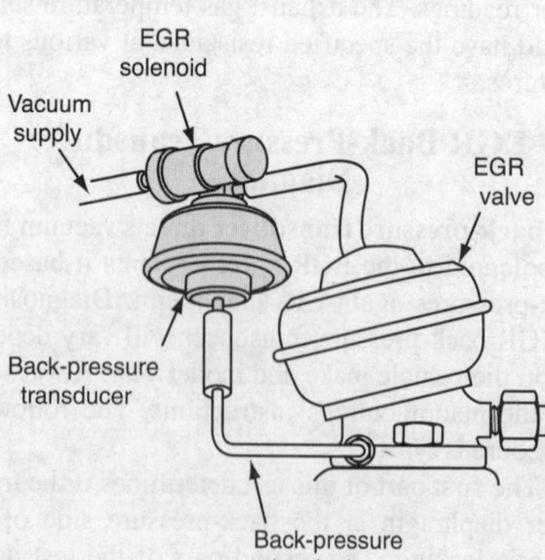


**Figure 14** *EGR valve with back-pressure transducer.*

If the above test passes, reconnect the hose to the vacuum inlet fitting on the EGR valve and disconnect the rubber hose at the vacuum outlet fitting on the EGR valve. Connect a vacuum gauge to this vacuum outlet fitting. Disconnect the electrical connector at the valve control. This will simulate an open circuit at the valve, activating the valve. A DTC will be set in the PCM that must be erased after testing is complete. Start the engine and warm it to normal operating temperature. While holding engine speed to approximately 2,000 rpm, observe the vacuum gauge. To allow full manifold vacuum to flow through the valve, exhaust back pressure must be present at the valve and it must be high enough to hold the bleed valve in the transducer portion of the valve closed. Have a helper hold a rag momentarily over the tailpipe opening to build some exhaust back pressure while observing the vacuum gauge.

**You Should Know**

*Do not cover the tailpipe opening for an extended period of time since damage to components or overheating may result.*

As temporary back pressure is built, full manifold vacuum should be observed on the vacuum gauge. Without back pressure and with the engine at approximately 2,000 rpm, the gauge reading should be low. At idle speed, the gauge reading may be erratic. If full manifold vacuum is not read on the gauge but was present at the inlet fitting, replace the valve.

## EGR Valve Position Sensor Diagnosis

**You Should Know**

*The EGR valve and EVP sensor may be very hot if the engine has been running. Wear protective gloves while servicing these components.*

Many EVP sensors have a 5-volt reference wire, a voltage signal wire, and a ground wire. The 5-volt reference wire and the ground wire can be checked by connecting the voltmeter test leads across the voltage signal wire to ground. With the ignition switch in the Run position, the voltage signal should be approximately 0.8 volt. Connect a handhold vacuum pump to the vacuum fitting on the EGR valve and slowly increase the vacuum from 0 to 20 in. Hg. The EVP sensor voltage signal should gradually increase to 4.5 volts at 20 in. Hg. Always use the EVP test procedure and specifications supplied by the vehicle's manufacturer. If the EVP sensor does not have the specified voltage, replace the sensor.

## Positive Crankcase Ventilation (PCV) System Diagnosis and Service

A PCV valve that is stuck in the open position causes excessive airflow through the valve. This results in a lean air-fuel ratio and possibly rough idle operation or engine stalling. A PCV valve that is plugged or a hose that is restricted results in excessive crankcase pressure. The excess pressure may force crankcase blowby gases through the clean air hose and into the filter in the air cleaner. This can cause the air filter to plug and/or the induction of oil into the combustion chamber (resulting in blue smoke coming out of the tailpipe). Also, a restricted PCV valve or hose may result in the accumulation of moisture and sludge in the engine and the engine oil.

Engine gasket and seal leaks not only result in potential oil leaks, but they also cause the escape of blowby gases to the atmosphere. In addition, the PCV system will draw unfiltered air through these leaks into the engine. This action could result in wear to the engine components, especially when the vehicle is operating in dusty conditions. Check all the engine gaskets for signs of oil leaks. Be sure the oil filler cap fits and seals properly.

Service to the PCV system involves a thorough inspection, testing the operation, and replacement of faulty parts. When replacing a PCV valve, match the part number on the valve with the vehicle manufacturer's specifications.

Some engines use a fixed orifice tube in place of a valve. Clean the tube periodically with a pipe cleaner soaked in carburetor cleaner. Although there is no PCV valve, this type of system is diagnosed in the same way as systems with a valve.

The first step in PCV servicing is a visual inspection. The PCV valve can be located in several places **(Figure 15)**. The most common location is in a rubber grommet in the valve or rocker arm cover (A). It can be installed in the middle of the hose connections (B), as well as directly in the intake manifold (C). Inspect the system using the following procedure:

1. Make sure all the PCV system hoses **(Figure 16)** are properly connected and that they are not cracked or broken.
2. Remove the air cleaner and inspect the carburetor or fuel-injector air filter for oil contamination. Clean or replace any dirty filters. Oil in the air cleaner assembly indicates that the PCV valve or hoses are plugged.
3. Check the crankcase inlet air filter, which is usually located in the air cleaner. If it is oil soaked, it is a good indication that the PCV system is not working properly. When the filter

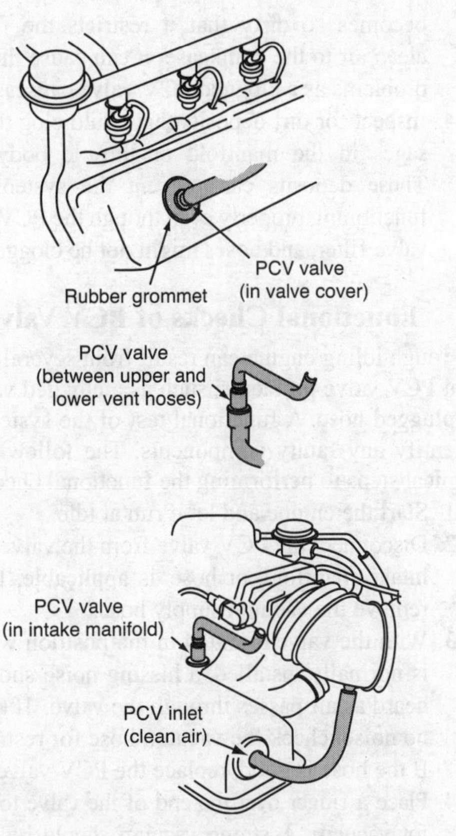


**Figure 15** *Various locations of PCV valves.*

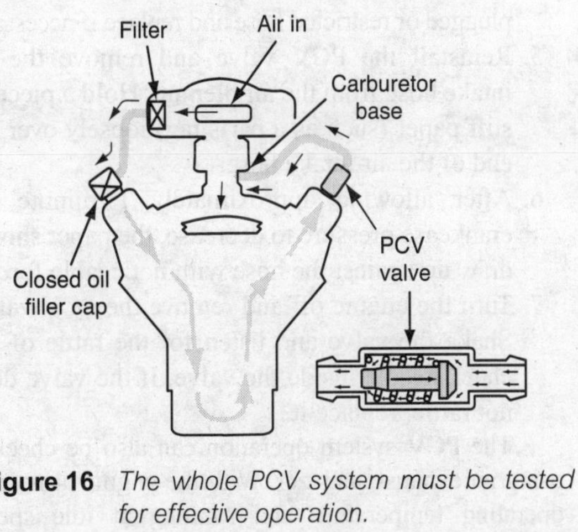


**Figure 16** *The whole PCV system must be tested for effective operation.*

becomes so dirty that it restricts the flow of clean air to the crankcase, it can cause the same problems as a clogged PCV valve causes.

4. Inspect for dirt deposits that could clog the passages in the manifold or throttle body base. These deposits can prevent the system from functioning properly even though the PCV valve, valve filter, and hoses might not be clogged.

## Functional Checks of PCV Valve

A rough idling engine can result from several different PCV valve problems, such as a clogged valve or a plugged hose. A functional test of the system will identify any faulty components. The following are typical steps to performing the functional check:

1. Start the engine and let it run at idle.
2. Disconnect the PCV valve from the valve cover, intake manifold, or hose as applicable. Do not remove the vacuum supply hose.
3. With the valve oriented in the position where it is normally installed, a hissing noise should be heard as air passes through the valve. If there is no noise, check the vacuum hose for restriction. If the hose is good, replace the PCV valve.
4. Place a finger over the end of the valve to check for vacuum. A strong vacuum should be felt. If there is little or no vacuum at the valve, check for a plugged or restricted hose and replace if necessary.
5. Reinstall the PCV valve and remove the air intake hose from the air plenum. Hold a piece of stiff paper (such as a parts tag) loosely over the end of the air intake hose.
6. After allowing approximately 1 minute for crankcase pressure to decrease, the paper should draw up against the hose with noticeable force.
7. Turn the engine off and remove the PCV valve. Shake the valve and listen for the rattle of the check needle inside the valve. If the valve does not rattle, replace it.

The PCV system operation can also be checked using an exhaust analyzer. With the engine at normal operating temperature and running at idle speed, observe the $O_2$ and CO readings. Remove the PCV valve from the rocker arm cover and allow this valve to pull in fresh air. The $O_2$ reading should increase and the CO reading should decrease. If the $O_2$ reading does not change, the PCV system is restricted. If the $O_2$ reading increases more than 1 percent or the CO reading decreases more than 1 percent, there are excessive blowby gases or the oil is contaminated with fuel.

Proper operation of the PCV system depends on a sealed engine. Remember that the crankcase is sealed by the dipstick, valve cover, gaskets, and sealed filler cap. If oil sludging or dilution is noted and the PCV system is functioning properly, check the engine for oil leaks and correct them to ensure that the PCV system functions as intended. Also, be aware that a worn engine may have more blowby than the PCV system can handle. If there are symptoms that indicate the PCV system is plugged (oil in air cleaner, saturated crankcase filter, and so forth) but no restrictions are found, check for engine wear.

Related to PCV is the issue of fuel in the crankcase. Some common causes of fuel in the crankcase include:

- The engine is in poor mechanical condition.
- The fuel injector is leaking.
- The fuel pressure regulator is defective.

Fuel in the crankcase leads to a rich air-fuel mixture. The following responses may occur:

- Fuel vapors are drawn into the intake through the PCV system.
- The mixture is rich because of an increase in fuel amount.
- The oxygen sensors detect a rich mixture and adjust accordingly.

One way to detect fuel in the crankcase is through exhaust gas testing. First test the exhaust gas ahead of the catalytic converter with the air pump (if equipped) disconnected. Then test the exhaust gas again with the PCV valve removed from the engine valve cover. If CO decreases by more than 1.5 percent, the crankcase is probably contaminated with fuel.

# Air-Injection System Diagnosis and Service

Not all engines are equipped with an air-injection system; however, in engines equipped with air-injection systems, these systems are vital to proper emissions. Each system has its own test procedure; always follow the manufacturer's recommendations for testing.

The first step to diagnosing an air-injection system is to check all vacuum hoses and electrical connections in the system. Many AIR system pumps have a centrifugal filter behind the pulley. Air flows through this filter into the pump and the filter keeps dirt out of the pump. The pulley and filter are bolted to the pump shaft and these two components are serviced separately **(Figure 17).** If the pulley or filter is bent, worn, or damaged, it should be replaced. The pump assembly is usually not serviced.

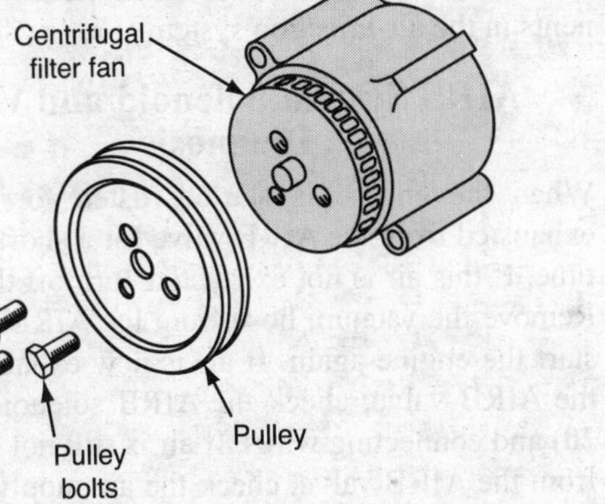


**Figure 17** *After the pulley is removed, the centrifugal filter can be inspected.*

The air pump belt must have the specified tension. A loose belt or a defective AIR system may result in high emission levels and/or excessive fuel consumption.

In some AIR systems, pressure-relief valves are mounted in the AIR bypass (AIRB) and AIR diverter (AIRD) valves **(Figure 18)**. Other AIR systems have a pressure-relief valve in the pump. If the pressure-relief valve is stuck open, airflow from the pump is continually exhausted through this valve, causing high tailpipe emissions.

If the hoses in the AIR system show evidence of burning, the one-way check valves are leaking. This allows exhaust to enter the system. Leaking air manifolds and pipes result in exhaust leaks and excessive noise.

Some AIR systems will set DTCs in the PCM if there is a fault in the AIRB or AIRD solenoids and related wiring. In some AIR systems, DTCs are set in the PCM memory if the airflow from the pump is continually upstream or downstream. Always use a scan tester to check for any DTCs related to the AIR system and correct the causes of these codes before proceeding with further system diagnosis.

If the AIR system does not pump air into the exhaust ports during engine warm-up, HC emissions will be high during this mode. In addition, the $O_2S$ will take longer to reach normal operating temperature. Under this condition, the PCM remains in open-loop mode longer. Since the air-fuel ratio is richer in open-loop mode, fuel economy is reduced.

When the AIR system pumps air into the exhaust ports with the engine at normal operating temperature, the additional air in the exhaust stream causes lean signals from the $O_2S$. The PCM responds to these lean signals by providing a rich air-fuel ratio from the injectors. This action increases fuel consumption. A vehicle can definitely fail an emissions test because of

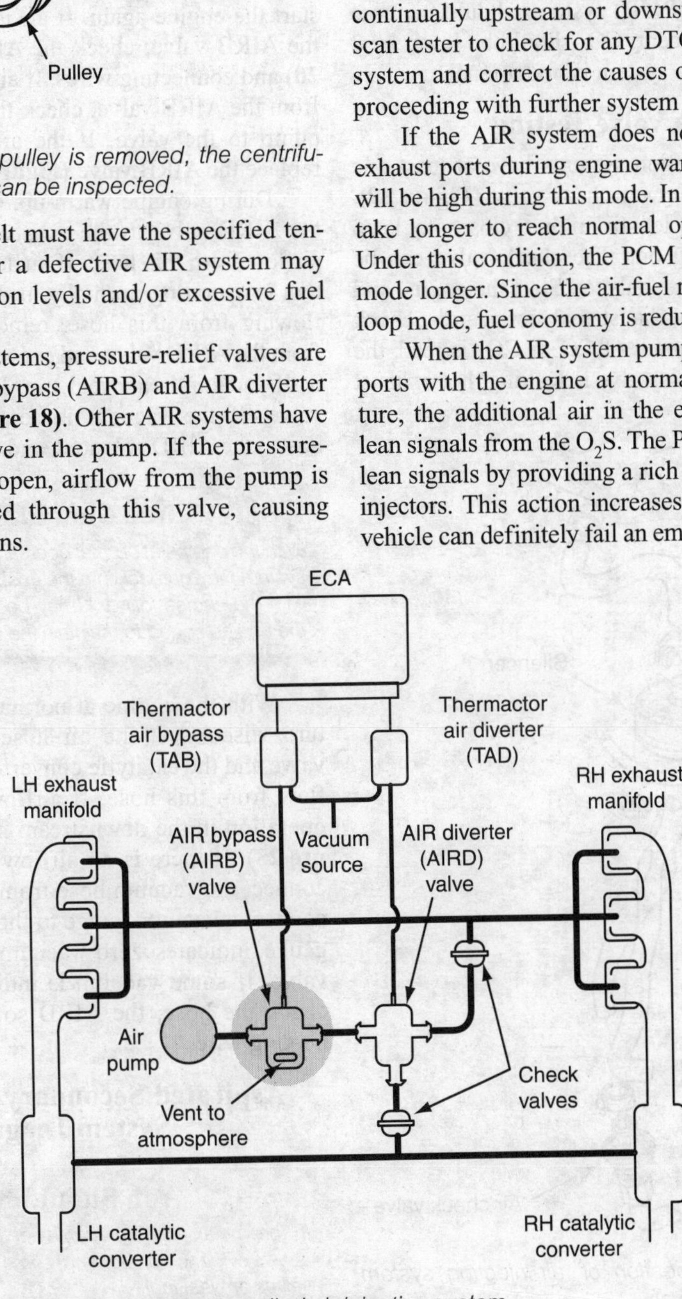


**Figure 18** *An example of a computer-controlled air-injection system.*

air flowing past the $O_2S$ when it should not be. If the $O_2S$ is always sending a lean signal back to the PCM, check the air-injection system.

As a general test of the air-injection system, operate the engine at normal temperature and idle speed. Connect an exhaust analyzer to the exhaust pipe and observe the $O_2$ and $CO_2$ readings. Next, disable the air-injection system and record the $O_2$ and $CO_2$ levels again. Compare the two readings; most AIR or pulsed AIR (PAIR) systems should increase the $O_2$ level by 2 percent; $CO_2$ should be lower by 2 to 4 percent when the system is operational.

## Check Valve Testing

All types of air-injection systems have at least one common aspect, a one-way check valve **(Figure 19)**. The valve opens to let air in but closes to keep exhaust from leaking out. The check valve can be tested with an exhaust analyzer. Start the engine and hold the probe of the exhaust analyzer near the check valve port. If any amount of CO or $CO_2$ is read, the valve is leaking. If this valve is leaking, hot exhaust is also leaking, which could ruin the other components in the air-injection system.

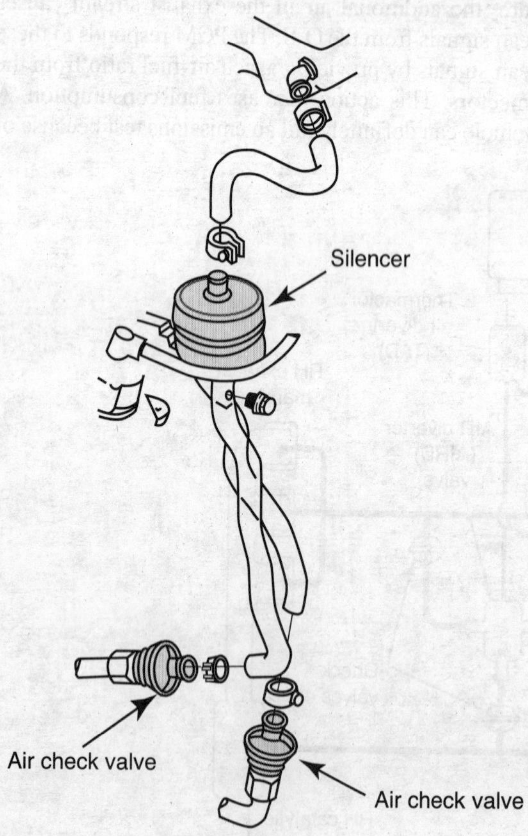


**Figure 19** *Typical location of air-injection system check valves.*

## AIRS System Solenoid and Valve Diagnosis

When the engine is started, listen for air being exhausted from the AIRB valve for a short period of time. If this air is not exhausted, turn off the engine. Remove the vacuum hose from the AIRB valve and start the engine again. If air is now exhausted from the AIRB valve, check the AIRB solenoid **(Figure 20)** and connecting wires. If air is still not exhausted from the AIRB valve, check the air supply from the pump to the valve. If the air supply is available, replace the AIRB valve **(Figure 21)**.

During engine warm-up, remove the hose from the AIRD valve to the exhaust ports and check for airflow from this hose. If airflow is present, the system is operating normally in this mode. If air is not flowing from this hose, remove the vacuum hose from the AIRD valve and connect a vacuum gauge to it. If vacuum is above 12 in. Hg, replace the AIRD valve. If the vacuum is zero, check the vacuum hoses, the AIRD solenoid, and the connecting wires.

> **You Should Know**
>
> *With the engine at normal operating temperature, the AIR system may go back into the upstream mode (Figure 22) with the engine idling. It may be necessary to increase the engine speed to maintain the downstream mode.*

With the engine at normal operating temperature, disconnect the air hose between the AIRD valve and the catalytic converters and check for airflow from this hose. If airflow is present, system operation in the downstream mode is normal **(Figure 23)**. If there is no airflow from this hose, disconnect the vacuum hose from the AIRD valve and connect a vacuum gauge to the hose. If the vacuum gauge indicates zero vacuum, replace the AIRD valve. If some vacuum is indicated on the gauge, check the hose, the AIRD solenoid, and the connecting wires.

## Aspirated Secondary Air-Injection System Diagnosis

> **You Should Know**
>
> *If the metal aspirator container or the clean air hose shows evidence of burning, some of the one-way check valves are leaking.*

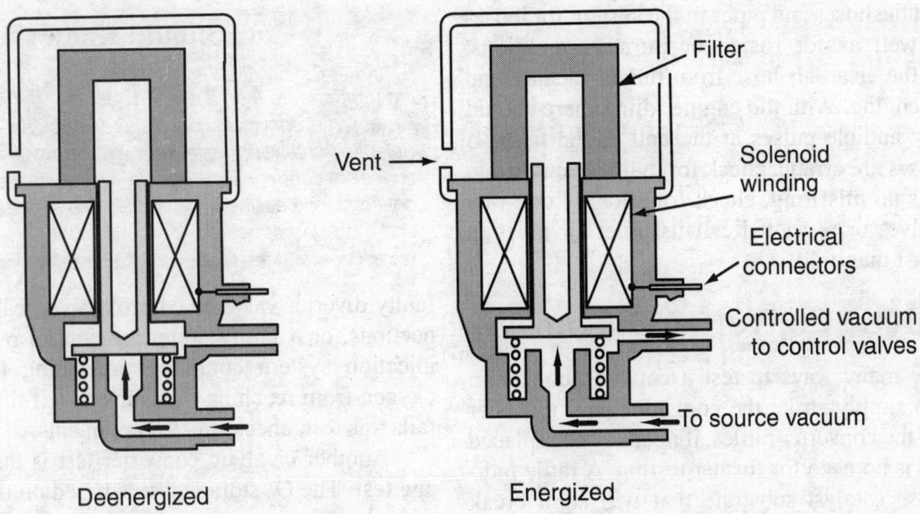


**Figure 20** *A TAB (AIRB) solenoid.*

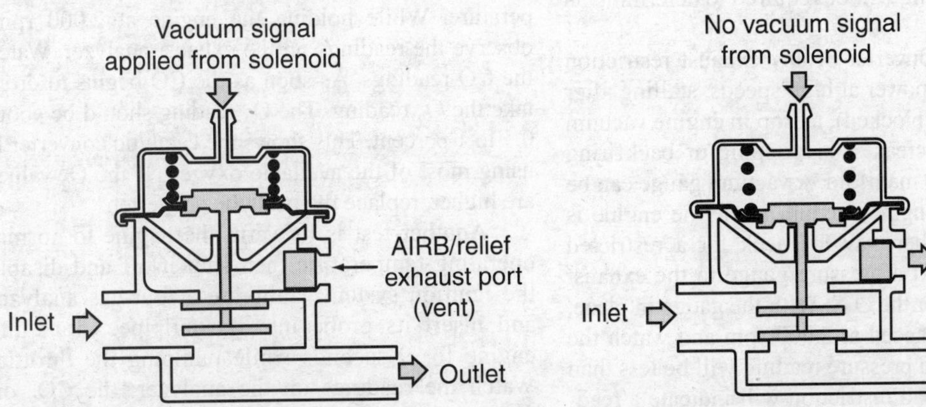


**Figure 21** *A TAB (AIRB) valve.*

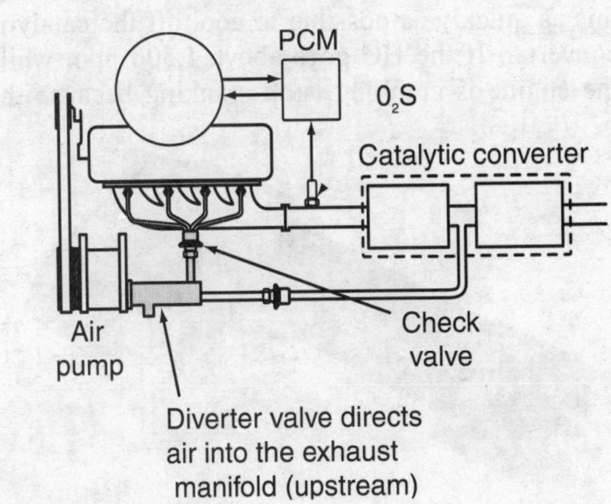


**Figure 22** *Upstream mode.*

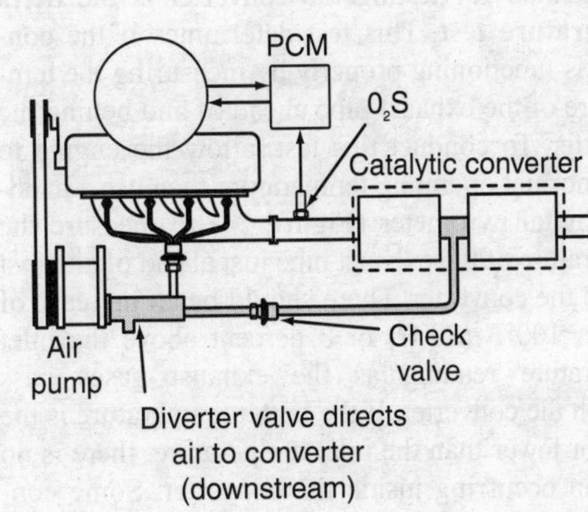


**Figure 23** *Downstream mode.*

Check all the hoses and pipes in the system for looseness as well as for rusted or burned conditions. Remove the clean air hose from the air cleaner and start the engine. With the engine idling there should be steady audible pulses at the end of the hose. If these pulses are erratic, check for cylinder misfiring. If there is no misfiring, check for sticking one-way check valves or restricted exhaust inlet air tubes in the exhaust manifold.

## Catalytic Converter Diagnosis

There are many ways to test a catalytic converter. One is to simply strike the converter with a rubber mallet. If the converter rattles, it needs to be replaced and there is no need for further testing. A rattle indicates loose catalyst substrate that will soon break into small pieces. Just because the catalyst does not rattle when stricken does not mean the converter is good; further testing will be required to determine its efficiency.

A plugged converter or any exhaust restriction can cause loss of power at high speeds, stalling after starting (if totally blocked), a drop in engine vacuum as engine rpm increases, or popping or backfiring through the intake manifold. A vacuum gauge can be used to observe engine vacuum while the engine is accelerated. Another way to check for a restricted exhaust is to insert a pressure gauge in the exhaust manifold's bore for the $O_2S$. With the gauge in place, hold the engine's speed at 2,000 rpm and watch the gauge. The desired pressure reading will be less than 1.25 psi. A very bad restriction will indicate a reading of more than 2:75 psi.

The catalytic converter should be checked for its ability to convert CO and HC into $CO_2$ and water. One method for testing the converter is the **delta temperature test**. This test determines if the converter is functioning properly by measuring the temperature of the exhaust pipe ahead of and behind the converter. To conduct this test, allow the engine to reach normal operating temperature then use a handheld digital pyrometer **(Figure 24)** to measure the temperature of the exhaust pipe just ahead of and just behind the converter. There should be an increase of at least 100°F (38°C) or 8 percent above the inlet temperature reading as the exhaust gases pass through the converter. If the outlet temperature is the same or lower than the inlet temperature, there is no reaction occurring inside the converter. Some converters use a supply of oxygen from the air pump to oxidize HC and CO emissions. A bad air pump, a faulty diverter valve or control valve, leaky air connections, or a faulty computer control over the air-injection system could be preventing the needed oxygen from reaching the converter. If the converter fails this test, check those components.

> **You Should Know**
>
> *A defective catalytic converter has a much lower operating temperature than a properly operating catalytic converter. When driving short distances in cold climates, a defective catalytic converter may fill up with condensation. This condensation may freeze when the engine is not running, resulting in a completely restricted converter and a no-start condition.*

Another catalytic converter test is the **O2 storage test**. The $O_2$ storage test is based on the fact that a good catalytic converter stores oxygen. To perform this test, disable the air-injection system (if equipped). Allow the engine to warm to normal operating temperature. While holding the engine at 2,000 rpm, observe the readings on the exhaust analyzer. Watch the CO readings. As soon as the CO begins to drop, take the $O_2$ reading. The $O_2$ reading should be about 0.5 to 1 percent. This shows the catalytic converter is using most of the available oxygen. If the $O_2$ values are higher, replace the catalytic converter.

Another test is to warm the engine to normal operating temperature then shut it off and disable the ignition system. Calibrate a four-gas analyzer and insert its probe into the tailpipe. Crank the engine for 9 seconds while pumping the throttle. Watch the readings on the analyzer; the $CO_2$ on fuel-injected vehicles should be more than 11 percent and carbureted vehicles should have a reading of more than 10 percent. After observing the readings, reconnect the ignition and start the engine. Do this as quickly as possible to cool off the catalytic converter. If the HC goes above 1,500 ppm while the engine is cranking, stop cranking because the

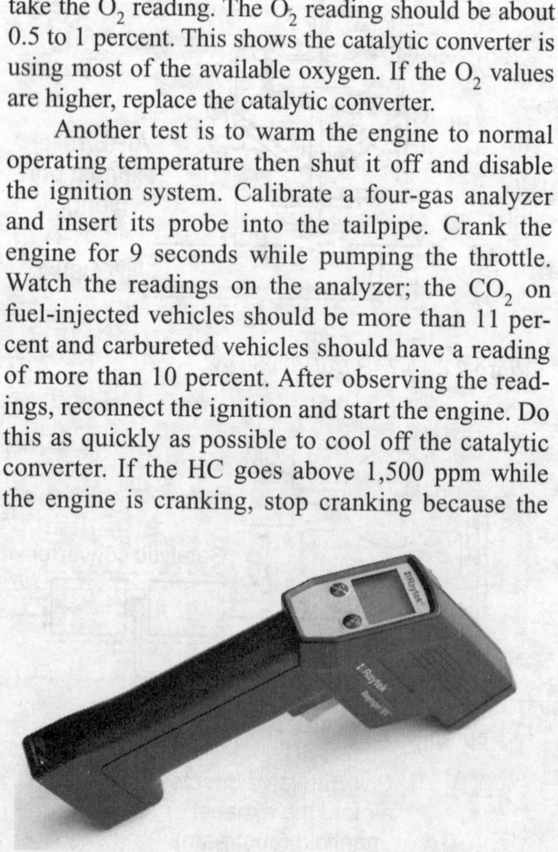

**Figure 24** *A digital pyrometer.*

converter is not working. Also stop cranking once the $CO_2$ reading reaches 10 or 11 percent. This indicates a good catalytic converter. If the catalytic converter is bad, there will be high HC and low $CO_2$ readings at the tailpipe. Do not repeat this test more than once without running the engine in between.

The last test method uses the four- or five-gas analyzer. Disable the air-injection system and warm the engine to normal operating temperature. Operate the engine at 2,000 rpm and observe the exhaust gas readings. On fuel-injected or computer-controlled carburetor engines the $O_2$ reading must be zero or very low and some CO must be present. If there is no CO present, add propane to the air intake with a propane enrichment kit to obtain some CO. On noncomputer-controlled fuel systems some $O_2$ is acceptable. Quickly accelerate the engine and observe the $O_2$ reading. If it goes above 1.2 percent, the catalytic converter is defective.

**You Should Know**

*Just changing a catalyst without correcting the real problem could result in damage to the new catalyst. Before replacing a catalyst, always evaluate all the other components that affect emissions. A defective catalyst cannot be repaired; it can only be replaced.*

## Review Questions

1. Technician A says the temperature should be the same at the inlet and outlet of the catalytic converter if it is working properly. Technician B says a defective belt-driven air pump may cause improper catalytic converter operation. Who is correct?
   A. Technician A only
   B. Technician B only
   C. Both Technician A and Technician B
   D. Neither Technician A nor Technician B
2. Technician A says if the EGR valve is open at idle and low speeds, the engine may surge during low-speed operation. Technician B says if the EGR valve is open at idle and low speeds, the engine may stall on deceleration. Who is correct?
   A. Technician A only
   B. Technician B only
   C. Both Technician A and Technician B
   D. Neither Technician A nor Technician B
3. Technician A says a defective throttle position sensor (TPS) may affect the EGR valve operation. Technician B says a defective engine coolant temperature (ECT) sensor may affect the EGR valve operation. Who is correct?
   A. Technician A only
   B. Technician B only
   C. Both Technician A and Technician B
   D. Neither Technician A nor Technician B
4. Technician A says that an engine with more than 14.5 percent $CO_2$ probably has a functioning catalytic converter. Technician B says that when there is a misfire, there is low $CO_2$ in the exhaust. Who is correct?
   A. Technician A only
   B. Technician B only
   C. Both Technician A and Technician B
   D. Neither Technician A nor Technician B
5. Technician A says a restricted PCV valve may result in oil contamination of the air cleaner element. Technician B says a restricted PCV valve causes high $NO_x$ emissions. Who is correct?
   A. Technician A only
   B. Technician B only
   C. Both Technician A and Technician B
   D. Neither Technician A nor Technician B
6. Technician A says if the EGR valve does not open, the engine may hesitate on acceleration. Technician B says if the EGR valve does not open, the engine may detonate on acceleration. Who is correct?
   A. Technician A only
   B. Technician B only
   C. Both Technician A and Technician B
   D. Neither Technician A nor Technician B
7. The diagnosis of a positive back-pressure EGR valve is being discussed. Technician A says with the engine running at idle speed, if a hand pump is used to supply vacuum to the EGR valve, the valve should open at 12 in. Hg of vacuum. Technician B says with the engine running at idle speed, any vacuum supplied to the EGR valve should be bled off and the valve should not open. Who is correct?
   A. Technician A only
   B. Technician B only
   C. Both Technician A and Technician B
   D. Neither Technician A nor Technician B

# SECTION 7 ON-BOARD DIAGNOSTICS SECOND GENERATION (OBD-II)

## Section Objectives

After you have read, studied, and practiced the contents of this section, you should be able to:

- Define the purpose of On-Board Diagnostics Second Generation (OBD-II).
- Define the terms used in reference to OBD-II.
- Explain the requirements to illuminate the malfunction indicator light (MIL) in an OBD-II system.
- Describe the data link connector (DLC) used in an OBD-II system.
- Describe how the MIL is extinguished after a fault is detected.
- Explain the procedure for self-erasure of DTCs.
- Define a trip in an OBD-II system.
- Explain the difference between one- and two-trip DTCs.
- Describe how the downstream heated oxygen sensors ($HO_2Ss$) monitor the catalytic converter efficiency.
- Describe the purpose and function of the comprehensive component monitor.
- Explain the purpose and function of the $HO_2S$ monitor.
- Explain how engine misfiring is monitored in an OBD-II system.
- Describe how the EGR flow is monitored in an OBD-II system.
- Describe the purpose and function of the EVAP monitor.
- Explain how OBD-II monitors are run in common systems.
- Describe the requirements of OBD-II.
- Access the different scan tool modes to obtain OBD-II service information.
- Evaluate DTC and freeze-frame data.
- Determine the enabling conditions for OBD-II monitors.
- Run and evaluate a comprehensive component monitor.
- Locate and interpret freeze-frame data.
- Perform a good trip after a comprehensive component monitor failure.
- Run and evaluate a fuel system monitor.
- Locate and interpret similar conditions window information.
- Perform a good trip after a fuel system monitor failure within the similar conditions window.

### Interesting Fact

*In 1961, the California Motor Vehicle State Bureau of Air Sanitation mandated the first automotive emissions-control technology in the nation, positive crankcase ventilation (PCV), to control hydrocarbon crankcase emissions.*

# CHAPTER 29 Introduction to OBD-II

## Introduction

As discussed in Section 6, the federal government over the years has passed legislation in an effort to maintain air quality standards **(Figure 1)**. In 1988, California required all automotive manufacturers to provide a system that identifies faults in the computer-controlled systems and notifies the driver via a malfunction indicator light (MIL) that a fault exists. This system monitored the functionality of components within the fuel metering system and the EGR system as well as additional emissions-related components. In addition to illumination of the Check Engine or Service Engine Soon lights, a DTC was stored in the computer's memory. The trouble code identifies the specific area of the fault.

Amendments to the Clean Air Act also included requirements for a second generation of OBD, referred to as **On-Board Diagnostics Second Generation (OBD-II)**. OBD-II is a legal regulation that requires vehicle manufacturers to have on board diagnostic routines to determine if the engine is polluting the environment. The California Air Resources Board (CARB) found that by the time a computer or emissions system component failure occurs and the MIL is illuminated, the vehicle emissions have been excessive for some time. The CARB developed requirements to monitor the performance of emissions systems as well as indicate component failure. The EPA accepted these requirements in 1990 and the federal government required manufacturers be compliant with OBD-II by model year 1996. The new requirement mandated that the monitoring results must be available to service personnel without the need for special test equipment marketed by the vehicle manufacturer and only available to franchised dealers.

Although the requirement did not mandate OBD-II compliance until the 1996 model year, vehicle manufacturers began installing OBD-II systems on some 1994 models. For example, in 1994 Ford installed this

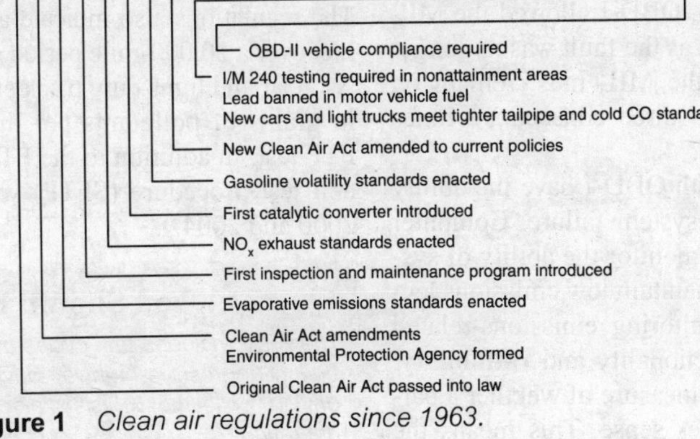


**Figure 1** *Clean air regulations since 1963.*

system on 3.8L Mustangs and 4.6L Thunderbirds and Cougars. In that same year, Chrysler installed OBD-II on the 2.0L Neon. Since OBD-II systems were not mandated until 1996, some of early OBD-II systems were partial systems that did not have all the monitors required on a complete OBD-II system.

OBD-II regulations dictate the function of the system but manufacturers are allowed to use their own methods to accomplish the task. This chapter details the purpose of OBD-II and what it means to the technician.

## OBD-I versus OBD-II

Hardware differences between OBD-I and OBD-II are trivial. Basically, OBD-II uses the same type of components as OBD-I. The only additional components needed are the postcatalyst heated oxygen sensors (downstream $HO_2S$) and an EVAP leak-detection pump or vacuum-control valve on some vehicles. In addition, some manufacturers had to install faster PCMs with additional storage capacity, drivers, and terminal connections.

From a manufacturing standpoint, the major changes required by OBD-II include the software used by the PCM, a standard J1962 DLC, and the J1850 communication protocol. Software changes were the most extensive since the PCM has a larger job to do. What this means for the technician, though, is that there is now more information stored in the PCM about a fault. Once the technician understands the purpose, function, and operation of OBD-II, this additional information should make it easier to diagnose faults.

There are some differences between OBD-I and OBD-II MIL illumination. On OBD-II vehicles, the MIL is illuminated only for emissions-related failures, whereas OBD-I allowed the MIL to be illuminated for nonemissions-related failures such as cruise control. In addition, OBD-I allowed the MIL to be extinguished as soon as the fault was no longer present. Under OBD-II, the MIL must remain on until a system test passes three consecutive times with no additional failures.

Computer systems with OBD-I have the ability to detect component and system failure. Computer systems with OBD-II can monitor the ability of systems and components to maintain low emissions levels. OBD-II requires monitoring emissions-related components for both functionality and **rationality**. Testing for rationality is a measure of whether a particular sensor signal makes sense. This means the OBD-II systems must illuminate the MIL if the vehicle emissions exceed one and one-half times the allowable standard for that model year based on a **federal test procedure (FTP)**. The FTP is a series of test procedures, including drive cycles, that is used to measure a vehicle's emissions. When a component or strategy fails or deteriorates to the point that it allows emissions to exceed the allowable standard level, the MIL is illuminated to inform the driver about the problem and a DTC is stored in the PCM.

## Federal Test Procedure Standards

As mentioned, OBD-II is the result of the Clean Air Act and its amendments. In recent years, additional mandated and voluntary guidelines for vehicle emissions have been put into place. These include the requirements for Tier 1, national low-emissions vehicles (NLEVs), and Tier 2. The FTP standard details the allowable emissions levels of a vehicle over its useful life. The allowable emissions are based on Tier 1 and Tier 2 requirements.

### Tier 1 Standards

Tier 1 standards apply to light-duty vehicles (LDVs) of less than 8,500 pounds gross vehicle weight rating (GVWR). LDVs are divided into the following categories:

- Passenger cars
- Light light-duty trucks (LLDTs), less than 6,000 pounds GVWR
- Heavy light-duty trucks (HLDTs), more than 6,000 pounds GVWR

Tier 1 regulations were finalized on June 5, 1991. The Tier 1 standards were phased in beginning in 1994 and were fully implemented by 1997. They apply to a full vehicle useful life of 100,000 miles. The regulations also indicate a standard that must be met over a 50,000-mile period **(Table 1)**.

Car and light-duty truck emissions are measured in grams of pollutants per mile (g/mile) using the FTP test. In addition to the FTP, a supplemental federal test procedure (SFTP) was phased in between 2000 and 2004.

**You Should Know**

*The SFTP includes test cycles that measure emissions during aggressive highway driving, and during urban driving while the vehicle's air-conditioning system is operating.*

## National LEV Program

On December 16, 1997, the EPA finalized the regulations for the NLEV program. The NLEV program provided the emissions standards to be used during the transitional period before Tier 2.

> **You Should Know**
>
> *The NLEV is a voluntary program that came into effect through an agreement of the northeastern states and the auto manufacturers.*

The NLEV requirements came in effect for the northeastern states in the 1999 model year 1999. Nationally they can into effect in the 2001 model year. Even though new cars and LLDTs had to meet tailpipe standards that were more stringent than the EPA could legally mandate before the 2004 model year, these standards are enforceable in any other federal new motor vehicle program.

## Tier 2 Standards

Tier 2 standards were adopted on December 21, 1999, and phased in beginning in 2004. Not only did the Tier 2 standards reduce the emissions level from that of Tier 1, but they also required more stringent emissions limits for larger vehicles. Under the Tier 2 standard, the same emissions standards apply to all vehicle weight categories. Also, the emissions standards are the same regardless of the fuel used. Tier 2 will be fully phased in by 2009.

The emissions standards for all pollutants for Tier 2 are shown in **Table 2**. Under Tier 2 standards, the vehicle "full useful life" period is extended to 120,000 miles.

# OBD-II Component Requirements

> **You Should Know**
>
> *Pinpointing exactly what the FTP standard is depends on the year of manufacturer, the weight classification of the vehicle, and the emissions classification. What is important to remember at this time is that OBD-II defines excessive emissions as exceeding one and one-half times the FTP standard.*

OBD-II requires that automotive manufacturers use standardized components and systems. These include DLCs, data circuits, diagnostic tests, and DTCs. In addition, the SAE in 1991 published standards for electrical/electronic system diagnostic terms, definitions, abbreviations, and acronyms. The resulting publication, J1930, applies to the following:

- Diagnostic, service, and repair manuals
- Bulletins and updates
- Training manuals
- Repair databases
- Underhood emissions labels
- Emissions certification applications

## Test Connector

OBD-II implemented the use of "generic" scan tools to access the emissions-related items and DTCs rather than relying on a manufacturer-specific scan tool. The standardized DLC, developed by the SAE, allows for these tools to communicate with the PCM. Standards require the DLC to be mounted in the passenger compartment out of sight of vehicle passengers. The DLC must be a 16-terminal connector with

| Category | 50,000 miles/5 years | | | | | | 100,000 miles/10 years[1] | | | | |
|---|---|---|---|---|---|---|---|---|---|---|---|
| | THC | NMHC | CO | $NO_x$ diesel | $NO_x$ *gasoline* | PM | THC | NMHC | CO | $NO_x$ diesel | $NO_x$ *gasoline* |
| Passenger car | 0.41 | 0.25 | 3.4 | 1.0 | *0.4* | 0.08 | — | 0.31 | 4.2 | 1.25 | *0.6* |
| LLDT, LVW <3,750 lbs. | — | 0.25 | 3.4 | 1.0 | *0.4* | 0.08 | 0.80 | 0.31 | 4.2 | 1.25 | *0.6* |
| LLDT, LVW >3,750 lbs. | — | 0.32 | 4.4 | — | *0.7* | 0.08 | 0.80 | 0.40 | 5.5 | 0.97 | *0.97* |
| HLDT, ALVW <5,750 lbs. | 0.32 | — | 4.4 | — | *0.7* | — | 0.80 | 0.46 | 6.4 | 0.98 | *0.98* |
| HLDT, ALVW >5,750 lbs. | 0.39 | — | 5.0 | — | *1.1* | — | 0.80 | 0.56 | 7.3 | 1.53 | *1.53* |

1—Useful life 120,000 miles/11 years for all HLDT standards and for THC standards for LDT

**Abbreviations**

LVW—Loaded vehicle weight (curb weight + 300 lbs).
ALVW—Adjusted LVW (the numerical average of the curb weight and the GVWR).
LLDT—Light light-duty truck (less than 6,000 lbs. GVWR).
HLDT—Heavy light-duty truck (more than 6,000 lbs. GVWR).
THC—Total hydrocarbon.
NMHC—Nonmethane hydrocarbon.
PM—Particulate matter.

**Table 1** *Tier 1 standards.*

| Bin # | 50,000 miles | | | | | 120,000 miles | | | | |
|---|---|---|---|---|---|---|---|---|---|---|
| | NMOG | CO | $NO_x$ | PM | HCHO | NMOG | CO | $NO_x$* | PM | HCHO |
| Temporary Bins | | | | | | | | | | |
| MDPV[c] | | | | | | 0.280 | 7.3 | 0.9 | 0.12 | 0.032 |
| 10[a,b,d,f] | 0.125 (0.160) | 3.4 (4.4) | 0.4 | — | 0.015 (0.018) | 0.156 (0.230) | 4.2 (6.4) | 0.6 | 0.08 | 0.018 (0.027) |
| 9[a,b,e] | 0.075 (0.140) | 3.4 | 0.2 | — | 0.015 | 0.090 (0.180) | 4.2 | 0.3 | 0.06 | 0.018 |
| Permanent Bins | | | | | | | | | | |
| 8[b] | 0.100 (0.125) | 3.4 | 0.14 | — | 0.015 | 0.125 (0.156) | 4.2 | 0.20 | 0.02 | 0.018 |
| 7 | 0.075 | 3.4 | 0.11 | — | 0.015 | 0.090 | 4.2 | 0.15 | 0.02 | 0.018 |
| 6 | 0.075 | 3.4 | 0.08 | — | 0.015 | 0.090 | 4.2 | 0.10 | 0.01 | 0.018 |
| 5 | 0.075 | 3.4 | 0.05 | — | 0.015 | 0.090 | 4.2 | 0.07 | 0.01 | 0.018 |
| 4 | — | — | — | — | — | 0.070 | 2.1 | 0.04 | 0.01 | 0.011 |
| 3 | — | — | — | — | — | 0.055 | 2.1 | 0.03 | 0.01 | 0.011 |
| 2 | — | — | — | — | — | 0.010 | 2.1 | 0.02 | 0.01 | 0.004 |
| 1 | — | — | — | — | — | 0.000 | 0.0 | 0.00 | 0.00 | 0.000 |

*—Average manufaturer fleet $NO_x$ standard is 0.07 g/mile.
a—Bin deleted at end of 2006 model year (2008 for HLDTs).
b—The higher temporary NMOG, CO, and HCHO values apply only to HLDTs and expire after 2008.
c—An additional temporary bin restricted to MDPVs expires after model year 2008.
d—Optional temporary NMOG standard of 0.195 g/mile (50,000) and 0.280 g/mile (120,000) applies for qualifying LDT4s and MDPVs only.
e—Optional temporary NMOG standard of 0.100 g/mile (50,000) and 0.130 g/mile (120,000) applies for qualifying LDT2s only.
f—50,000 mile standard optional for diesels certified to bin 10.

**Abbreviations**
NMOG—Nonmethane organic gases.
HCHO—Formaldehyde (chemical equation, not acronym).
PM—Particulate matter.

**Table 2** *Tier 2 standards.*

specific terminals defined by the SAE **(Figure 2)**. This connector is referred to as J1962 **(Figure 3)**.

> **You Should Know**
> *Just because a vehicle has a 16-terminal DLC does not necessarily mean it is equipped with an OBD-II system.*

A new requirement for automotive manufacturers started being implemented in the 2004 model year. It is the use of the controller area network (CAN) bus for diagnostics. The requirement (J2284) is for this bus system be connected to pins 6 and 14. Previously these pins were discretionary. Thus, some manufacturers that were using these two pins for their own purposes have had to start migrating to a new DLC configuration.

## Serial Data Circuits

As part of the legislation adopted by the federal government, the Clean Air Act of 1990 required PCMs to be able to communicate with any generic scan tool using a common protocol for vehicle communication. These requirements were to be implemented no later than the 1996 model year.

However, a manufacturer can use any communication method it prefers when using its own scan tool. For this reason, there may be more than one data link connected to the DLC on OBD-II systems.

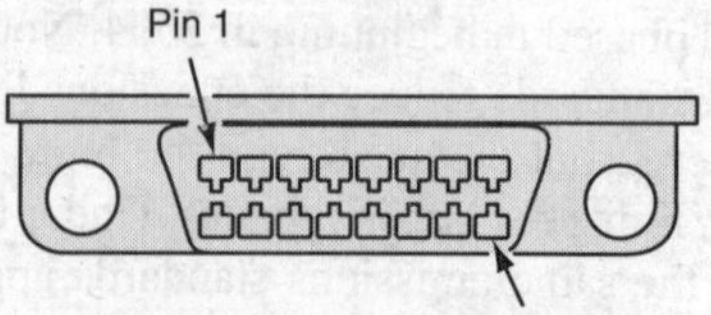


**Figure 2** *DLC terminal identification.*

| Cavity | General Assignment |
|---|---|
| 1 | Ignition control |
| 2 | BUS (+) SCP |
| 3 | Discretionary |
| 4 | chassis ground |
| 5 | Signal ground (SIG RTN) |
| 6 | Discretionary |
| 7 | K line of ISO 9141 |
| 8 | Discretionary |
| 9 | Discretionary |
| 10 | BUS (–) SCP |
| 11 | Discretionary |
| 12 | Discretionary |
| 13 | FEPS (flash EEPROM) |
| 14 | Discretionary |
| 15 | L line of ISO 9141 |
| 16 | Battery power |

**Figure 3** *J1962 16-terminal DLC.*

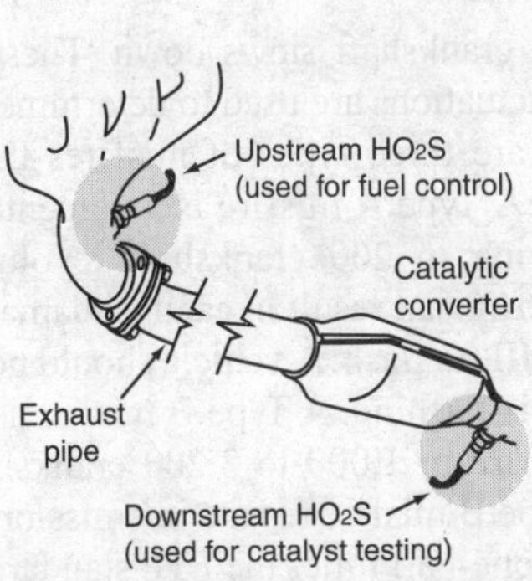


**Figure 4** *The efficiency of the catalytic converter is monitored by the use of upstream and downstream oxygen sensors.*

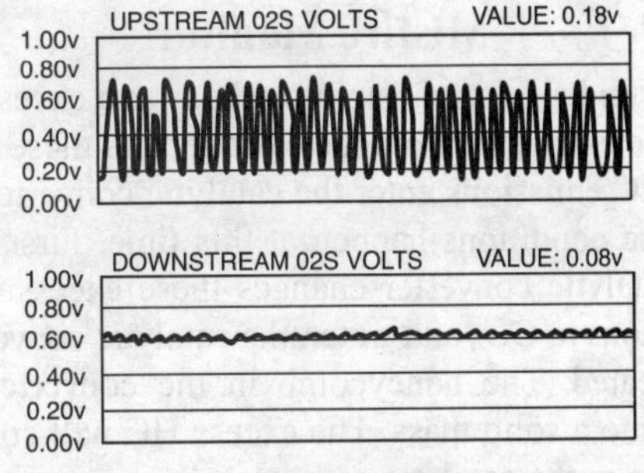


**Figure 5** *Normal upstream and downstream $O_2S$ activity with a good catalytic converter.*

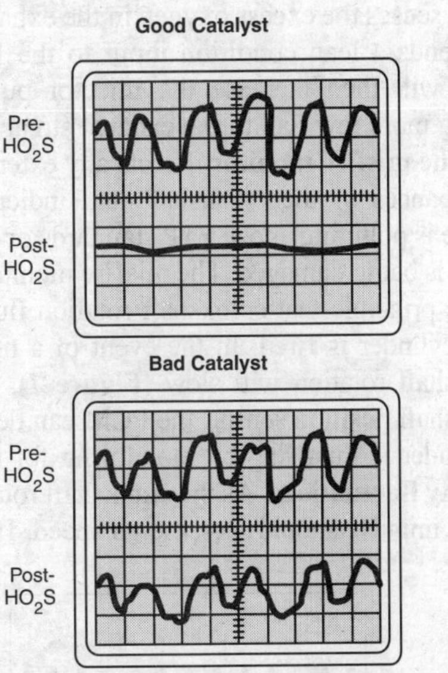


**Figure 6** *Comparison of $O_2S$ activity for a good and bad catalyst.*

## OBD-II Monitors

The OBD-II monitors discussed here provide a general description of each monitor. Monitor testing requirements and methods are discussed to provide an overall understanding of each monitor's operation. In following chapters, detailed monitor information for each system is discussed.

OBD-II monitors include:

1. Catalyst efficiency monitor
2. Engine misfire monitor
3. Fuel system monitor
4. Heated oxygen sensor ($HO_2S$) monitor
5. Evaporative system monitor
6. EGR monitor
7. Secondary air-injection monitor
8. Comprehensive component monitor
9. Chlorofluorocarbon (CFC) monitor

### Catalyst Efficiency Monitor

When a lean air-fuel ratio is present for an extended length of time, the oxygen content in the catalytic converter will reach maximum. When a rich air-fuel ratio is present, the oxygen will be depleted. When this occurs, the catalyst fails to convert the harmful emissions gases. The **catalyst monitor** detects the catalyst's ability to store and give off oxygen through the use of a **postcatalyst oxygen sensor (Figure 4)**. The postcatalyst oxygen sensor is used to measure the $O_2$ content in the exhaust after the catalyst to determine the converter's storage capacity for $O_2$. A good catalyst should have 95 percent HC conversion efficiency. A catalytic converter stores oxygen during lean engine operation and gives up this stored oxygen during rich operation to oxidize excessive HC. Catalytic converter efficiency is measured by monitoring the oxygen storage capacity of the converter during closed-loop operation.

> **You Should Know**
>
> *The postcatalyst $O_2S$ is also called the downstream $O_2S$ or catalyst monitor sensor.*

> **You Should Know**
>
> *Depletion of oxygen in the converter is called catalyst punch-through.*

The downstream $O_2S$ will show a relatively flat output voltage compared with the upstream $O_2S$ **(Figure 5)**. A high oxygen storage capacity indicates a good catalyst, while low oxygen storage capacity indicates a deteriorating catalyst. Voltage spikes that match the upstream $O_2S$ output **(Figure 6)** indicate a deteriorated catalyst. Peaks on the downstream $O_2S$ indicate the catalyst has lost its ability to store oxygen.

## Misfire Monitor

If a cylinder misfires, unburned gas and excess oxygen are exhausted from the cylinder and these excessive HC emissions enter the catalytic converter. Two adverse conditions happen at this time. First, when the catalytic converter changes these excessive HC emissions to $CO_2$ and water, the catalytic converter is overheated. The honeycomb in the converter may melt into a solid mass. The excess HC will continue to burn in the catalytic converter, increasing its temperature. If this action occurs, the converter is no longer efficient in reducing emissions. Second, the $HO_2S$ senses the excess oxygen in the exhaust stream and sends a lean condition input to the PCM. The PCM will then increase the injector pulse width, adding more raw fuel to the exhaust stream.

The **misfire monitor** detects any external torque disturbances of the crankshaft and indicates if the engine is polluting above FTP standards or if the converter is being damaged. The misfire monitor is based on the principle that crankshaft rotation fluctuates as each cylinder is fired. In the event of a misfire, the crankshaft rotation will slow **(Figure 7)**. Using the crankshaft position sensor, the PCM can determine if a cylinder is misfiring by monitoring for **rotational velocity fluctuations**. As the crankshaft rotates, every power impulse should increase its speed. If a misfire occurs, the crankshaft slows down. These rotational velocity fluctuations are used to determine misfire.

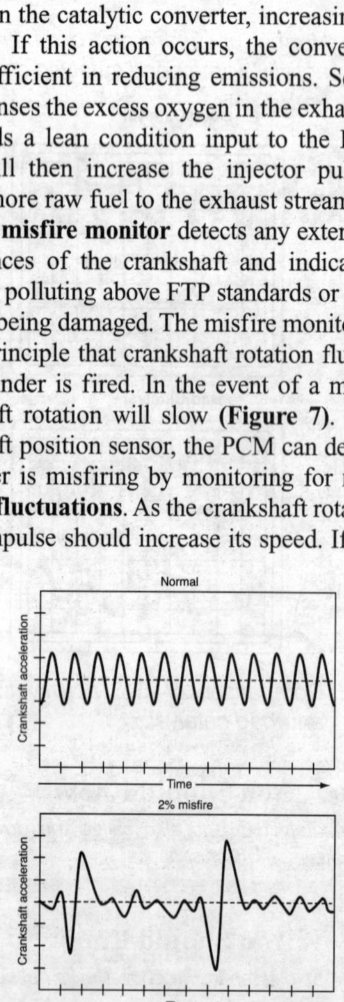


**Figure 7** *The crankshaft position sensor indicates to the PCM when a misfire occurs.*

There are three types of misfires that must be monitored. A Type A misfire is a percentage of misfire occurring in 200 crankshaft revolutions. This type of misfire can result in catalyst damage and will cause the MIL to flash. A vehicle should not be driven if the MIL is flashing. A Type B misfire is a percentage of misfire in 1,000 to 3,200 crankshaft revolutions. A Type B misfire results in emissions in excess of one and one-half times the FTP standard. With this type of misfire, the MIL is on but not flashing. A Type C misfire is identical to a Type B misfire except it indicates failure of the California I/M test.

### You Should Know

*The manufacturers predetermine the percentage of misfire for each test. The percentage of misfire for malfunction criteria varies based on engine rpm and load. Failure percentages vary from engine to engine.*

In addition, the system can detect which cylinder is misfiring and set individual DTCs for each cylinder. All systems must have a DTC for random or multiple cylinder misfires.

**Crank Sensor Learning.** Before the misfire monitor can run, the PCM software must learn and correct for mechanical inaccuracies in the crankshaft position tone wheel tooth spacing. Since a crankshaft tone wheel will not be manufactured with perfect spacing between teeth, the PCM will learn what the normal pattern of the sensor input is as compared to a perfect pattern. The difference that is learned then allows the PCM to correct for inaccuracies during manufacturing of the tone wheel. The sum of all the angles between crankshaft teeth equals 360 degrees. This constant means that a correction factor can be calculated for each misfire sample interval that makes all the angles between individual teeth equal. To prevent any fueling or combustion differences from affecting the correction factors, crankshaft sensor pattern learning is done during fuel cut-off decelerations. During this time, the crankshaft is still rotating but the engine is not producing any power pulses.

The misfire monitor is not active until the crank sensor corrections have been learned. In the event of battery disconnection or loss of keep alive memory (KAM), the correction factors are lost and they must be relearned by the PCM.

## Fuel System Monitor

The **fuel system monitor** determines if the running condition of the engine is too rich or too lean to be controlled and brought back to stoichiometric, resulting in excessive emissions (more than one and one-half times the FTP standard). The OBD-II regulations require that the fuel-delivery system be monitored continuously to comply with emissions standards. The regulations require monitoring of the long-term fuel trim limits.

As the engine is running, the PCM monitors $HO_2S$ activity to determine the quality of the combustion burn. If the $HO_2S$ indicates a lean air-fuel condition, the PCM will initiate a fuel trim program to add an additional amount of injector pulse width to richen the mixture until the $HO_2S$ indicates normal activity. If the $HO_2S$ indicates a rich air-fuel condition, the fuel trim program will reduce the injector pulse width to correct the reading. The monitor watches changes in the fuel trim to determine if the engine is running too rich or too lean. Most manufacturers use short-term and long-term fuel trim values. These values are displayed on the scan tool as a percentage of correction. Different cells are used based on engine rpm and MAP sensor inputs. If the long-term and short-term trim values reach and stay at their maximum limits for a period of time, a malfunction is indicated.

## $HO_2S$ Monitor

OBD-II regulations require the monitoring of the $HO_2S$. The PCM runs at least two tests on the $HO_2S$—one on the sensor operation and the other on the heater operation. For both monitors, upstream and downstream oxygen sensors are tested.

## EVAP Monitor

The EVAP system collects fuel vapor from the fuel tanks and then controls the flow of those vapors into the engine to be burned. Since a leak in this system as small as 0.020 in. (0.5 mm) in diameter can result in a yield of 1.35 grams of HC per mile (more than thirty times the allowable limits), OBD-II requires the EVAP system to be monitored for proper operation. OBD-II regulations have been modified to require systems to detect leaks of 0.020 in. (0.5 mm) in diameter or larger. Vacuum tubing and connections located between the intake manifold and the purge valve are excluded from these regulations. There are two types of EVAP systems that can be used on OBD-II vehicles. Each is monitored differently.

To monitor an **enhanced EVAP** system, the system must be pressurized or put under vacuum to detect a leak. Enhanced EVAP systems use either positive or negative pressure to detect leaks. The manufacturer is left to its discretion as to how the system will be tested but it must detect leaks of 0.020 in. (0.5 mm) in diameter or larger and be able to verify airflow through the system.

**Nonenhanced EVAP** systems are used if the vehicle is not equipped with a leak-detection system. Changes in short-term fuel trim shifts, movement in target IAC at idle, or idle speed change are used to monitor the system. The manufacturer must certify with the EPA that nonenhanced EVAP systems have an expected-use life of 100,000 miles.

> **You Should Know**
>
> *Nonenhanced EVAP systems are also referred to as stricter EVAP systems.*

Since the EPA is more in favor of leak detection than stricter EVAP, manufacturers wishing to install stricter EVAP may also be required to have leak-detection systems installed on a sampling of their vehicles to prove system integrity. For example, Chrysler minivans built in November 1995 (for the 1996 model year) have both leak-detection pumps and stricter EVAP.

The EVAP system monitor will not run unless specific fill requirements in the tank are met. As a general rule, the monitor will not run if the tank is filled more than 80 percent or less than 20 percent.

## EGR Monitor

OBD-II requires the EGR system be monitored for low and high flow rates that can result in excessive emissions levels. The EGR system is considered failed if an EGR system component fails or a change in the EGR flow rate results in the vehicle exceeding one and one-half times the FTP standard. The DTC must be set within two vehicle trips once the malfunction has been detected.

## AIR System Monitor

If a vehicle is equipped with an AIR system, OBD-II requires monitoring for the presence of airflow in the exhaust stream and functional monitoring of the pump and any switches or solenoids.

## Comprehensive Component Monitor

While the input signal of **comprehensive components** is constantly being monitored for electrical opens and shorts, the components are also tested for rationality. Comprehensive components are any components that may affect emissions. These are generally components that were tested under OBD-I.

One strategy for monitoring inputs involves checking certain inputs for electrical defects and out-of-range values by checking the input signals at the A/D converter. The comprehensive component monitor also checks inputs by performing rationality checks. During a rationality check, the monitor uses other sensor readings and calculations to determine if a sensor reading is proper for the present conditions. The following are examples of sensors that would be monitored by this system:

- Engine coolant temperature sensor
- Manifold absolute pressure sensor
- Throttle position sensor
- Mass airflow sensor
- Idle control
- Injector control

In addition, comprehensive components that are outputs are monitored for functionality along with electrical testing. When the PCM provides a voltage to an output component, it can verify that the command was carried out by monitoring specific input sensors for expected changes.

### Exponentially Weighted Moving Averaging

Some manufacturers use **exponentially weighted moving average (EWMA)** monitors. EWMA is a process of using statistical data to reduce the variability of data. EWMA is used to determine the decay of a component's function and is commonly used for EGR and catalyst monitors. Based on the EWMA, it is possible to determine if a failure is the result of aging or catastrophic failure of a component. EWMA uses a percentage of instantaneous and historical data to monitor a component's function. By filtering the inputs that may experience variability, this prevents normal voltage variances from causing false DTCs from being set.

A typical fast response average may use 90 percent of instantaneous input and 10 percent of historical data. This average can then be used to determine catastrophic failure. On the other hand, a slow moving average will use a lower percentage of instantaneous input and a larger percentage of historical data to determine the failing trend. Usually, an EWMA monitor uses the result from the last time it ran and a new average is determined every time the monitor runs.

## OBD-II Terms

There are specific OBD-II terms used to describe a test, a driving method, or monitor criteria. The following identifies some of these terms.

### Drive Cycle

A **drive cycle** is a specific driving method to verify a symptom or the repair for the symptom. A drive cycle is also a specific driving method to begin and complete a specific OBD-II monitor. A minimum drive cycle is from engine start-up and vehicle operation until after the PCM enters closed-loop mode. To complete a drive cycle, all five trip monitors must be completed. The catalyst monitor must then be completed after the other five monitors are completed in a trip. A steady throttle opening between 40 to 60 mph (64 to 96 kph) for 80 seconds is required to complete the catalyst efficiency monitor.

The following conditions must occur to complete all OBD-II monitors:

- The misfire, comprehensive component, and fuel monitors are checked continuously from engine warm-up and can complete at any time.
- The misfire monitor, on applications with fuel deceleration shutoff, requires a deceleration at closed throttle for 10 seconds following the acceleration to 55 mph at one-quarter to one-half throttle. Decelerations following acceleration must be performed at least twice consecutively to satisfy this misfire requirement.
- A transmission component functional verification in the comprehensive component monitor requires at least six complete stops in the normal city portion of the drive cycle.
- The EGR and secondary air-injection monitors require a series of idles and accelerations.
- The $HO_2S$ monitor requires a steady speed drive for approximately 1 minute at 30 to 40 mph.
- The secondary air-injection monitor requires almost 12 minutes of vehicle operating time from initial start-up.
- The catalyst efficiency monitor requires a steady speed drive for 5 minutes at 40 to 60 mph, followed by a normal city drive between 25 and 40 mph for 10 minutes.
- The EVAP monitor requires at least 3 minutes of the steady throttle part of the drive cycle (10 minutes) between 45 to 60 mph to test the EVAP system.
- The OBD-II readiness function is available on all scan tools to indicate the status of the once-per-trip monitors.

### Trip

A **trip** is a set of vehicle operating conditions that must be met for a specific monitor to run. All trips require a key cycle. The term *trip* is a difficult concept to define since the requirements for a trip vary depending on the test being run. The minimum requirement for a trip is that an ignition switch-off period must precede an OBD-II drive cycle. Basically, in order to have a trip, there must be an ignition-off period and then the engine must be started and the vehicle must be driven.

Vehicle tests vary in length and may be either a **once-per-trip monitor** or a **continuous monitor**. Once-per-trip monitors run once for every driving trip defined by an engine start/stop cycle. A continuous monitor is one that runs all the time to determine the readiness of the OBD system to set DTCs. For example, the misfire, comprehensive, and fuel system monitors are checked continually once enabling conditions are met. During a trip, these monitors can complete at any time. A series of engine idle and accelerations is required for the EGR monitor to complete. A steady speed of between 20 to 45 mph (32 to 72 kph) for 20 seconds after engine warm-up is required to complete the $HO_2S$ monitor **(Figure 8)**. The catalyst efficiency monitor is not required to define an OBD-II trip.

What constitutes a particular OBD-II monitor is important since it is possible that the vehicle may need to fail more than once in order to illuminate the MIL and record a DTC. Monitors can be either **one-trip faults** or **two-trip faults**. One-trip faults are monitors that illuminate the MIL after one failure is recorded. These are also called A monitors. Two-trip faults require two failures in a row before the MIL is illuminated. This allows the system to double-check itself to prevent the MIL from illuminating when a fault does not exist. These are also called B monitors.

There are times when a monitor may be stopped. A monitor is **pended** if the diagnostic routine has found fault in another monitor that will cause this monitor to fail. The monitor is stopped pending repair of the first fault. For example, if the MIL is illuminated due to an $O_2S$ fault, the catalyst monitor will not run since it requires signals from the $O_2S$. Running the monitor would cause inaccurate results so the monitor is not run pending the resolution of the related problem.

At times there may be other monitors running or existing faults that **conflict** with the operation of a monitor. A monitor conflict occurs when two monitors have met the enabling conditions to run; one of the monitors will stop running since the testing of the other monitor may cause false failures. A conflict can also occur in some monitors when the air-conditioner compressor is turned on. In this event, the system will not run the monitor.

In addition, a two-trip fault may not be allowed to mature by having the monitor's result **suspended** until the results from another monitor are received. A suspended monitor means it has recorded a failure when the monitor ran the first time and now the second run of the monitor is indicating a failure. The fault code maturing may be suspended until other components are run again to check their function since these components may be causing this monitor to fail.

## Warm-up Cycle

A **warm-up cycle** is defined by vehicle operation after an engine shutdown period. During the vehicle operation, the engine coolant temperature must rise at least 40°F (22°C) and the coolant temperature must reach at least 160°F (71°C). The warm-up cycle counts are used to erase the DTC from the PCM's memory. The number of warm-up cycles does not increase until the MIL is turned off by three good trips. Most DTCs are erased after forty warm-up cycles.

## Diagnostic Trouble Codes

Once a monitor has been run, the OBD-II software must determine whether the system passed or failed.

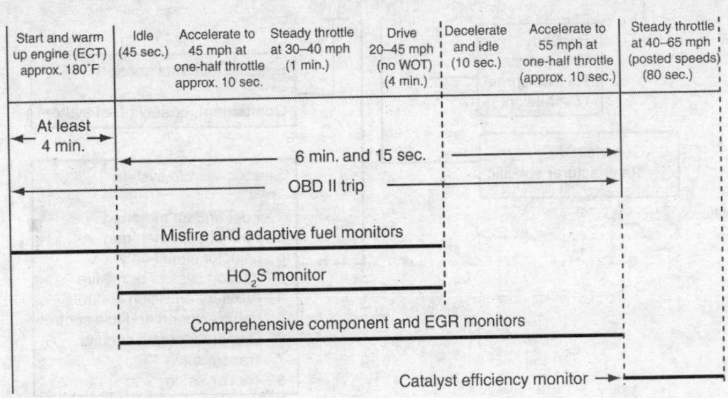


**Figure 8** *Drive cycle chart.*

If the system failed, the OBD-II software must then determine if it failed enough times to illuminate the MIL. If not, the PCM will store a maturing code. When this monitor is run again on the next trip, the results will once again indicate pass or fail. If the monitor fails, the code will mature, a DTC will be set, and the MIL will be illuminated. The MIL will be turned off after the system passes the monitor three consecutive times. After forty warm-up cycles with no other failures, the DTC is erased.

**You Should Know**

*A maturing code is also referred to as a one-trip failure or pending code.*

The SAE published the J2012 trouble code definition to describe industrywide standards for a uniform DTC format. This format allows a generic scan tool to access any OBD-II system. The OBD-II codes contain five characters: one letter and four numbers **(Figure 9)**. The letter identifies the function of the device that has generated the fault code:

- P = Power train
- C = Chassis
- B = Body
- U = Network or DLC

The first digit of the code indicates whether the DTC is generic or manufacturer-specific. A "0" here indicates the code is generic while a "1" indicates it is manufacturer-specific. The second digit indicates the vehicle system that generated the fault code:

- 1 = Fuel and air metering
- 2 = Injector circuit malfunctions
- 3 = Ignition system or misfire
- 4 = Auxiliary emissions controls
- 5 = Vehicle speed-control and idle-control systems
- 6 = Computer output circuit
- 7 = Transmission
- 8 = Transmission

The last two digits indicate the specific fault designation.

## Freeze-Frame

All OBD-II-monitored systems provide **freeze-frame** data on a vehicle's operating conditions when a maturing code is set. A freeze-frame is a snapshot of the vehicle's operating conditions when a fault is first detected. This data is retrievable by the use of a scan tool. The data is for the conditions at the time the system *first* failed, not necessarily at the time the MIL was illuminated. In the event of multiple monitor failures, the first failure to occur is stored in the freeze-frame. Since only one freeze-frame is stored, the monitor with the highest priority will overwrite a lower-priority freeze-frame. Information contained within the freeze-frame includes:

- DTC
- Engine rpm
- Engine load
- Fuel trim
- ECT
- MAP
- Operation mode (open- or closed-loop)
- Vehicle speed
- Freeze-frame priority

Freeze-frame priorities are based on the severity of the malfunction. For example, when a fault in the system is detected, the maturing code or DTC is prioritized for freeze-frame data. Maturing codes or DTCs with higher-priority overwrite lower-priority codes in a freeze-frame. The freeze-frame priority DTCs are as follows:

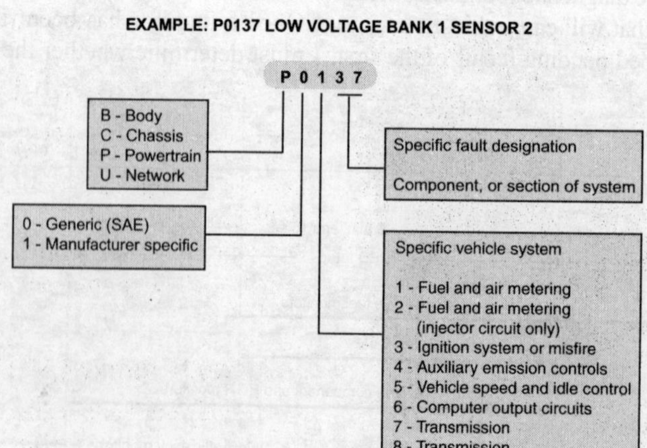


**Figure 9** *DTC structure as dictated by J2012.*

- Priority 0—Nonemissions-related trouble codes
- Priority 1—One-trip failure of a two-trip fault for nonfuel system and nonmisfire
- Priority 2—One-trip failure of a two-trip fault for fuel system or misfire
- Priority 3—Two-trip failure of a two-trip fault for nonfuel system and nonmisfire
- Priority 4—Two-trip failure of a two-trip fault for fuel system or misfire

## Similar Conditions Window

If a DTC is stored due to misfire or fuel-system-related problems, the PCM requires the engine be returned to a **similar conditions window (SCW)** to retest for a pass of the monitor. The SCW is a stored frame of data that records the operating conditions at the time the fault originated. After a misfire or fuel system fault is detected and the MIL is illuminated, the PCM must perform the tests once the SCW is entered. The monitor must pass three consecutive times within the SCW to turn off the MIL. Similar conditions are defined as being within 375 rpm and within 20 percent of the MAP reading at the time of the fault. In addition, the engine must be in the same fuel-control mode and close to the same temperature.

> **You Should Know**
>
> *Although the priority levels are the same, some manufacturers use numbers to distinguish the priority while others use letters.*

> **You Should Know**
>
> *It is important to note that the vehicle does not need to be driven in the SCW in order to mature a code. However, the vehicle must be in the SCW and pass to increment the good trip counter for turning off the MIL.*

## CARB Readiness Indicator

The **CARB readiness indicator** identifies whether all of the once-per-trip monitors have been run. The CARB readiness indicator is used when emissions I/M test is performed. EPA regulations mandate that certain cities perform these tests before a license registration is issued. One part of the regulations states that in order to pass the I/M test, the vehicle must have had the entire CARB readiness indicator monitors run at least once. If any monitor fails, the MIL is illuminated and the vehicle cannot be registered until it is repaired. The readiness indicator only flags once-per-trip monitors. Some cities use the CARB readiness indicator in lieu of sampling emissions from the tail pipe.

> **You Should Know**
>
> *The CARB readiness indicator indicates that the monitor ran, not necessarily that it passed the test. The emissions inspector will check to see if the monitors ran and if the MIL is on.*

## OBD-II Scan Tool

The SAE document J1978 describes the minimum requirements for an OBD-II scan tool. This document covers the required capabilities of and conformance criteria for OBD-II scan tools. Tool manufacturers are allowed to add additional capabilities if they desire. The following are basic OBD-II requirements:

- Automatic determination of the communication interface used
- Automatic determination and display of inspection and maintenance readiness information
- Display of the emissions-related DTC, current data, freeze-frame data, and oxygen sensor data
- Ability to perform the Expanded Diagnostic Protocol function described in J2205

## Review Questions

1. Describe an OBD-II warm-up cycle.
2. The downstream $HO_2S$ monitors __________ efficiency.
3. All OBD-II-monitored systems provide __________ __________ data on the vehicle's operating conditions when a maturing code is set.
4. Technician A says that in an OBD-II system, the PCM illuminates the MIL if a defect causes emissions levels to exceed two and one-half times the emissions standards for that model year. Technician B says that if a misfire condition threatens to cause engine or catalyst damage, the PCM flashes the MIL. Who is correct?
   A. Technician A only
   B. Technician B only
   C. Both Technician A and Technician B
   D. Neither Technician A nor Technician B
5. Technician A says that there is no leak-detection pump in an enhanced EVAP system. Technician

B says that in a nonenhanced EVAP system there may be a leak-detection pump. Who is correct?

A. Technician A only

B. Technician B only

C. Both Technician A and Technician B

D. Neither Technician A nor Technician B

6. Explain trip and drive cycle in an OBD-II system.
7. A ______________ is a specific driving method to verify a symptom or the repair for the symptom and to begin and complete a specific OBD-II monitor.
8. Technician A says that a maturing fault code is a one-trip failure of a two-trip monitor. Technician B says that a freeze-frame is stored when a maturing fault is detected. Who is correct?

   A. Technician A only

   B. Technician B only

   C. Both Technician A and Technician B

   D. Neither Technician A nor Technician B

9. A technician is using a digital storage oscilloscope (DSO) to diagnose an OBD-II system. The downstream $HO_2S$ has the same voltage waveform as the upstream $HO_2S$. Technician A says that the catalytic converter is defective. Technician B says that the air-fuel ratio is too rich. Who is correct?

   A. Technician A only

   B. Technician B only

   C. Both Technician A and Technician B

   D. Neither Technician A nor Technician B

10. Technician A says that the purpose of the warm-up cycle is to turn off the MIL. Technician B says that the purpose of a trip is to erase the fault code. Who is correct?

    A. Technician A only

    B. Technician B only

    C. Both Technician A and Technician B

    D. Neither Technician A nor Technician B

CHAPTER

# 30 DaimlerChrysler's OBD-II

## Introduction

The Chrysler group of DaimlerChrysler Motors Corporation uses three different PCMs: Single Board Engine Controller (SBEC) for passenger cars, Jeep/Truck Engine controller (JTEC) for Jeep and Dodge truck vehicles, and Next Generation Controller (NGC) starting in the 2002 model year. The NGC replaces the SBEC and JTEC as of the 2005 model year. All three OBD-II systems are similar. This chapter discusses the SBEC and JTEC systems. The NGC and how it differs from the SBEC and JTEC systems is discussed in Chapter 31.

## Overview

The Chrysler group of DaimlerChrysler Motors Corporation uses a **task manager** software program within the PCM to organize and prioritize the diagnostic procedures and the protocol for recording and displaying the results. During the course of the vehicle operation, many diagnostic steps must be performed. Most of these steps must be performed under separate operating conditions to be accurate. The task manager is responsible for determining when these diagnostic tests will run. The following is a list of the responsibilities of the task manager:

- Trip indicator
- Test sequence
- Readiness indicator
- DTC identification, maturation, and erasure
- Freeze-frame data storage and erasure
- Freeze-frame priority
- MIL illumination
- Test status

In summary, the task manager determines if the conditions are appropriate for a test to be run, knows the definition of a trip for each test, and records the results of each test when it is completed.

Trips are criteria used to turn the MIL off. The scan tool displays the number of good trips under a trip counter. Chrysler has defined four good trip counters:

- Global good trip
- Fuel system good trip
- Misfire good trip
- Alternate good trip

Basically, to increment a **global good trip** all monitors that run once per trip must have run and passed their tests. Usually this means the $HO_2S$ and catalyst efficiency monitors must run and pass the tests. A **fuel system good trip** is used to turn the MIL off after a fuel system DTC has been set. It requires three good trips. The fuel system good trip counter increments a good trip if the following conditions are met and the test passes:

- Engine in closed-loop mode
- Operating in similar conditions window
- Short-term trim multiplied by long-term trim is less than threshold
- Less than threshold for predetermined time

A **misfire good trip** is used to turn the MIL off after a misfire DTC has been set. It also requires three good trips. The misfire good trip counter is incremented if the vehicle is operating in the similar conditions window and there are 1,000 engine revolutions with no misfire.

An **alternate good trip** is used in place of a global good trip for comprehensive components and major monitors. Since a global good trip could not occur if there were a fault with a comprehensive component (because the test is pended), the MIL could never be extinguished. To prevent this from happening, an alternate good trip is counted if there are 2 minutes of engine run time without any other faults occurring. The task manager counts an alternate good trip for a major monitor when the monitor runs and passes its test. Only the major monitor that failed needs to pass to count an alternate good trip.

If a fault is detected, the maturing code or DTC is prioritized. Maturing codes or DTCs with higher priority overwrite lower-priority codes. This is true concerning freeze-frame data only. Chrysler systems can store several different DTCs; however, pre-2000 model year OBD-II systems only stored one freeze-frame. The MIL is illuminated for only Priority 3 and Priority 4 DTCs. A freeze-frame is stored when a fault is *first* detected. If the fault matures and turns on the MIL, only DTC priority changes, and the other data remains the same.

## Comprehensive Components Monitor

While the input signal of comprehensive components is constantly being monitored for electrical opens and shorts, the components are also tested for rationality. In addition, comprehensive components that are outputs are monitored for functionality along with electrical testing. When the PCM provides a voltage to an output component, it can verify that the command was carried out by monitoring specific input sensors for expected changes. On Chrysler-built vehicles, these are one-trip faults that carry a Priority 3 DTC, thus the MIL is illuminated after the first failure. The only exception is on JTEC vehicles where the EVAP purge solenoid is a two-trip fault. On Mitsubishi-built vehicles, the comprehensive components are all two-trip faults. The first time a failure is detected, a Priority 1 freeze-frame is stored. At this time, the MIL is off and no DTC is stored. If the component fails on the next test, the freeze-frame is updated to a Priority 3, the MIL is illuminated, and a DTC is stored.

## Fuel System Monitor

The fuel system monitor is a two-trip monitor. The tests must fail twice in succession before the MIL is illuminated. Once the MIL is on, the fuel system monitor tests must pass three times within a similar conditions window.

The PCM monitors the short-term and long-term adaptive memories on a constant basis. If a lean air-fuel condition exists, then both adaptive memory values will increase until normal $HO_2S$ inputs are received again. If during this lean condition the combined short-term and long-term fuel-correction values exceed a predetermined threshold (which varies according to engine), the PCM sets a fuel system failure for that trip and a freeze-frame is entered.

If a rich air-fuel condition exists, the PCM combines short-term and long-term fuel correction values. If the result is less than a predetermined value, the PCM checks the purge-free cells. **Purge-free cells** are values placed in adaptive memory cells when the EVAP purge solenoid is off. At least two purge-free cells are used; one corresponds to an adaptive memory cell at idle in drive, the other(s) to a cell at off/idle or cruise. If all purge-free cells are less than a certain percentage, the PCM sets a fuel system rich fault for that trip and a freeze-frame is entered. If the purge-free values are above the threshold, the fault code maturing is suspended until the EVAP system is tested.

## Heated Oxygen Sensor Response Monitor

For the PCM to detect a shift in the air-fuel ratio, the voltage input from the heated oxygen sensor ($HO_2S$) must change beyond its threshold. These thresholds are the lean and rich **switch points (Figure 1)**. As the voltage crosses the switch points, the PCM alters the air-fuel mixture accordingly.

The $HO_2S$ is tested for **big slope** and **half-cycle counts.** Big slope is the rate of change that an $O_2S$ experiences. It is determined by voltage change over time. The big slope is determined by the formula:

$$\text{Slope} = \Delta \text{ voltage}/\Delta \text{ time}$$

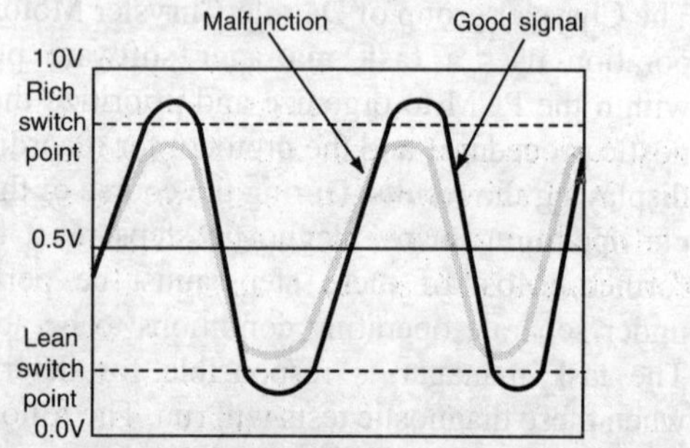


**Figure 1** *Heated oxygen sensor switch points.*

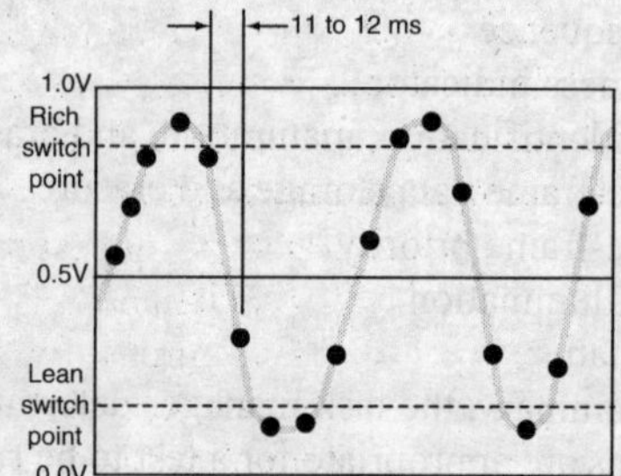


**Figure 2** *Big slope is the change in voltage over time.*

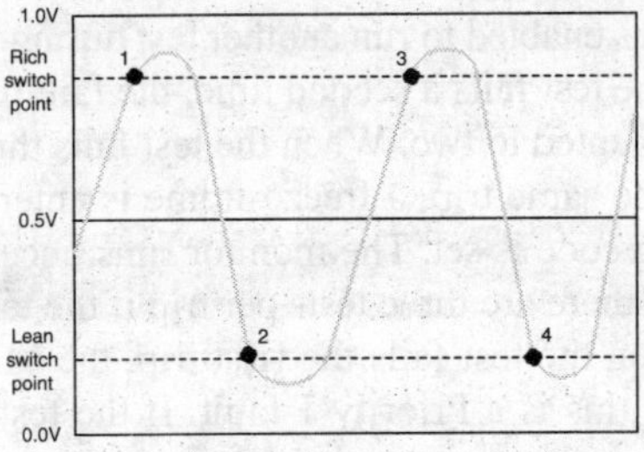


**Figure 3** *The half-cycle counter counts the number of times the oxygen sensor voltage crosses the switch points.*

The PCM checks the voltage level of the $HO_2S$ every 11 to 12 milliseconds **(Figure 2).** If the slope is great enough, the counter is incremented by one. If the count reaches a predetermined value within a specified length of time, the monitor passes.

The half-cycle counts indicate the frequency of the $O_2S$. The PCM monitors the number of times the $HO_2S$ signal goes beyond the switch points **(Figure 3)**. Each time the voltage surpasses the threshold, a counter is incre-mented by one. If it counts a sufficient number of half cycles within a defined time frame, the monitor passes its test.

### You Should Know

*Both the big slope and half-cycle counters do not have to pass for the monitor to pass its test. If either of the two tests passes, the monitor passes. The $HO_2S$ monitor is a two-trip monitor that is tested once per trip.*

## Oxygen Sensor Heater Monitor

In order for the $HO_2S$ to operate properly, it is heated to approximately 572°F to 662°F (300°C to 350°C). OBD-II requirements mandate the monitoring of the oxygen sensor's heater circuit. Most of Chrysler's oxygen sensors contain heater elements that are positive temperature coefficient (PTC) devices. The PTCs in the sensor are fed direct battery voltage through the automatic shutdown (ASD) relay. As the PTCs receive voltage, their internal structure is changed and the temperature increases.

Chrysler does not check the heater element directly. Instead, the $O_2S$ is heated while its bias voltage is monitored. Basically, the resistance in the $O_2S$ output circuit is tested to determine heater operation. The internal circuit of the $O_2S$ itself is a negative temperature coefficient (NTC) device. At room temperature, it has about 4.5M ohms of resistance, but as it warms the resistance drops to about 100 ohms. The PCM sends a 5-volt bias voltage through the $O_2S$ to chassis ground to monitor this circuit. The voltage value received by the PCM during the pulsing of the 5 volts will depend on the voltage drop over the internal resistance of the sensor, which depends on its temperature.

On Chrysler passenger cars that use the SBEC PCM, the $O_2S$ heater monitor is run after the vehicle has been driven and then shut off. After shutoff, the $O_2S$ is still warm enough to produce a voltage. The voltage will be low due to high $O_2$ content in the exhaust pipe. A 5-volt signal is pulsed every 1.6 seconds and is on for 35 milliseconds each time. As the sensor cools, the internal resistance increases and the pulsed bias voltage begins to increase **(Figure 4)**. The PCM uses the difference between high-bias voltage and low-bias voltage to determine if the sensor was heated sufficiently during the drive cycle. The difference must be less than 1.57 volts.

Next, the PCM checks to see if the $HO_2S$ is cooled enough to allow the monitor to run. This is done in two steps. First, the PCM monitors the low-bias voltage. As the $HO_2S$ cools, the bias voltage increases since the sensor's internal resistance increases. The low-bias level must increase more than 0.49 volt. Once low-bias voltage exceeds this threshold, the PCM enters the second step. During

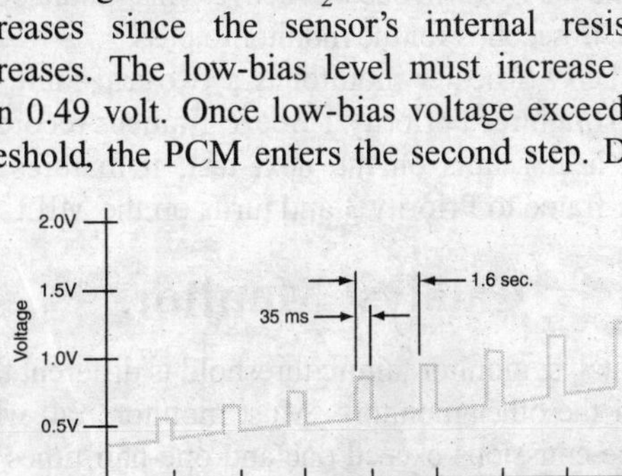


**Figure 4** *Checking to see if the $HO_2S$ is cooled enough by measuring the difference between high-bias voltage and low-bias voltage.*

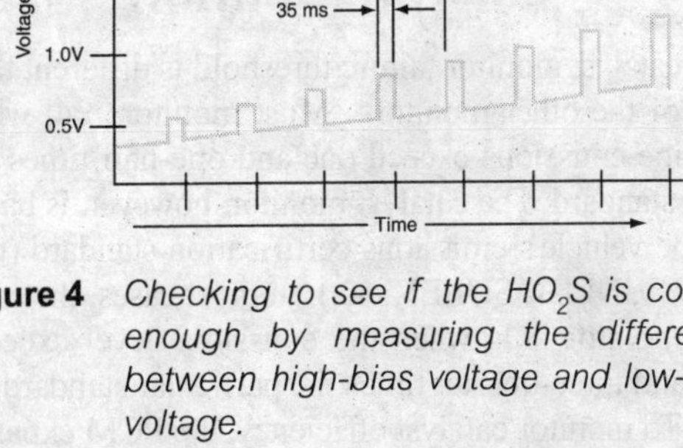


**Figure 5** *The PCM turns on the $O_2S$ heater and monitors the difference between pulses to see if the low-bias voltage is dropping.*

this step, the PCM monitors the difference between high-bias voltage and low-bias voltage as the circuit is pulsed. Sufficient sensor cooldown is indicated when the difference between these voltages is greater than the parameters.

If these two tests pass, the $O_2S$ is cool enough to run the heater monitor. The PCM energizes the ASD relay, turning on the heater element. The heater element begins to increase the sensor's temperature. The PCM will continue to bias the 5-volt signal to the sensor. The PCM takes a low-bias voltage reading before and after each pulse **(Figure 5)**. For the monitor to pass, the voltage difference of the low-bias voltage must be greater than 0.16 volt within 32 pulses.

Jeep and Dodge trucks that use the JTEC-type PCM run the heater monitor on initial engine start-up. The engine coolant temperature must be below 140°F (60°C) in order for the monitor to run. Also, the engine temperature and the outside ambient temperature must be within 20°F (11°C) of each other for the monitor to run. After the engine is started, 5 volts is applied to the $O_2S$ circuit. As the $O_2S$ heats, this voltage will drop to zero and the $O_2S$ will become active. This should occur within 45 seconds for the monitor to pass.

The $O_2S$ heater monitor is a two-trip fault. On the first failure, a priority 1 freeze-frame is recorded. If the heater fails on the next test, it matures the freeze-frame to Priority 3 and turns on the MIL.

## Catalyst Monitor

The catalyst monitor failure threshold is different than that of the other monitors. Most monitors fail when tailpipe emissions exceed one and one-half times the FTP standard. The catalyst monitor, however, is based on the vehicle's emissions certification standard (that is, LEV, NLEV, SULEV, etc.). In most cases, the catalyst monitor fails when the emissions level exceeds one and three-quarter times the particular standard.

To monitor catalyst efficiency, the PCM expands the rich and lean switch points of the $HO_2S$. The expanded switch point causes the air-fuel mixture to be richer and leaner than normal. This will overburden the catalytic converter. Once the test is initiated, the PCM counts the number of upstream and downstream $HO_2S$ switches. The **switch ratio** is calculated by dividing the number of downstream switches by the number of upstream switches. If at any point during the test the switch ratio reaches 90 percent for vehicles equipped with automatic transmissions or 70 percent for vehicles equipped with manual transmissions, a failure counter is incremented by one. The monitor is enabled to run another test during the same trip. If the test fails a second time, the failure counter is incremented to two. When the test fails three times within the same trip, a freeze-frame is entered and a maturing code is set. The monitor runs once per trip, however there are three tests per trip if the tests fail.

When the test fails the first trip, the MIL is not on since this is a Priority 1 fault. If the test fails on the second consecutive trip, the freeze-frame priority is matured to 3 and the MIL is illuminated.

In order to increment a good trip after MIL illumination, the downstream $O_2S$ switch rate must be less than 80 percent of the upstream sensor for vehicles equipped with automatic transmissions and 60 percent for vehicles equipped with manual transmissions.

Later-year SBEC and JTEC controllers use an EWMA method for determining catalyst failures. The EWMA catalyst monitor differs from earlier catalyst monitoring technology in that it requires six separate test sequences to run before a failure is recorded. The methods for running the tests differ between SBEC and JTEC vehicles. On SBEC vehicles, all six tests are conducted within the same trip. On JTEC vehicles, testing is suspended until the next trip if the PCM detects three failures. Once enabling conditions are met during the next trip, the PCM resumes testing. If three additional failures are detected, the PCM stores a DTC and illuminates the MIL. The JTEC uses this strategy because if the underfloor catalytic converters become wet enough during a rainstorm, a cooling effect and a false failure may result.

This monitor uses a **slow-moving average** and a **fast-moving average** along with **instantaneous switching frequency**. Slow-moving average is used to detect a naturally aging high-mileage catalyst. The fast-moving average is used to detect a major catalyst failure, such as a severe misfire that has melted the catalyst substrate. The instantaneous switching frequency is the rate at which both the upstream and downstream oxygen sensors cross the rich to lean switching points.

The slow-moving average is determined by adding 90 percent of the "old" slow-moving average and 10 percent of the "new" instantaneous switching frequency. The fast-moving average is determined by adding 75 percent of the "old" fast-moving average and 25 percent of the "new" instantaneous switching frequency.

A suspicious counter is incremented when the slow-moving average is more than 20 percent higher than the instantaneous switching frequency. In SBEC vehicles, when the suspicious counter increments six times and the slow-or fast-moving average is above a calibrated frequency, a DTC sets and the MIL illuminates. On JTEC vehicles, the third suspicious count sets a one-trip failure. Counting resumes on the next trip. After three more suspicious counts, a DTC is declared and the MIL is illuminated.

When the catalyst monitor is about to fail, the fault code maturing is suspended and the $O_2S$ response monitor is retested during the catalyst monitor. This is done to verify that a lazy $O_2S$ is not corrupting the results of the catalyst monitor.

## Misfire Monitor

Chrysler uses a crankshaft speed fluctuation method to monitor for misfire. The crankshaft position sensor can detect slight variations in engine speed that result from cylinder misfire. The monitor must take into account such variables as component wear, sensor fatigue, and machining tolerances. In a perfect design, the crankshaft signal produces a series of 0- to 5-volt pulses to identify the position of two cylinders. In a six-cylinder engine, there are three groups of slots. Each group is 120 degrees apart and each slot within a group is 20 degrees apart **(Figure 6)**. This perfect pattern is programmed into the PCM. Also, the PCM is programmed to know it will never see a perfect design. In order for the misfire monitor to run, it must first learn the **adaptive numerator**. The adaptive numerator is the learned value of the crankshaft tone wheel. To learn this, the PCM uses the first signal set as a point of reference. It then measures where the second set of signals is compared with where engineering data has determined it should be **(Figure 7)**. The PCM calculates the adaptive numerator during engine full fuel cutoff. It may take several attempts before the adaptive numerator is learned.

The PCM also checks the machining tolerances for each group of slots. By monitoring the speed of the crankshaft from the first slot to the last slot in a group, the PCM can calculate engine rpm. The variance between groups of crankshaft tone wheel slots is known as the rpm error **(Figure 8)**. In order for the PCM to run the misfire monitor, the rpm error must be less than approximately 5 percent.

To detect misfire, the PCM monitors the crankshaft speed. If a misfire occurs, the PCM will detect a decrease in crankshaft speed and count the misfire. Every 1,000-revolution window contains five 200-revolution windows **(Figure 9)**. The PCM counts the number of misfires in every 200 revolutions of the crankshaft. The value of the 200-revolution

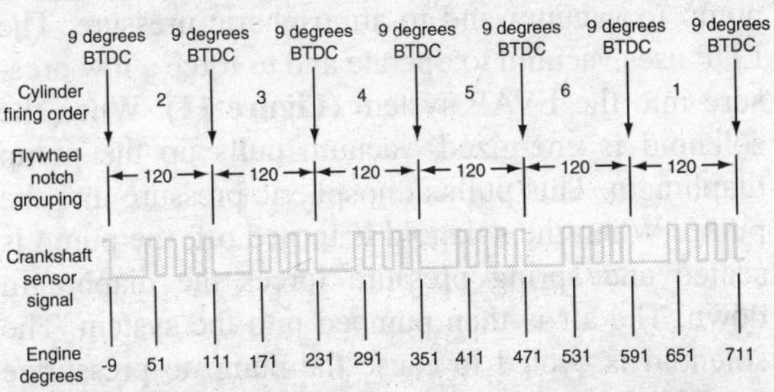


**Figure 6** *The PCM monitors the crankshaft position sensor to determine the position of the pistons and rotational speed.*

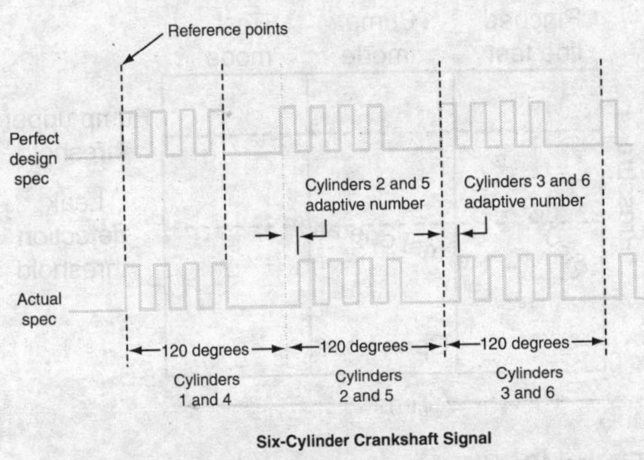


**Figure 7** *The adaptive numerator is the variance between a perfect crankshaft position sensor signal and the actual signal.*

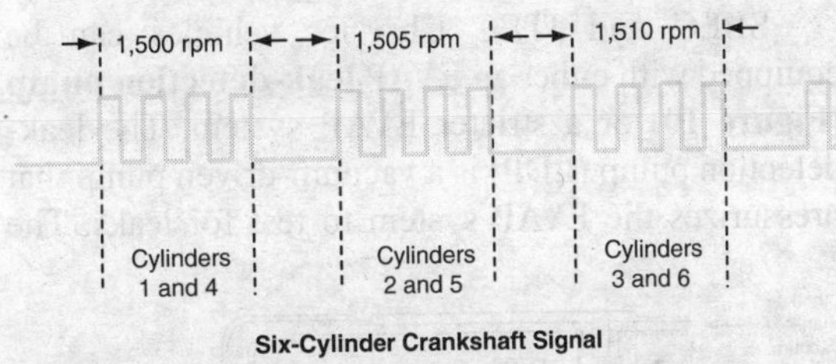


**Figure 8** *The rpm error is the amount of speed fluctuations within a group of crankshaft tone wheel slots.*

| 200 Revolutions | 400 Revolutions | 600 Revolutions | 800 Revolutions | 1,000 Revolutions | |
|---|---|---|---|---|---|
| 5 | 5 | 15 | 20 | 25 | 1,000-revolution counter |
| 5 | 0 | 10 | 5 | 10 | 200-revolution counter |

**Figure 9** *The 1,000-revolution window contains five 200-revolution windows. The PCM counts the misfires for each 200-revolution window and carries that value over for the 1,000-revolution window.*

window is carried over for the 1,000-revolution window. If during five 200 counters (1,000 revolutions) the misfire percentage exceeds the threshold, a maturing fault is set and a freeze-frame is entered. The data in the freeze- frame corresponds to the last 200 revolutions of the 1,000-revolution period. Also, a similar conditions window is recorded. A second failure on the next trip will mature the code, set a DTC, and illuminate the MIL. The PCM will attempt to identify which cylinder is misfiring at the time of the fault.

Since the 1,000-revolution counters are two-trip monitors, the MIL is not illuminated on the first trip even if the misfire continues. However, if the misfire exceeds the malfunction percentage in any 200-revolution period, this can result in catalyst damage and the MIL will be flashed. The Type A misfire is a one-trip monitor. While the MIL is flashing, the PCM converts to an open-loop fuel strategy to prevent additional fuel flow to the cylinders. In order to extinguish the MIL, the test must pass under the similar conditions window three consecutive times.

## EVAP Monitor

Emissions tests have determined that EVAP system leaks of 0.020 in. (0.5mm) in diameter can result in the release of about 1.35 grams of HC per mile. This is thirty times the allowable exhaust emissions standard. OBD-II standards have dictated that the EVAP system be tested for leaks of 0.02 in. (0.5mm) in diameter or larger.

SBEC and JTEC Chrysler vehicles can be equipped with either an EVAP **leak-detection pump (Figure 10)** or a stricter EVAP system. The leak-detection pump (LDP) is a vacuum-driven pump that pressurizes the EVAP system to test for leaks. The

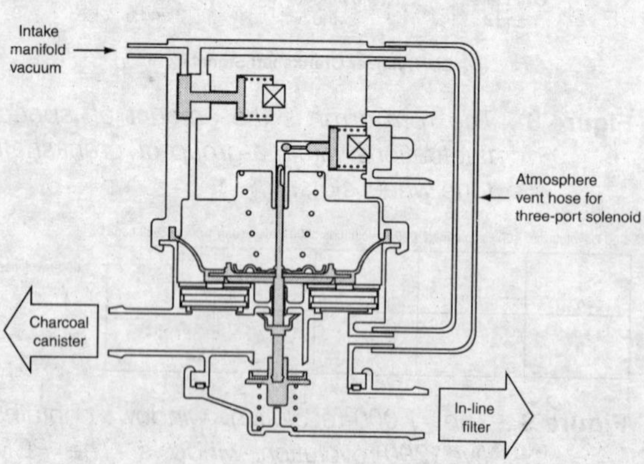


**Figure 10** *The leak-detection pump pressurizes the EVAP system to test for leaks.*

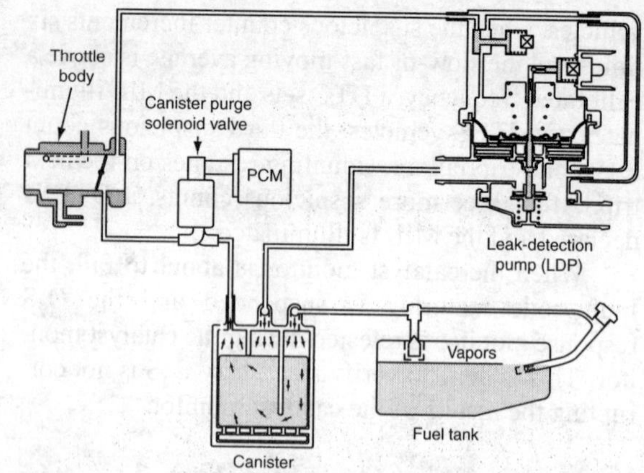


**Figure 11** *Leak-detection pump system.*

LDP performs two functions. First, it pressurizes the EVAP system for testing. Second, it seals the charcoal canister. The leak-detection portion of the EVAP monitor is run immediately after a cold start if the ambient temperature is within predetermined parameters. The PCM uses the LDP solenoid to operate the system. The solenoid does this by connecting the pump to vacuum and to atmospheric pressure. The LDP uses vacuum to operate and to force a low pressure into the EVAP system **(Figure 11)**. When the solenoid is energized, vacuum pulls up the pump diaphragm. This pulls atmospheric pressure into the pump. When the solenoid is turned off, the pump is sealed and spring pressure forces the diaphragm down. The air is then pumped into the system. The solenoid is cycled to cause the pump to pressurize the EVAP system to about 7.5 in. of water column pressure. If no leaks are present in the system, pressure will equalize and the pump stops.

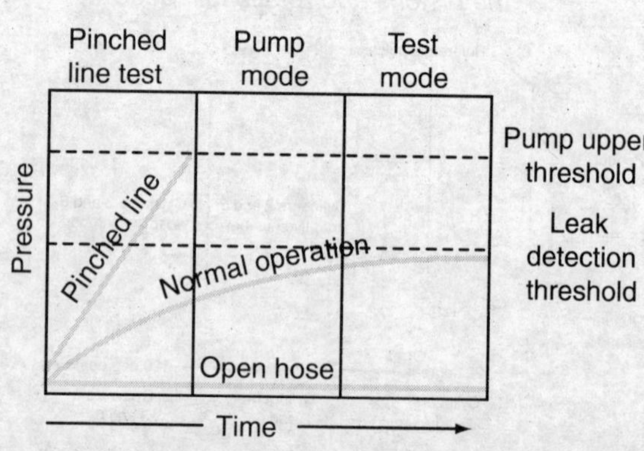


**Figure 12** *The rate of pressure buildup determines if there is a pinched line, a leak, or normal operation.*

The first portion of the leak-detection section of the monitor is a test to determine if a line is pinched. This test is done by monitoring the pump rate. The PCM monitors the pump rate through the use of a reed switch. If the pump rate falls to zero too soon, the PCM determines there is insufficient space to pressurize **(Figure 12)**. This indicates the possibility of a pinched line. Before the PCM can store a one-trip failure for the pinched line, the LDP monitor test and the purge flow test must both be run.

Next, the PCM checks the system for leaks. During this test, the LDP is controlled in two modes: **pump mode** and **test mode**. In the pump mode the pump's on and off periods are cycled at a fixed rate to pressurize the system to 7.5 in. of water. During the pump mode, the PCM can determine the amount of time required to pressurize the system based on fuel-level inputs. If it takes longer to pressurize the system than determined, the PCM enters a low pressure check mode. The **pressure check mode** means the PCM will loop back and run in the pump mode again.

With the system pressurized, the PCM switches the pump to test mode to determine if there are any leaks. In test mode, the on period only is cycled at a fixed rate. The pump is operated until the system is pressurized and the pump stalls. If a leak larger in diameter than 0.02 in. (0.5mm) is in the system, the pump rate falls to a rate proportional to the size of the leak. When the PCM determines that the pump rate has revealed the presence of a leak, a one-trip failure occurs and a freeze-frame is entered. If no leaks are detected, the purge solenoid is activated to depressurize the system.

If a vehicle is not equipped with a LDP, it is certified by the stricter EVAP standards. This EVAP monitor uses two stages to test the system. Stage 1 is a **passive test**. The passive test monitors the system or components during normal operation. The PCM compares adaptive memory values between purge and purge-free cells. The PCM uses these values to determine the amount of fuel vapors entering the system. In order to pass this stage of the test, the difference between purge and purge-free cells must be greater than a predetermined value. This would indicate that vapors are entering the combustion chamber, so the EVAP system must be intact. If this stage fails, the monitor progresses to Stage 2.

During Stage 2, the PCM deenergizes the purge solenoid and then waits until engine rpm, short-term adaptive fuel trim, and idle air control have all stabilized. When this is accomplished, the PCM increments up the purge solenoid duty cycle about 6 percent every eight engine revolutions. If during this test any one of the following three conditions occurs prior to the purge cycle reaching 100 percent, the EVAP test passes. The three conditions are:

- rpm rises by a predetermined value
- Short-term adaptive value drops by a predetermined value
- Idle air control closes by a predetermined value

If none of these conditions occur, the test fails and the failure counter is incremented by one. The PCM runs the test three times during a trip. If the failure counter increments to three failures within the trip, the monitor fails and a freeze-frame is entered.

## EGR Monitor

During this test, the PCM activates the solenoid, allowing vacuum to be applied to the transducer. With negative back pressure, the manifold vacuum is vented; however, with positive pressure, the transducer modulates the vacuum supplied to the EGR valve. While the monitor is running, the EGR solenoid is turned off and the PCM monitors the short-term adaptive memory values. If the PCM does not detect a change in short-term memory within a given percentage range, the monitor will fail and a freeze-frame will be stored.

## Review Questions

1. Describe the half-cycle counter.
2. Technician A says that OBD-II systems have two heated oxygen sensors downstream from the catalytic converter to monitor converter operation. Technician B says that the PCM monitors exhaust temperature to determine ignition misfire. Who is correct?
   A. Technician A only
   B. Technician B only
   C. Both Technician A and Technician B
   D. Neither Technician A nor Technician B
3. What is the purpose of the task manager?
4. Technician A says that if a lean condition exists and the combined short-term and long-term fuel correction values exceed a predetermined

threshold, the PCM sets a fuel system lean failure for that trip and a freeze-frame is entered. Technician B says that if a rich condition exists, the PCM combines short-term and long-term fuel correction values and if the result is less than a predetermined value, the PCM sets a fuel system rich failure for that trip and a freeze-frame is entered. Who is correct?

A. Technician A only
B. Technician B only
C. Both Technician A and Technician B
D. Neither Technician A nor Technician B

5. Describe how the $O_2S$ heater monitor runs on a vehicle equipped with an SBEC system.
6. Technician A says that to detect misfire, the PCM monitors the crankshaft speed. Technician B says that the 1,000-revolution counters are two-trip monitors. Who is correct?
   A. Technician A only
   B. Technician B only
   C. Both Technician A and Technician B
   D. Neither Technician A nor Technician B
7. Explain the term *adaptive numerator*.
8. Technician A says that if the catalytic converter is not reducing emissions properly, the voltage frequency increases on the downstream $HO_2S$. Technician B says that if a fault occurs in the catalyst monitor system on three drive cycles, the MIL is illuminated. Who is correct?
   A. Technician A only
   B. Technician B only
   C. Both Technician A and Technician B
   D. Neither Technician A nor Technician B
9. What constitutes an alternate good trip?
10. Technician A says that for the heated $HO_2S$ monitor to fail, both big slope and half-cycle counters must be below the threshold. Technician B says that big slope is the number of times the $O_2S$ voltage crosses over the upper and lower switch points. Who is correct?
    A. Technician A only
    B. Technician B only
    C. Both Technician A and Technician B
    D. Neither Technician A nor Technician B
11. Explain the operation of the EWMA catalyst monitor.
12. Technician A says that if the LDP stalls too early, a pinched line may be the cause. Technician B says that with the system pressurized, the PCM switches the pump to test mode to determine if there are any leaks. Who is correct?
    A. Technician A only
    B. Technician B only
    C. Both Technician A and Technician B
    D. Neither Technician A nor Technician B

CHAPTER

# 31 DaimlerChrysler's OBD-II with Next- Generation Controller

## Introduction

Beginning in the 2002 model year, Chrysler began to phase in its Next Generation Controller (NGC). This controller offers faster computer speed than the SBEC or the JTEC and uses high-side quad and dual drivers for many output functions. Another unique feature of this system is that the oxygen sensor's return circuit is biased to 2.5 volts. Some of the OBD-II monitors run differently than those of the SBEC and JTEC. These differences are discussed in this chapter.

## Freeze-Frame

During the 2000 model year, SBEC and JTEC controllers began to support two freeze-frames. Freeze-frame 1 is the CARB-mandated freeze-frame. Rules of priority and how long the failure has been present apply to this freeze-frame. With two freeze-frames, the failure snapshot first attempts to occupy freeze-frame 1. If freeze-frame 1 is occupied with an equal or greater priority, then the snapshot moves to freeze-frame 2. There is no data transfer possible between the two freeze-frames.

The NGC has five freeze-frames **(Table 1)**. The first one is still the CARB-mandated freeze-frame but it has been expanded to include more data than originally provided. This freeze-frame is titled

| CARB Freeze-Frame | Freeze-Frame 1 | Freeze-Frame 2 | Freeze-Frame 3 | Most Recent Freeze-Frame |
|---|---|---|---|---|
| Order of occurance does not matter | First failure | Second failure | Third failure | Last failure |
| Priority does matter | Priority does not matter | Priority does not matter | Priority does not matter | Priority does not matter |
| Could be one-trip failure or DTC | Could be one-trip failure or DTC | Could be one-trip failure or DTC | Could be one-trip failure or DTC | Could be one-trip failure or DTC |
| Highest-priority failure | | | | |

**Table 1** *Rules of order for multiple freeze-frames.*

"CARB Freeze-Frame." Freeze-frame 1 is the first failure regardless of priority. Freeze-frame 2 is the second failure, freeze-frame 3 is the third failure, and the "most recent freeze-frame" is the last failure. An intermittent or chronic condition has the potential to fill all five freeze-frames with the same "P" code, but the snapshot condition and priority could vary. The current rules dictate that for a two-trip fault, only the priority is updated, not the data. EPA/CARB rules change often, so consult the manufacturer's service information for the latest rules.

> **You Should Know**
>
> *Since the SAE requires all powertrain DTCs to be prefixed with the letter "P" powertrain DTCs are often referred to as "P codes."*

The priority of the faults has also changed. The fault code priority for the NGC system is:

Priority 1—The first occurrence of a two-trip fault, except fuel system rich/lean or misfire

Priority 3—The second consecutive occurrence of a two-trip fault (matured fault), except fuel system rich/lean and misfire or a one-trip fault of a comprehensive component

Priority 4—The first occurrence of a two-trip fault for fuel system rich/lean or misfire

Priority 6—The second consecutive occurrence of a two-trip fault (matured fault) for fuel system rich/lean or misfire

This change in priority was made so the technician can still have the freeze-frame for the first occurrence of a fuel system or misfire fault. On the SBEC and STEC systems, the Priority 2 freeze-frame would be erased if a Priority 3 fault matured. This meant the first occurrence of the Priority 2 failure would be lost. On NGC systems, the MIL is illuminated on Priority 3 or Priority 6 faults, not on Priority 2 or Priority 4 faults.

## Good Trip Counter

As with previous controllers, a good trip means that the monitor that failed has since been tested and passed. On NGC systems, 2 minutes of run time is required in order for the good trip to count up (as seen on the scan tool). This 2-minute run time is required for all failures, not just the comprehensive component monitor. In addition, the NGC processes a good trip on shutdown. This means that the ignition key must be cycled to see a good trip increment on the scan tool.

## $H0_2S$ Response Monitor

Like the SBEC and the JTEC, the $HO_2S$ response monitor is a two-trip fault and a once-per-trip monitor. On the NGC system, the oxygen sensor voltage range is 2.5 to 3.5 volts. The $HO_2S$ response monitor runs under light engine load at approximately 25 to 50 mph (40 to 80 km/h). Big slope is not used on NGC; the method of judging pass/fail is a voltage switch counter that replaces the half-cycle and big slope methods. The monitor takes 20 seconds to run. However, there is a fast-pass mode if the specification is met within 10 seconds. This would indicate that the trend is good enough for a pass condition. Two consecutive failures result in an $O_2S$ slow response failure DTC.

## Oxygen Sensor Heater Monitor

The $HO_2S$ heater monitor can fail in two ways: an electrical circuit failure and a monitor failure. An electrical circuit failure results in a heater circuit fault. The electrical circuit failure is a one-trip fault and results in a DTC and illumination of the MIL on the first occurrence.

A monitor failure results in a performance failure. The $O_2S$ heater monitor is a two-trip fault and a once-per-trip-monitor. Unlike previous controllers, the NGC measures the temperature in the heater circuit. This monitor runs on cold start-up (coolant temperature less than 122°F, or 50°C).

## Fuel System Lean/Rich Monitor

The NGC system does not have to update the purge-free cells during a rich failure before maturing a fault code as SBEC and JTEC controllers did. All long-term cells of the NGC are purge-free. This means the cells do not update while EVAP purge is occurring. A saturated EVAP system will not set a fuel system rich fault. The NGC updates the speed density pulse-width equation through purge vapor ratio and purge adaptive instead of long-term/short-term adaptive memory values.

## Misfire Monitor

The NGC misfire monitor is a continuous monitor that is a two-trip fault. As with previous controllers, the misfire monitor must learn the adaptive numerator in order to detect any misfires. There are two misfire categories: the 200-revolution counter looks for catalyst damage from misfire and the 1,000-revolution counter looks for FTP emissions failure due to misfire.

If misfire is at a rate high enough to cause catalyst damage, the MIL will flash. The percentage of misfire within 200 revolutions needed to log a misfire is dependent on NGC calibration, which is different for each application. If the vehicle is operating at less than 3,000 rpm, a group of three 200-revolution blocks is necessary to set a one-trip failure. If the vehicle is operating at more than 3,000 rpm, one 200-revolution block with misfire will set a one-trip failure. To set a DTC, a second consecutive trip with misfire is required. The time to set the DTC is dependent on the second trip's rpm operating range. For these reasons, it is possible for a vehicle to have a flashing MIL without a set DTC.

The FTP emissions misfire is set if a percentage misfire is exceeded in a 1,000-revolution block. A misfire that is present upon start-up will set a one-trip failure after 1,000 revolutions. A misfire that occurs after the vehicle has run 1,000 revolutions without a misfire takes four blocks of 1,000 revolutions to set a one-trip failure. To set a misfire, the DTC requires two consecutive trips with a misfire present. On the second trip, regardless of whether the misfire was present at start-up, it will only take one 1,000-revolution block to set a misfire DTC.

A good trip takes at least 2 minutes of run time with a pass through the similar conditions window with no misfire occurring.

## EVAP and Purge Monitors

The NGC system uses natural vacuum leak detection (NVLD) instead of a leak-detection pump. This system relies on the gas law principal. The **gas law principle** states that in a sealed container, a vacuum will be created as the temperature of the gas decreases. Even small leaks can cause the pressure to equalize with the outside ambient pressure.

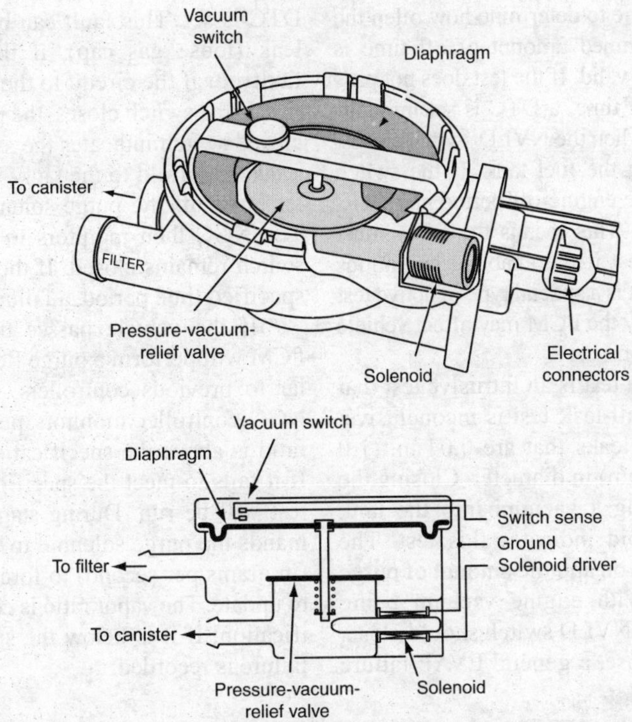


**Figure 1** *Components of the natural vacuum leak-detection (NVLD) assembly.*

A vent valve seals the canister vent during engine-off conditions. If the evaporative system has a leak of less than the failure threshold, the evaporative system will be pulled into a vacuum due to a cooldown from ambient temperatures. When the vacuum in the system exceeds about 1 in. $H_2O$ (0.25 kPa), a vacuum switch closes. If the PCM detects the switch is closed, the small leak monitor will record a pass condition on the next start. If the switch state does not change, either the system has a leak or the proper temperature change did not occur. This will result in the test being flagged as "inconclusive."

The NVLD assembly **(Figure 1)** is designed with a normally open vacuum switch, a normally closed (deenergized) solenoid, and a pressure-vacuum-relief valve. Both the solenoid and a diaphragm actuate the pressure-vacuum-relief valve.

> **You Should Know**
>
> *The NVLD is located on the atmospheric vent side of the canister. The NVLD assembly may be mounted on top of the canister outlet or in line between the canister and atmospheric vent filter.*

The normally open vacuum switch closes when about 1 in. $H_2O$ (0.25 kPa) vacuum lifts the diaphragm. The normally closed pressure-vacuum-relief valve in the NVLD is intended to maintain the seal on the evaporative system during engine-off conditions. If vacuum in the evaporative system exceeds 3 to 6 in. $H_2O$ (0.75 to 1.5 kPa), the valve will be pulled off the seat, opening the seal. This protects the system from excessive vacuum as well as allows sufficient purge flow in the event that the solenoid becomes inoperative.

The PCM activates the solenoid when the engine is running so the valve will unseat and allow the canister to vent. The PCM deenergizes the solenoid to close the vent during the medium- and large-leak tests and during the purge flow test. The diaphragm will open the seal in the NVLD if the pressure in the evaporative system is too great (about 0.5 in. $H_2O$ or 9.012 kPa). This permits the venting of vapors through the canister during refueling or if there is an increase in temperature.

The NVLD assembly has three wires: switch sense, solenoid driver, and ground. The NGC utilizes a high-side driver to energize and duty-cycle the solenoid.

The small-leak test is a passive test performed while the engine is off. The NGC logs ignition-off time as well as ignition-on time to determine how often the test is run. A predetermined amount of off time is required to make the test valid. If the test does not pass in a specified amount of time, a DTC is set immediately. The test is passed when the NVLD switch closes; this indicates vacuum in the fuel tank. If the switch does not close during the engine-off period, the NGC has inconclusive results. This means the PCM must perform the **intrusive test** if the enabling conditions are met. The intrusive test is a special type of active test in that the action taken by the PCM may affect vehicle performance and emissions.

The engine-running test is an intrusive test that will only run if the small-leak test is inconclusive. It is used to check for leaks that are 0.04 in. (1.0 mm) and 0.06 in. (1.5 mm) in diameter. Closing the EVAP vent and drawing a vacuum into the tank using the purge solenoid indicates this test. The purge solenoid is turned on and the amount of purge is slowly increased. With engine vacuum being drawn into the tank, the NVLD switch should close. If the switch fails to close, a general EVAP failure DTC is set. This fault can be set if there is a large leak (loose gas cap), if the switch contacts are faulty, or if the circuit to the switch is open.

If the switch closes, the purge monitor passes. A closed switch indicates the system is intact from the intake manifold to the tank. Next, the NGC controller turns off the purge solenoid to seal the system. The NGC then monitors in seconds how long the switch remains closed. If the switch opens before a specified time period, a failure is recorded.

If the vehicle passes the small-leak test, the PCM will perform a purge flow monitor that is similar to previous controllers. During stage one, the NGC controller monitors purge vapor ratio. If the ratio is above the specification, the monitor passes. If it fails to meet the specification, stage two of the test will be run. During stage two, the NGC commands the purge solenoid to flow at a specified rate (in grams per second) to force the purge vapor ratio to update. The vapor ratio is compared with the specification; if it is below the specification, a one-trip failure is recorded.

## Review Questions

1. List the DTC priorities of the NGC and their respective fault types.
2. Technician A says that NGC uses the big slope counter for the $O_2S$ response monitor. Technician B says that a normal pass for the $O_2S$ response monitor is 20 seconds. Who is correct?
   A. Technician A only
   B. Technician B only
   C. Both Technician A and Technician B
   D. Neither Technician A nor Technician B
3. Technician A says that a misfire that occurs after the vehicle has run 1,000 revolutions without a misfire takes six blocks of 200 revolutions to set a one-trip failure. Technician B says that a good trip takes at least 2 minutes of run time with a pass through the similar conditions window with no misfire occurring. Who is correct?
   A. Technician A only
   B. Technician B only
   C. Both Technician A and Technician B
   D. Neither Technician A nor Technician B
4. Technician A says that if the natural vacuum switch does not close after the engine is turned off, a one-trip failure will be set. Technician B says that if the switch closes during the intrusive test, the purge monitor passes. Who is correct?
   A. Technician A only
   B. Technician B only
   C. Both Technician A and Technician B
   D. Neither Technician A nor Technician B
5. What additional requirements are included with the NGC to record a good trip?
6. Technician A says that if the vehicle is operating at less than 3,000 rpm, a group of three 200-revolution blocks is necessary to set a one-trip misfire failure. Technician B says that a misfire that is present upon start-up will set a one-trip failure after 1,000 revolutions. Who is correct?
   A. Technician A only
   B. Technician B only
   C. Both Technician A and Technician B
   D. Neither Technician A nor Technician B
7. Technician A says that natural vacuum leak detection (NVLD) is used to detect a medium or large leak. Technician B says that vacuum decay is used to detect a small leak. Who is correct?
   A. Technician A only
   B. Technician B only
   C. Both Technician A and Technician B
   D. Neither Technician A nor Technician B

CHAPTER

# 32 General Motors's OBD-II

## Introduction

This chapter covers the OBD-II system used by General Motors. Although the intent of OBD-II is the same for all manufacturers, each manufacturer can determine how it will accomplish the stated goals. This chapter discusses those functions of the OBD-II system that are different than those discussed in previous chapters.

## Diagnostic Executive

The PCM performs two functions using different management systems. One controls vehicle systems using sensor input and software programming and the other performs system diagnostics. Both of these functions communicate with each other. The PCM uses the **diagnostic management system** to perform diagnostic testing, record the results of the tests, and request the test fail actions. The different types of diagnostic tests performed by the diagnostic management system include passive, **active**, and intrusive tests **(Figure 1)**. Active testing requires the PCM to control the system or components in a specific manner while monitoring the results.

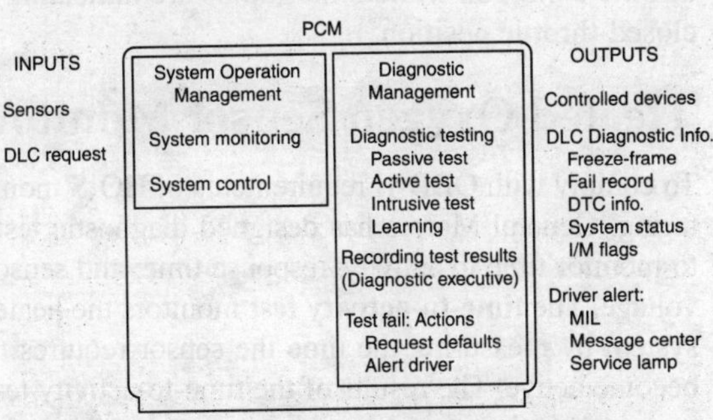


**Figure 1** *The different types of diagnostic tests performed by the diagnostic management system.*

The diagnostic tests occur during a trip and the PCM "learns" from the results of the tests. The PCM charts the results and creates baseline (normal) test results. This allows the PCM to filter out information that could result in erroneous DTCs. The diagnostic management system stores the test results in a software program called the **diagnostic executive**. The diagnostic executive stores the following information:

- DTC information
- Freeze-frame data
- Fail records
- System status
- Warm-up cycles

DTC information indicates the status of the diagnostic testing for the specified DTC. This information includes the pass/fail status of the test along with information concerning when the test failed and if the DTC requires any diagnostic lights to be illuminated. The diagnostic executive records freeze-frame records on one frame of data about the operating conditions at the time for the *first* failed test of an emissions-related system or component. A history DTC is stored and the MIL is illuminated. The freeze-frame will not be updated if the test fails a second time. However, fuel trim and misfire DTCs have a higher priority than other DTCs and will overwrite lower-priority freeze-frames. Fail records store up to five DTCs whether they are emissions related or not. The fail records contain the same type of information that would be stored in the freeze-frame. The system status program stores data on which diagnostic tests have been run.

### Test Fail Action

If a diagnostic test fails and a DTC is set, the diagnostic management system performs a series of **test fail actions**. Test fail actions are the responses the PCM has to perform when a test fails. These actions are dependent upon the DTC. The diagnostic management system may perform the following actions:

- Illuminate the MIL
- Illuminate the service lamp
- Send data message to the drive information center (DIC)
- Initiate usage of default values stored in program
- Request PCM default operations
- Store a freeze-frame
- Store a fail record

The MIL is illuminated only if the DTCs impact vehicle emissions. Upon a failure, the diagnostic executive turns the MIL on as requested by the diagnostic management system. In the event of a misfire malfunction resulting in possible catalytic converter damage, the diagnostic executive will flash the MIL once per second. The MIL will continue to flash until the engine operating conditions are such that catalytic damage is no longer imminent; however, at this time the MIL will be on steady.

If there are three consecutive trips in which the diagnostic test passes (for the test that turned the MIL on originally), the MIL will be turned off. However, for fuel trim or misfire DTCs, the test must run and pass in similar operating conditions to which the DTC was originally set under. The window is within 375 rpm and within 10 percent of the engine load and similar engine temperature.

On some General Motors vehicles, a service lamp is also available. This lamp alerts the driver to nonemissions-related malfunctions, such as air conditioning and cruise control faults. This lamp only illuminates if the DTC is a Type C. In addition, some vehicles are equipped with a DIC that alerts the driver to Type C failures.

## DTC Types

If there is a failure during a trip, the diagnostic management system sets a DTC that is directly related to the diagnostic test. Some DTCs will not set until there have been two consecutive failures. The following are the four types of DTCs and the characteristics of the code:

1. Type A
   - Emissions-related failure
   - MIL illuminated on the first trip with a failure
   - Stores a historical DTC on the first trip with a failure
   - Stores a freeze-frame if empty
   - Stores a fail record
   - Updates the fail record each time the test fails
2. Type B
   - Emissions-related failure
   - "Armed" after one failure
   - Cleared after one trip with a pass
   - MIL illumination after two consecutive trips that fail
   - Stores a historical DTC on the second trip failure
   - Stores a freeze-frame after the second trip failure (if empty)
   - Stores fail record on the first trip failure
   - Updates the fail record each time the test fails
3. Type C
   - Nonemissions-related fault
   - Requests illumination of the service lamp or DIC (not the MIL) on the first failure
   - Stores a history DTC on the first failure
   - No freeze-frame stored
   - Stores fail record at first failure
   - Updates the fail record each time the test fails
4. Type D
   - Nonemissions-related fault
   - No illumination of any lamps
   - Stores a history DTC on the first failure
   - No freeze-frame stored
   - Stores fail record at first failure
   - Updates the fail record each time the test fails

## Comprehensive Components Monitor

Input and output components are monitored for minimum circuit continuity and out-of-range values. Comprehensive components are also monitored for functionality, such as the TPS showing a wide-open throttle condition while other inputs are indicating a closed-throttle position.

## Heated Oxygen Sensor Monitor

To comply with OBD-II requirements of $HO_2S$ monitoring, General Motors has designed diagnostic tests to monitor time to activity, response time, and sensor voltage. The **time-to-activity test** monitors the heater system by measuring the time the sensor requires to become active. The results of the time-to-activity test are compared with calibrated parameters within the PCM. The heater circuit indicates a failure when the time to activity increases to a point outside the parameters. This test is only run following a cold start.

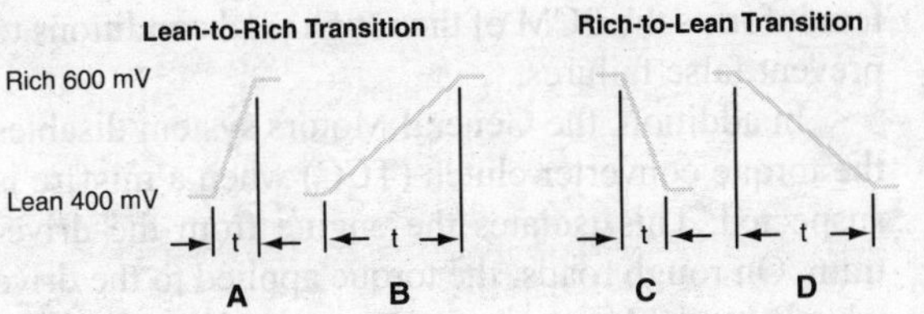


**Figure 2** *Both chart A and chart C illustrate correct $HO_2S$ transitions. Chart B and chart D illustrate $HO_2S$ transitions that may cause a test to fail.*

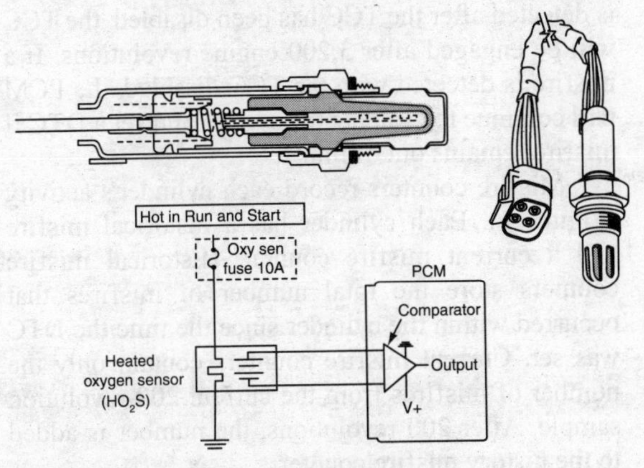


**Figure 3** *$HO_2S$ circuit.*

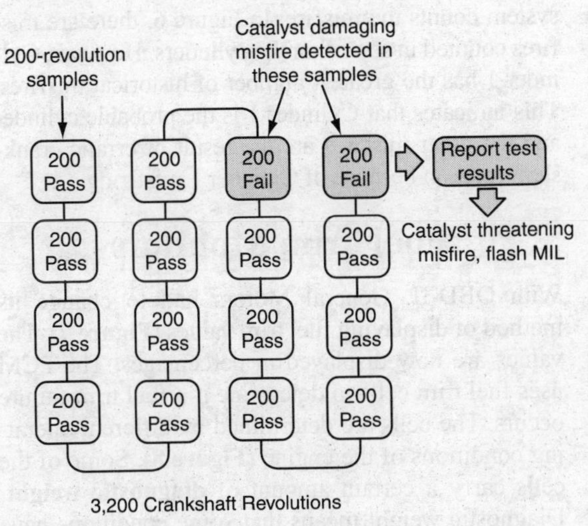


**Figure 4** *The PCM maintains a record of the previous 3,200 revolutions, which are divided into sixteen 200-revolution counters. If a misfire occurs during any 200-revolution sample that could damage the catalyst, the MIL flashes.*

The **response time test** monitors the transition times between lean-to-rich and rich-to-lean switches of the upstream oxygen sensors. If the response time test indicates too much time has elapsed between the lean-to-rich and rich-to-lean transitions, the test fails **(Figure 2)**.

The **sensor voltage test** monitors upstream and downstream $HO_2S$ for voltages that are out of range. The sensor voltage test looks for sensors that are inactive, shorted to voltage, shorted to ground, or have rich-to-lean shifts of the fuel-control system. Factors such as fuel quality, vehicle operating conditions, and the condition of the catalytic converter can affect $HO_2S$ operation.

General Motors uses an $HO_2S$ with a high signal wire and a low signal wire that are connected to a comparator in the PCM **(Figure 3)**. The low signal wire provides the comparator with reference low (ground). This is different from the $HO_2S$ used in OBD-I vehicles in which the return low circuit went directly to ground.

## Catalyst Monitor

The catalyst monitor determines the oxygen storage capacity of the catalytic converter. To test this capacity, the PCM runs a two-stage diagnostic test. In the first stage, the PCM calculates the oxygen storage capacity of the catalytic converter and compares it with calculated values stored in the PCM. This stage is done when the engine and catalyst have reached normal operating temperatures and the vehicle is being driven at a steady cruising speed. By monitoring the upstream and downstream $HO_2S$, the PCM determines if the activity of the two is similar. Similar activity indicates the storage capacity of the catalytic converter has deteriorated. If the calculated storage capacity is within the fail parameters, the test fails Stage 1. This does not necessarily indicate a failed catalyst.

If the catalyst monitor fails Stage 1, Stage 2 is entered. Stage 2 is additional monitoring of the storage capacity. If the monitor fails Stage 2, a DTC is set. Staged testing of the catalyst allows the PCM to statistically filter the test data to prevent erroneous passes or fails while still passing or failing a marginal catalyst.

## Misfire Monitor

To detect engine misfire, the PCM stores a record of the previous 3,200 engine revolutions. The 3,200 revolutions are then divided into sixteen 200-revolution counters. If a misfire occurs during any one of the 200 revolution samples, a DTC is set and the

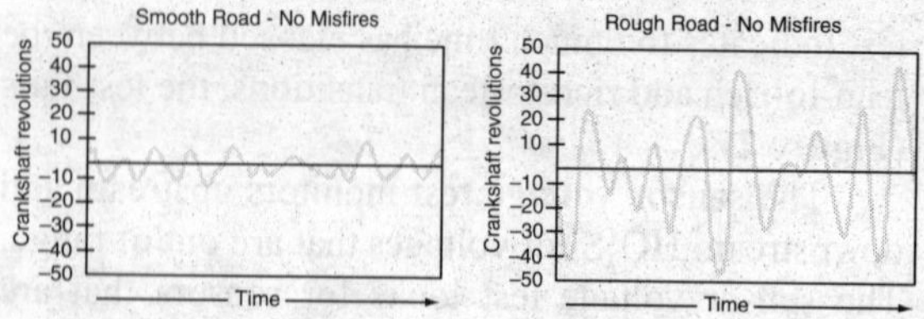


**Figure 5** *Rough road conditions can alter the crankshaft position sensor input to look like a misfire.*

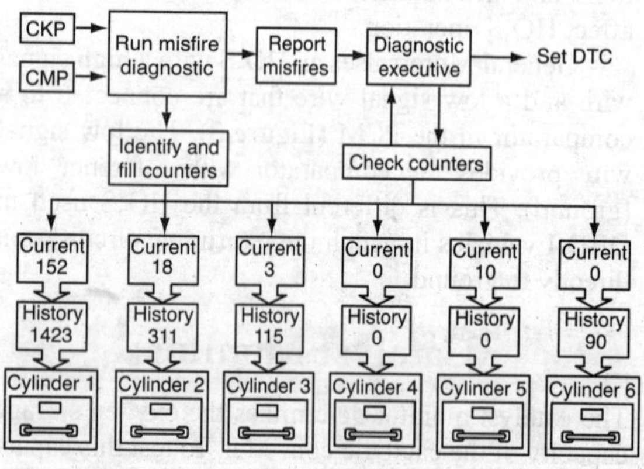


**Figure 6** *Misfire counters shown for each cylinder.*

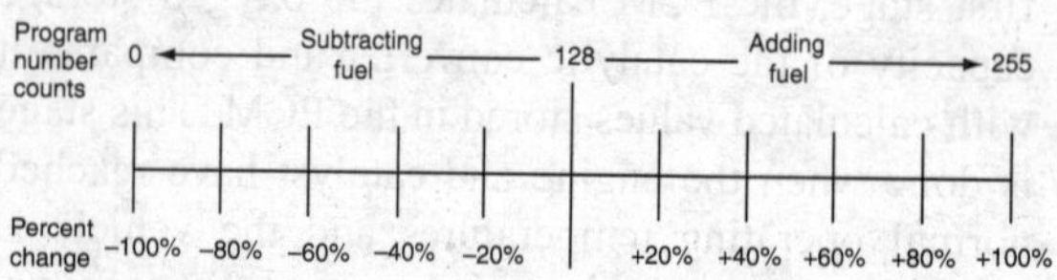


**Figure 7** *Fuel trim conversion to percentage.*

MIL flashes **(Figure 4)**. If the misfire is inconsistent, the misfire must fail five of the test samples. If the test fails in a second trip, a DTC is stored and the MIL is turned on. This type of misfire will not result in catalyst damage. Noncatalyst damaging misfires will set a DTC and turn on the MIL under the following conditions:

1. A second consecutive trip contains a failed test.
2. A second nonconsecutive trip contains a failed test that occurred under the same operating conditions window as the first test within eighty trips of the first failure.

Since rough roads can cause false misfire failures, the software in the PCM is enhanced to detect most rough road conditions using the crankshaft sensor **(Figure 5)**. General Motors also uses the antilock brake system (ABS) to detect rough road conditions. Monitoring the wheel speed sensors can indicate rough roads. As the wheels go over the rough surfaces, variations in wheel speed occur. The ABS system informs the PCM of the rough road conditions to prevent false failures.

In addition, the General Motors system disables the torque converter clutch (TCC) when a misfire is suspected. This isolates the engine from the drivetrain. On rough roads, the torque applied to the drive wheels and drivetrain can temporarily decrease engine speed. The PCM can detect this decrease and falsely indicate a misfire. By isolating the engine from the drivetrain, the PCM will not fail the misfire test because of road conditions. If no misfire activity is detected after the TCC has been disabled, the TCC will be engaged after 3,200 engine revolutions. If a misfire is detected with the TCC disabled, the PCM will continue to evaluate the system and set a DTC if misfire remains out of limits.

Misfire counters record each cylinder's activity **(Figure 6)**. Each cylinder has a historical misfire and a current misfire counter. Historical misfire counters store the total number of misfires that occurred within the cylinder since the time the DTC was set. Current misfire counters contain only the number of misfires from the current 200-revolution sample. After 200 revolutions, the number is added to the history misfire counter.

At the time of a cylinder misfire, the detection system counts the misfire. In Figure 6, there are misfires counted in several of the cylinders. However, Cylinder 1 has the greatest number of historical misfires. This indicates that Cylinder 1 is the probable cylinder and the other misfires are the result of erratic crankshaft rotation because of Cylinder 1 misfiring.

## Fuel Trim Monitors

With OBD-II, General Motors had to change its method of displaying fuel trim values **(Figure 7)**. The values are now displayed in percentages. The PCM uses fuel trim cells to determine if a fuel trim failure occurs. The cells are determined by different operating conditions of the engine **(Figure 8)**. Some of the cells carry a certain amount of **diagnostic weight**. Diagnostic weight means that some conditions have higher precedence. Events that happen within these conditions are weighted (have more value). Only the cells that are assigned a diagnostic weight value are used to set DTCs. A cell that has a higher diagnostic weight value can set a DTC easier than one with a lower value. For example, in **Figure 8**, fuel trim cells 4, 5, and 9 are the weighted cells. No fuel trim DTC will set regardless of the fuel trim count unless that count is located in one of the weighted cells.

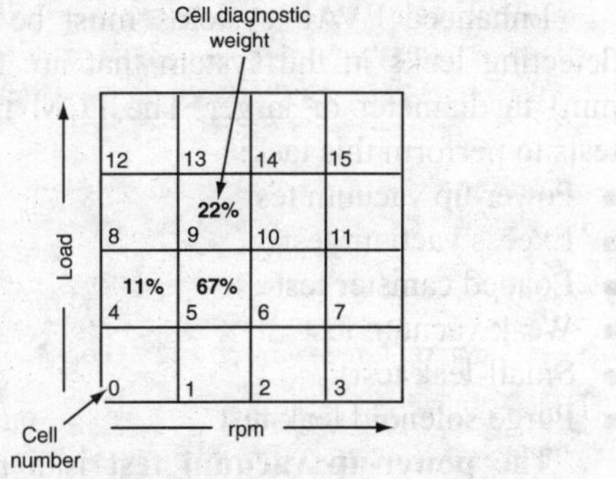


**Figure 8** *Fuel trim cells are determined by the operating conditions of the engine. Some cells carry a higher diagnostic weight than others.*

The fuel trim diagnostic test compares an average of long-term trim values and short-term trim values with rich and lean limits that are the calibrated fail thresholds for the test. If either value is within the fail thresholds, a pass is recorded. If both values are outside the thresholds, then a failure exists.

## Exhaust Gas Recirculation Monitor

Changes in the EGR valve will have an effect on MAP and rpm inputs. Opening the valve increases manifold pressure while closing the valve decreases pressure **(Figure 9)**. The amount of MAP sensor input change is correlated to the amount of EGR flow.

The EGR monitoring system runs an intrusive test by forcing the EGR valve open during a closed-throttle deceleration. Closing the valve during a steady cruise condition can also run the test. The PCM records the average MAP changes over several EGR valve activations. If the resulting average is outside parameter limits, the test fails.

## Secondary Air-Injection Monitor

General Motors's AIR monitoring system uses both passive and active tests to evaluate the AIR system. The AIR diagnostic system indicates a test pass if the passive portion of the AIR system observes the correct lean reading from the upstream $HO_2S$ after start-up and prior to closed-loop operation. The $HO_2S$ will read low voltage if the AIR pump is delivering air to the exhaust system. When the AIR system is turned off, the test looks for $HO_2S$ voltage to toggle.

If the AIR monitor passes during the passive test, the test is completed. However, if the passive test fails, the active test is initiated. The active test activates the AIR pump during closed-loop operation and observes the response from the upstream $HO_2S$. It may also monitor short-term fuel trim values. A lean $HO_2S$ voltage or a fuel trim value that increases indicates the AIR system is operating properly.

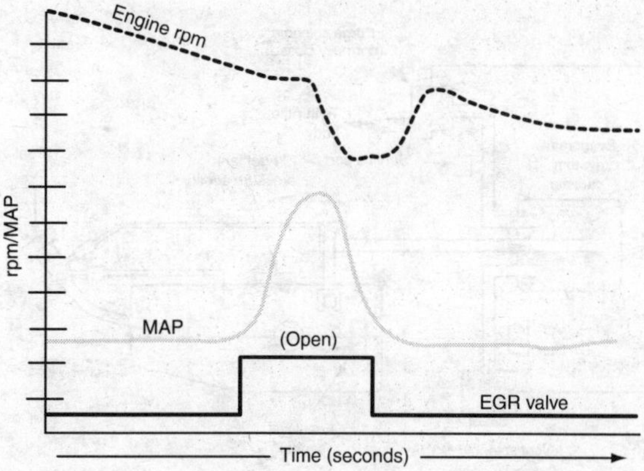


**Figure 9** *Changes in MAP values are monitored as the EGR operates.*

## Evaporative Emissions System Monitor

General Motors vehicles can be equipped with either a nonenhanced or an enhanced EVAP system. The nonenhanced system includes the following components:

- EVAP emissions pressure-control valve
- Vented canister
- Diagnostic switch
- EVAP purge solenoid

The EVAP solenoid is pulse-width modulated and controls the amount of canister purge. When the solenoid is energized, vapors are allowed to flow from the canister into the intake manifold. The **diagnostic switch** is normally closed and will open when the solenoid is energized and there is sufficient intake manifold vacuum. The diagnostic switch is used to determine correct operation of the solenoid. If the solenoid is stuck in the open position, vacuum will be present at the switch and it will remain in the open position. A solenoid that is stuck in the closed position constantly indicates atmospheric pressure. In addition, a clogged vent could trap vacuum between the switch and the canister, causing the switch to open even with the solenoid deenergized.

The enhanced EVAP system includes the following components **(Figure 10)**:

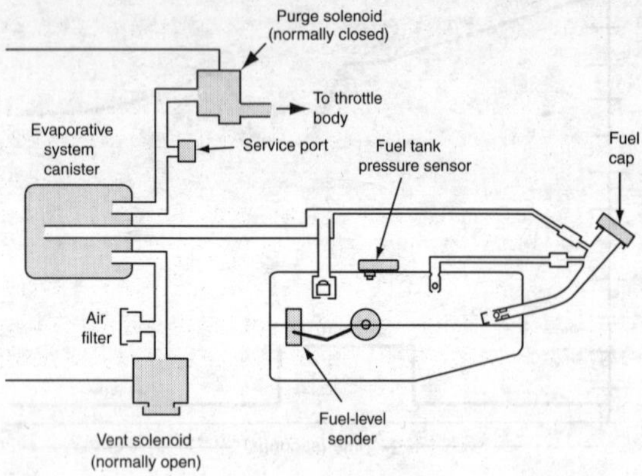


**Figure 10** *Typical enhanced EVAP system.*

- EVAP canister
- Fuel tank pressure sensor
- Canister purge solenoid
- Canister vent solenoid
- Service port
- Fuel-level sensor
- Fuel cap

The purge solenoid operates in the same manner as in the nonenhanced system. The vent solenoid is a normally open solenoid that allows fresh air to the canister during purge and allows the diagnostics to pull a vacuum on the fuel tank when closed. The fuel tank pressure sensor is a three-wire strain gauge that measures the difference between the air pressure in the tank and the atmospheric pressure **(Figure 11)**.

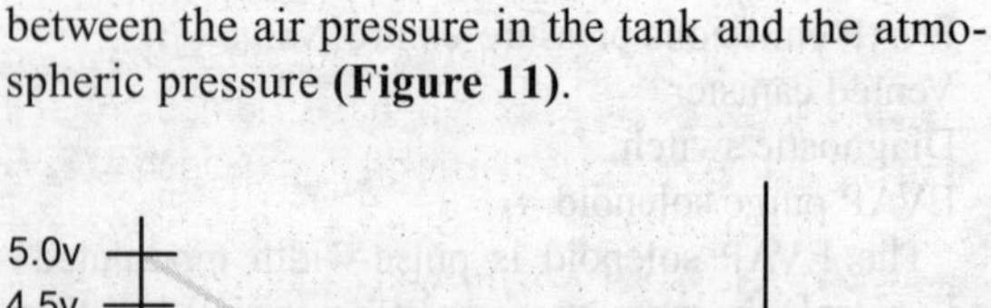


**Figure 11** *A fuel pressure sensor that is mounted on top of the fuel sending unit measures the difference between the air pressure in the tank and the atmospheric pressure.*

Enhanced EVAP systems must be capable of detecting leaks in the system that are 0.2 in. (0.5 mm) in diameter or larger. The PCM uses several tests to perform this task:

- Power-up vacuum test
- Excess vacuum test
- Loaded canister test
- Weak vacuum test
- Small-leak test
- Purge solenoid leak test

The **power-up vacuum test** is a passive test used to detect restrictions in the vent path. The power-up vacuum test is run when the vent solenoid is open with the purge solenoid closed. The engine must also be cold and off with the ignition in the run position. Under these conditions, the fuel tank pressure sensor should not indicate pressure or vacuum. This test will not run if the fuel-level sensor indicates more than 85 percent full or 15 percent empty.

The **excess vacuum test** is also a passive test designed to detect restrictions in the vent path. The excess vacuum test runs during normal purge operations with the solenoid valve open. During this test, the pressure sensor in the tank should not indicate excessive vacuum.

The **loaded canister test** is a passive test designed to detect leaks in the entire EVAP system. The loaded canister test monitors the $HO_2S$ voltage levels and fuel trim values during normal EVAP operation. If large amounts of fuel vapors are being purged, the canister is sufficiently loaded and the system is functioning properly.

The **weak vacuum test** is an active test designed to detect large leaks. The weak vacuum test only runs if the loaded canister test fails. During this test, the vent solenoid is closed during normal EVAP operation. The fuel tank pressure sensor should indicate the presence of vacuum during this test.

If the weak vacuum test passes, the PCM then initiates the small-leak test while vacuum is still present in the fuel tank. The vent and purge solenoids are closed to seal the system. With the system sealed, the tank should hold vacuum. The fuel tank pressure sensor is monitored for a decay rate. If the decay is too rapid, the test fails.

The **purge solenoid leak test** is an active test designed to detect a leak through the purge solenoid. During the purge solenoid leak test, the vent and purge solenoids are closed when EVAP purge is not occurring. The fuel tank pressure sensor is monitored; if vacuum is present the purge solenoid is leaking.

# Review Questions

1. Heated $O_2S$ monitor function is being discussed. Technician A says that the time-to-activity test monitors the transition times between lean-to-rich and rich-to-lean switches of the upstream oxygen sensors. Technician B says that the response time test monitors the time the sensor requires to become active. Who is correct?
   A. Technician A only
   B. Technician B only
   C. Both Technician A and Technician B
   D. Neither Technician A nor Technician B
2. Explain the operation of the misfire monitor.
3. Explain how proper EGR operation is monitored.
4. Technician A says that during the first stage of the catalyst monitor the PCM calculates the oxygen storage capacity of the catalytic converter and compares it with calculated values stored in the PCM. Technician B says that similar activity of the upstream and downstream oxygen sensors indicates the storage capacity of the catalytic converter is efficient. Who is correct?
   A. Technician A only
   B. Technician B only
   C. Both Technician A and Technician B
   D. Neither Technician A nor Technician B
5. Technician A says that a diagnostic weight means that some conditions have higher precedence and events that happen in these conditions are weighted. Technician B says that the fuel trim diagnostic compares an average of long-term trim values and short-term trim values with rich and lean limits that are the calibrated fail thresholds for the test. Who is correct?
   A. Technician A only
   B. Technician B only
   C. Both Technician A and Technician B
   D. Neither Technician A nor Technician B
6. Explain the operation of the nonenhanced EVAP system monitor.
7. Explain the operation of the enhanced EVAP system monitor.

CHAPTER

# 33 Ford's OBD-II

## Introduction

This chapter discusses Ford's OBD-II system. Ford began installing the ODB-II system in the 3.8L Mustang and the 4.6L Thunderbird and Cougar in 1994 with the introduction of the Electronic Engine Control V (EEC V). The EEC V system hardware includes the PCM, sensors, switches, actuators, solenoids, wires, terminals, and the constant control relay module (CCRM). The main hardware difference between the EEC V system and previous systems is the addition of downstream heated oxygen sensors mounted behind the catalytic converter. The conventional $HO_2S$ is mounted in the exhaust manifold.

The major software difference between EEC V with OBD-II and previous systems is the software containing the monitoring strategies inside the PCM. The **OBD Executive** is the PCM software program that manages the sequencing and execution of all OBD-II diagnostic tests. The components in the EEC V with OBD-II are designed to perform within specifications for 100,000 miles.

## Catalyst Efficiency Monitor—Steady State

The 20-second steady-state catalyst efficiency monitor (used in 1994 through some 1996 vehicles) uses an $HO_2S$ located downstream of the catalytic converter to infer catalyst efficiency based on oxygen storage capacity. A high-efficiency catalyst will result in a lower downstream $HO_2S$ switch rate as compared with the upstream sensors.

The downstream $HO_2Ss$ have additional protection to prevent the collection of condensation on the ceramic shell. The internal heater is not turned on until the ECT sensor signal indicates a warmed-up engine. This action prevents cracking of the ceramic shell. Gold-plated pins and sockets are used in the $HO_2Ss$ and the downstream and upstream sensors have different wiring harness connectors.

When the catalytic converter is storing oxygen properly, the downstream $HO_2Ss$ provide low-frequency voltage signals. If the catalytic converter is not storing oxygen properly, the voltage signal frequency increases on the downstream $HO_2Ss$ until it approaches the frequency of the upstream $HO_2Ss$.

The monitor runs after all enabling conditions have been met. The catalyst efficiency monitor enabling conditions are based on input from the engine coolant temperature (ECT), intake air temperature (IAT), throttle position sensor (TPS), crankshaft position (CKP), and vehicle speed sensor (VSS) along with a calculated length of engine run time in closed-loop mode. Once the monitor is activated, closed-loop fuel control is temporarily transferred from the upstream $HO_2S$ to the downstream $HO_2S$. The monitor then analyzes the downstream $HO_2S$ signal switching frequency to determine if the catalytic converter has failed. This actual measured output frequency of the downstream $HO_2S$ is called the **test frequency**. High catalytic converter efficiency is determined by a slow test frequency.

A second **calibrated frequency** is determined based on the engine speed and load. The calibrated frequency serves as a high-limit threshold for the test frequency and is determined by the PCM. If the test frequency is less than the calibrated frequency, the catalytic converter passes the monitor test. If the test frequency is higher than the calculated frequency, the test fails and a pending DTC is stored. If the test fails on two consecutive drive cycles, the MIL is turned on.

## Catalyst Efficiency Monitor—Federal Test Procedure

The federal test procedure (FTP) switch ratio method (used beginning in the 1996 model year) to determine catalyst efficiency is activated during part-throttle (cruise), closed-loop fuel conditions. The engine must be warm and the inferred catalyst temperature must be within specified limits. When this monitor is activated, the front and rear $HO_2S$ switches are counted and then a switch ratio is calculated. The front $HO_2S$ switches are accumulated in up to nine different air mass regions (cells). A single cell is used for rear switch counting. Once the front sensor has accumulated enough switches in the cell the total number of rear switches accumulated is divided by the total number of front switches. The result is the switch ratio.

A switch ratio of 0:1 means the catalyst has high oxygen storage capacity and will oxidize HC emissions. A switch ratio near 1:1 means the catalyst has low oxygen storage capacity and will not oxidize HC effectively. The switch ratio is compared with an emissions limit. If the switch ratio is greater than the limit, the catalyst has failed. It may require up to six consecutive drive cycles with a failure for a DTC to set and turn on the MIL.

Beginning in 2001, Ford began to use the **index ratio** method to monitor the catalyst. The index ratio uses the ratio of signal lengths to determine proper catalyst activity. The conditions for this monitor are similar to those of the switch ratio method. Front $HO_2S$ switches are accumulated in up to three cells. Once the front sensor has accumulated enough switches in each cell, the total signal length of the rear $HO_2S$ is divided by the total signal length of the front $HO_2S$. The result is the catalyst index ratio.

An index ratio near 0:1 means the catalyst has high oxygen storage capacity and will oxidize HC emissions. An index ratio near 1:1 means the catalyst has low oxygen storage capacity and will not oxidize HC effectively. The index is compared with an emissions limit. If the index ratio exceeds the limit, the catalyst has failed.

If the catalyst monitor does not complete during a particular driving cycle, the data that has been accumulated is retained and is used during the next driving cycle. This gives the monitor a better opportunity to complete.

Low-emission vehicles (LEVs) and ultralow-emission vehicles (ULEVs) will not monitor all of the catalyst volume. Usually they only monitor the light-off catalyst. This is done so these vehicles can meet the one and three-quarter times the FTP standard for this monitor. The test is done in this way because the catalyst nearest the engine will usually deteriorate first. Monitoring this catalyst increases monitor sensitivity so the MIL will illuminate correctly for the lower-emissions standards.

The FTP catalyst monitor uses EWMA. Usually three to six drive cycles must record a failure before the MIL is illuminated. **Table 1** shows examples for a typical calibration.

## Evaporative (EVAP) Emissions System Monitor

Some EVAP systems have a purge flow sensor (PFS) connected in the vacuum hose between the canister purge (CANP) solenoid and the intake manifold

| Monitor evaluation (new data) | EWMA filter calculation, "normal" filter constant Set to 0.4 Malfunction threshold = 0.75 | Weighted average (new average) | Driving cycle number | Comments |
|---|---|---|---|---|
| 0.15 | 0.15 (0.4) + 0.15 (1.0 - 0.4) | 0.15 | | Normal 100K system |
| 1.0 | 1.0 (0.4) + 0.15 (1.0 - 0.4) | 0.49 | 1 | Catastrophic failure |
| 1.0 | 1.0 (0.4) + 0.49 (1.0 - 0.4) | 0.69 | 2 | |
| 1.0 | 1.0 (0.4) + 0.69 (1.0 - 0.4) | 0.82 | 3 | Exceeds threshold |
| 1.0 | 1.0 (0.4) + 0.82 (1.0 - 0.4) | 0.89 | 4 | MIL "on" |
| 0.15 | 0.15 (0.4) + 0.15 (1.0 - 0.4) | 0.15 | | |
| 0.8 | 0.8 (0.4) + 0.15 (1.0 - 0.4) | 0.41 | 1 | Normal 100K system |
| 0.8 | 0.8 (0.4) + 41.0 (1.0 - 0.4) | 0.57 | 2 | threshold failure |
| 0.8 | 0.8 (0.4) + 0.57 (1.0 - 0.4) | 0.66 | 3 | |
| 0.8 | 0.8 (0.4) + 0.66 (1.0 - 0.4) | 0.72 | 4 | |
| 0.8 | 0.8 (0.4) + 0.72 (1.0 - 0.4) | 0.75 | 5 | Exceeds threshold |
| 0.8 | 0.8 (0.4) + 0.75 (1.0 - 0.4) | 0.77 | 6 | MIL "on" |

**Table 1** *EWMA catalyst monitor calibration table.*

**(Figure 1)**. The PCM monitors the PFS signal once per drive cycle to determine if there is vapor flow through the solenoid to the intake manifold.

Some vehicles that meet the enhanced evaporative requirements use a vacuum-based evaporative system check to test for 0.02 in. (0.5 mm) diameter leaks. This test uses a fuel tank pressure transducer (FTPT), a canister vent solenoid (CVS), and fuel-level input (FLI) and vapor management valve (VMV) to find EVAP system leaks with a diameter of 0.02 in. (0.5 mm), 0.04 in. (1.0 mm), or larger **(Figure 2)**.

> **You Should Know**
>
> *Early vehicles with enhanced EVAP monitoring only checked for leaks that were 0.04 in. (1.0 mm) in diameter or larger. This parameter was used prior to the stricter requirements of testing for leaks of 0.02 in. (0.5 mm) in diameter.*

The VMV is a normally closed valve used to control the flow of fuel vapors into the engine. The CVS is a normally open solenoid used to control the evaporative flow between the carbon canister and the atmosphere. The monitor relies on the CVS to seal the entire EVAP system from the atmosphere and it relies on the VMV to pull engine vacuum on the fuel tank. The CVS is closed and the VMV is opened a calibrated amount. If the FTPT does not sense a target vacuum within a given time, then there is a leak or flow fault. If the target vacuum is obtained, then both solenoids are closed to lock the vacuum for a specified length of time. If the vacuum bleeds above a fault threshold within that time period, the EVAP monitor fails the test. DTCs are set after three unsuccessful attempts to hold vacuum.

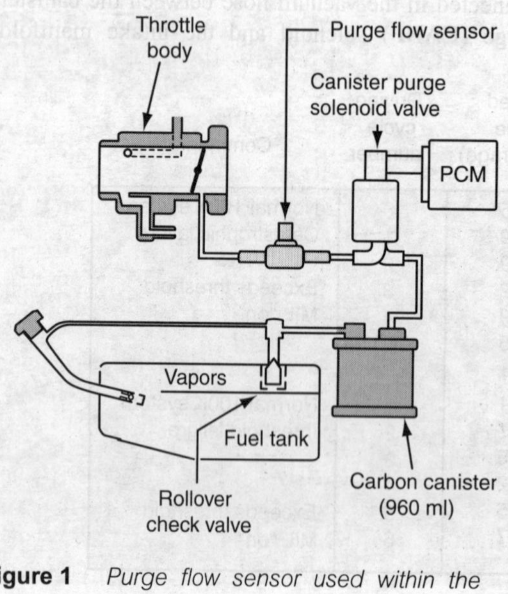


**Figure 1** *Purge flow sensor used within the EVAP system.*

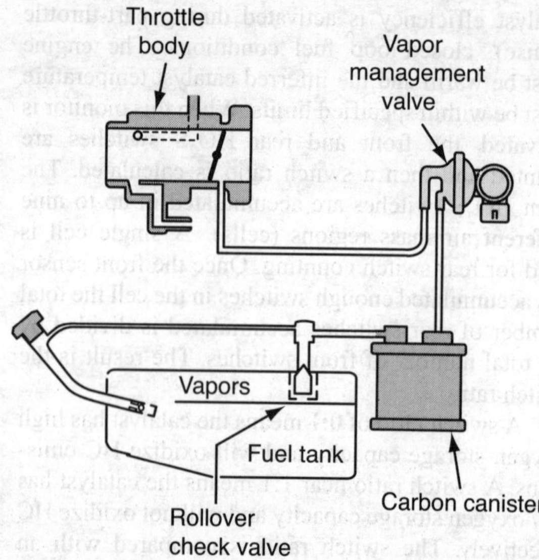


**Figure 2** *Vapor management valve used within an EVAP system.*

The EVAP system leak test is done under two different sets of conditions. The first is a cruise test that is used to detect leaks of 0.04 in. (1.0 mm) in diameter and to screen for leaks of 0.02 in. (0.5 mm) in diameter. If a 0.02 in. (0.5 mm) diameter leak is suspected during the cruise test, the idle test is performed to verify the leak under a more restrictive cold start condition. This test is performed with the system idling.

The cruise test is run after a six-hour engine-off time and then the vehicle is driven at steady highway speeds. Also, the test requires the ambient air temperature to be between 40°F and 100°F (4°C and 37°C).

The cruise test is done in the following phases:

- **Phase 0**. First, the CVS is closed to seal the entire evaporative system and then the VMV is opened to pull a 7 in. $H_2O$ vacuum. If this pull-down vacuum cannot be reached, a large leak is indicated. In addition, at every engine start-up, a check for refueling events is done. If the fuel level at start-up is at least 20 percent greater than the fuel level when the engine was shut off, a **refueling flag** is set in the PCM executive. The refueling flag is used to prevent the 0.02 in.

(0.5 mm) idle test from running until the cruise test for a large leak is done first. The refueling flag remains set until the EVAP monitor completes Phase 0.

If a vacuum of 7 in. H2O could not be achieved after a refueling flag is set, a large leak is indicated. In this case it is possible the fuel cap is loose or off. The minimum fuel tank pressure obtained during the pulldown test is stored and a "Check Fuel Cap" light may be illuminated.

If the initial vacuum is excessive, a vacuum malfunction is indicated. Plugged or pinched vapor lines between the FTPT and the CVS or a stuck open VMV could cause this. If there are large leak faults detected during this phase of the test, the EVAP test will not run the other phases.

- **Phase 1**. This portion of the test is referred to as vacuum stabilization. At this time, the VMV is closed and vacuum is allowed to stabilize for a fixed time. If the pressure in the tank rises immediately, this phase of the test is aborted and Phase 2 is started.
- **Phase 2**. During this portion of the test, the vacuum is held for a calibrated time. Determined by the fuel level in the tank, two test times are used. The shorter timer is used to detect 0.04 in. (1.0 mm) diameter leaks. The longer timer is used to detect 0.02 in. (0.5 mm) diameter leaks.

When Phase 2 is entered, the initial vacuum is recorded. Once the timer for the 0.04 in. (1.0 mm) diameter leak test has expired the vacuum value is recorded again. If the difference between the two readings exceeds the 0.04 in. (1.0 mm) diameter leak threshold, a failure is logged. If three successive test attempts indicate a 0.04 in. (1.0 mm) diameter leak is likely, then a final vapor generation check is done to verify the leak. This is performed in Phase 3 and Phase 4.

If the difference in vacuum reading does not indicate a medium leak, the test continues until the 0.02 in. (0.5 mm) diameter leak test timer expires. The difference between the vacuum levels at the beginning of this phase and at the completion of the small-leak timer are used to determine if a 0.02 in. (0.5 mm) diameter leak is present. If the difference in readings is above the threshold on a single monitoring attempt, a 0.02 in. (0.5 mm) diameter leak is likely and a final vapor generation check is done to verify the leak.

If neither the medium- nor the small-leak threshold is exceeded, the leak test passes.

**You Should Know**

*Fuel level and IAT are used to determine the appropriate fuel tank vapor volume and temperature. Steady state conditions must be maintained throughout this portion of the test. The test will abort if there is an excessive change in engine load, fuel tank pressure, or fuel-level input. These changes may indicate that fuel slosh is occurring. If the monitor does abort, it will attempt to run the test again (up to twenty times) until a maximum time limit after start is reached.*

- **Phase 3**. The CVS is opened for a fixed period of time and releases any vacuum in the system. The VMV remains closed.
- **Phase 4.** During this time, the system is sealed by closing the CVS and is monitored for a rise in tank pressure that results from excessive vapor generation. At the conclusion of Phase 3, the tank pressure reading is recorded. The pressure is monitored for a change from the initial pressure and for absolute pressure.

If the pressure rise is below the threshold value for both the absolute pressure reading and the change in pressure (and a 0.04 in. (1.0 mm) diameter leak was indicated in Phase 2), a DTC is stored.

If the pressure rise is below the threshold value for both the absolute pressure and the change in pressure (and a 0.02 in. (0.5 mm) diameter leak was indicated in Phase 2), the PCM will run the 0.02 in. (0.5 mm) diameter leak test during idle conditions.

Some 2003 model year PCMs have a test to identify if a VMV is leaking during Phase 1. If Phase 1 of the test develops more than 1 in. $H_2O$ of additional vacuum, Phase 2 is skipped and the monitor goes to Phase 3. At this time the PCM will open the CVS to release the vacuum and begin Phase 4. The PCM closes the CVS and measures the vacuum that develops. If the vacuum exceeds approximately 4 in. $H_2O$, a DTC is set.

The long period of time required to detect a 0.02 in. (0.5 mm) diameter leak coupled with the changes in driving conditions may lead to false leak indications while the vehicle is in motion. To confirm that a leak is really present, the **idle check** repeats Phase 0 and Phase 2 with the vehicle stationary. The idle check is done on cold engine start-up to ensure that the fuel is cool. This test is run only during the first 10 minutes after the engine is started. The test is aborted if vehicle speed exceeds 10 mph. However, to minimize the amount of idle time to complete the test, the initial vacuum pulldown (Phase 0) can begin with the vehicle in motion.

Results from the idle check are:

- Loss of vacuum greater than the 0.02 in. (0.5 mm) diameter leak threshold is recorded as a failure.
- Vacuum loss below the vacuum threshold is recorded as a pass. The pending code resulting from the cruise test is erased.
- If the vacuum loss is greater than the pass value but below the failure limit, the test is considered indeterminate and does not count as a pass or fail.

EVAP bypass logic is programmed into the PCM executive in the event that the EVAP monitor cannot be completed. Lack of completion may be due to ambient temperature conditions encountered outside of the 40°F to 100°F (4°C to 37°C ) range or by the elevation being above 8,000 feet (2,438 meters) during a drive cycle. In the 1997 through 1999 model years, the bypass logic considers the EVAP monitor as completed.

Model year 2000 and newer vehicles utilize a timer that is incremented if the EVAP monitor conditions are met (with the exception of the 40°F to 100°F [4°C to 37°C] ambient temperatures or elevation). The timer value represents conditions whereby the EVAP monitor could have run but did not because the temperature or altitude did not allow it. If the timer accumulates over 30 seconds during a drive cycle that all monitors ran, the EVAP monitor is then considered complete.

Another method Ford uses is to measure air-fuel ratio shift between no purge flow and purge flow. The monitor runs when the engine speed is below 900 rpm, the engine is in closed-loop fuel strategy, the engine coolant is between 104°F and 215°F (40°C and 102°C), and vehicle speed is less than 3.76 mph (60 km/h). When the monitor runs, the air-fuel ratio is sampled by the $O_2S$ and averaged before fuel vapors are allowed to flow. The EVAP solenoid is then progressively opened in steps beginning with 10 percent to 40 percent duty cycle, then from a 40 percent to 100 percent duty cycle. During this time, the resulting air-fuel ratio value is compared with the averaged no-flow value. If there is not a sufficient shift in values, the system fails the first attempt (Test 1) of the EVAP monitor. Upon completion of Test 1, the monitor returns to normal operation of the EVAP system. During this time, the monitor determines when most of the vapors should have been drawn from the canister. The monitor then makes a second attempt (Test 2) to check for sufficient vapor flow. If this test fails, a fault is set. The MIL is illuminated after one of the two tests fails on two consecutive drive cycles.

## Misfire Monitor

Cylinder misfire monitoring requires measuring the contribution of each cylinder to engine power. The misfire monitoring system uses a crankshaft angle measurement to determine crankshaft acceleration each time a cylinder is fired. A high data rate crankshaft sensor (eighteen position references per crankshaft revolution) is usually used for this function. While the engine is running, the PCM monitors the crankshaft acceleration time for each cylinder firing. If a cylinder is contributing normal power, a specific amount of crankshaft acceleration occurs. In the event of a misfire, the cylinder does not contribute normal power and the crankshaft acceleration for that cylinder is slowed.

The high data rate input signal is processed using two different programming procedures in the PCM. The first procedure, **pattern cancellation**, is designed to detect low rates of misfire. A special chip in the PCM is used to perform the high data processing. Results of the misfire test are then sent to the PCM's microprocessor over a dedicated serial communication link.

To detect misfire, first the PCM programming learns the normal pattern of cylinder accelerations. This procedure, called **profile correction**, identifies the normal pattern of the crankshaft sensor input. Since factors such as the tone wheel tooth spacing inaccuracies result in different crankshaft input patterns, the PCM cannot be preprogrammed with normal firing events. Profile correction is learned during fuel cutoff deceleration mode. During this time, there are no power impulses occurring. The misfire monitor will not run until the profile of the crankshaft sensor is learned.

Normal crankshaft accelerations are learned from proper firing events. Once the PCM learns what is normal, it can then detect deviations from this pattern. The second procedure is designed to detect if one or more cylinders are continuously misfiring. Since crankshaft tensional vibrations can degrade the sensor signal, this program filters the signal. By filtering the signal the PCM is better able to detect cylinders that are continuously misfiring. Both the low-rate and high-rate misfire programs identify crankshaft acceleration that has deviated from the learned normal pattern.

During the 2003 model year a low data rate system was introduced on vehicles using four-cylinder engines. The high data rate system is still used for misfire monitoring on six- and eight-cylinder engines.

The low data rate system uses only one position reference signal at 10 degrees BTDC for each cylinder. In addition, PCM programming for the 2003 model year was enhanced to provide for misfire detection that occurs in six engine revolutions after initial start-up. This enhancement was done to meet the requirement of identifying misfires within two engine revolutions once the warm drive and idle rpm conditions have been met.

When monitoring for Type A misfires, the monitor checks cylinder misfiring over a period of 200 rpm. This monitor is used to determine if the rate of misfire is enough to result in catalyst damage. The rate of misfire that indicates Type A is rpm and load dependent. The rate ranges from 40 percent at idle to 4 percent at higher rpm and load. Misfire under this condition may result in the PCM shutting off the fuel to the misfiring cylinder. This will help to prevent the catalytic converter from overheating due to raw fuel being burned inside of it. The PCM may turn off up to two injectors at the same time on misfiring cylinders. This fuel shut-off strategy is used on many eight-cylinder engines and some six-cylinder engines. Injector shutdown is not used on four-cylinder engines.

If a Type A misfire is detected and the PCM does not shut off the injectors, the MIL flashes to indicate severe misfire is occurring to the vehicle driver. If a Type A misfire is detected and the PCM shuts off an injector or injectors, the MIL light is illuminated continually indicating that the engine is polluting more than the FTP standard permits.

In current vehicles, the catalyst temperature model that has been used to infer the temperature of the catalyst for the catalyst and oxygen sensor monitors is also used by the misfire monitor. By using the catalyst temperature model, the PCM can determine the temperature of the catalyst while the misfire is occurring. With engine packages that include this strategy, both the catalyst damage misfire rate and the catalyst damage temperature must be over limits before the MIL is illuminated. This prevents fault detection of transit misfire that may result from outside influences other than real misfire, such as driving on a wash-out dirt road. The logic does ensure that the MIL is illuminated for misfires that truly damage the catalyst.

When monitoring for a Type B misfire, the PCM checks cylinder misfire over a period of 1,000 rpm. This monitor is used to determine if the misfire is causing the engine to pollute more than the FTP standard permits. If cylinder misfiring exceeds 1 to 2 percent during this period, the monitor considers the misfiring to be excessive. This amount of cylinder misfiring may not overheat the catalytic converter but it may cause excessive emissions levels. If a misfire is determined in either the first 1,000 engine revolutions since engine start-up, or four groups of 1,000 engine revolutions during the drive cycle after start-up, then a pending DTC is set. If a misfire is detected on the second consecutive drive cycle, the MIL is illuminated and a DTC is stored. Beginning with the 2003 model year, many engine packages will set a specific DTC if the Type B misfire limit is exceeded during the first 1,000 revolutions after engine start-up.

## Fuel System Monitoring

The fuel system monitor checks short-term fuel trim (SFT) and long-term fuel trim (LFT) while the PCM is operating in closed-loop mode. A correction factor is added to the fuel injector pulse-width calculation according to the SFT and the LFT as needed to compensate for variations in the fuel system. The fuel system monitor operates continually when the PCM is in closed-loop mode. Once the monitor is activated, it looks for the adaptive tables to reach a corrective limit and for the air-fuel ratio to exceed a calibrated threshold. When deviation is excessive, the PCM is not capable of proper air-fuel control and emissions increase. The fuel system monitor sets a DTC and illuminates the MIL if the fault occurs during two consecutive drive cycles.

A fuel system lean fault is recorded if LFT is at +25 percent and SFT is greater than +5 percent. A rich malfunction is recorded if LFT is at −25 percent and SFT is less than −10 percent **(Figure 3)**.

## Heated Oxygen Sensor Monitor

All $HO_2Ss$ are monitored once per drive cycle. However, there are separate tests for the upstream and downstream sensors. The upstream $HO_2S$ monitor checks the voltage signal frequency of the upstream $HO_2S$. If the signal voltage frequencies are below the threshold, a faulty sensor is indicated. During the test, the $HO_2S$ monitor varies the fuel delivery and checks for $HO_2S$ response. The response rate is monitored when a special fuel control routine is entered. This routine uses a 1.5 Hz square wave control that will drive the air-fuel ratio around stoichiometry at a specific frequency and magnitude. This results in a $HO_2S$ signal amplitude that can be

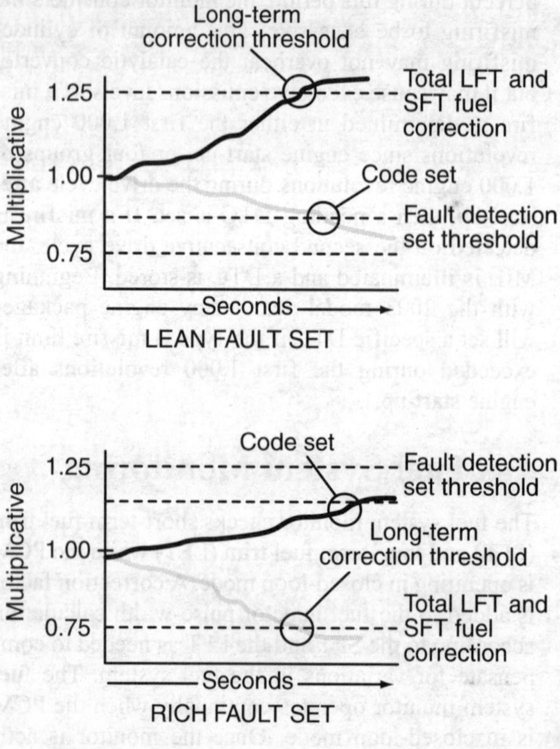


**Figure 3** *Fuel system monitor failure thresholds.*

predetermined. An amplitude lower than that predicted indicates a slow sensor.

The downstream $HO_2S$ function test is performed during normal vehicle operation. The PCM continually monitors the peak rich and lean voltages of the sensors. If the voltages exceed the adjustable rich and lean thresholds, the sensor is considered to be functioning properly. If the voltages do not exceed the rich and lean thresholds after an extended period of vehicle operation, the PCM modifies the air-fuel ratio toward rich or lean in an attempt to force a response from the downstream sensors. If the peak rich and lean thresholds are not exceed, a failure is recorded.

The fuel system and misfire monitors must complete successfully before the $HO_2S$ monitor is enabled.

## Oxygen Sensor Heater Monitor

The oxygen sensor heater circuit is monitored separately from the response monitor just discussed. The heater monitor uses a separate current-monitoring circuit. This is a once-per-trip monitor; however, the heater current is tested three times. If the current value for two of the three tests is lower or higher than the failure thresholds, the sensor heater function is not operating properly. The failure limits are determined by the type of $HO_2S$ used:

- NTK (standard)—Less than 220 mA or greater than 3 amperes
- NTK (fast light off)—Less than 465 mA or greater than 3 amperes
- Bosch (standard)—Less than 400 mA or greater than 3 amperes
- Bosch (fast light off)—Less than 230 mA or greater than 3 amperes

**You Should Know**

*NTK and Bosch are manufacturers of $O_2S$ used by many vehicle manufacturers.*

## EGR System Monitoring

The EGR system contains a **delta pressure feedback EGR (DPFE) sensor (Figure 4)**. The DPFE sensor is a capacitive-type pressure transducer that monitors the pressure difference across a metering orifice. The DPFE sensor receives this signal through two hoses. An orifice is located under the EGR valve and small exhaust pressure hoses are connected from each side of this orifice to the DPFE sensor. During the EGR test, the PCM first checks the DPFE signal. If this sensor signal is within the normal range, the monitor proceeds with the tests.

With the engine idling and the EGR valve closed, the PCM checks for pressure difference at the two pressure hoses connected to the DPFE sensor. When the EGR valve is closed and there is no EGR flow, the pressure should be the same at both pipes. If the pressure is different at these two hoses, the EGR valve is stuck open.

The PCM commands the EGR valve to open and then checks the pressure at the two exhaust hoses connected to the DPFE sensor. With the EGR valve open and EGR flow through the orifice, there should be higher pressure at the upstream hose than at the downstream hose.

The PCM checks the EGR flow by checking the DPFE signal value against an expected DPFE value for the engine operating conditions at steady throttle within a specific rpm range. If a fault is detected in any of the EGR monitor tests, a DTC is set in the PCM memory. If the fault occurs during two consecutive drive cycles, the MIL is illuminated. The EGR monitor operates once per drive cycle.

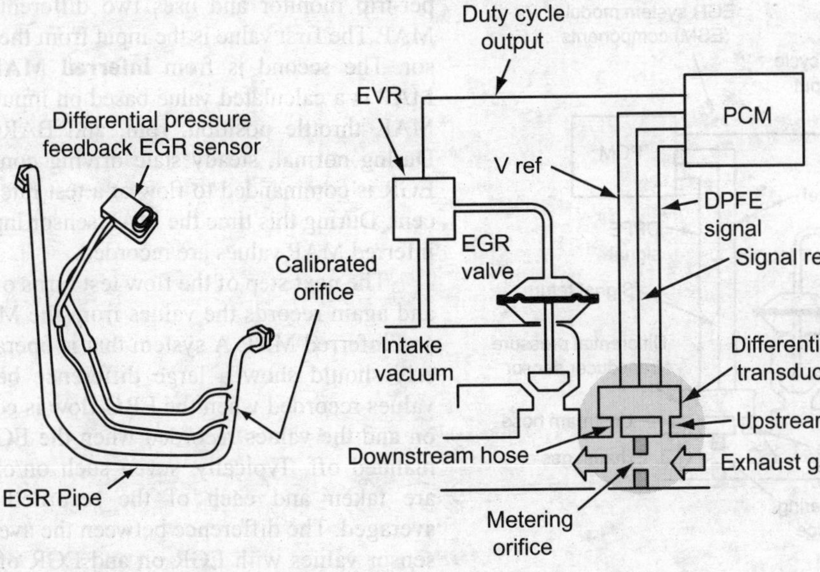


**Figure 4** *The EGR system incorporates a DPFE sensor.*

Since the EGR system normally has water vapors present as the result of the engine combustion process, these vapors may freeze if ambient temperature is too cold. To prevent MIL illumination due to freezing, the PCM uses the following logic:

- If an EGR system failure is determined with ambient temperature above 32°F (0°C), both the EGR system and the EGR monitor are disabled for the remainder of the ignition cycle. On the second consecutive failure a DTC is stored and the MIL is illuminated.
- If an EGR system failure is determined with ambient temperature below 32°F (0°C), only the EGR system is disabled for the remainder of the ignition cycle. At this time a DTC is not stored nor is the I/M readiness status changed. The EGR monitor continues to run and if it determines that the failure is no longer present, the EGR system is enabled again to allow normal system operation.

During the EGR test, the differential pressure indicated by the DPFE sensor is monitored. The high flow check is performed at idle with zero EGR flow. If the differential pressure exceeds a failure threshold, a stuck-open EGR valve is indicated. During vehicle acceleration, the pressure difference input from the DPFE sensor is zero. At this time EGR flow is checked to ensure that both hoses to the DPFE sensor are connected. During this check the pressure difference should be zero. If the pressure difference exceeds a maximum positive threshold then the downstream DPFE hose has a failure. If the pressure difference falls below a minimum negative threshold then the upstream DPFE hose has a failure.

After the vehicle has warmed sufficiently to allow EGR system operation, the low flow check is performed. Whenever the EVR duty cycle is greater than 80 percent the pressure difference is monitored to determine the amount of EGR restriction. If the pressure difference is below the threshold, a low flow is indicated and a failure is recorded.

Beginning with the 2002.5 model year, Ford began to use a revised DPFE system **(Figure 5)**. This system operates much the same as the conventional DPFE system. The difference is that the components of the system have been combined into a single component called the EGR system module (ESM). With the EGR orifice relocated to the intake, the downstream pressure signal now monitors manifold absolute pressure (MAP). The ESM provides a differential DPFE input that is the same as that of the conventional DPFE system.

The electric stepper motor EGR system **(Figure 6)** directly activates the EGR system through the use of a stepper motor. This type of system does not rely on engine vacuum and diaphragm activation within the EGR valve. The stepper motor range is from zero to fifty-two steps. Zero steps indicates the EGR valve is fully closed and fifty-two steps indicates the valve is fully open. A MAP sensor is used to determine proper EGR flow.

The EGR monitor in vehicles using the stepper motor system consists of an electrical and a

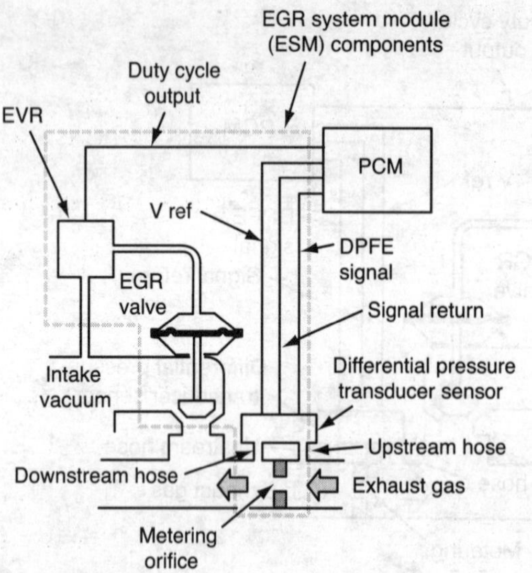


**Figure 5** *Revised DPFE system.*

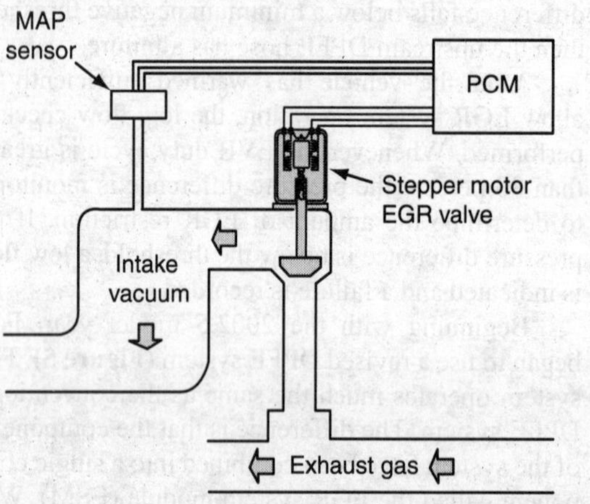


**Figure 6** *Some Ford engines are equipped with an electric stepper motor EGR system.*

functional test. The electrical test continually checks for continuity-type faults and the functional test checks for flow rate. If the electrical test determines an open circuit, short to power, or short to ground has occurred in one or more of the stepper motor coils, the EGR system is disabled and the monitor stops running for the remainder of the ignition cycle. In addition, the MAP sensor is checked for electrical failures. If a failure is determined, the monitor suspends running the EGR flow test.

The functional test is performed once the engine has warmed to an operating temperature that can support normal EGR flow rates. This test is a once-per-trip monitor and uses two different values of MAP. The first value is the input from the MAP sensor. The second is from **inferred MAP.** Inferred MAP is a calculated value based on inputs from the MAF, throttle position, rpm, and BARO sensors. During normal, steady state driving conditions the EGR is commanded to flow at a test rate of 10 percent. During this time the MAP sensor input and the inferred MAP values are recorded.

The next step of the flow test turns off the EGR and again records the values from the MAP sensor and inferred MAP. A system that is operating properly should show a large difference between the values recorded when the ERG flow is commanded on and the values recorded when the EGR is commanded off. Typically, seven such on/off samples are taken and each of the recorded values is averaged. The difference between the average MAP sensor values with EGR on and EGR off is calculated. Also, the difference between the average inferred MAP with EGR on and EGR off is calculated. The resulting differences between MAP sensor and inferred MAP are added together. If the sum of these two values is greater than 8.00 in. Hg or less than 0.62 in. Hg, a flow malfunction is indicated.

## Secondary Air-Injection (AIR) System Monitor

If equipped on the vehicle, the AIR system is monitored with both passive and active tests. During the passive test, the voltage of the precatalyst $HO_2S$ is monitored from start-up to closed-loop operation. The AIR pump is normally on during this time. Once the $HO_2S$ is warm enough to produce a voltage signal, the voltage should be low if the AIR pump is delivering air to the exhaust manifolds. The secondary AIR monitor will indicate a pass if the $HO_2S$ voltage is low at this time. The passive test also looks for a higher $HO_2S$ voltage when the AIR flow to the exhaust manifolds is turned off by the PCM. If the AIR system passes the passive test, no further testing is done. If the AIR system fails the passive test or the test is inconclusive, the AIR monitor proceeds with the active test.

During the active test, the PCM cycles the AIR flow to the exhaust manifolds on and off during closed-loop operation and monitors the precatalyst $HO_2S$ voltage and the SFT value. When the AIR flow to the exhaust manifolds is turned on, the $HO_2S$ voltage should decrease and the SFT value should indicate a leaner condition. The secondary AIR system monitor illuminates the MIL and stores a DTC in the PCM memory if the AIR system fails the active test on two consecutive trips.

Some Ford vehicles have an electric air pump (EAP) system. In this system, the EAP is controlled by a solid-state relay (SSR). A signal from the PCM operates the SSR. The PCM also operates an air-injection bypass solenoid. This solenoid supplies vacuum to dual air diverter valves **(Figure 7)**.

When the engine is started, the PCM signals the SSR to start the EAP. The SSR supplies the high current required by the EAP. The EAP may provide a 10-second delay in pump operation after the engine is started. The PCM also energizes the air-injection bypass solenoid. When this solenoid is energized, it supplies vacuum to the dual air diverter valves. This action opens the normally closed air diverter valves and air from the pump is now delivered to the exhaust manifolds. After the oxygen sensors warm up, the secondary air pump continues to be energized while the fuel system goes into closed-loop mode. The secondary air system continues to run in closed-loop mode until the air pump is deenergized. The purpose of the EAP system is to oxidize HC and CO in the exhaust manifolds for 20 to 120 seconds after the engine is started. The length of EAP operation depends on engine temperature. Cold engine temperatures provide longer EAP operation. Once the catalytic converter is operating properly, the PCM signals the SSR to shut off the EAP. The PCM also deenergizes the air-injection bypass solenoid, which allows the air diverter valves to close.

The PCM monitors the SSR and the EAP to determine if secondary air is present. The AIR pump flow check monitors the $HO_2S$ signal at idle to determine if secondary air is being delivered to the exhaust system. The air-fuel ratio is commanded to open-loop rich mode and the AIR pump is turned on. The time required for the $HO_2S$ signal to go low is monitored. If the $HO_2S$ signal does not go low within the allowable time, a low/no-flow malfunction is indicated.

This PCM monitor for the EAP system runs once per drive cycle. When a malfunction occurs in the EAP system on two consecutive drive cycles, a DTC is stored in the PCM memory and the MIL is illuminated.

## Comprehensive Components Monitor

The comprehensive components monitor (CCM) uses two strategies to monitor inputs and two strategies to monitor outputs. One strategy involves checking inputs for electrical defects and out-of-range values by monitoring the input signals at the A/D converter. The inputs that are monitored in this way include:

1. Rear $HO_2S$ inputs
2. $HO_2S$ inputs
3. MAF sensor
4. Manual lever position (MLP) sensor
5. Throttle position sensor (TPS)
6. ECT sensor
7. IAT sensor

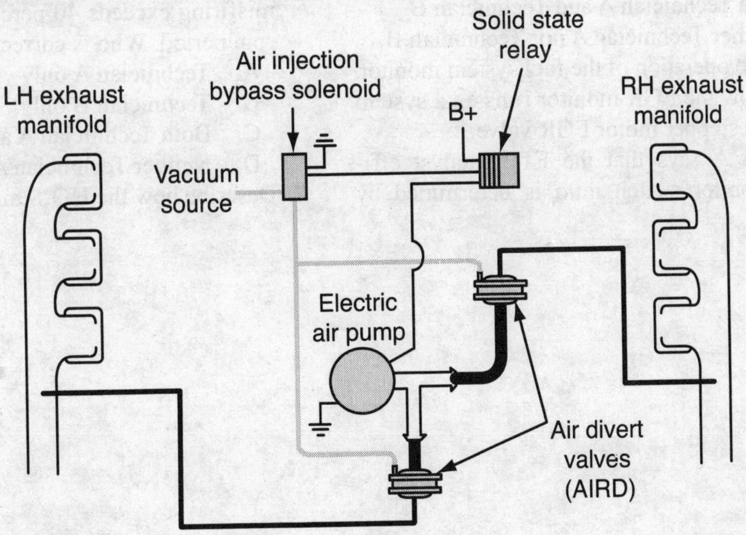


**Figure 7** *Some Ford vehicles are equipped with an electric air pump (EAP) system.*

The second strategy of the CCM is to perform rationality checks. During a rationality check, the monitor uses other sensor readings and calculations to determine if a sensor reading is correct for the present conditions. The rationality checks are:

1. Profile ignition pickup (PIP)
2. Output shaft speed (OSS) sensor
3. Ignition diagnostic monitor (IDM)
4. Cylinder identification (CID) sensor
5. Vehicle speed sensor (VSS)

Functionality checks of outputs are also performed. For example, the PCM output that controls the idle air control (IAC) motor is monitored by checking the idle speed demanded by the inputs against the closed-loop idle speed correction supplied by the PCM to the IAC motor.

In the CCM checks most of the outputs by monitoring the voltage of each output solenoid, relay, or actuator at the output driver in the PCM. If the output is off, this voltage should be high. This voltage is pulled low when the output is on.

Monitored outputs include:

1. Wide-open throttle A/C cutoff (WAC)
2. Shift solenoid 1 (SS1)
3. Shift solenoid 2 (SS2)
4. Torque converter clutch (TCC) solenoid
5. HO2S heaters
6. High fan control (HFC)
7. Fan control (FC)
8. Electronic pressure control (EPC) solenoid

## Review Questions

1. Describe the operation of the steady-state catalyst efficiency monitor.
2. The evaporative emissions system monitor is being discussed. Technician A says that the cruise test is run after a six-hour engine-off time and then the vehicle is driven at steady highway speeds. Technician B says that one method of EVAP monitoring is to measure air-fuel ratio shift with no purge flow as compared to purge flow. Who is correct?
   A. Technician A only
   B. Technician B only
   C. Both Technician A and Technician B
   D. Neither Technician A nor Technician B
3. Explain the operation of the fuel system monitor.
4. Explain how the EGR monitor runs on a system that uses a stepper motor EGR valve.
5. Technician A says that the FTP catalyst efficiency monitor switch ratio is determined by dividing the total number of rear switches by the total number of front switches. Technician B says that if the switch ratio is less than the emissions threshold, the catalyst has failed. Who is correct?
   A. Technician A only
   B. Technician B only
   C. Both Technician A and Technician B
   D. Neither Technician A nor Technician B
6. Technician A says that a Type A misfire is detected when cylinder misfire exceeds 1 to 2 percent over a 1,000 rpm period. Technician B says that a type B misfire is detected if cylinder misfiring exceeds 40 percent at idle over a 200 rpm period. Who is correct?
   A. Technician A only
   B. Technician B only
   C. Both Technician A and Technician B
   D. Neither Technician A nor Technician B
7. Describe how the $HO_2S$ monitor runs.

CHAPTER

# 34 Import OBD-II

## Introduction

This chapter provides an example of how import manufacturers have approached OBD-II requirements. Many European countries also have an OBD-II system similar to the United States. Their version is called Europe On-Board Diagnostics (EOBD).

> **You Should Know**
>
> *It is beyond the scope of this book to describe every import OBD-II system. The following Toyota system is offered as an example.*

## Evaporative Systems Monitor

There are two different methods used to detect EVAP system leaks. These are classified as non-intrusive pressure detection and as intrusive monitor systems. Early EVAP systems had to detect leaks of 0.040 in. (1.0 mm) in diameter and greater. Beginning with the 2000 model year, the EVAP monitor had to meet the mandated standard of detecting 0.020 in. (0.5 mm) diameter leaks. In addition, the EVAP monitor has to determine proper vapor purge and component operation.

### Nonintrusive Pressure Detection

In the nonintrusive pressure detection monitor system, the PCM uses a signal from the **vapor pressure sensor (VPS)** to measure the pressures in the EVAP system and in the purge side of the charcoal canister **(Figure 1)**. This sensor uses a silicon chip with a calibrated reference pressure on one side and the other side is exposed to vapor pressure. Changes in vapor pressure cause the chip to flex and vary the voltage signal to the PCM. As vapor pressure increases, the voltage signal increases. The changes in the pressures being measured can be less than 8.3 in.$H_2O$ (0.3 psi).

To separate canister side pressure from tank pressure the two pressures, a **three-way vacuum solenoid valve (VSV)** is connected to the VPS, fuel tank and lines, and charcoal canister **(Figure 2)**. The VSV allows the VPS to detect either canister or tank pressure. When the three-way VSV is off, the VPS measures the pressure in the canister purge system.

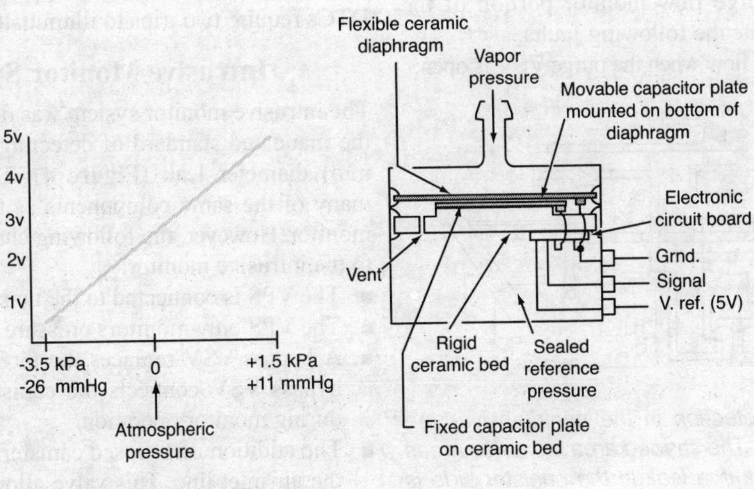


**Figure 1** *Vapor pressure sensor.*

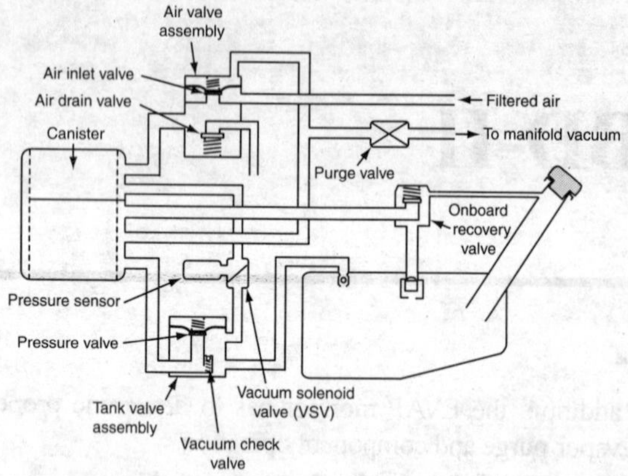


**Figure 2** *Early-type pressure-detection system.*

When the PCM activates the VSV, the VPS measures the pressure in the fuel tank.

When the VPS detects pressure that is higher or lower than atmospheric pressure, the PCM determines that there are no leaks in that portion of the system. If the VPS detects that either the tank or canister purge side of the system is at atmospheric pressure under specific conditions, the PCM determines there is a leak in that portion of the system. If a leak is detected on two consecutive completed trips, the MIL is illuminated and a DTC is stored. The PCM may take more than 20 minutes to complete testing the fuel tank side.

Separate DTCs are logged based on the location of the leak. One DTC is for leaks on the fuel tank side of the EVAP system and another is for the canister side. The fuel tank side of the test includes the lines between the fuel tank and *part* of the canister. **Figure 3** illustrates possible locations for a leak on the canister side.

The EVAP purge flow monitor portion of the test is used to isolate the following faults:

- Restricted purge flow when the purge VSV is open.

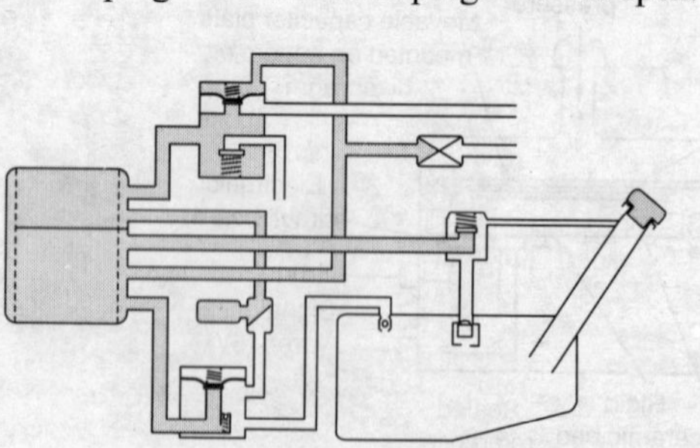

**Figure 3** *Leak detection in the nonintensive-type system. The shaded area indicates areas to check if a leak in the canister side is suspected.*

- Low purge flow when the purge VSV is closed.
- Pressure pulsations during normal purge conditions.

To test for restricted vapor purge flow, the PCM uses the VPS to monitor pressure pulsations. When the EVAP system is being purged the cycling of the purge VSV should generates pressure pulsations as the pressure in the canister side drops. If the pulsations are not detected the PCM determines the EVAP system is not functioning properly.

When the engine is first started, purge flow should not occur. During this time low vapor purge flow is tested. At start-up the canister side pressure should be equal to atmospheric pressure. If the canister pressure drops to a value that is close to intake manifold pressure (vacuum) than this indicates that the purge VSV is stuck open or leaking.

The three-way VSV is monitored for component performance using a test with two modes. The three-way VSV is considered to be functioning properly if there is a pressure difference between the tank and canister when the three-way VSV is activated. If the pressures are equal, then the PCM looks for the following conditions:

- If during purging, pressure pulsations generated by the purge VSV are not detected in the canister side, the three-way VSV is determined to be faulty.
- If pressure pulsations detected in the fuel tank than the three-way VSV is determined to be leaking or did not activate.

Two separate DTCs can set depending on the outcome of the test. These DTCs indicate a faulty VPS or a fault in the VPS circuit. Since the PCM monitors the VPS signal under several different conditions maturing of DTCs can be suspended until the EVAP monitor has fully completed testing. Both DTCs require two trips to illuminate the MIL.

## Intrusive Monitor System

The intrusive monitor system was developed to meet the mandated standard of detecting a 0.020 in. (0.5 mm) diameter leak **(Figure 4)**. This system uses many of the same components as the non-intrusive monitor. However, the following changes were made to the intrusive monitor:

- The VPS is connected to the tank side.
- The VPS only monitors pressure in the tank.
- A bypass VSV replaces the three-way VSV. The bypass VSV connects the canister and the tank during monitor operation.
- The addition of a closed canister valve (CCV) on the air inlet line. This valve allows the system to be sealed during the test.

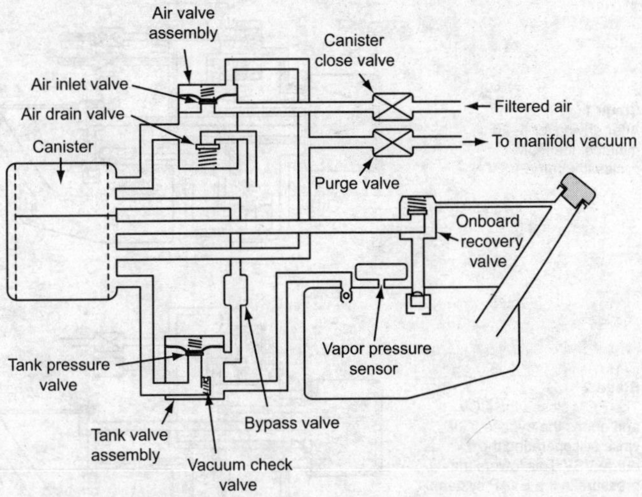


**Figure 4** *Intrusive-type pressure-detection system.*

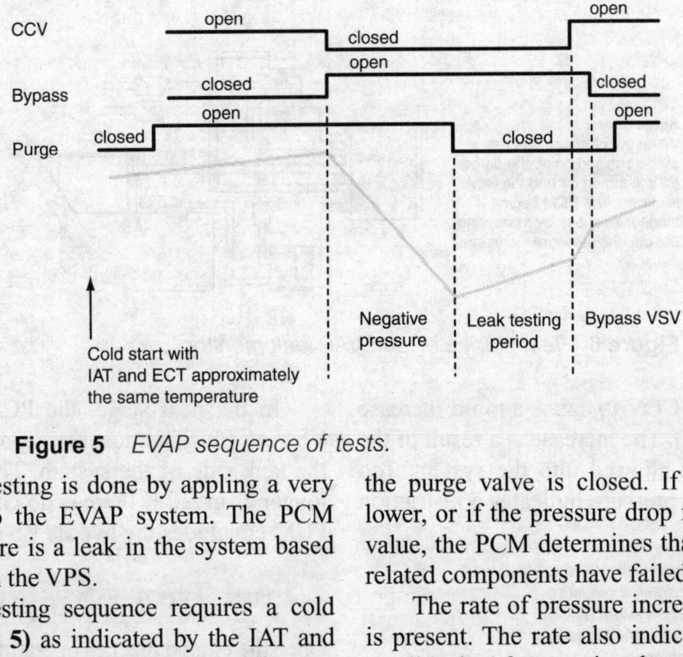


**Figure 5** *EVAP sequence of tests.*

- Leak detection testing is done by appling a very small vacuum to the EVAP system. The PCM determines if there is a leak in the system based on the input from the VPS.

The monitor testing sequence requires a cold engine start **(Figure 5)** as indicated by the IAT and ECT sensors having approximately the same temperature value. With the engine running, the PCM monitors the fuel tank pressure. The "gas law" principle states that as the temperature of the fuel increases, pressure increases. The PCM will purge the canister and then the bypass VSV is closed. With a sealed system, pressure will increase in the fuel tank **(Figure 6)**.

At a predetermined point, the PCM closes the CCV and opens the bypass VSV. This causing the pressure to drop in the EVAP system. The PCM will continue to operate the purge valve until the pressure is lowered to a specified value. Once this is obtained the purge valve is closed. If the pressure fails to lower, or if the pressure drop is more than the limit value, the PCM determines that the purge VSV and related components have failed.

The rate of pressure increase indicates if a leak is present. The rate also indicates if it is a large or small leak. After purging the system, the purge VSV is turned off and a vacuum is sealed in the EVAP system. The PCM will now monitor for a pressure increase. An increase in pressure is expected due to temperature. However, the pressure increase should be slow. A rapid increase in pressure indicates that there is a leak in the EVAP system, since atmospheric pressure is entering the system. A pressure rise just above normal indicate a very small leak.

To check the CCV and vent for proper operation of the air inlet side, the PCM monitors the VPS for a vapor pressure increase to a specified point. The

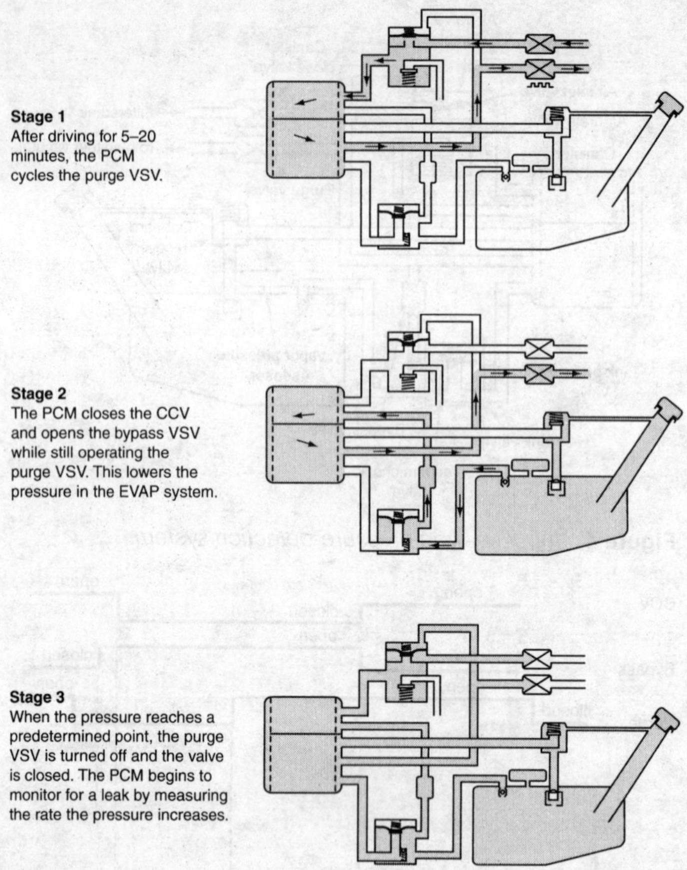


**Figure 6** *Test stages of an EVAP leak monitor.*

PCM then opens the CCV to cause a rapid increase in pressure **(Figure 7)**. The increase is a result of the atmospheric pressure allowed into the system. Too low of an increase in pressure indicates a restriction on the air inlet side.

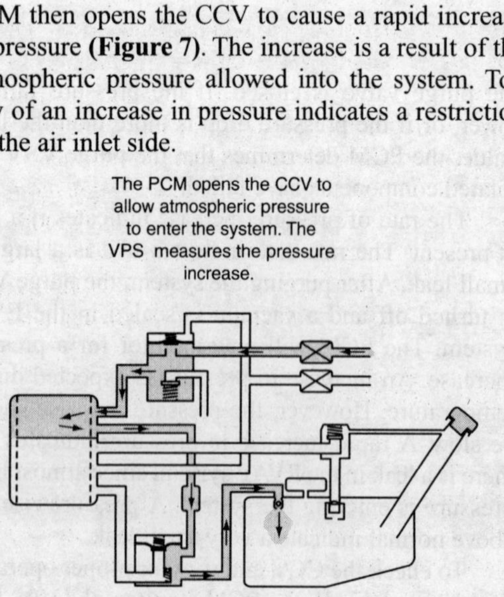


**Figure 7** *Operation of the CCV.*

In the next stage, the PCM closes the bypass VSV to block atmospheric pressure from entering the tank side of the system. The pressure rise is no longer as great. If there is no change in pressure, the PCM concludes the bypass VSV did not close.

## Fuel Trim—System Too Lean

Once the engine reaches operating temperature and the air-fuel ratio feedback is stabilized, the fuel trim reaches its limit of correction to the rich side if a fuel system lean fault is detected. The fuel system lean monitor requires two consecutive failures to turn on the MIL.

## Fuel Trim—System Too Rich

Once the engine reaches operating temperature and the air-fuel ratio feedback is stabilized, the fuel trim reaches its limit of correction to the lean side if a fuel system rich fault is detected. The fuel system rich monitor requires two consecutive failures to turn on the MIL.

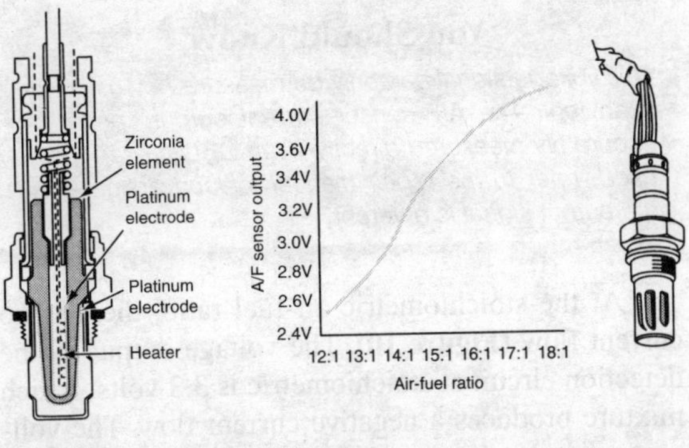


**Figure 8** *Air-fuel ratio sensor.*

## Misfire Monitor

The PCM measures crankshaft speed fluctuations by monitoring the crankshaft position sensor. The crankshaft will accelerate each time a cylinder is on the power stroke. If a cylinder has a misfire, the crankshaft speed slows down. The PCM detects the change in frequency in the crankshaft position sensor signal. The PCM uses the camshaft position sensor to identify the cylinder that has experienced the misfire. Severe rough road conditions may temporarily suspend misfire monitor operation.

The Type A misfire indicates that the misfire severe enough to damage the catalytic converter. This is indicated to the driver by a blinking MIL as long as catalyst temperatures of 1,832°F (1,000°C) or more is inferred. The PCM calculates the catalyst temperature based on driving conditions and the percentage of misfire.

The Type B misfire is less severe but will increase emissions above FTP standards. This type of misfire is determined by the percentage of misfire compared with PCM emission specifications. Type B misfire is a two-trip monitor.

## Heated Oxygen Sensor Monitor

The $HO_2S$ monitor checks for sensor circuit malfunctions, slow response rate, and for a malfunction of the sensor's heater circuit. There are separate DTCs for each of these conditions and for each of the sensors except for the downstream sensor, which is not monitored for response rate.

When the PCM sees the right conditions, it tests the heated oxygen sensors for performance by measuring the signal response as the fuel injected into the cylinder is varied. The faster the $O_2S$ responds, the better the sensor. A DTC is stored when there is little or no signal response from the $O_2S$. This DTC is a one-trip code. A fault is determined if the engine is at normal operating temperature and oxygen sensor output does not indicate rich air-fuel mixture even when the following conditions are present for at least 1.5 minutes:

- Engine speed is 1,500 rpm or more.
- Vehicle speed is 25–62 mph (40–100 km/h).
- Throttle valve does not fully close.
- At least 140 seconds has elapsed since starting engine.

The oxygen sensor's output voltage portion of the monitor determines if the signal to the PCM stays high or low during the test period. The faults are detected if the voltage output of the oxygen sensor remains at 0.40 volt or more, or 0.55 volt or less during idling after the engine is warmed up. This is a two-trip fault.

The PCM also monitors the response of the $O_2S$. Response is judged by the time it takes for the $O_2$ sensor to switch from 350 mV and 550 mV. Response time for the heated oxygen sensor's voltage output to change from high to low, or from low to high is 1.1 seconds or more during idling after the engine is warmed up. This is also a two-trip failure.

## Oxygen Sensor Heater Monitor

The PCM monitors the current required to heat the sensor. If the current is above 2 amperes or below 200 mA, a DTC is set. This is a two-trip failure.

## A/F Sensor Monitor

The air-fuel (A/F) sensor monitor is similar to the $O_2S$ monitor. Since the A/F sensor has characteristics that differ from those of the oxygen sensor, the operating parameters of the monitor are different.

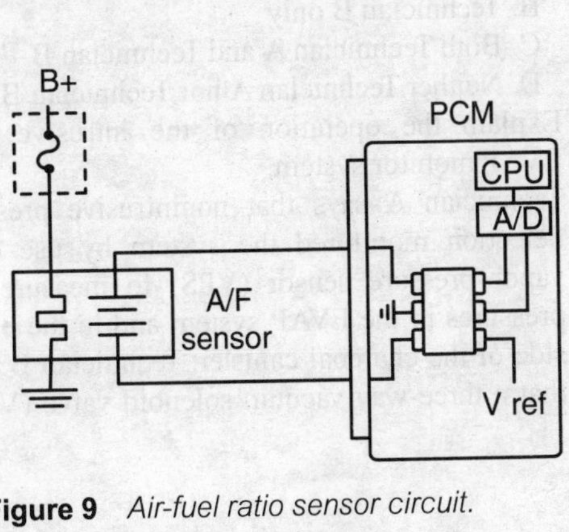


**Figure 9** *Air-fuel ratio sensor circuit.*

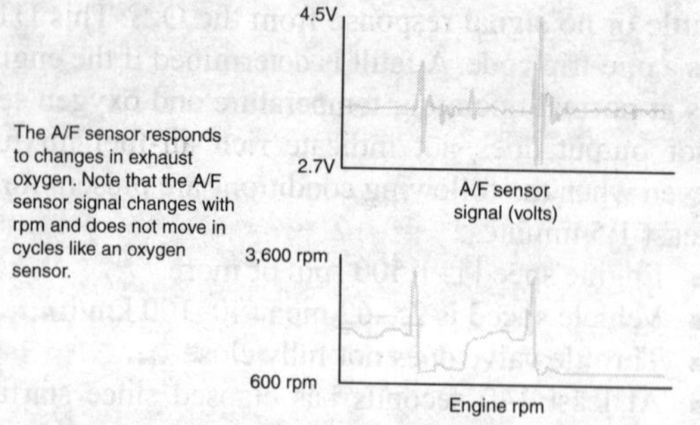


**Figure 10** *Air-fuel ratio sensor output signals.*

### You Should Know

*The A/F sensor is also referred to as a wide-range or wide-ratio sensor because of its ability to detect air-fuel ratios over a wide range.*

The A/F sensor **(Figure 8)** operates at approximately 1,200°F (648°C) which is hotter than the typical oxygen sensor that operates at 750°F (400°C). The biggest difference is that the A/F sensor changes its current output in relation to the amount of oxygen in the exhaust stream. A detection circuit in the PCM detects the change and strength of current flow and puts out a voltage signal that is relatively proportional to exhaust oxygen content **(Figure 9)**.

### You Should Know

*This voltage signal can only be measured by using the scan tool. The A/F sensor current output cannot be accurately measured directly. If an OBD-II scan tool is used, refer to the repair manual for conversion since the output signal is different.*

At the stoichiometric air-fuel ratio, there is no current flow **(Figure 10)**. The voltage output by the detection circuit at stoichiometric is 3.3 volts. A rich mixture produces a negative current flow. The voltage output by the detection circuit will be below 3.3 volts. A lean mixture produces a positive current flow. The detection circuit will now produce a voltage signal above 3.3 volts.

DTCs are related to voltage output. This DTC will set when a sensor output has very little or no activity. The A/F sensor is monitored for voltage change activity when the following conditions are met:

- Engine speed is 1,500 rpm or more.
- Vehicle speed is 25 to 62 mph (40 to 100 k/hr).
- TPS does not indicate idle.
- The enabling conditions continue for at least 90 seconds.
- At least 140 seconds must have passed since the engine was started.

## Review Questions

1. Technician A says that the nonintrusive EVAP system monitor detects leaks of 0.020 in. (0.5 mm) in diameter. Technician B says that the nonintrusive EVAP system monitor determines leaks by measuring pressures in both sides of the EVAP system. Who is correct?
   A. Technician A only
   B. Technician B only
   C. Both Technician A and Technician B
   D. Neither Technician A nor Technician B
2. Explain the operation of the intrusive type EVAP monitor system.
3. Technician A says that nonintrusive pressure detection monitored the system by use of a vapor pressure sensor (VPS) to measure the pressures in the EVAP system and in the purge side of the charcoal canister. Technician B says that a three-way vacuum solenoid valve (VSV) is used to detect either canister or tank pressure. Who is correct?
   A. Technician A only
   B. Technician B only
   C. Both Technician A and Technician B
   D. Neither Technician A nor Technician B
4. Technician A says that the PCM test the oxygen sensors for performance by measuring the signal response as the fuel injected into the cylinder is varied. Technician B says that the slower the oxygen sensor responds, the better the sensor. Who is correct?
   A. Technician A only
   B. Technician B only
   C. Both Technician A and Technician B
   D. Neither Technician A nor Technician B
5. Explain the differences between an A/F sensor and an oxygen sensor.

CHAPTER

# 35 Introduction to OBD-II Diagnostics

## Introduction

Because all manufacturers have continually updated, expanded, and improved their engine control systems, there are now hundreds of different domestic and import systems on the road. Methods to read onboard diagnostic data and troubleshoot systems can vary greatly among manufacturers and among model lines and years of the same manufacturer. This is why it is extremely important that the service manual be referred to during diagnostics.

Fortunately, regulations have been passed that call for a standardized test procedure for all electronic control systems. As a result of this mandate, known as OBD-II, vehicles use the same terms, acronyms, and definitions to describe their components. They also have the same type of diagnostic connector, the same basic test sequences, and they display the same trouble codes. OBD-II began in 1994 and was required to be on all passenger cars sold in North America after the 1996 model year. This chapter discusses OBD-II diagnostics.

**You Should Know**

*Standardized test procedures, terms, acronyms, and definitions are referred to as the SAE J1930 standard.*

The purpose of OBD-II was to provide standard service procedures without the use of dedicated special tools. To accomplish this goal, manufacturers were required to change many aspects of their electronic control systems. According to the guidelines of OBD-II, all vehicles will have:

- A universal diagnostic test connector, known as the data link connector (DLC), with dedicated pin assignments
- A standard location for the DLC—under the dash on the driver's side of the vehicle
- A standard list of diagnostic trouble codes (DTCs)
- Vehicle identification that is automatically transmitted to the scan tool
- Stored trouble codes that can be cleared from the computer's memory with the scan tool
- The ability to record and store in memory a snapshot of the operating conditions that existed when a fault occurred
- The ability to store a code whenever a fault is detected that affects exhaust quality
- A standard glossary of terms, acronyms, and definitions used for all components in the electronic control systems

The standard DLC **(Figure 1)** is a 16-pin, D-shaped connector with guide keys that allow the scan tool to be installed only one way. Use of a standard connector design and pin designation allows data retrieval with any scan tool designed for OBD-II.

Many of the trouble codes from an OBD-II system will indicate the same malfunction regardless of manufacturer. However, some of the codes will pertain only to a particular system or will mean something different with each system. The DTC is a five-character code with both letters and numbers **(Figure 2)**.

**You Should Know**

*The DTC layout is called the alphanumeric system.*

## OBD-I versus OBD-II Diagnostics

In OBD-I-equipped vehicles, the MIL would come on if a fault was detected. If the fault was not found on the next test of the circuit, the MIL would go out. If the light came on for a few seconds then never came back on, most drivers would not bother bringing the vehicle to the shop for diagnosis. With OBD-II, there are not any intermittent MILs. Once the MIL is turned on, it takes a series of trips without a

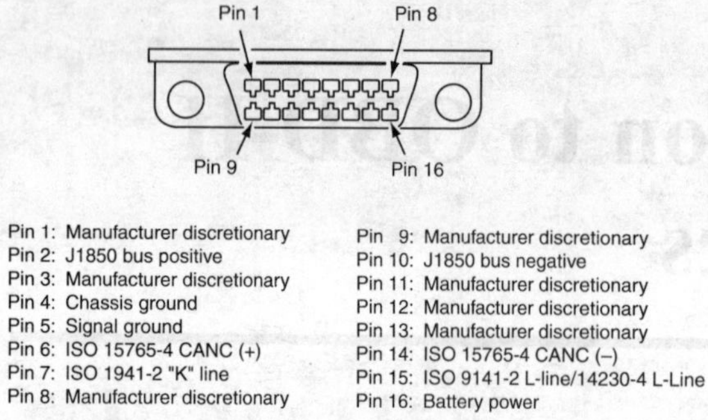


**Figure 1** *Standard OBD-II data link connector.*

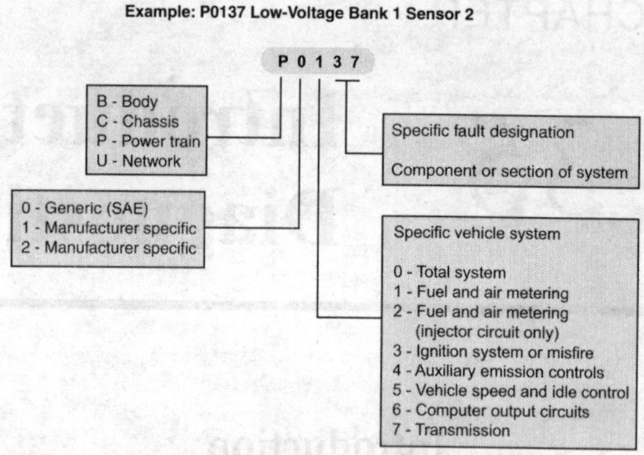


**Figure 2** *OBD-II trouble code format.*

fault to turn it off. Because of this, customers are more likely to bring their vehicles in for diagnosis.

Once the vehicle is in the shop, other differences exist between OBD-I and OBD-II systems. With OBD-I systems, it is a common practice to read all fault codes, record them, and then erase them. After the DTCs are erased, the vehicle is driven to see which DTCs reappear. These codes are then diagnosed. On OBD-II-equipped vehicles, most manufacturers recommend that the fault codes be erased *after* the repair is verified. This means that after the repair is done, allow the monitor to run again to verify that there are no more faults. It is recommended that fault codes be erased only *after* the repair is verified since erasing the fault codes on OBD-II-equipped vehicles clears all OBD-II data. This includes freeze-frames and similar conditions windows that are helpful for diagnostics. For an OBD-II-equipped vehicle that comes in with the MIL on, the following is a typical diagnostic procedure:

1. Read and record all DTCs **(Figure 3)**.
2. Examine the freeze-frame data and determine which DTC corresponds with the freeze-frame **(Figure 4)**.

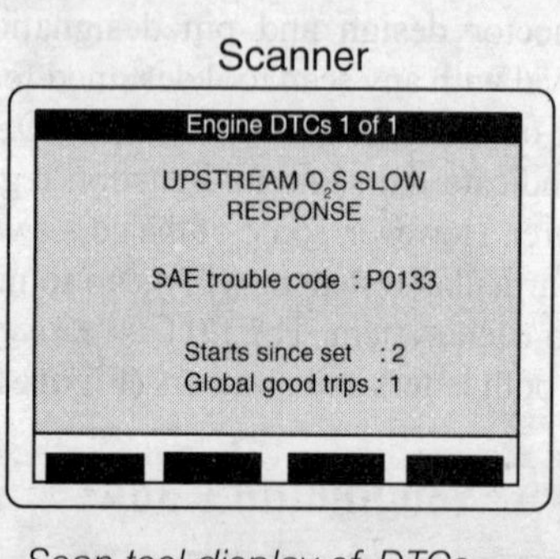


**Figure 3** *Scan tool display of DTCs.*

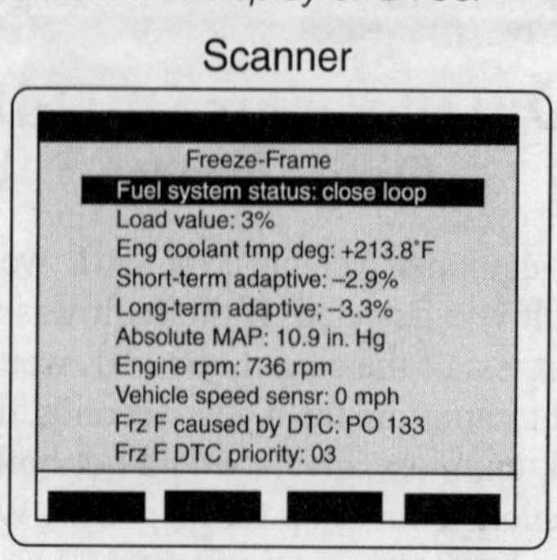


**Figure 4** *Freeze-frame data is stored when the PCM first detects a failure.*

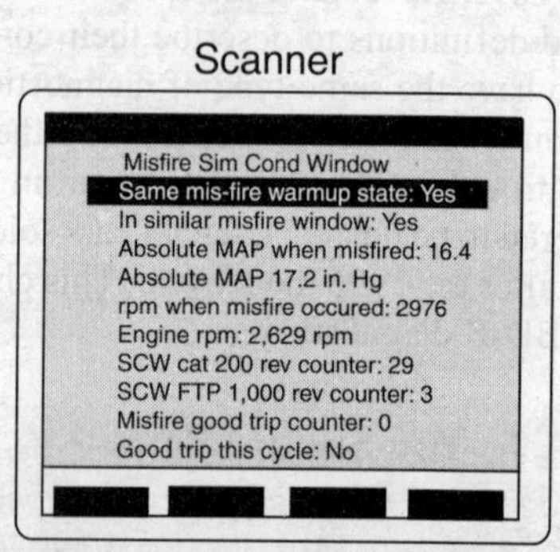


**Figure 5** *Similar conditions window.*

**Figure 6** *Good trip counter.*

3. If applicable, examine the similar conditions window **(Figure 5)**.
4. Determine the cause of the fault using the diagnostic procedures outlined in the service manual.
5. Perform the needed repairs.
6. Test drive the vehicle while staying in the similar conditions window (if applicable) or in the same conditions as shown in the freeze-frame.
7. Observe the **good trip counter**. The good trip counter increments when the monitor has run in the required parameters and passed. Some counters do not record a good trip until the ignition key is cycled after the monitor has run. If it increments by one **(Figure 6)**, pull over to the side of the road and shut off the engine.
8. Restart the engine and drive back to the shop. During this time, the good trip counter should increment to two good trips.
9. At this time, you have confirmed the repair was successful. If no good trips were incremented, there is still a failure in the system.
10. If the repair is confirmed as successful, erase the fault codes and return the vehicle to the customer.

> **You Should Know**
>
> *With some OBD-II systems, a good trip requires the monitor to meet all conditions* and *the engine to run for 2 minutes. In addition, the good trip counter display on the scan tool may not increment until the ignition key is cycled.*

Also, keep in mind that the freeze-frame data and similar conditions window is recorded when the PCM first detects the failure. The driving conditions explained by the customer may not match those found in these two recordings. If the scan tool has the capabilities, it is a good practice to read any one-trip failures that are recorded.

## Generic Scan Tool Modes

J1979 OBD-II regulations allow "generic" scan tools to be used to retrieve emissions-related information. Some scan tools may provide additional features and functions than what is presented here. The following is a brief description of each test mode supported by the generic scan tool application. The information is characterized by requesting information using a **parameter ID (PID)**. A PID is a type of request/response that is legislated as part of SAE J1979 support. Not all PID-requested information is applicable or supported by all systems. Consequently, the scan tool software can only support what is available for a given vehicle in a given model year.

### Mode 1—Request Current Power-Train Data

The purpose of this mode is to allow access to current emissions-related data values including inputs and outputs, system status information, and calculated parameter values. The request for information includes a PID value that indicates to the onboard system the specific information requested. The onboard module will respond to this message by transmitting a module ID and the requested data values.

### Mode 2—Request Power-Train Freeze-Frame Data

The purpose of this mode is to allow access to emissions-related data values that are reported at the time the fault matures (MIL on). The request for information includes a PID value that indicates to the onboard system the specific information requested. The onboard module will respond to this request by transmitting the requested data value stored by the system. This mode also allows expansion to meet manufacturer-specific requirements not necessarily related to the required freeze-frame and not necessarily containing the same data values as the required freeze-frame.

Trip failures are not stored in the J1979 Mode 2 Freeze-Frame; only matured faults (DTCs) are stored. However, the data stored in J1979 Mode 2 Freeze-Frame will be from the first occurrence of the first emissions-related failure unless overwritten by a higher-priority DTC.

### Mode 3—Emissions-Related DTCs

The purpose of this mode is to obtain stored emissions-related power-train DTCs. If there are no stored codes, then the module will respond with a message indicating zero codes are stored. If a manufacturer-specific DTC is found, the message "No text available for DTC code" or a similar message will be displayed.

### Mode 4—Clear/Reset DTCs

The purpose of this test mode is to provide a means to command onboard modules to clear all emissions-related diagnostic information. This includes the following:

- Clear number of diagnostic trouble codes (Mode 1)
- Clear number of diagnostic trouble codes (Mode 3)

- Clear trouble code for freeze-frame data (Mode 1)
- Clear freeze-frame data (Mode 2)
- Clear $O_2S$ test data (Mode 5)
- Reset status of system monitoring tests (Mode 1)
- Clear onboard monitoring test results (Mode 6 and Mode 7)
- Clear odometer mileage counter for accumulated distance traveled with MIL on (Mode 1)

## Mode 5—$O_2S$ Monitoring Results

The purpose of this test mode is to allow access to the onboard $O_2S$ monitoring test results as required by OBD-II regulations. Use of this mode is optional depending on the method used by the vehicle manufacturer to comply with the requirement for $O_2S$ monitoring. For example, systems that use an A/F sensor will not use Mode 5.

## Mode 6—Last Monitor Results

The purpose of this test mode is to allow access to the results from onboard diagnostic monitoring tests of specific components/systems that are not continuously monitored—except fuel system and misfire monitors. The latest test results must be retained, even over several ignition cycles. The data stored in Mode 6 can be replaced by more recent test results. Test results are reported only for supported combinations of test limit type and component ID and are reported as positive values. Only one test limit is included in a response message but that limit could be either a minimum or a maximum limit. If both minimum and maximum test limits are to be reported, then two response messages will be transmitted in any order. Last result data will not store if the PCM software is not calibrated for certain monitors, if the monitors have not run, or if the monitors have not successfully been completed.

Information on this screen may include the following:

- **Test ID**. This value is equal to the mode test identification number and is an identifier for the module being queried or responding.
- **Value**. This column is the actual data stored from the last time the monitors ran. This data represents whether the monitor passed or failed when compared with the "Threshold Spec for Failure" column.
- **Spec**. This column represents the **Threshold Spec for Failure**. The Threshold Spec for Failure is either the Low Limit Specification or the High Limit Specification. The technician must refer to the limit column to use this specification appropriately.
- **Limit.** Either the appropriate "High" or appropriate "Low" or "Both" limit specifications will be displayed.

This test mode is used as an alternative to replace the discretionary Mode 5 for reporting $O_2S$ monitor test results.

## Mode 7—One-Trip Failures

The purpose of this test mode is to obtain test status results for emissions-related power-train components or systems that are continuously monitored. The intended use of this data is to assist the service technician after a vehicle repair and after clearing diagnostic information. This mode will report the test results after a single driving cycle. If the test failed during the driving cycle, the DTC associated with that test will be reported.

Test results reported by this mode do not necessarily indicate a faulty component or system. If the test results indicate a failure after additional driving, then the MIL is illuminated and a DTC is set and reported with Mode 3 indicating a faulty component or system. Also, once Mode 3 indicates a fault Mode 7, it may or may not continue to report it as a one-trip failure.

**You Should Know**

*Test results for Mode 7 are reported in the same format as the DTCs reported in Test Mode 3.*

## Mode 8—Onboard System Test

The purpose of this mode is to control the operation of an onboard system, test, or component. This is basically an activation test of devices and circuits that can set emissions-related faults. This mode can also be used to run diagnostic tests of some systems (the evaporative system leak test, for example). The scan tool can display three different messages. The first states that the test is denied. This may be displayed if the test conditions are not met or the test is not available. The second message provides further instructions as needed; for example, "engine must be running." The third message states that the task is being performed.

## Mode 9—Request Vehicle Information

The purpose of this mode is to enable the scan tool to request vehicle-specific information such as the vehicle identification number (VIN), the calibration identifier, or the calibration verification number (CVN) used in the validation of software integrity. Support for

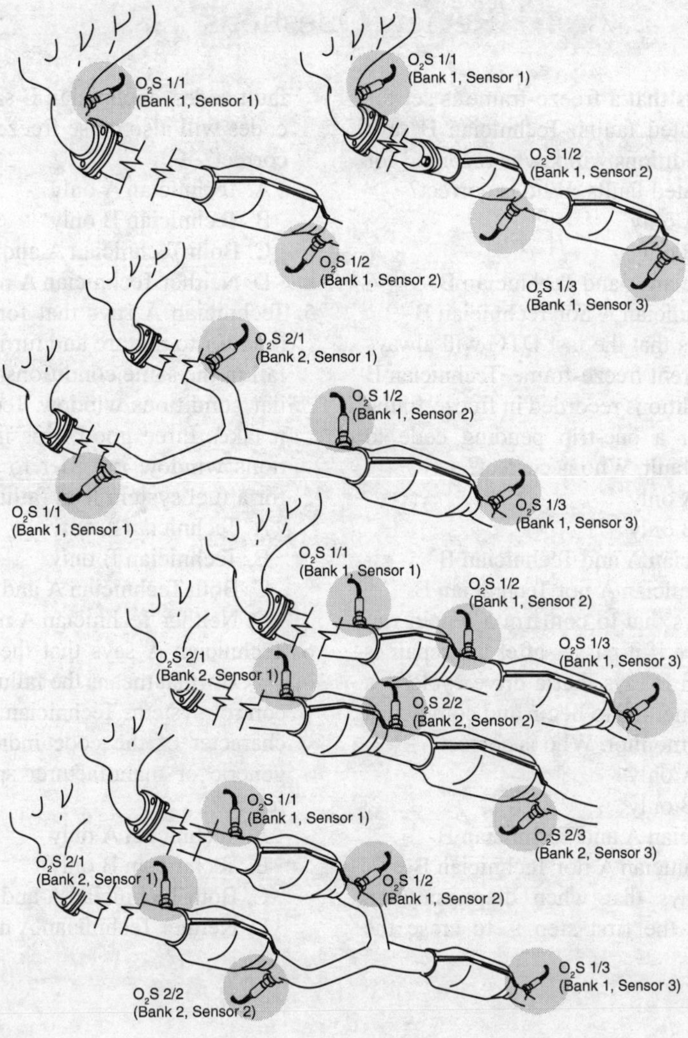


**Figure 7** *Oxygen sensor configurations an nomenclature.*

Mode 9 generally applies only to control modules with reprogrammable flash memory. Therefore, some of this information may be required by federal regulations and other information may be simply reported in a standard format if desired by the vehicle manufacturer. The calibration identifier and the CVN are intended as a means to verify the contents of the PCM's memory. These values must "identically match" what the original equipment manufacturer released.

These values are compared against a reference specification that documents the unique numbers corresponding to each released product (cross-referenced by VIN or part number). This is to protect against aftermarket alterations to the control module memory that could affect emissions quality.

## OXYGEN Sensor and Catalyst System Configurations

To satisfy OBD-II regulations for the common identification of $O_2S$ components located within the exhaust flow of various catalyst system configurations, a standard was developed for consistent placement and reference of these components **(Figure 7)**. These conventions are used for identifying and referencing $O_2S$ placement in multiple-bank systems and are documented with full concurrence in the following recommended practices: SAE J1930 "E/E Systems Diagnostic Terms, Definitions, Abbreviations, and Acronyms," SAE J1979 "Legislative Diagnostic Protocol," and SAE J2012 "Diagnostic Trouble Code Definition."

# Review Questions

1. Technician A says that a freeze-frame is set for all emissions-related faults. Technician B says that a similar conditions window is recorded for all emissions-related faults. Who is correct?
   A. Technician A only
   B. Technician B only
   C. Both Technician A and Technician B
   D. Neither Technician A nor Technician B
2. Technician A says that the last DTC will always overwrite the current freeze-frame. Technician B says that the conditions recorded in freeze-frame must be met for a one-trip pending code to mature to a hard fault. Who is correct?
   A. Technician A only
   B. Technician B only
   C. Both Technician A and Technician B
   D. Neither Technician A nor Technician B
3. Technician A says that to confirm a repair, run the monitor to see if it passes after the repair is made. Technician B says that a drive cycle is a specific driving method to begin and complete a specific OBD-II monitor. Who is correct?
   A. Technician A only
   B. Technician B only
   C. Both Technician A and Technician B
   D. Neither Technician A nor Technician B
4. Technician A says that when diagnosing an OBD-II vehicle, the first step is to erase the fault codes. Technician B says that erasing fault codes will also erase freeze-frame data. Who is correct?
   A. Technician A only
   B. Technician B only
   C. Both Technician A and Technician B
   D. Neither Technician A nor Technician B
5. Technician A says that for a fuel system lean monitor to mature and turn on the MIL, it must fail in the same conditions recorded in the similar conditions window. Technician B says that it takes three good trips in the similar conditions window in order to extinguish the MIL for a fuel system lean failure. Who is correct?
   A. Technician A only
   B. Technician B only
   C. Both Technician A and Technician B
   D. Neither Technician A nor Technician B
6. Technician A says that the "P" in the uniform DTC format means the failure is within the PCM control system. Technician B says that the first character of the code indicates if the DTC is generic or manufacturer specific. Who is correct?
   A. Technician A only
   B. Technician B only
   C. Both Technician A and Technician B
   D. Neither Technician A nor Technician B

CHAPTER

# 36 OBD-II Diagnostics—Drive Cycles

## Introduction

A **drive cycle** is a specific driving method to verify a symptom or the repair for the symptom. A drive cycle is also a specific driving method to begin and complete a specific OBD-II monitor. Often, a complete drive cycle is not required after performing a repair. To increment a good trip, many manufacturers only require the monitor that failed to be rerun and pass. However, some manufacturers do recommend erasing the fault code and running a complete drive cycle to confirm the repair. These differences in procedures require the technician to follow the proper service manual information.

Also, erasing the fault codes will clear the readiness indicators. The **readiness indicator** identifies whether or not all of the once-per-trip monitors have been run. This means that if the customer were to go directly to an I/M testing station, the vehicle may not pass inspection since all of the readiness indicators are not showing a completed status. A completed drive cycle will ensure that all readiness indicators show a completion. This chapter provides a guide to performing a drive cycle. Remember that things change, so always refer to the latest service information.

**You Should Know**

*It is also important to realize that the readiness indicator will show the monitor is completed even if a fault is detected. The readiness indicator only shows that the monitor ran.*

## Executing General Motors's Drive Cycle

A completed drive cycle will perform the diagnostic routines on all monitors. Most General Motors drive cycles can be completed within 15 minutes. The following is the procedure for performing the drive cycle:

1. **Cold start**. This requires that the engine coolant temperature be below 122°F (50°C) and within 11°F (6°C) of the ambient air temperature when the engine is started.
2. **Idle**. Allow the engine to run for 2.5 minutes. The air conditioner and the rear defroster should be turned on to provide additional electrical load. During this stage of the drive cycle, the more electrical load applied will ensure valid results. During the idle stage of the drive cycle the following monitors will run: the O2S heater, passive air, the no-flow portion of purge, misfire, and fuel trim (if in closed-loop) mode.
3. **Accelerate**. During this stage of the drive cycle, turn off all electrical loads applied during the idle stage. Accelerate using about half-throttle until a speed of 55 mph (88 km/h) is obtained. During this stage of the drive cycle the misfire, fuel trim, and purge flow monitors will run.
4. **Hold steady speed**. Maintain a constant speed of 55 mph (88 km/h) for 3 minutes. During this stage of the drive cycle the $O_2$ response, air intrusive, EGR, purge, misfire, and fuel trim monitors will run.
5. **Decelerate**. Without shifting or applying the brake, ease off the accelerator pedal and allow the vehicle to coast to a speed of 20 mph (32 km/h). This stage of the drive cycle will run the EGR, purge, and fuel trim monitors.
6. **Accelerate**. Accelerate at three-quarter throttle until a speed between 55 to 60 mph (88 to 96 km/h) is reached. This stage of the drive cycle will perform the same diagnostics as during the acceleration stage.
7. **Hold steady speed**. During this stage vehicle speed needs to be held steady at 55 mph (88 km/h) for 5 minutes. This stage of the drive cycle will run the same monitors as those in the first steady speed stage (Step 4) plus the catalyst monitor. It may take five complete driving cycles

to determine the state of the catalyst if it is marginal or if the battery has been disconnected.

8. **Decelerate**. This stage is a repeat of Step 5 and will run the same monitors.

After the drive cycle is complete, use a scan tool to check for any pending or mature fault codes.

## Executing Ford's Drive Cycle

Ford's drive cycle will run all of the monitors on the vehicle and will clear the I/M readiness code (P1000). However, not all of the drive cycle needs to be completed to test and verify a repair. In the following drive cycle procedure, perform Steps 1 through 4 first, then the step required for the specific monitor.

The following practices will help to ensure that the monitors can run and complete the drive cycle:

- Operate the accelerator pedal smoothly. During any cruising conditions maintain a steady pedal.
- The fuel level should be between one-half and three-quarters full.

To perform the drive cycle, follow these steps:

1. **Prepare the engine**. This stage of the drive cycle is performed to bypass the engine soak timer and reset the status of the monitors. Begin by installing the scan tool and cycling the ignition switch to the Run position two times (do not start the engine). Use the scan tool to clear any DTCs. With the ignition switch still in the Run position (be sure not to return it to the Off position), navigate the scan tool so engine coolant temperature, intake or idle air temperature (IAT), EVAP duty cycle, fuel-level input (FLI) and throttle position inputs can be monitored. Start the engine and allow it to idle for 15 seconds and then accelerate smoothly to a speed of 40 mph (64 km/h). Drive at this speed until the engine temperature reaches at least 170°F (76.7°C).
2. **Prepare for entering monitors**. This stage allows the engine to fully warm up and for the IAT to provide input to the PCM. Use the scan tool to confirm that the IAT is between 40°F and 100°F (4.4°C and 37.8°C). If the IAT is not within this range, you can complete the following steps but you will be required to perform Step 12 to bypass the EVAP monitor and clear code P1000.
3. **Run the oxygen sensor monitor**. This stage requires a steady cruise speed of 40 mph (64 km/h) for up to 4 minutes. Monitor the scan tool to determine when the monitor is complete.
4. **Run the EVAP monitor.** This stage can only be performed if the IAT is between 40°F and 100°F (4.4°C and 37.8°C). The EVAP monitor requires a steady cruise speed of between 45 and 65 mph (72 and 104 km/h) for 10 minutes. During this cruise time, avoid any maneuvers that may cause the fuel to slosh around in the tank since this can contaminate the results. For this monitor to run the following is required:
   - Throttle position Mode indicating part-throttle
   - EVAPDC greater than 75 percent
   - FLI between 15 and 85 percent
5. **Run the catalyst monitor**. This stage requires simulating stop-and-go traffic conditions. Included in this stage is the need to perform five separate steady cruise speeds ranging from 25 to 45 mph (40 to 72 km/h) over a 10-minute period.
6. **Run the EGR monitor**. This stage requires acceleration from a complete stop to 45 mph (72 km/h). During this operation accelerate using one-half throttle. This step needs to be repeated three times.
7. **Run the idle speed control (ISC) portion of the comprehensive components monitor.** This stage requires bringing the vehicle to a stop and allowing the engine to idle with the transmission in drive (neutral for manual transmissions) for 2 minutes.
8. **Run the transmission portion of the comprehensive components monitor**. This stage depends on if the vehicle is equipped with an automatic or a manual transmission. For vehicles equipped with automatic transmissions, begin by accelerating from a stop to 50 mph (80 km/h) and cruise for at least 15 seconds. Next, stop the vehicle and accelerate to 40 mph (64 km/h) but this time without allowing the transmission to go into overdrive. Cruise at this speed for at least 30 seconds. While maintaining the 40 mph (64 km/h) speed, activate overdrive and accelerate to 50 mph (80 km/h). Cruise at this speed for at least 15 seconds. Bring the vehicle to a complete stop and allow it to idle in drive for at least 20 seconds. Repeat these steps five times. For vehicles equipped with manual transmissions, accelerate from a stop to 50 mph (80 km/h), then go to Step 9.
9. **Run the misfire and fuel monitors**. This stage requires the vehicle be accelerated from a complete stop to a speed of 65 mph (104 km/h). Without using the brakes, allow the vehicle to decelerate at closed throttle until a speed of 40 mph (64 km/h) is obtained. Repeat this procedure three times.

10. **Check readiness**. This function determines if any monitor has not completed. Using the scan tool, navigate to Mode 1. This screen indicates if all noncontinuous monitors have completed. If they are not all completed, continue to Step 11.
11. **Check pending code and EVAP monitor bypass**. Use the scan tool to check for any **pending codes** that are preventing the clearing of code P1000. A pending code indicates a potential problem was detected but a second failure of the diagnostic test is required for it to mature to a DTC. If there are pending codes present, perform the diagnostic procedures for the code and repair the cause. If there are no pending codes, the monitors that are not complete will need to be run again. The only exception is if the EVAP monitor is not complete. If the IAT was out of range in Step 2 (or altitude is over 8,000 feet (2,438 m)), then the EVAP bypass procedure in Step 12 is necessary to clear code P1000.
12. **Bypass the EVAP monitor**. Since the EVAP monitor will not run if the IAT in Step 2 is outside the 40°F to 100°F (4.4°C to 37.8°C) range (or the altitude is above 8,000 feet (2,438 m)), the PCM must detect these conditions two times to allow the EVAP monitor to be bypassed so code P1000 can be cleared. This function allows the bypass counter to increment to two. This will require a cold soak of the engine for at least 8 hours. After the cold soak, perform Steps 2 through 10.

## Executing Toyota's Drive Cycle

Toyota monitors do not need to run in a particular order. Once the conditions to run a monitor are met, it will run. The following describes how to make each of the monitors run.

### EGR Monitors

The EGR monitor has several preconditions: the MIL is off, altitude is 7,800 feet (2,377 m) or less, and the IAT is 14°F (−10°C) or higher. The status of these conditions can be confirmed with the scan tool. To run the monitor **(Figure 1)**, start the engine and allow it to idle until the IAT sensor reading increases to higher than 50°F (10°C). Next, drive the vehicle at between 43 to 56 mph (70 to 90 km/h) for a period of 3 to 5 minutes without allowing the TPS to exceed 30 percent. Maintain a steady engine speed with smooth throttle operation and avoid sudden accelerations. After this drive time is accomplished, stop the vehicle and let the engine idle for 3 to 5 minutes.

After the idle time is completed, repeat the drive and idle patterns one more time. If the readiness status does not indicate "complete," use the scan tool to see if a pending code is set. Since this is a two-trip DTC the readiness status may not switch to "complete" after the first drive pattern trip if a pending code has been set. If a pending code is not set, turn the ignition off and repeat the two drive and idle patterns again.

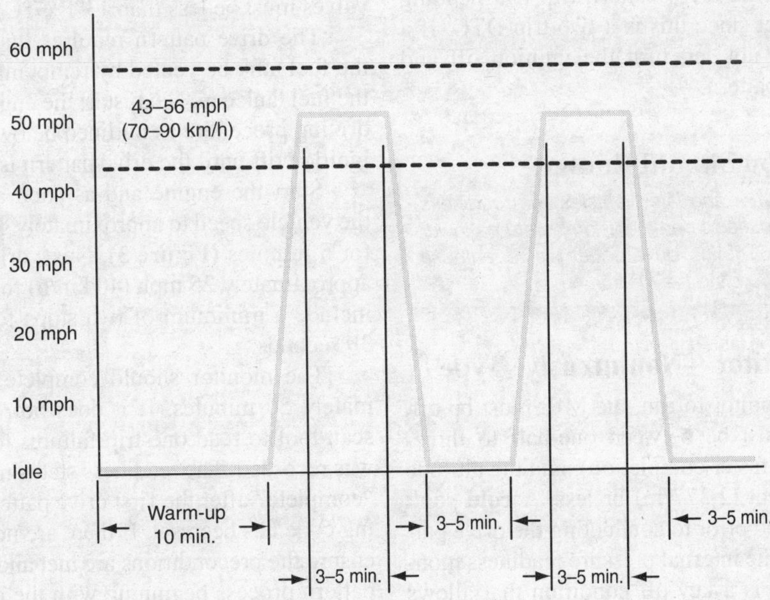


**Figure 1** *EGR monitor drive pattern.*

## Catalyst Monitor

The precondition requirements for this monitor are as follows: the MIL is off, the ECT temperature is higher than 176°F (80°C), and the IAT is higher than 14°F (−10°C). The status of these conditions can be confirmed with the scan tool. During engine start-up, note the IAT sensor reading. The driving time will be adjusted based on this reading. If the ECT sensor indicates a temperature below 176°F (80°C), allow the engine to idle until the temperature crosses over this value. To run this monitor **(Figure 2)**, drive the vehicle with smooth throttle operation (avoiding sudden acceleration) at a speed of 40 to 55 mph (64 to 88 km/h) for the time listed below:

- If IAT was lower than 50°F (10°C) when the engine was started, drive for 7 minutes.
- If IAT was higher than 50°F (10°C) when the engine was started, drive for 3 minutes.

For $O_2$S-equipped vehicles, after the above drive time is accomplished, decelerate the vehicle with a smooth release of the throttle to its closed position. Maintain a speed of 35 to 45 mph (56 to 72 km/h). Avoid sudden deceleration. Maintain a steady speed for approximately 7 minutes.

If the vehicle is equipped with A/F sensors, drive the vehicle and allow the speed to fluctuate between 35 and 45 mph (56 and 72 km/h) for 16 minutes.

If the readiness status does not indicate "complete," use the scan tool to see if a pending code is set. The readiness status may not switch to "complete" after the first drive pattern trip if a pending code has been set since this is a two-trip DTC. If a pending code is not set, turn the ignition off and repeat the drive pattern.

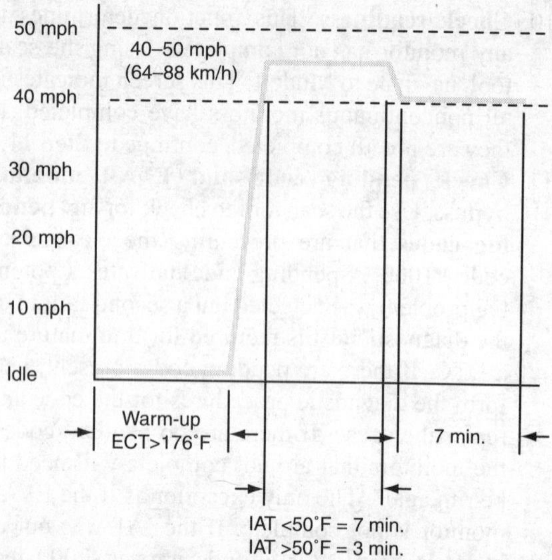


**Figure 2** *Catalyst monitor drive pattern.*

### You Should Know

*For 2002 model year and later vehicles, the readiness test can be completed in ambient temperatures lower than 14°F (−10°C) if the drive pattern is repeated a second time after cycling the ignition off.*

## EVAP Monitor—Nonintrusive Type

For the EVAP monitor to run, the MIL must be off, the fuel level must be between one-half to three-quarters full (for faster completion), and the altitude must be 7,800 feet (2,377 m) or less. A **cold soak** must be performed prior to conducting the drive pattern to complete the internal pressure readiness monitor. A cold soak is a key-off condition that allows the ECT to equal ambient temperature. To accomplish the cold soak requirement, start the engine and allow it to warm until the ECT sensor indicates 176°F (80°C) or higher. Once the ECT temperature meets this requirement, let the vehicle cold soak for 8 hours or until the difference between IAT and ECT is less than 13°F (7°C).

After the cold soak conditions are satisfied, the same conditions for the presoak must be met for the monitor to run. In addition, the ECT and the IAT must both be between 40°F and 95°F (4.4°C and 35°C) and the difference between the ECT and IAT valves must be less than 13°F (7°C).

The drive pattern requires that the pressure in the fuel tank be vented by removing and reinstalling the fuel tank cap. Next, start the engine and begin the driving procedure as outlined below. Do not turn the ignition off until the drive pattern is completed.

Start the engine and as soon as possible bring the vehicle speed to approximately 45 mph (72 km/h) for 5 minutes **(Figure 3)**. Next, drive the vehicle at approximately 25 mph (40 km/h) for 15 minutes and include a minimum of two stops for approximately 30 seconds.

The monitor should complete within approximately 20 minutes. If it does not, use the OBD-II scan tool to read one-trip failures to see if a failure was recorded. The readiness status may not switch to "complete" after the first drive pattern trip if a pending code has been set. If there are no pending codes, ensure the preconditions are met and repeat the drive pattern process beginning with the cold soak procedure.

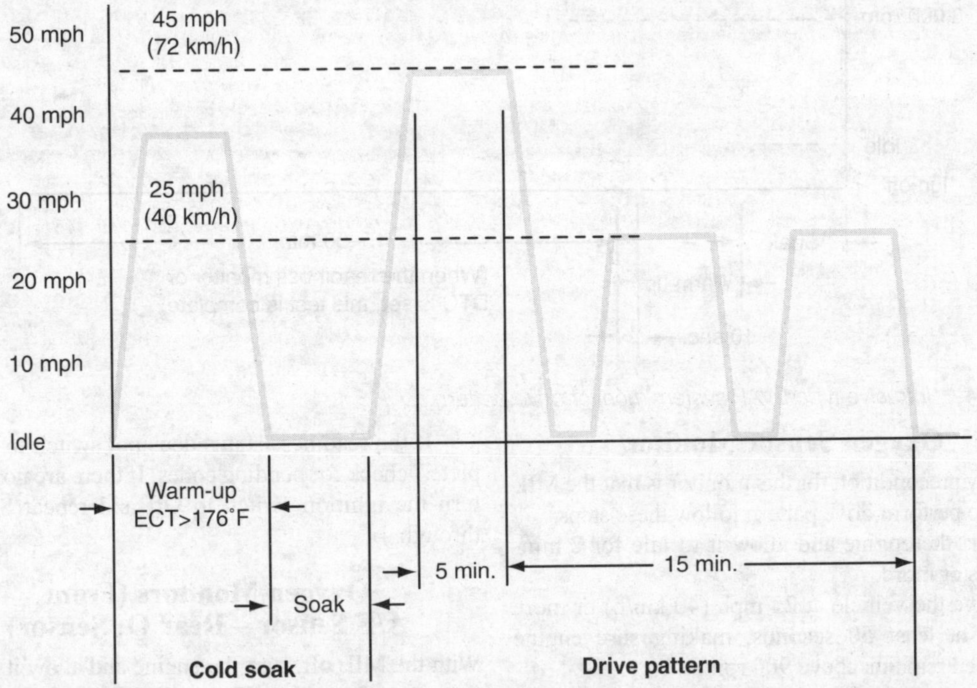


**Figure 3** *Nonintrusive-type EVAP system monitor drive pattern.*

## EVAP Monitor—Intrusive Type

For this type of EVAP monitor, the preconditions for the cold soak procedure are the same as for the nonintrusive type. The vehicle must be allowed to cold soak for 8 hours or until the difference between IAT and ECT is less than 13°F (7°C). There is no requirement for an engine warm-up first.

After the cold soak condition has been satisfied and before starting the engine, use the OBD-II scan tool to read the ECT and IAT values. Both values must be between 40°F and 95°F (4.4°C and 35°C) and the difference between ECT and IAT must be less than 13°F (7°C). Also, vent the pressure in the fuel tank by removing and reinstalling the fuel tank cap.

Next, start the engine and allow it to idle until the ECT is 167°F (75°C) or higher **(Figure 4)**. With the transmission in neutral, rapidly increase the engine speed to 3,000 rpm and hold for approximately 10 seconds. Allow the engine to idle with the A/C on for 15 to 50 minutes.

### You Should Know

*The reason for turning on the A/C system is to create a slight load against the engine. If the vehicle is not equipped with A/C, allow the engine to idle with the transmission in drive for at least 15 minutes.*

### You Should Know

*Make sure the parking brake is set and the wheels are chocked properly before placing the transmission in drive. Otherwise, personal injury or property damage may result.*

### You Should Know

*Make sure to connect a properly operating shop exhaust system to the vehicle's tailpipe to prevent the possibility of carbon monoxide poisoning.*

If the readiness status does not switch to "complete" after the first drive pattern, check for one-trip failures that have been set. If there are no faults recorded, rerun the test again (including cold soak).

### You Should Know

*For 2002 model year and later vehicles, the readiness test can be completed in cold ambient conditions (less than 40°F/4.4°C) and/or at high altitudes (above 7,800 feet/2,377 m) if the complete drive pattern (including cold soak) is repeated a second time after cycling the ignition off.*

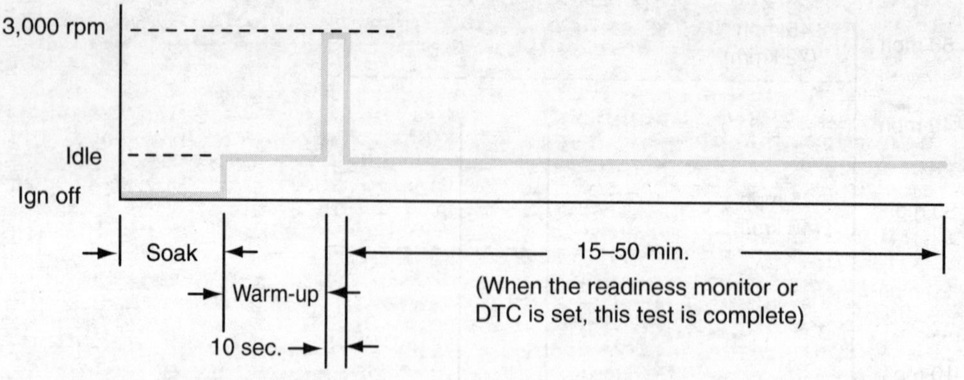


**Figure 4** *Intrusive-type EVAP system monitor drive pattern.*

## Oxygen Sensor Monitor

The only precondition for this monitor is that the MIL is off. To perform drive pattern follow these steps:

1. Start the engine and allow it to idle for 2 minutes or more.
2. Drive the vehicle at 25 mph (40 km/h) or more for at least 50 seconds, making sure engine speed remains above 900 rpm.
3. Stop the vehicle and allow the engine to idle for 40 seconds or more.
4. Performing Steps 2 and 3 ten times.

If the readiness status does not switch to "complete," check for pending codes. If there are no faults, turn the ignition switch to Off and repeat Steps 1 through 4.

## Oxygen Monitors (Front A/F Sensor – Rear $O_2$ Sensor)

With the MIL off, start the engine and allow it to idle for at least 2 minutes **(Figure 5).** Next, drive the vehicle at 40 to 70 mph (64 to 112 km/h) or more for at least 3 minutes. Be sure to maintain engine speed

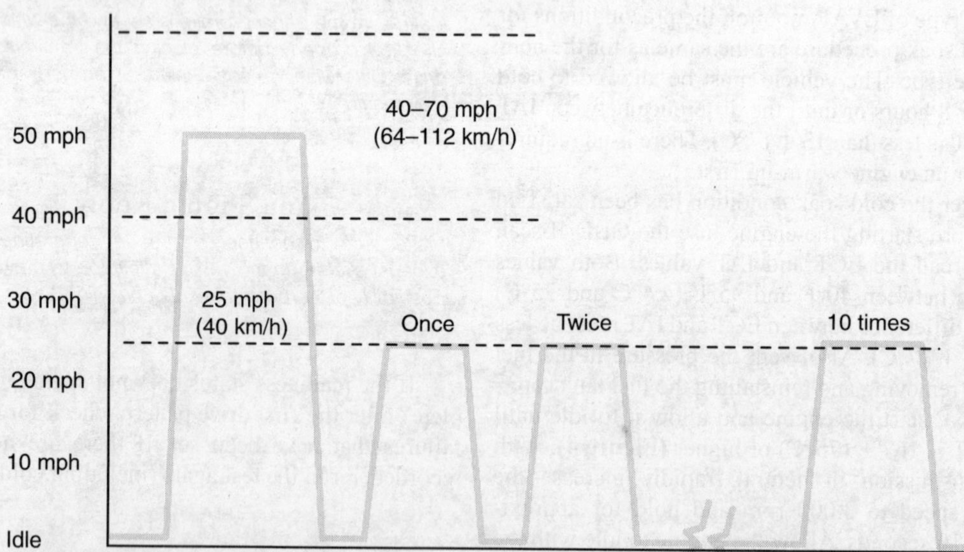


**Figure 5** *$O_2$S monitor drive pattern.*

between 900 to 3,200 rpm. Stop the vehicle and allow the engine to idle for 10 seconds or more, then drive the vehicle at 25 mph (40 km/h) for at least 40 seconds, making sure to maintain engine speed above 900 rpm.

Stop the vehicle and allow the engine to idle for at least 10 seconds. Drive the vehicle at 25 mph (40 km/h) for at least 40 seconds (maintaining engine speed above 900 rpm). Stop the vehicle and allow the engine to idle for at least 10 seconds ten times.

If the readiness status does not switch to "complete," check for pending codes. If no faults are present, ensure preconditions are met, turn the ignition switch off, and repeat all the steps.

### Oxygen Sensor Heater Monitors

The $O_2S$ and A/F sensor monitors run similarly. For the monitors to run, be sure the MIL is off. Start the engine and allow it to idle for 9 minutes. Next, drive the vehicle at least 25 mph (40 km/h) for at least 2 minutes. If the readiness status does not switch to "complete," check for pending codes. If there are no faults, turn the ignition off and repeat the pattern.

## Review Questions

1. Define the term *drive cycle*.
2. Executing a General Motors drive cycle is being discussed. Technician A says that if the catalyst is marginal or the battery has been disconnected, it may take five complete driving cycles to determine the state of the catalyst. Technician B says that in order to be classified as a cold start, the ECT must be between 40°F to 180°F (4°C to 82°C). Who is correct?
   A. Technician A only
   B. Technician B only
   C. Both Technician A and Technician B
   D. Neither Technician A nor Technician B
3. For a Toyota drive cycle, explain the requirements for the EGR monitor.
4. Executing a General Motors drive cycle is being discussed. Technician A says that the $O_2S$ may require the air conditioner and the rear defroster to be turned on. Technician B says that for the misfire, fuel trim, and purge flow diagnostics to be performed, the air conditioner and all the other loads need to be turned off. Who is correct?
   A. Technician A only
   B. Technician B only
   C. Both Technician A and Technician B
   D. Neither Technician A nor Technician B
5. For a Ford drive cycle, explain how to run the EGR monitor.

# CHAPTER 37 EVAP Leak Diagnosis

## Introduction

When a vehicle is brought to the shop for an MIL illumination concern, diagnosing of the fault typically only requires that normal procedures be followed. For example, when the misfire monitor has set a fault, the technician follows the same procedures to locate the cause of the misfire for both OBD-I- and OBD-II-equipped vehicles. This chapter focuses on diagnosing the EVAP monitor. Studies have indicated that the major cause of an MIL illumination is the EVAP leak-detection monitor.

> **You Should Know**
>
> *Most gross leak fault codes are the result of a loose filler neck cap. Make sure the gas cap is properly installed before testing the rest of the system.*

## Leak-Detection Pump Systems

Test Mode 8 of the scan tool enables the conditions required to conduct an evaporative system leak test but it does not actually run the test. An example is to close a purge solenoid to prevent leakage if the system is pressurized. The vehicle manufacturer is responsible for determining the criteria to automatically stop the test (open the solenoid in this example). The criteria may include a running engine, vehicle speed greater than zero, or exceeding a specified time period. Mode 8 supports closing the canister vent with an external command. Upon receipt of a valid J1979 Mode 8 request from a scan tool, the PCM energizes the leak-detection pump (LDP) solenoid **(Figure 1)** and closes the purge solenoid. The PCM replies with a positive response to indicate the request was carried out.

Energizing the LDP and purge solenoids seals the evaporative system. The service technician can now pressurize the system using a pressure source applied at the evaporative service port **(Figure 2)** or at the gas tank cap **(Figure 3)**. Once the system is pressurized, the technician can detect for leaks using leak-detection equipment **(Figure 4)**.

Most leak detectors can visually detect leaks smaller than 0.02 in. (0.5 mm) in diameter in the evaporative system. This is done by using shop air and a smoke generator. The food grade oil used to make the smoke includes a UV trace dye that will leave telltale signs of the leak under a black light. This is helpful when components have to be removed to determine the exact leak location. The following is a typical procedure for using the leak-detection equipment.

The first part of the test is to confirm that an actual leak is present. Keep in mind that anything that prevents the system from being pressurized while the leak monitor is running will cause a failure. This can include a loose gas cap.

The following steps are to prepare the leak-detection machine:

1. Connect the scan tool and access Mode 3 and record the DTC.
2. Connect the red power lead to a 12V DC power supply and the black lead to a chassis ground.
3. Connect shop air to the leak detector.
4. Set the control knob to "AIR."
5. To calibrate the machine, insert the tester's air supply tip into the appropriate calibration orifice on the tester's control panel based on the size of leak indicated by the DTC **(Figure 5)**.
6. Press and hold the remote start button while positioning the flag on the flow meter so it is aligned with the indicator ball **(Figure 6)**.

Once the flow meter is calibrated for the size of the leak being diagnosed, install the leak detector's air hose to the green service port to the fuel tank cap adapter. It will be necessary to apply a constant source of vacuum to the LDP solenoid. One way of doing this is to use an A/C evacuation pump. Connect

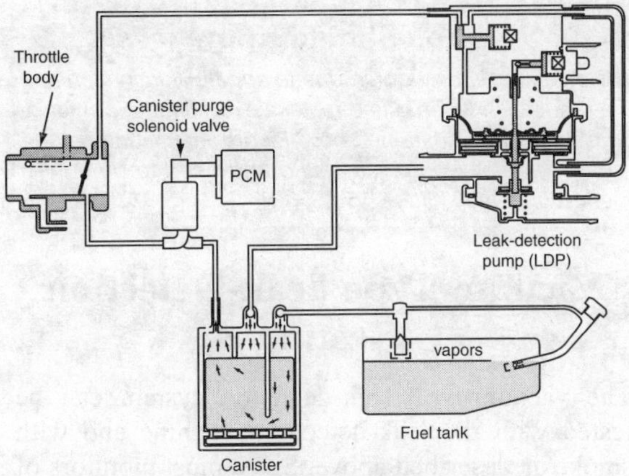


**Figure 1** *The leak-detection pump pressurizes the EVAP system to test for leaks.*

**Figure 2** *Pressure line from leak-detection equipment connected to the service "T."*

**Figure 3** *Pressure line from leak-detection equipment connected to the gas cap adapter.*

**Figure 4** *The evaporative emissions leak detector makes the job of finding small leaks easier.*

**Figure 5** *Calibrate the machine for the size of the leak.*

**Figure 6** *Adjust the indicator to the flow meter ball.*

> **You Should Know**
>
> *Some vehicles are equipped with a "T" in the vacuum line for connecting the vacuum source.*

the vacuum pump to the line going from the intake manifold to the LDP solenoid. This is necessary to allow the pump to operate. Using a generic scan tool may not allow the use of engine vacuum since the engine rpm signal may deny the test.

Using Mode 8, initiate the LDP test to close the system. This energizes the LDP solenoid and allows the vacuum from the constant vacuum source to apply vacuum to the LDP pump diaphragm. This lifts the diaphragm up and seals the atmospheric vent at the bottom of the LDP. The evaporative system is now ready to be pressurized.

Press the remote button to activate airflow. Large-volume fuel tanks or low fuel levels may

### You Should Know

*Another location that may be used to connect the air source, and later the smoke, is at the fresh-air filter connection at the LDP.*

require 4 or 5 minutes to completely fill the system with air. Compare the flow meter indicator ball reading with the flag on the flow meter. If the ball is below the flag, it indicates that the system integrity is acceptable and the test is complete. If the ball is above the flag, a leak is indicated and further testing is required.

### You Should Know

*If the fuel level is more than 85 percent on vehicles with ORVR, it may be necessary to perform this procedure from both the service valve and the fuel tank cap adapter due to closure of the ORVR flow-control valve.*

With the air flowing into the system, listen for air leaks at all components including the gas cap. If you are unable to locate the leak, remove the air supply hose and connect the smoke supply tip to the EVAP system. Set the control switch to smoke and press the remote start button.

If inputting the smoke through the service port adapter, verify that smoke has filled the evaporative system by removing the fuel cap until the smoke begins to escape, then reinstall the cap. If inputting smoke through the fuel tank adapter, verify that smoke has filled the evaporative system by depressing the Schrader valve on the service port adapter until smoke begins to escape from the adapter. For best results, continue to input smoke into the system for an additional 60 seconds. Continue introducing smoke at 15-second intervals as necessary.

Using the white light, follow the evaporative system path and look for the source of the leak by locating exiting smoke. Make sure to check that the normally closed purge solenoid is not leaking into the intake manifold. If the leak is in a concealed area, stop the smoke and use the ultraviolet black light and the yellow glasses to locate the residual ultraviolet dye left behind by the escaping smoke. The leak deposits a residual fluid that is either bright green or yellow in color when viewed with a UV light source.

Repair the leak and retest the system by performing the air pressure test again to verify the repair. Follow up with an EVAP monitor test and view the monitor's last results in Mode 6.

### You Should Know

*If a vehicle is brought in due to an MIL illumination as a result of a loose gas cap, take extra time to demonstrate to the customer how to properly tighten the cap. Most caps are ratcheting and need to be tightened until they click.*

## Vacuum-Type Leak-Detection Testing

The vacuum-type leak-detection system can be tested with the leak-detection machine and with smoke as described above. Since most monitors of this type require extended cold soak times, the following is offered as an alternative method to confirm a repair.

System integrity and function can be confirmed using a scan tool to monitor the vacuum solenoid assembly switch while applying vacuum to the EVAP system. Connecting the fuel cap adapter to the fill neck and attaching a hand vacuum pump to it **(Figure 7)** does this. With the ignition switch in the Run position, monitor the switch status while pumping vacuum into the tank. Once the switch indicates that vacuum has been applied, monitor the switch and see if it holds for at least 10 minutes. If the switch remains closed, there is no leak.

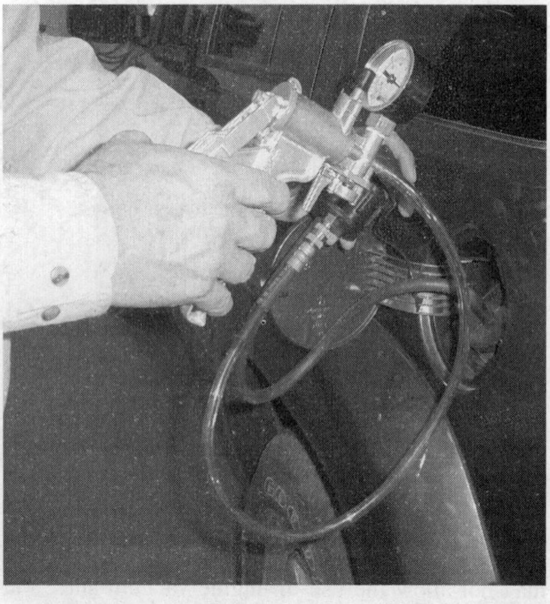

**Figure 7** *Use a hand vacuum pump and gas cap adapter to apply vacuum to the EVAP system.*

## Review Questions

1. Technician A says that systems with natural vacuum leak detection cannot be pressurized to check for leaks. Technician B says that a loose gas cap can cause a large-leak failure. Who is correct?
   A. Technician A only
   B. Technician B only
   C. Both Technician A and Technician B
   D. Neither Technician A nor Technician B
2. Explain why it may be necessary to connect an external source of vacuum to the LDP in order to pressurize the system.
3. Technician A says that Mode 8 runs the EVAP leak monitor test. Technician B says that during Mode 8 the PCM energizes the LDP solenoid. Who is correct?
   A. Technician A only
   B. Technician B only
   C. Both Technician A and Technician B
   D. Neither A Technician nor Technician B
4. Technician A says that the leak-detection machine's flow indicator is used to identify if a leak exists. Technician B says that smoke can be used to help locate a small leak. Who is correct?
   A. Technician A only
   B. Technician B only
   C. Both Technician A and Technician B
   D. Neither Technician A nor Technician B
5. Describe how to test the system integrity and function of a vacuum-type leak-detection system.

SECTION

# 8 Alternative Fuel and Power Vehicles

## Section Objectives

After you have read, studied, and practiced the contents of this section, you should be able to:

- Explain the purpose of alternative fuel vehicles.
- Describe the flow of the fuel-delivery system of a liquid-propane-injection vehicle.
- List and describe the function of the propane fuel system components.
- List and describe the function of typical compressed natural gas (CNG) fuel system components.
- Explain the purpose of propane and CNG tank codes and the regulations concerning these tanks.
- Explain the operation of the flex fuel vehicle.
- Detail the modes of operation of CNG fuel systems.
- Demonstrate a full understanding of the safety concerns associated with CNG vehicle service.
- Test the CNG system for gas leaks.
- Properly and safely purge the natural gas from the tubes and cylinders.
- Properly perform fuel tube fitting replacement.
- Perform a fuel system pressure test and determine needed repairs.
- Diagnose common CNG electrical components including the fuel low-pressure sensor, the fuel temperature sensor, the fuel gauge pressure sensor, the high-pressure fuel shutoff solenoid, the low-pressure fuel shutoff solenoid, and the fuel injectors.
- Explain the basic operation of electric vehicles.
- Describe the typical operation of hybrid vehicles.
- Explain the difference between parallel and series hybrids.
- Explain the purpose of regenerative braking.
- Describe how a proton exchange membrane produces electricity in a fuel-cell system.
- List and describe the different fuels that can be used in a fuel-cell system.
- Describe the purpose of the reformer.
- Explain how different types of reformers operate.

### Interesting Fact

*In 1993, President Clinton signed the Partnership for a New Generation of Vehicles (PNGV). This initiated a cooperative program between automotive manufacturers and government agencies to develop a new generation of cars by 2004. One of the criteria for the new vehicles was fuel economy of 80 miles per gallon (mpg) of gasoline or gasoline equivalent. In 2002, the Bush administration replaced the PNGV program with the FreedomCAR (Freedom Cooperative Automobile Research) partnership established with DaimlerChrysler, Ford, and General Motors. The new purpose of the program is to reduce the country's dependence on foreign oil while also reducing tailpipe pollution. In addition, the program focuses research on the long-term transition from petroleum-based vehicles to the use of clean, renewable hydrogen energy. Overall goals include the development of technology that will enable mass production of affordable hydrogen-powered fuel-cell vehicles and the infrastructure to support them.*

CHAPTER

# 38 Alternative Fuel Vehicles

## Introduction

Due to the increase in regulations concerning emissions and the public's desire to becoming less dependent on foreign oil, most major automotive manufacturers have developed alternative fuel or alternative power vehicles. This chapter explores many of the most common methods used in these fields. It also includes a look at alternative fuel vehicles that run on propane, compressed natural gas (CNG), and flex fuel.

## Alternative Gaseous Fuels

Recent federal energy and environmental legislation affects the fuel and vehicle choices of some U.S. fleet operators. The Energy Policy Act of 1992 (EPACT) requires certain federal and state fleet operators to acquire alternative fuel vehicles (AFVs). Some private, municipal, and other fleet operators may also be required to obtain AFVs in the future. The Clean Air Act introduced the Clean Fuel Fleet program, which requires that some fleet operators in cities with the greatest air pollution acquire vehicles that meet special clean fuel vehicle emissions standards. State and local requirements also affect the fuel and vehicle choices of fleet operators. The requirements vary from area to area and depend on the location of a fleet.

Original equipment manufacturer (OEM) AFVs are designed specifically to use alternative fuels such as propane, CNG, methanol, or ethanol. The engine, emissions, and performance of the vehicle are optimized for that fuel. Such a vehicle features a unique engine, catalyst, and single- or multipoint injection (MPI) system that meters fuel more precisely. The onboard computer is designed specifically for the fuel. In addition, the suspension system and shocks may be designed to withstand the added weight of propane or CNG tanks. Often, the fuel system parts are designed and installed specifically for that type of vehicle.

Because the supply of OEM vehicles is limited, some fleet operators may choose to use **converted vehicles** as a way to meet vehicle acquisition requirements. Converted vehicles are those that have been changed to use an alternative fuel source instead of gasoline fuel after the vehicle was produced. Compressed natural gas and propane are the two most common types of fuel for such vehicles.

### Propane

Liquefied petroleum gas (LPG), or liquid propane, is a byproduct of natural gas processing and crude oil refining. About 65 percent of the propane supply is extracted from natural gas production while 35 percent is a byproduct of crude oil refining. The United States is one of the world's largest producers of propane. Almost 92 percent of the U.S. demand for propane gas is supplied domestically.

> **You Should Know**
>
> *The terms* liquefied petroleum gas *and* propane *are often used interchangeably.*

The automotive propane fuel system is a completely closed system that contains a supply of pressurized LPG. The gas is stored in liquefied form in tanks that are pressurized to around 160 psi (110.3 KPa). The fuel remains a liquid as long as it is under pressure.

Propane gas has the potential to produce less CO and HC than gasoline does. It is clean burning, producing virtually no particulate or sulfur emissions. CARB and U.S. EPA tests have shown that emissions of HC and CO from propane-fueled engines are consistently reduced 50 to 80 percent, and $NO_x$ emissions are reduced 5 to 30 percent. In addition, since propane is contained in a sealed system there are no evaporative emissions.

On a per-unit weight basis, liquid propane contains about 5 percent more energy than gasoline; however, the density is 30 percent less. The net result is that propane contains less energy than gasoline for the same quantity. Propane has an octane of about 110. With propane, a greater quantity of fuel is required to travel the same distance. However, because of its gaseous form, the engine burns the fuel more efficiently, which helps offset the extra fuel required. Since propane has about 80 percent of the energy of gasoline on a volumetric basis, fuel consumption is slightly higher with propane as compared with gasoline. However, engines running on propane require significantly less maintenance than engines running on gasoline because of its clean burning characteristics.

Propane is readily available with more than 5,000 refueling stations across Canada, and more than 10,000 refueling stations in the United States. Propane, composed primarily of propane and butane with smaller amounts of propylene and butylene, is supplied in four grades of different composition. **HD-5** is the only grade appropriate for automotive applications. HD-5 is 95 percent propane and 5 percent butanes.

**LPG Fuel-Injection Systems**. In recent years there has been a decline in propane fuel use. This is partly due to the automobile industry's change from carburetor or throttle body injection (TBI) fuel systems to MPI systems. The propane conversion companies have failed to keep up with changing technologies. Today there are a few companies that provide injection systems for converting from gasoline to propane and a few years ago automotive manufacturers returned to the propane vehicle market. Several automotive manufacturers offer different models with propane fuel systems. Ford offers a number of models through its Qualified Vehicle Modifier (QVM) program with factory-certified conversions. The Chrysler group of DaimlerChrysler Motors offered a completely engineered propane vehicle (not a conversion) including liquid fuel injection and an integrated fuel tank that conforms to the body design. However, even interest on the part of automotive manufacturers has declined and in the 2004 model year only the Ford F150 series trucks was offered with propane. Although manufacturers are not currently offering propane fueled light duty vehicles, a number of manufacturers continue to produce medium and heavy duty vehicles that are fueled by propane. These vehicles include transit buses, shuttles, and trolleys.

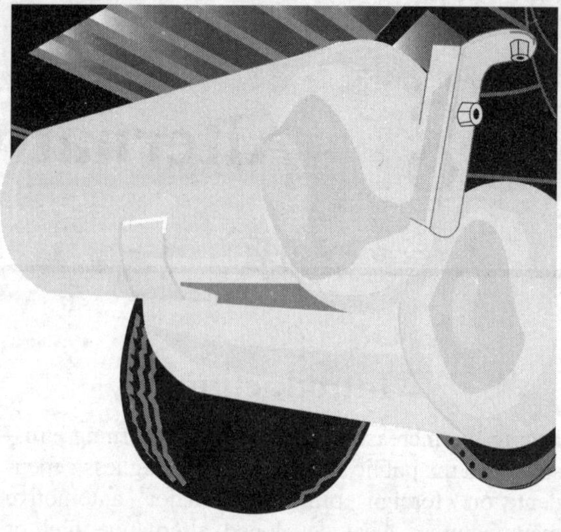

**Figure 1** *LPG fuel-injected system tanks.*

The LPG fuel-injection systems operate similarly to gasoline fuel-injection systems. The propane electronic fuel-injection system uses a special LPG tank **(Figure 1)**. The tank also contains an internal pump **(Figure 2)**. The fuel hose performs two functions **(Figure 3)**. First, it delivers the liquid fuel to the fuel rail through the center section of the hose. Second, it is the return for vapor and unused fuel back to the tank through the outer section of the hose.

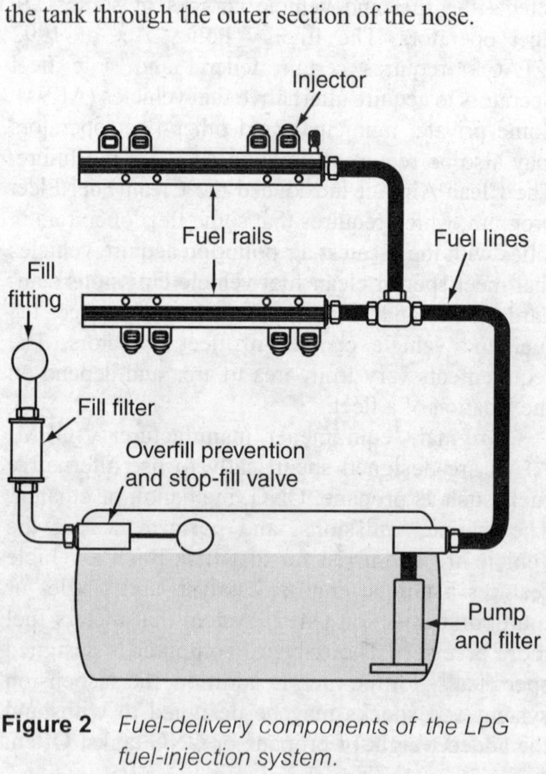


**Figure 2** *Fuel-delivery components of the LPG fuel-injection system.*

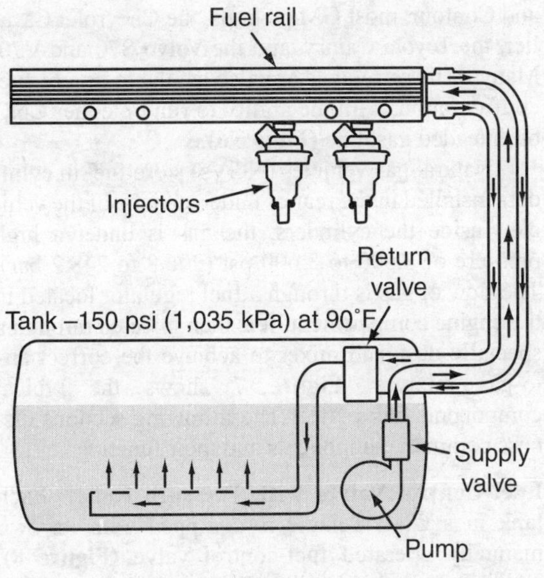


**Figure 3** *A single hose performs the functions of delivery and return.*

The system is similar enough to gasoline fuel injection that the engine can be designed as a **bi-fuel system (Figure 4)**. A vehicle equipped with a bi-fuel system has two separate fuel systems, with the capability to switch from one to the other. A fuel selector switch allows the driver to change between gasoline and propane fuel. In addition, when the fuel level of the propane tank drops to a predetermined point, the PCM automatically switches to the gasoline system.

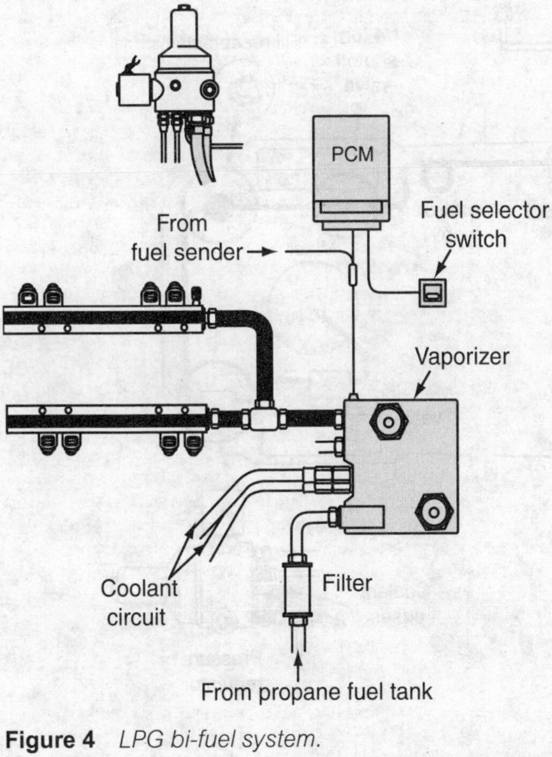


**Figure 4** *LPG bi-fuel system.*

## Compressed Natural Gas (CNG)

CNG offers several advantages over gasoline and other alternative fuels. Some of these advantages include:

- Lower CO emissions
- Lower HC emissions
- Lower $NO_x$ emissions
- Cleaner burning with fewer deposits
- Higher octane, eliminating engine knock

Natural gas is composed mainly of methane and has an octane rating from 105 to 130. Unlike LPG, natural gas remains a gas even when it is compressed to 3,000 to 3,600 psi (206.8 to 248.2 bar). Natural gas is lighter than air and when released it rises and dissipates into the atmosphere. CNG also is a plentiful resource and distribution networks are already in place making it more economical than many other alternative fuels **(Figure 5)**.

CNG has a lower British thermal unit (Btu) rating per cubic inch than gasoline. Due to its lower energy output, larger storage tanks are required to carry the fuel. As a result, traveling distance per tankful is somewhat reduced from that of gasoline vehicles. There are a number of natural gas vehicles available, including the Honda Civic GX, the Dodge Caravan and Dodge B-Van, the Ford Crown Victoria

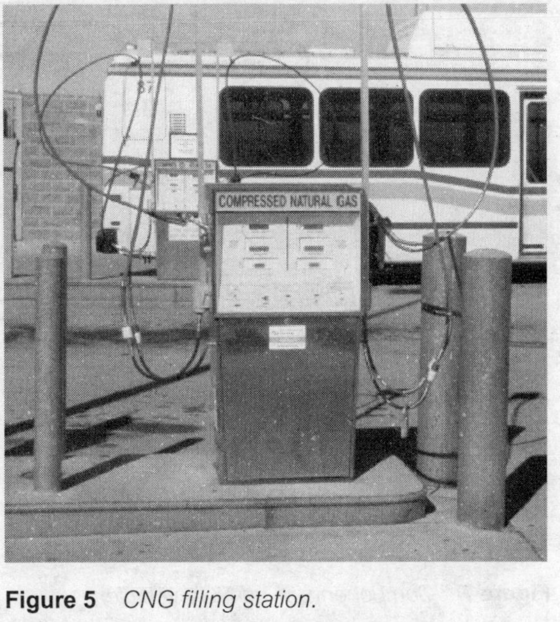


**Figure 5** *CNG filling station.*

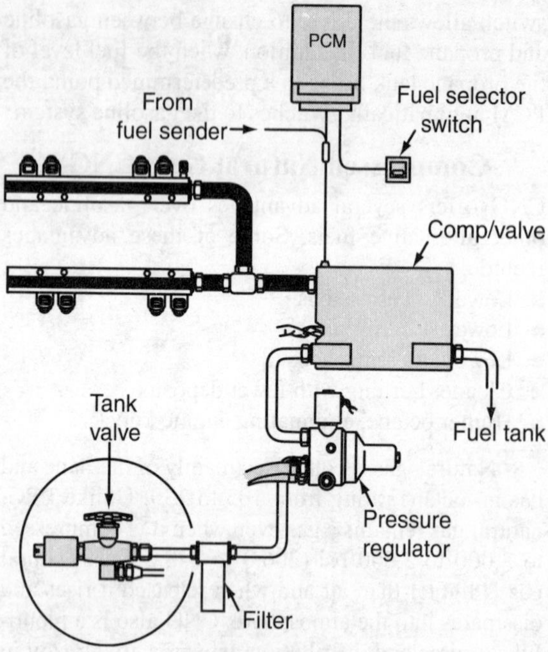


**Figure 6** *CNG bi-fuel system.*

and Contour, most GM pickups, the Chevrolet Cavalier, the Toyota Camry, and the Volvo S70 and V70. Many of these models are also available in a bi-fuel configuration, with the ability to run on either CNG or unleaded gasoline **(Figure 6)**.

Natural gas vehicles (NGVs) store fuel in cylinders installed in the rear or undercarriage of the vehicle. Inside the cylinders, the gas is under a high pressure of 3,000 to 3,600 psi (206.8 to 248.2 bar). The flow of gas is through a fuel regulator located in the engine compartment. It is then injected through a specially designed mixer to achieve the correct air-to-gas mixture. **Figure 7** shows the typical components of an NGV. The following sections discuss common components and their functions.

**Fuel-Control Valves with Pressure Relief**. Each tank in a CNG vehicle is equipped with its own manually operated fuel-control valve **(Figure 8)**. Turning the valve handle all the way in the clockwise direction stops gas flow. Turning the valve handle all the way in the counterclockwise direction provides gas flow.

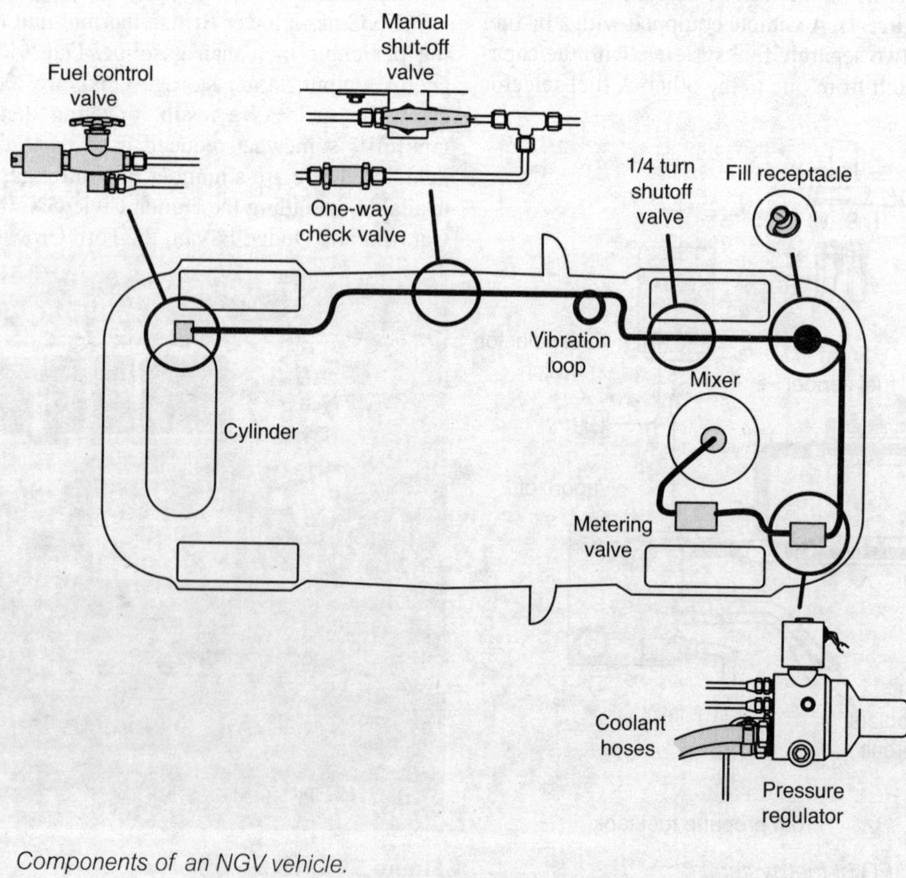


**Figure 7** *Components of an NGV vehicle.*

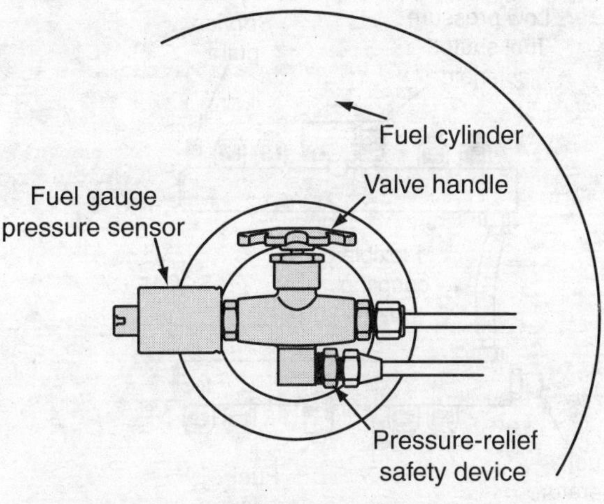


**Figure 8** *Typical fuel-control valve.*

Each control valve is equipped with a pressure-relief device for safety. The device releases excess cylinder pressure to the atmosphere if the temperature of the cylinder rises above approximately 217ºF (103ºC). When the re-lief device opens, the CNG from the entire system (all cylinders) is released into the atmosphere. The relief device is not reusable.

**Quarter-Turn Manual Shutoff Valve**. A hand-operated manual shutoff valve is provided to stop fuel flow from all cylinders **(Figure 9)**. When closed, the valve prevents fuel from going to the delivery system. A "Manual Shutoff Valve" label is affixed to the exterior body panel of the vehicle.

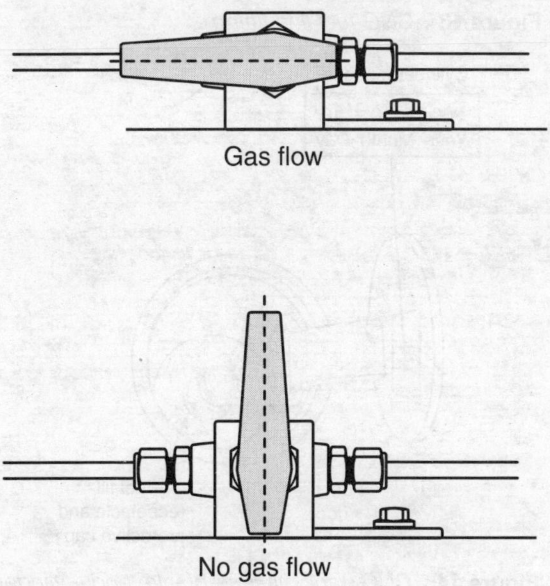


**Figure 9** *Typical manual valve operation.*

When the manual shutoff knob is turned parallel to the fuel line, the valve is open. When the valve is turned perpendicular to the fuel line, the valve is closed and fuel cannot reach the delivery system.

**Fuel Pressure Regulator**. The fuel pressure regulator reduces the fuel cylinder pressure to 90 to 140 psi (6.2 to 9.6 bar). The regulator is located downstream from the manual shutoff valve **(Figure 10)**. The regulator contains a built-in pressure-relief device and a fuel filter. The pressure-relief device is a safety item that vents CNG to the atmosphere. It is on the regulator's low-pressure side and vents excess pressure above approximately 225 psi (15.5 bar).

Expanding the fuel to a lower pressure has a thermodynamic cooling effect. To counteract this effect, engine coolant is routed to the regulator to prevent freezing **(Figure 11)**. The hoses are tied into the vehicle's heater hoses in the engine compartment.

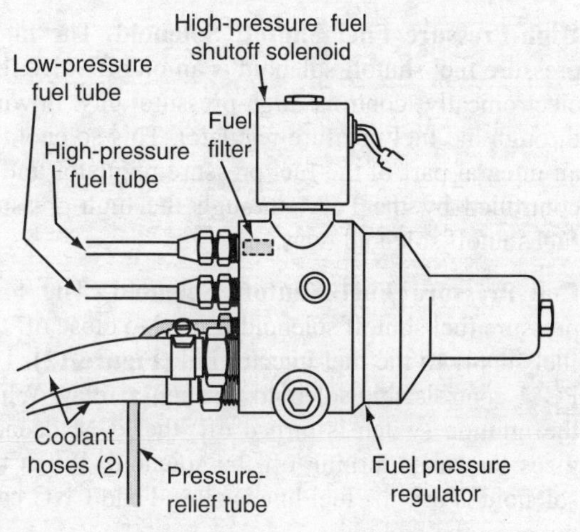


**Figure 10** *CNG fuel pressure regulator.*

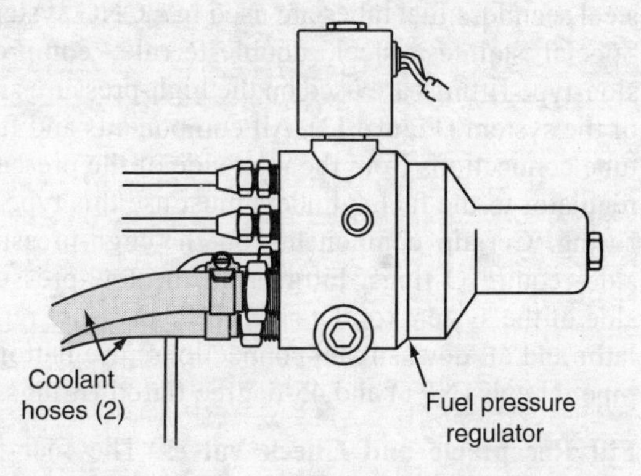


**Figure 11** *Engine coolant hoses routed to the regulator to prevent freezing.*

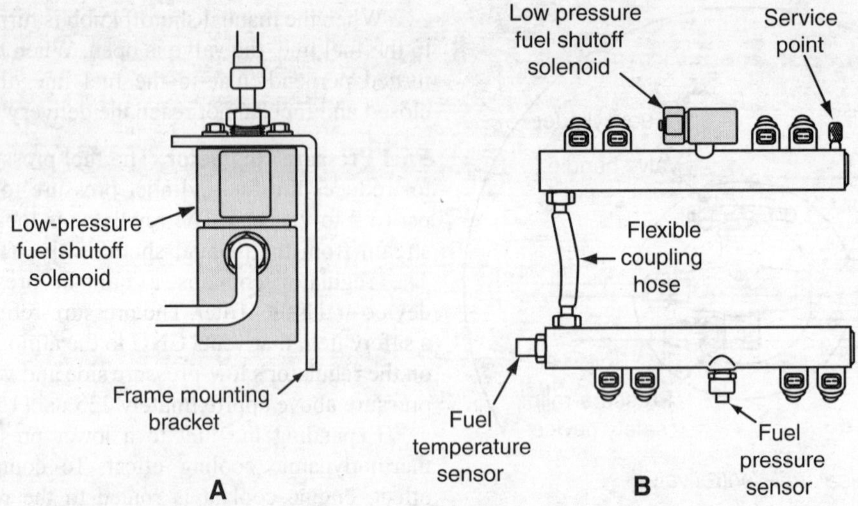


**Figure 12** *The low-pressure fuel shutoff solenoid can be located in the fuel rail feed line (A) or on the fuel rail (B).*

**High-Pressure Fuel Shutoff Solenoid**. The high-pressure fuel shutoff solenoid is an on/off valve that electronically controls high-pressure gas flowing through the fuel pressure regulator. This solenoid is an integral part of the fuel pressure regulator and is controlled by the PCM through the high-pressure fuel shutoff solenoid relay.

**Low-Pressure Fuel Shutoff Solenoid**. The low-pressure fuel shutoff solenoid is used to close off the fuel supply to the fuel-injector rail **(Figure 12)**. The PCM controls this solenoid through a relay. When the ignition switch is turned off, the PCM deenergizes the relay, turning off the solenoid. When the solenoid is off, the fuel line is closed and CNG cannot enter the fuel rail.

**Fuel Lines and Fittings**. High-pressure stainless steel seamless fuel tubes are used in a CNG system. Special stainless steel, double-ferrule, compression-type fittings are used on the high-pressure side of the system **(Figure 13)**. All components and fuel tube connections from the inlet side of the pressure regulator to the fuel cylinders must use this type of fitting. Certain components on the high-pressure side require O-rings. Fittings on the low-pressure side of the system (outlet side of the pressure regulator and all downstream connections) use national pipe threads (NPT) and 45-degree flared fittings.

**Fill Receptacle and Check Valves**. The fuel fill receptacle is mounted behind the fuel filler door **(Figure 14)**. The receptacle contains an integral one-

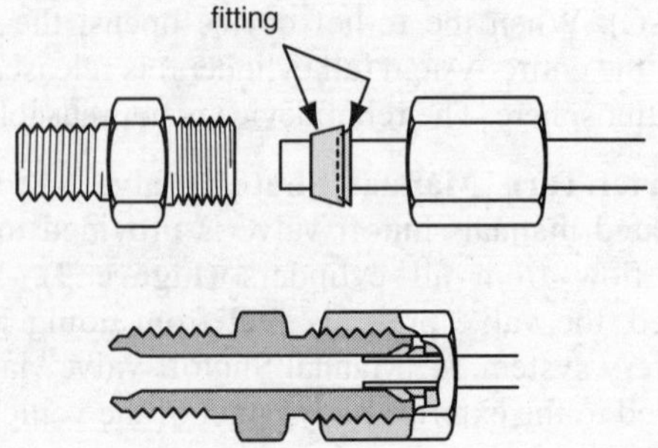


**Figure 13** *CNG fuel line fitting.*

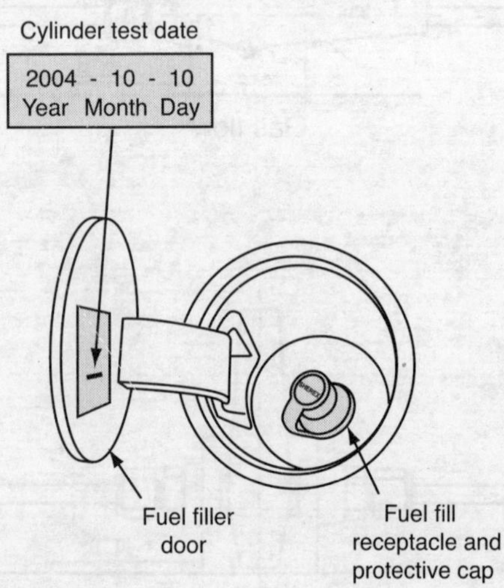


**Figure 14** *CNG tank fill receptacle and cylinder test date sticker.*

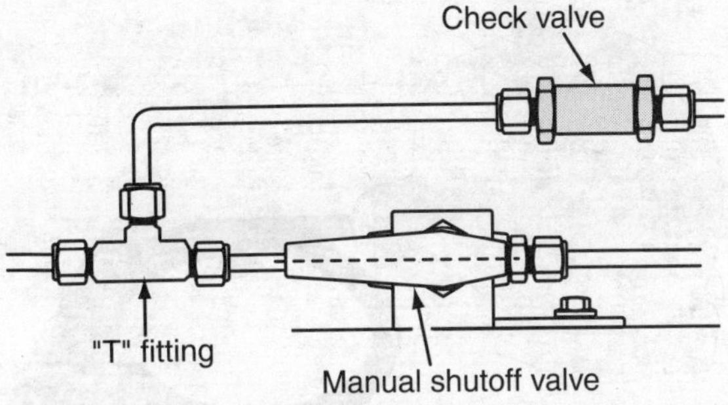


**Figure 15** *One-way check valve.*

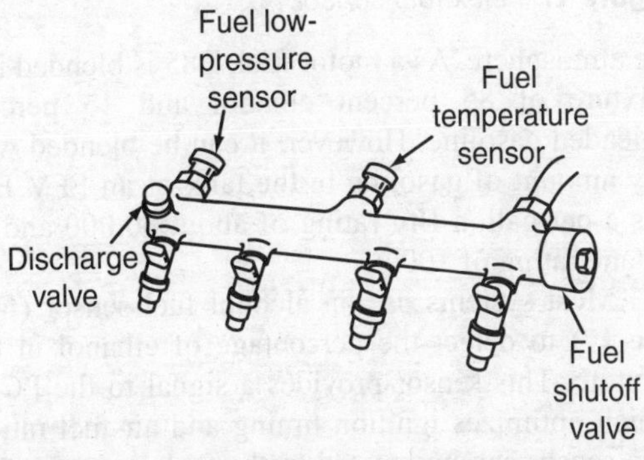


**Figure 16** *Fuel low-pressure sensor and fuel temperature sensor.*

way check valve. An additional one-way check valve is installed in the fuel fill line between the fuel pressure regulator and the fuel fill receptacle **(Figure 15)**. Both the receptacle check valve and the in-line check valve prevent fuel from escaping through the filling receptacle.

**Fuel Low-Pressure Sensor**. The fuel low-pressure sensor provides an input to the PCM based on the pressure of the CNG fuel at the fuel rail **(Figure 16)**. The PCM uses this input along with various other sensors to calculate fuel-injector timing.

**Fuel Temperature Sensor**. The PCM uses input from the fuel-temperature sensor to help calculate fuel-injector timing.

## CNG and Propane Cylinder Codes

CNG is stored in specially approved cylinders at pressures up to 3,600 psi (248.2 bar). The cylinders can be made of steel, steel composite, or aluminum composite. Many are constructed of aluminum liners wrapped in fiberglass. The exterior of the cylinders is coated with an epoxy paint and clear polyurethane to provide additional environmental protection.

Government regulations have standardized the testing and use of CNG fuel tanks throughout the automotive industry. The U.S. Department of Transportation (DOT) requires that each CNG fuel cylinder manufactured on or after March 27, 1995, be removed and reinspected every 3 years in accordance with Federal Motor Vehicle Safety Standard 304. A qualified person must perform the inspection in accordance with the cylinder manufacturer's established reinspection criteria and the appropriate Compressed Gas Association guideline. Retest markings must be stamped on the cylinder neck or marked on a label securely affixed to the cylinder near the original test date and coated with epoxy. Reheat treatment or repair of rejected cylinders is not allowed. The fuel cylinder expires and must be removed from service 15 years from the date of manufacture. A label can also be attached to the fuel filler door stating the cylinder retest date and expiration date.

Canadian requirements state that the cylinder must be reinspected and hydrostatically retested every 3 years in accordance with the Canadian Standards Association (CSA) CAN/CSA-B339, as prescribed for TC-3FCM containers. Retest dates must be stamped on the exposed metallic surface of the cylinder neck or marked on a label securely affixed to the cylinder near the original test date and coated with epoxy. Reheat treatment or repair of rejected cylinders is not allowed. The fuel cylinder expires and must be removed from service 15 years from the date of manufacture. A label can also be attached to the fuel filler door stating the first cylinder retest date and expiration date.

Because propane is not stored at such high pressures, the testing for propane cylinders is not as rigorous as that for CNG cylinders. Every propane cylinder must be exposed to twice its service pressure and one out of every 500 is exposed to four times its pressure, or about 960 psi (66 bar). Two types of containers are authorized for propane: DOT cylinders and American Society of Mechanical Engineers (ASME) tanks.

DOT cylinders are manufactured under the provisions of DOT Hazardous Materials Regulations. They must be requalified for continued use 12 years from the date of manufacture. If the DOT-authorized visual inspection procedure is used, the cylinders must be requalified every 5 years thereafter. They must be checked for physical wear or damage every time they are refilled and the paint must be kept in good condition. DOT engine fuel cylinders are usually removed from the vehicle and refilled elsewhere, though they may be refilled in place if they are properly installed and equipped for that purpose.

ASME engine fuel tanks are always refilled in the vehicle. These containers are manufactured under the provsions of the ASME Pressure Vessel Code. Periodic requalification is not required but the tanks should be inspected for unusual wear or physical damage and the paint must be kept in good condition.

## Flex Fuel Vehicles

A **flex fuel vehicle (FFV)** has one fuel tank and a single fuel system. The vehicle is designed to run on regular unleaded gasoline and an alcohol fuel (either ethanol or methanol) in any mixture. Most manufacturers produce a version of the FFV. Initially, most of these vehicles were designed to run on M85 (85 percent methanol). Methanol is typically made from natural gas, though it is possible to produce it by fermenting biomass. M85 has a per-gallon Btu rating between 56,000 and 66,000 with an octane rating of 100. However, methanol is more corrosive than gasoline and requires changes in some of the materials in the fuel-handling systems of both the vehicle and the refueling station.

Due to a lack of availability (due to storage and handling concerns) of M85, all domestic manufacturers now make their FFVs to run on E85. Ethanol, or grain alcohol, is produced by fermenting biomass. Ethanol is a renewable resource and contributes nothing in itself to greenhouse gas loading of the atmosphere. As a motor fuel, E85 is blended in a mixture of 85 percent ethanol and 15 percent unleaded gasoline. However, it can be blended with any amount of gasoline in the tank of an FFV. E85 has a per-gallon Btu rating of about 80,000 and an octane rating of 100.

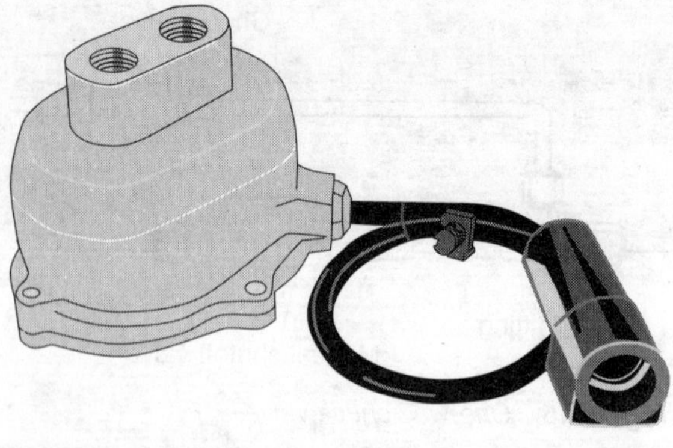

**Figure 17** *Flex fuel sensor.*

Most systems use an alcohol fuel sensor **(Figure 17)** to detect the percentage of ethanol in the mixture. This sensor provides a signal to the PCM, which optimizes ignition timing and air-fuel ratios. The sensor can be located in the tank or in the fuel line between the tank and the injectors **(Figure 18)**. Some manufacturers do not use the sensor. Instead, they use the $O_2S$ to indicate that the engine is being run on E85. No matter which sensor is used, the end result is the PCM will use different strategies based on the detection of gasoline or E85.

Other PCM changes that occur when E85 is detected include how OBD-II monitors will be run.

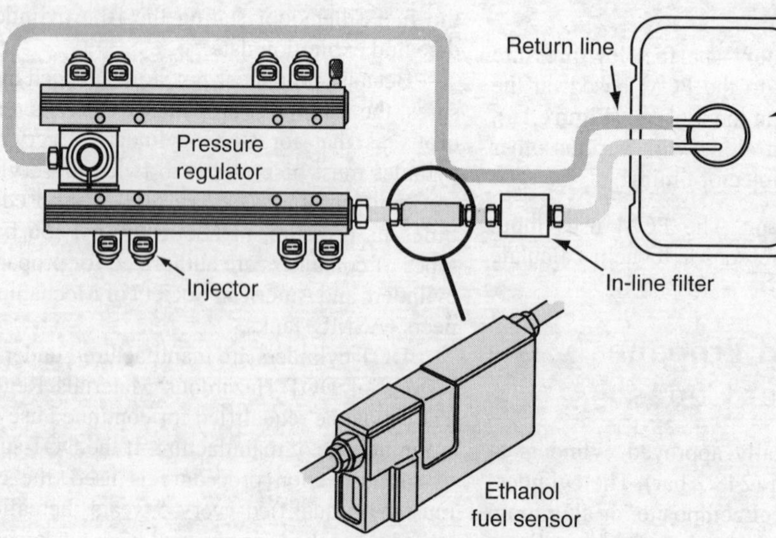


**Figure 18** *E85 flex fuel system with ethanol sensor.*

For example, catalyst index ratios for ethanol vary based on the changing concentration of alcohol in the fuel. The malfunction threshold typically increases as the percentage of alcohol increases. A malfunction threshold of 0.5 may be used at E10 (10 percent ethanol) and 0.9 may be used at E85 (85 percent ethanol). The malfunction thresholds are therefore adjusted based on the percentage of alcohol in the fuel.

E85 fuel systems do not look any different from most sequential fuel-injection (SFI) systems. However, the components of the E85 system usually differ from those of a standard gasoline system. The components must be able to withstand the actions of ethanol. In most cases, different engine oil is also recommended when running E85 exclusively.

## Review Questions

1. Describe the flow of the fuel-delivery system of an LPG fuel-injection vehicle.
2. The automotive propane fuel system is a completely ________ system that contains a supply of pressurized liquefied petroleum gas (LPG).
3. Special stainless steel, ________, compression-type fittings are used on the high-pressure side of a CNG system.
4. Flex fuel vehicles run on M85, E85, or ________.
5. The ________ has a single fuel tank, fuel system, and engine. The vehicle is designed to run on regular unleaded gasoline and an alcohol fuel (either ethanol or methanol) in any mixture.
6. Natural gas has an octane rating between:
   A. 105 and 130
   B. 85 and 91
   C. 120 and 160
   D. 45 and 65
7. Explain the difference between a bi-fuel and a flex fuel system.
8. Natural gas vehicles (NGVs) store their fuel in cylinders under a high pressure of ________ to ________ psi.
9. CNG tanks must be inspected every ________ years and replaced every ________ years.
10. The ________ converts the liquid propane to a gaseous vapor form.
11. The LPG fuel-injection system tanks contain:
    A. A rollover valve
    B. A pump
    C. A regulator/vaporizer
    D. A float valve
12. Flex fuel vehicles can run on gasoline and:
    A. Natural gas
    B. Propane
    C. Hydrogen
    D. E85

CHAPTER

# 39 Alternative Fuel System Service

## Introduction

This chapter covers typical procedures for diagnosing and servicing vehicles with alternative fuel systems. Coverage is limited to propane and compressed natural gas (CNG) vehicles. Due to safety concerns, the manufacturers of electric and hybrid vehicles only allow factory-trained master technicians to service these vehicles. Fuel-cell systems are not yet readily available and service of these vehicles is restricted to manufacturer-assigned personnel.

## CNG Fuel System Modes of Operation

The PCM controls all functions of a CNG system by using input signals and controlling the various output devices. Similar to how it operates in gasoline systems, the PCM in a CNG fuel system operates in open- and closed-loop modes depending on operating conditions.

The PCM monitors various inputs and recognizes the following modes of operation:

Open loop:

- Ignition switch on (engine not running)
- Engine start-up (crank)
- Engine warm-up
- Deceleration
- Ignition switch off

Closed loop (with engine at operating temperature):

- Idle
- Cruise
- Acceleration
- Wide-open throttle (WOT)

All modes of operation are the same as for the gasoline system except for WOT and deceleration modes. WOT mode is normally an open-loop mode on gasoline-powered vehicles. In CNG vehicles WOT is a closed-loop mode of operation. During WOT with CNG vehicles, the PCM operates the injectors as it continues fuel injection and adjusts the injector pulse width and spark advance.

When the vehicle is under hard deceleration with proper rpm and closed-throttle conditions, the PCM enters open-loop mode. It then enters fuel cutoff until engine rpm approaches idle speed and then the PCM adjusts the IAC motor.

When the vehicle is under light deceleration, the PCM continues fuel injection but adjusts injection pulse width accordingly. Based on all of the inputs, the PCM adjusts engine idle speed through the IAC motor. This provides a smooth transition to idle mode.

## CNG Fuel-Delivery System Service

This section covers diagnosis and service of typical CNG fuel systems. Prior to servicing these vehicles, you must become familiar with all the safety warnings and cautions. Do not skip this section. It is imperative that you understand the safety concerns associated with service of this type of system.

### CNG Service Precautions

The technician must observe the following warnings when servicing a CNG system:

- A CNG system should be serviced only by qualified natural gas vehicle technicians. All work performed in the school's shop must be supervised by an instructor at all times.
- Natural gas vapors at atmospheric pressure are lighter than air and will rise and disperse in open areas. In enclosed areas, natural gas vapor may collect and form a combustible mixture.
- If the vehicle is routinely placed in an enclosed area, the area should be provided with adequate ventilation and/or a natural-gas-detection system.

- For long-term storage, the manual shutoff valve and each individual fuel control valve should be closed.
- A fire or explosion may result if a source of ignition or flame is present with natural gas in a confined, unventilated area. Provide proper ventilation when servicing CNG vehicles.
- Natural gas contains an odor additive; occasionally mild natural gas odors may be noticed during engine start-up and after engine shutdown because of small quantities of unburned gas in the engine intake and exhaust manifolds. While the odors are normal, a persistent odor of natural gas indicates a leak is present. Check the vehicle immediately for the source of the leak. If a leak is detected, do not bring the vehicle into an unventilated area.
- Do not attempt to weld any fuel system component.
- Do not attempt to refill the fuel cylinders if any of the outer composite covering is damaged.
- Do not attempt to modify any fuel system component or replace any component with parts that do not meet the manufacturer's requirements.
- Do not apply any paint or undercoating material to any fuel system component.
- Do not attempt to force open or tamper with any fuel tubes or valves. A sudden release of high-pressure gas may occur, resulting in bodily injury.
- When a vehicle has been involved in an accident that may have caused damage or dislocation to the CNG system, the system should be inspected and pressure tested.
- Any fuel system components that have been subjected to fire may not be returned to service due to reduced pressure capability.
- U.S. regulations require that CNG fuel cylinders manufactured on or after March 27, 1995, be removed and reinspected for damage and deterioration after a motor vehicle accident or fire and at least every 36 months or 36,000 miles (whichever occurs first), in accordance with Motor Vehicle Safety Standard No. 304 "Compressed Natural Gas Fuel Containers," and the Compressed Gas Association (CGA) guideline C-6.4. Retest markings must be stamped on the cylinder neck or marked on a label securely affixed to the cylinder near the original test date and overcoated with epoxy.
  –Reheat treatment or repair of rejected cylinders is not authorized.
  –The fuel cylinder expires and must be removed from service 15 years from the date of manufacture. A label on the fuel filler door states the cylinder retest date and cylinder expiration date. A similar label is attached to each fuel cylinder.
  –If there is a concern about the proper inspection of the CNG cylinder, contact the manufacturer as identified on the cylinder label.
- Canadian regulations require that CNG cylinders must be reinspected for damage and deterioration after a motor vehicle accident or fire and at least every 36 months or 36,000 miles (58,000 km) (whichever occurs first), in accordance with Canadian Standards Association (CSA) B51-97, Part 2 "High-Pressure Cylinders for the Onboard Storage of Natural Gas as a Fuel for Automotive Vehicles" and the CGA guideline C-6.4. Retest dates must be stamped on the exposed metallic surface of the cylinder neck or marked on a label securely affixed to the cylinder near the original test date and coated with epoxy.
  –Reheat treatment or repair of rejected cylinders is not authorized.
  –The fuel cylinder expires and must be removed from service 15 years from the date of manufacture. A label on the fuel filler door states the first cylinder retest date and cylinder expiration date.
- It is recommended that the fuel system components be inspected periodically for leaks and/or excessive wear.
- Fuel the vehicle using an NGV-certified fill nozzle as specified on the label inside the fuel filler door. Do not attempt to fuel the vehicle with any other type of nozzle.
- Before opening any part of a CNG fuel system for repair, purge the system of all natural gas.

## CNG Fuel-Delivery Visual Inspection

A visual check can reveal many faults and save unnecessary diagnostic time. Always perform a thorough visual inspection for loose, damaged, disconnected, or incorrectly routed wires and fuel lines. To perform the visual inspection, begin by verifying that the **manual shutoff valve (Figure 1)** is open by turning the valve handle parallel to the gas line. Then ensure the **fuel-control valves (Figure 2)** are open by turning the valve handle fully counterclockwise. The manual shutoff valve provides a single point that allows for stopping the flow from all cylinders. Fuel-control valves are manually operated valves located on each cylinder. They are used to stop the flow of gas from the

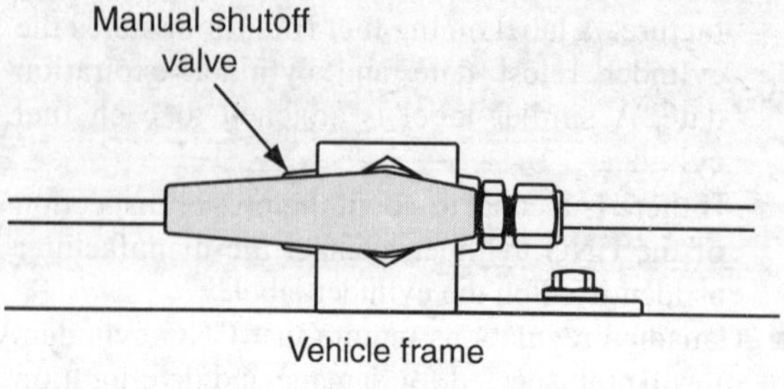


**Figure 1** *Manual shutoff valve.*

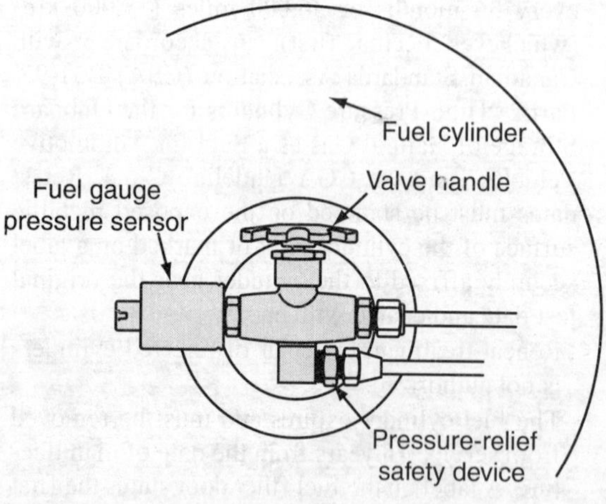


**Figure 2** *A fuel-control valve is located on each cylinder.*

cylinder. Visually inspect the following:

1. Check electrical connections at the following components:
   - Battery cables
   - PCM connector
   - Low-pressure fuel shutoff solenoid
   - High-pressure fuel shutoff solenoid
   - Fuel temperature and fuel pressure sensors
   - High-pressure fuel shutoff solenoid relay
   - Injector driver module (if equipped)
2. Inspect the fuel tubes for damage or kinking.
3. Inspect all fuel tube connections for leaks. Additional items can be visually inspected. Refer to the appropriate service manual for further information on visual inspections.

> **You Should Know**
>
> *Any time a customer brings in a CNG vehicle for service, inspect the tanks and note when the next inspection is needed. U.S. regulations require CNG fuel cylinders be removed and inspected at least every 36 months or 36,000 miles (whichever occurs first). Inspection of the tanks is very important and should not be overlooked.*

## Testing for Gas Leaks

If a gas leak is suspected at a fuel tube, fuel tube fitting, or fuel system component, obtain a commercially available leak detector. A tube fitting supplier is a good source for this. When inspecting for leaks, be sure the fuel is under operating pressure. Apply a generous amount of detecting fluid to the suspected leak area(s). Watch the area for about 3 to 5 minutes. In most cases, a leak will cause a steady flow of bubbles to appear at the leak source.

## System Purging

Purge the fuel system of natural gas before any of the following components are removed or repaired **(Figure 3):**

- High- and low-pressure fuel lines
- Fuel injector rail
- Fuel injectors
- Fuel temperature sensor
- Fuel low-pressure sensor
- Fuel gauge pressure sensor
- Fuel fill receptacle
- Fuel cylinder
- Fuel-control valve
- Manual shutoff valve
- Fuel filter
- Low-pressure fuel shutoff solenoid
- High-pressure fuel shutoff solenoid
- Fuel pressure regulator
- One-way check valve

> **You Should Know**
>
> *Before any part of a CNG fuel system is opened for repair or replacement, purge the system of all natural gas.*

**Fuel Tube Purging**. Tube purging can be used if the fuel tanks are not being serviced. There are two separate procedures for purging the tubes of natural gas. The procedure used depends on whether or not the engine can be started.

If the engine will run, close the fuel-control valves by turning the valve handle fully clockwise on all fuel cylinders. Verify that the manual shutoff valve is in the open position to allow gas flow by

> **You Should Know**
>
> *Although the fuel tubes are now purged of gas, the fuel cylinders are still under high pressure. Do not open the closed valves until all high-pressure components have been resealed.*

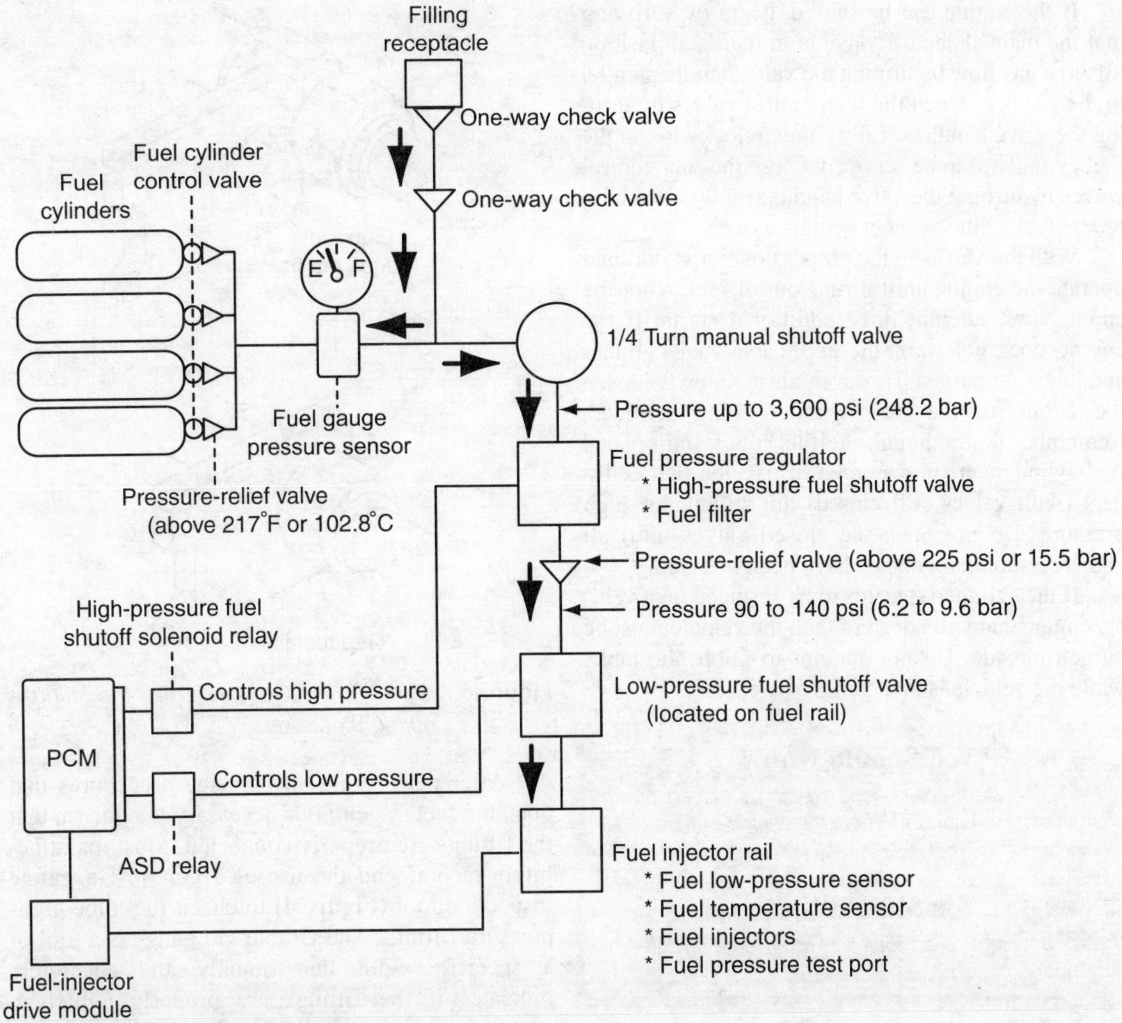


**Figure 3** *CNG fuel flow diagram.*

turning the valve handle parallel to the gas line. Start and run the engine until it runs out of fuel. After the engine dies, attempt three additional starts. If the engine does not start, the fuel tube is purged. However, always slowly loosen any fittings being serviced.

If the tubes need to be purged on a vehicle engine that will not start, move the vehicle outdoors. Do not attempt to purge the tubes with the vehicle inside a confined area.

**You Should Know**

*Wear approved eye protection.*

With the vehicle safely outside, close the fuel-control valves by turning the valve handles fully clockwise on all fuel cylinders. Verify that the manual shutoff valve is in the open position by turning the valve handle parallel to the gas line. Next, cycle the ignition switch to the Run position a few times to activate the high-pressure fuel shutoff solenoid. With the ignition switch in the Off position, slowly loosen the fuel line at the low-pressure fuel shutoff solenoid. Pressure will be observed. Leave the line loosened until all pressure is relieved. Although the fuel tubes are now purged of gas, the fuel cylinders are still under high pressure. Do not open the closed valves until all high-pressure components have been resealed.

**Fuel Cylinder Purging**. To service the fuel cylinders, they must be purged. There are two procedures for purging the fuel cylinders. The procedure used depends on whether or not the engine can be started.

If the engine can be started, begin by verifying that the manual shutoff valve is in the open position to allow gas flow by turning the valve handle parallel to the gas line. Open the fuel-control valves by turning the valve handle(s) fully counterclockwise on the fuel cylinder(s) to be serviced. Close the fuel-control valves by turning the valve handle(s) fully clockwise on the fuel cylinder(s) not being serviced.

With the valves in the proper positions, start and operate the engine until it runs out of fuel. After the engine dies, attempt three additional starts. If the engine does not start, the applicable fuel cylinder and tubes are purged. However, always slowly loosen the fitting to the fuel cylinder being serviced. Remember that although the fuel tubes and opened fuel cylinders are now purged of gas, the fuel cylinders (with valves still closed) are still under high pressure. Do not open the closed valves until all high-pressure components have been resealed.

If the cylinder(s) needs to be serviced on a vehicle engine that will not start, then the vehicle must be purged outside. Do not attempt to purge the tanks while the vehicle is in a confined area.

**You Should Know**

*If a fuel system component is being serviced on an engine that will not start, move the vehicle outdoors.*

**You Should Know**

*Wear approved eye protection when purging the fuel cylinder.*

With the vehicle outside, open the manual shutoff valve by turning the valve handle parallel to the gas line. Open the fuel-control valves by turning the valve handle(s) fully counterclockwise for the fuel cylinder(s) to be serviced. Close the fuel-control valves by turning the valve handle(s) fully clockwise on the fuel cylinder(s) not being serviced. Very slowly loosen the fuel tube fitting at the fuel-control valve of the fuel cylinder being serviced. High pressure will be observed. Leave the line loosened until all pressure is relieved.

## Fuel Tube Fitting Replacement

**You Should Know**

*Do not attempt to remove or tighten a fitting with the fuel system under pressure. Always purge the system prior to servicing the fittings.*

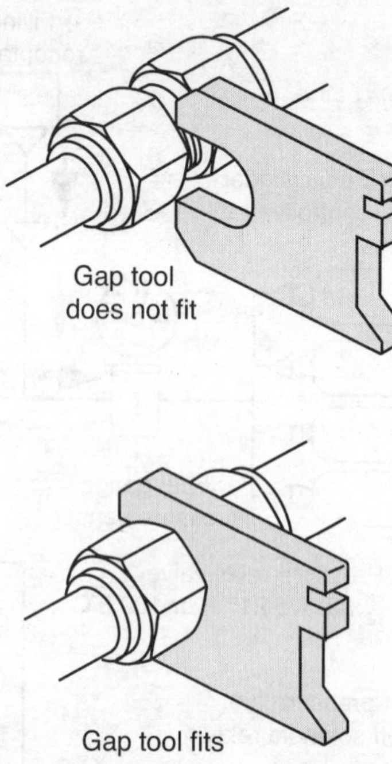


**Figure 4** *If the Go-No-Go gauge fits, the fitting is not tight enough.*

After performing any service procedures that open the fuel system, it is necessary to confirm that the fittings are properly connected. Most manufacturers recommend the use of a **Go-No-Go gauge** inspection tool **(Figure 4)** to check fuel tube high-pressure fittings. The Go-No-Go gauge is a tool of a specific width that quickly and accurately indicates if the fittings are properly tightened. Always use the correct size Go-No-Go gauge to ensure the fitting has been sufficiently pulled up (tightened) on the fitting body. Different size gauges are available for ¼ in. and ¾ in. fuel tubes.

**You Should Know**

*To prevent damage to the fitting and the tool, do not install the Go-No-Go gauge inspection tool to the fitting while tightening. It is to be used only as a test gauge after the fitting has been tightened.*

High-pressure stainless steel, seamless fuel tubes are used in a CNG system. Special stainless steel, **double-ferrule**, compression-type fittings are used on the high-pressure side of the system **(Figure 5)**. The single ferrule incorporates one ferrule that must both grip and seal. The double-ferrule design splits these functions by using the front ferrule to seal and the rear ferrule to grip. All components and

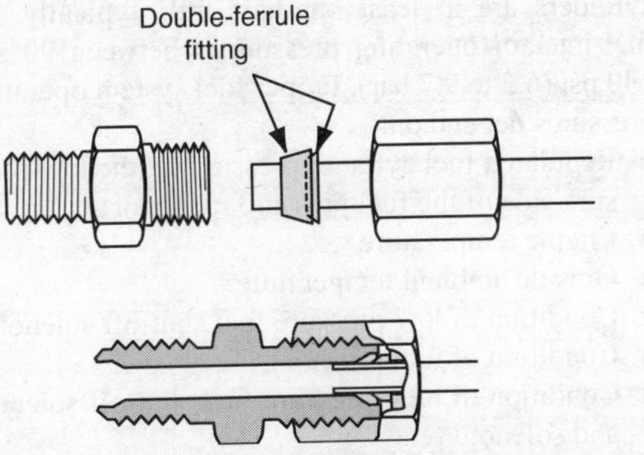


**Figure 5** *Double-ferrule fitting provides a positive seal and grip of the tube.*

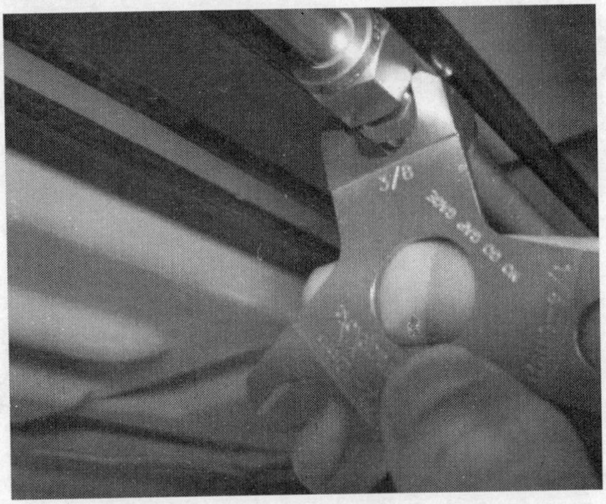


**Figure 6** *Slowly continue to tighten the fitting until the gauge does not fit between the fitting and the hex on the fitting body. This indicates the fitting has been properly tightened. Do not overtighten the fitting.*

**Figure 7** *Leak test the component.*

fuel tube connections from the inlet side of the pressure regulator to the fuel cylinders (high pressure) must use this type of fitting. Fittings on the low-pressure side of the system (outlet side of the pressure regulator and all downstream connections) use NPT pipe threads and 45-degree flared fittings.

If any fuel tube, fitting, or component connected to a fuel tube must be removed, thoroughly clean the fitting and tube before disassembly. This helps to prevent dirt or foreign material from entering the fuel system. New high-pressure tube fittings are delivered completely assembled with two ferrules installed. Disassembly is not recommended because dirt or foreign material could get into the fitting and cause a leak. To install a new high-pressure fuel tube fitting, insert the fuel tube into the fitting. Ensure that the fuel tube is completely through the ferrules and resting firmly on the shoulder of the fitting body. Tighten the fitting finger. Next, use a soft-tip marker to place a reference mark on the fitting at the six o'clock position. Attach a backup wrench to the fitting body and with a second wrench rotate the fitting one-half turn. The reference mark should now be at the twelve o'clock position. Inspect the fitting with a Go-No-Go gauge **(Figure 6)**. If the gauge fits between the fitting and the hex on the fitting body, additional tightening is required. Remove the Go-No-Go tool from the fitting before proceeding. If necessary, slowly continue to tighten the fitting until the gauge does not fit between the fitting and the hex on the fitting body. This indicates the fitting has been properly tightened. Do not overtighten the fitting. Finally, leak test the component **(Figure 7)**.

Used fittings can be used several times. When making a high-pressure connection with used fittings, make sure to thoroughly clean the tube and the fitting before installation and tightening. Insert the fuel tube with ferrules that have been pre-swaged and collapsed into the fitting body. Fully seat the ferrule into the fitting body. Tighten the fitting nut by hand, then attach a backup wrench to the fitting body. Rotate the fitting nut to its original position with a second wrench. An increase in resistance will be felt when the fitting nut has reached its original position. Slowly tighten the fitting nut an additional one-twelfth of a turn. Do not overtighten the fitting. After the installation and tightening, check the fitting with a Go-No-Go gauge tool. After the fitting is properly tightened, restore fuel flow through the system and check for leaks.

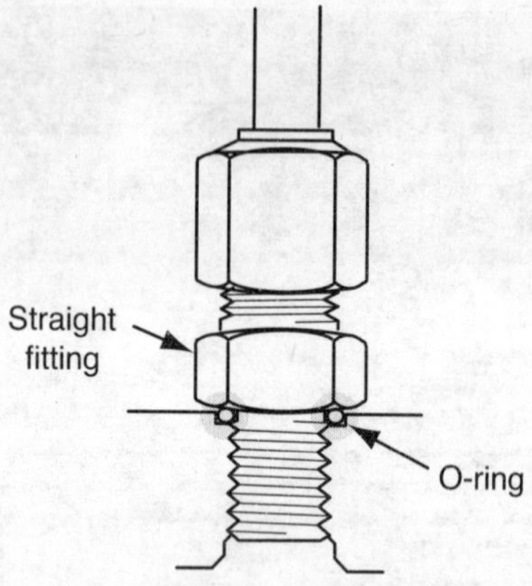


**Figure 8** *Straight-type fitting with an O-ring seal.*

Certain components on the high-pressure side require O-rings. Special precautions must be taken when removing and installing this type of fitting:

- Make sure the tube and fitting are thoroughly clean before removing the fitting. Inspect both the threads and the sealing surface to confirm they are free of nicks, burrs, scratches, or any foreign material.
- Lubricate the O-ring with special O-ring lubricant.
- Tighten straight-type fittings **(Figure 8)** to 245 in.-lbs. (28 N•m) torque.
- Check the fitting for leaks.

When installing standard NPT tapered fittings to the low-pressure side of the fuel system, apply a small amount of anaerobic thread sealant to the pipe threads *only*. Be extremely careful not to let any of the thread sealant into any fuel system opening.

### You Should Know

*Use only an anaerobic thread sealant containing Teflon™.*

## Fuel System Pressure Testing

If diagnostics indicate that the fuel system pressure needs to be checked and confirmed, be sure that the cylinders are at least one-half full. Typically, the fuel-injector operating pressure is between 90 and 140 psi (6.2 to 9.7 bar). Proper fuel system operating pressures depend on:

- Regulated fuel cylinder pressure (at the low-pressure side of the fuel pressure regulator)
- Engine temperature
- Outside ambient temperature
- Condition of low-pressure fuel shut-off solenoid
- Condition of fuel injectors
- Condition of high-pressure fuel shut-off solenoid and solenoid relay
- Engine load requirements

To test system pressure, perform the procedure outlined below.

### You Should Know

*Use the correct gauge to measure the expected pressures. Otherwise, damage to equipment may result.*

Remove the fuel pressure service port cap on the fuel rail (if equipped) and attach a pressure test gauge to the service port. Start the engine and bring it to normal operating temperature. Check the pressure reading and compare the specifications.

If the fuel pressure is above the specifications and the fuel cylinders are one-half full or more, replace the fuel pressure regulator **(Figure 9).** If the fuel pressure is lower than specifications, check the following:

- Verify that the fuel cylinder control valves are open.
- Verify that the manual shutoff valve is open.
- Verify fuel volume.
- Check for damaged fuel tubes.
- Perform low-pressure fuel shutoff solenoid electrical tests.
- Perform high-pressure fuel shutoff solenoid electrical tests.
- Refer to fuel-injector tests.

After the tests have been completed, remove the pressure test gauge and check the system for leaks.

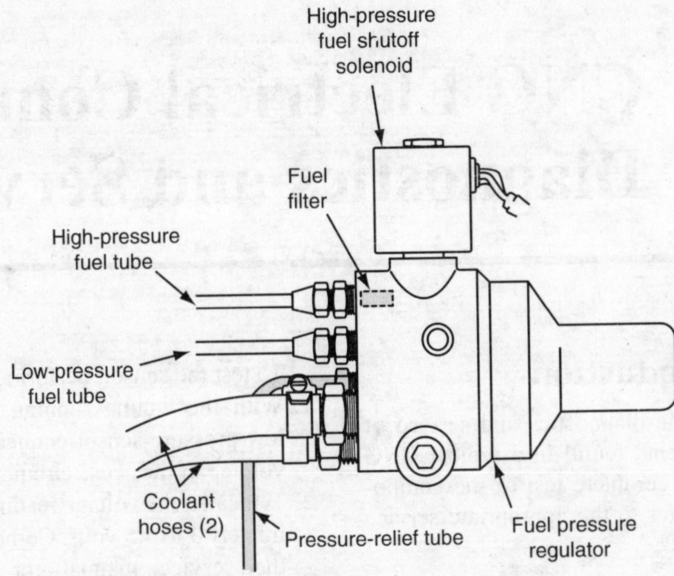


**Figure 9** *Fuel pressure regulator assembly with high-pressure shutoff solenoid and fuel filter.*

## Review Questions

1. Technician A says that to perform a pressure test on a propane-injected system, the cylinders need to be at least one-half full. Technician B says that typically the fuel-injector operating pressure in a propane system is between 90 to 140 psi. Who is correct?
   A. Technician A only
   B. Technician B only
   C. Both Technician A and Technician B
   D. Neither Technician A nor Technician B
2. Technician A says that any fuel system components that have been subjected to fire may not be returned to service. Technician B says that the fuel cylinder expires and must be removed from service 15 years from the date of manufacture. Who is correct?
   A. Technician A only
   B. Technician B only
   C. Both Technician A and Technician B
   D. Neither Technician A nor Technician B
3. Technician A says that when purging the tubes of a CNG vehicle when the engine will not start, place the shop ventilation hoses over the fitting being loosened. Technician B says that the Go-No-Go gauge should be installed between the fitting body and the fitting nut and the nut should be tightened until it contacts the gauge. Who is correct?
   A. Technician A only
   B. Technician B only
   C. Both Technician A and Technician B
   D. Neither Technician A nor Technician B
4. Technician A says that a CNG system enters open-loop mode at wide-open throttle. Technician B says that during WOT, the PCM operates the injectors as it continues fuel injection. Who is correct?
   A. Technician A only
   B. Technician B only
   C. Both Technician A and Technician B
   D. Neither Technician A nor Technician B
5. Technician A says that a CNG system must be purged before servicing the fuel line fittings. Technician B says that the high-pressure fittings are 45-degree flare fittings. Who is correct?
   A. Technician A only
   B. Technician B only
   C. Both Technician A and Technician B
   D. Neither Technician A nor Technician B

CHAPTER

# 40 CNG Electrical Component Diagnostics and Service

## Introduction

This chapter details the diagnostics and service of the electrical components found in a typical CNG vehicle. To perform a complete test of the components and circuitry, refer to the appropriate service manual.

## Fuel Low-Pressure Sensor Testing

The **fuel low-pressure sensor** is a transducer that sends the pressure signal to the PCM to be used with other inputs to calculate fuel delivery through the injectors. The fuel low-pressure sensor is located on the fuel-injector rail **(Figure 1)**. The sensor has a three-wire connector. For pin location, refer to the wiring diagrams of the appropriate service manual. The first circuit is a reference signal from the PCM to the sensor (usually 5 volts). The second circuit is a ground from the sensor back to the PCM. The third circuit is the sense circuit to the PCM. The PCM monitors the pressure by measuring the voltage on this circuit.

To test the sensor, perform the following steps:

1. With the engine running, backprobe the fuel low-pressure sensor connector and measure the voltage of the sense circuit.
2. Typically, the voltage reading should be approximately 3 to 4.5 volts. Compare the reading with the service manual for the vehicle being serviced.
3. Disconnect the low-pressure solenoid.
4. Measure the voltage of the sense circuit after the engine runs out of fuel. The voltage reading should be approximately 0.5 volt.
5. If 0 volt is measured on this circuit, the sensor has a short circuit. If 5 volts are measured, the sensor has an open circuit.

## Fuel Temperature Sensor Testing

The PCM uses the input from the **fuel temperature sensor** to determine injector timing. The fuel temperature sensor is located on the fuel-injector rail. The sensor has a two-wire connector. For pin location,

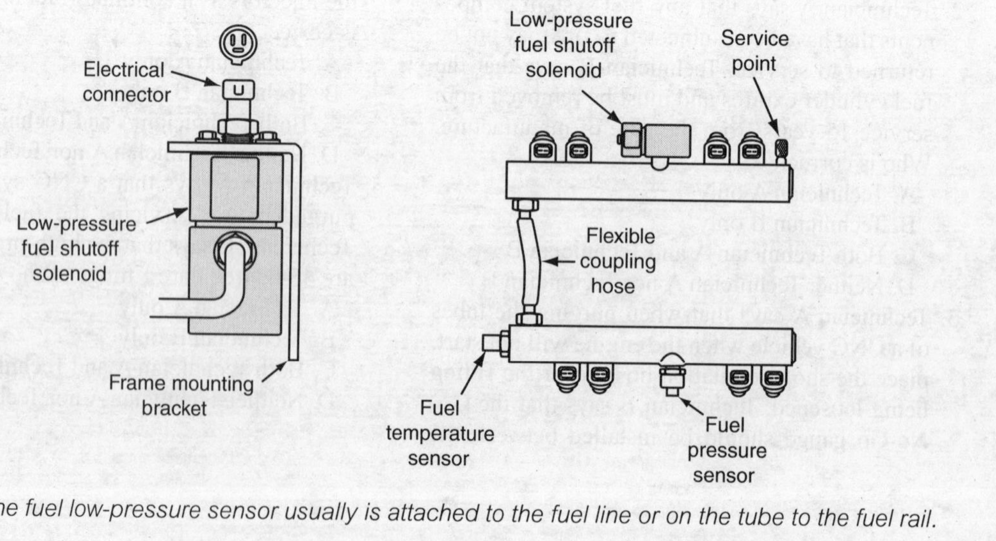


**Figure 1** *The fuel low-pressure sensor usually is attached to the fuel line or on the tube to the fuel rail.*

refer to the wiring diagrams of the appropriate service manual. The first circuit provides the fuel temperature sensor signal to the PCM. The second circuit is a ground from the sensor back to the PCM. To test the sensor, perform the following steps:

1. Run the engine until operating temperature is reached.
2. Disconnect the fuel temperature sensor wiring connector.
3. Using a DVOM, check the internal resistance of the sensor. Resistance should be less than 4,000 ohms with the engine at operating temperature. **Table 1** provides typical resistance values at different temperatures. Replace the sensor if the resistance reading is out of the given range.
4. Test the resistance of the wire harness between the PCM connector and the fuel temperature sensor connector. Repair the wire harness as necessary if resistance is greater than 1 ohm.

## Fuel Gauge Pressure Sensor Testing

The **fuel gauge pressure sensor** is a temperature-compensated transducer used to measure the pressure of the cylinder. The voltage signal is translated to fuel level and displayed on the vehicle's instrument cluster. The fuel gauge pressure sensor is mounted to the fuel-control valve of one of the fuel

| Fuel Temperature | | Resistance (ohms) | |
|---|---|---|---|
| C° | F° | Minimum | Maximum |
| –40° | –40° | 291,490 | 381,710 |
| –20° | –4° | 85,850 | 108,390 |
| –10° | 14° | 49,250 | 61,430 |
| 0° | 32° | 29,330 | 35,990 |
| 10° | 50° | 17,990 | 21,810 |
| 20° | 68° | 11,370 | 13,610 |
| 25° | 77° | 9,120 | 10,880 |
| 30° | 86° | 7,370 | 8,750 |
| 40° | 104° | 4,900 | 5,750 |
| 50° | 122° | 3,330 | 3,880 |
| 60° | 140° | 2,310 | 2,670 |
| 70° | 158° | 1,630 | 1,870 |
| 80° | 176° | 1,170 | 1,340 |
| 90° | 194° | 860 | 970 |
| 100° | 212° | 640 | 720 |
| 110° | 230° | 480 | 540 |
| 120° | 248° | 370 | 410 |

**Table 1** *Fuel temperature sensor resistance values.*

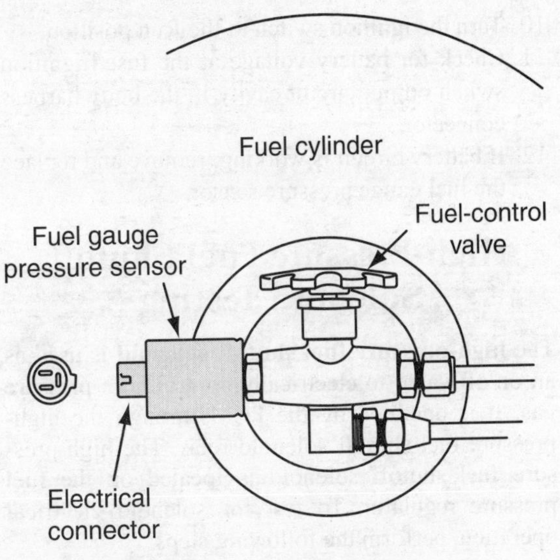


**Figure 2** *Fuel gauge pressure sensor.*

cylinders **(Figure 2)**. The sensor has a three-wire connector. For pin location, refer to the wiring diagrams of the appropriate service manual. The first circuit is typically a 12-volt battery supply. The second circuit is a ground return. The third circuit is the sense circuit to the fuel gauge. The PCM monitors the pressure by measuring the voltage on this circuit. To test the fuel gauge pressure sensor, perform the following steps:

1. Turn the ignition switch to Run (early models may require the engine to be running).
2. Disconnect the electrical connector at the fuel gauge pressure sensor.
3. The fuel gauge should move to Empty. If it does so, proceed to Step 6.
4. If the fuel gauge did not move to Empty, check the signal circuit between the connector and the gauge.
5. If the signal circuit is working, remove and replace the fuel gauge.
6. Install a jumper wire between the fuel gauge pressure sensor signal circuit and the ground circuit cavities in the body harness.
7. The fuel gauge should move to Full. If it does so, proceed to Step 10.
8. If the fuel gauge did not move to Full, turn the ignition switch to off; check for continuity between the ground circuit cavity in the body harness and a good ground.
9. If the ground circuit is all right, remove and replace the fuel gauge.

10. Turn the ignition switch to the Run position.
11. Check for battery voltage at the fused ignition switch output circuit cavity in the body harness connector.
12. If battery circuit is working, remove and replace the fuel gauge pressure sensor.

## High-Pressure Fuel Shutoff Solenoid Testing

The **high-pressure fuel shutoff solenoid** is used as an on/off valve to electrically control high-pressure gas. It is operated by the PCM through the high-pressure fuel shutoff solenoid relay. The high-pressure fuel shutoff solenoid is located on the fuel pressure regulator. To test for solenoid electrical operation, perform the following steps:

1. Unplug the two-way electrical connector from the fuel pressure regulator.
2. Use a set of jumper wires (18 gauge) to momentarily apply battery voltage across the solenoid terminals.
3. The solenoid should cycle and a "click" should be heard as the battery voltage is applied and removed.
4. If the solenoid does not cycle, replace the fuel pressure regulator assembly.

## Low-Pressure Fuel Shutoff Solenoid Testing

After the cylinder pressure has been lowered by the fuel pressure regulator, the **low-pressure fuel shutoff solenoid** is used as an on/off valve to electrically control CNG entering the fuel rail assembly. Testing the electrical operation of the low-pressure fuel shutoff solenoid begins by unplugging the two-way electrical connector at the solenoid. Using a set of jumper wires (18 gauge), momentarily apply battery voltage across terminals (polarity is not important). The solenoid should cycle and a "click" should be heard as battery voltage is applied and removed. If the solenoid does not cycle, replace the solenoid.

## Fuel-Injector Testing

### You Should Know

*Before performing the following electrical test, be sure the ignition switch is in the Off position. Disconnect the electrical connector at the low-pressure fuel shutoff solenoid to prevent natural gas from entering the engine.*

There are no tests for injector spray patterns since CNG is not a liquid fuel and is not readily visible. Pressure testing of the fuel injectors consists of checking the pressure at the fuel rail as discussed earlier. Fuel-injector electrical testing is done by disconnecting the fuel-injector electrical connector and jumping battery voltage and ground to the injector terminals. The injector should click as battery voltage is applied and removed.

### You Should Know

*Since liquid is not passing through the injector to help muffle sound, CNC injectors may seem to click louder than normal.*

With the injector still disconnected, use an ohmmeter to measure the resistance values of the injector coil. Check the results with the service manual specifications. If the injector does not click or if resistance values cannot be met, replace the fuel injector.

## Review Questions

1. Technician A says that if the fuel pressure is too high on a properly filled CNG system, replace the fuel pressure regulator. Technician B says if the pressure is too low, the low-pressure fuel shutoff solenoid may be faulty. Who is correct?
   A. Technician A only
   B. Technician B only
   C. Both Technician A and Technician B
   D. Neither Technician A nor Technician B
2. Technician A says that an ohmmeter can be used to test the fuel temperature sensor. Technician B says to jump 12 volts to the sensor while monitoring the scan tool. Who is correct?
   A. Technician A only
   B. Technician B only
   C. Both Technician A and Technician B
   D. Neither Technician A nor Technician B
3. Technician A says that to test the injectors of a CNG engine, remove them and perform a spray pattern test. Technician B says that fuel-injector electrical testing is done by disconnecting the fuel-injector electrical connector and jumping

battery voltage and ground to the injector terminals. Who is correct?

A. Technician A only
B. Technician B only
C. Both Technician A and Technician B
D. Neither Technician A nor Technician B

4. Technician A says that the fuel low-pressure sensor is a three-wire transducer. Technician B says that if 0 volt is measured on the sense circuit, there may be a short to ground. Who is correct?

A. Technician A only
B. Technician B only
C. Both Technician A and Technician B
D. Neither Technician A nor Technician B

5. Technician A says that when testing the fuel gauge pressure sensor, if the sensor is unplugged the gauge should read Full. Technician B says that if a jumper wire is installed between the fuel gauge pressure sensor signal circuit and the ground circuit, the fuel gauge should read Empty. Who is correct?

A. Technician A only
B. Technician B only
C. Both Technician A and Technician B
D. Neither Technician A nor Technician B

CHAPTER

# 41 Vehicles with Alternative Power Sources

## Introduction

This chapter provides a brief description of several alternative power sources. This includes a study of common hybrid systems. In addition, the final section of this chapter covers fuel-cell theories and some of the methods that manufacturers are using to approach this alternative power source. These power sources are either being sold in limited numbers or they are still in the research and development stage.

## Electric Vehicles

Since the 1990s, most major automobile manufacturers have developed an **electric vehicle (EV)**. The EV powers its motor from a battery pack. The primary advantage of an EV is a drastic reduction in noise and emissions levels. The CARB established a low-emission vehicles/clean fuel program to further reduce mobile source emissions in California during the late 1990s. This program established emissions standards for five vehicle types **(Figure 1)**: conventional vehicle (CV), transitional low-emission vehicle (TLEV), low-emission vehicle (LEV), ultralow-emission vehicle (ULEV), and zero-emission vehicle (ZEV). The electric vehicle meets ZEV standards. **Figure 2** shows the basic components of an electric vehicle.

| | CV | TLEV | LEV | ULEV | ZEV |
|---|---|---|---|---|---|
| Nonmethane organic gas (NMOG) | 0.250 * | 0.125 | 0.075 | 0.040 | 0.000 |
| Carbon monoxide (CO) | 3.400 | 3.400 | 3.400 | 1.700 | 0.000 |
| Nitrogen oxide (NOx) | 0.400 | 0.400 | 0.200 | 0.200 | 0.000 |

(*) Emissions standards of nonmethane hydrocarbon (NMHC)

**Figure 1** *California tailpipe emissions standards in grams per mile at 50,000 miles.*

General Motors introduced the EV1 electric car to the market in 1996. The original battery pack in this car contained twenty-six 12-volt batteries that delivered electrical energy to a three-phase 102-kilowatt (kW) AC electric motor. The electric motor is used to drive the front wheels. The driving range is about 70 miles (113 km) of city driving or 90 miles (145 km) of highway driving. Temperature, vehicle load, and speed affect this range. A 1.2 kW charger in the vehicle's trunk can be used to recharge the batteries. This charger takes about 15 hours to fully recharge the batteries. An external Delco Electronics's Magne Charge 6.6 kW inductive charger operating on 220 volts/30 amperes can recharge the batteries in 3 to 4 hours **(Figure 3)**. The weatherproof

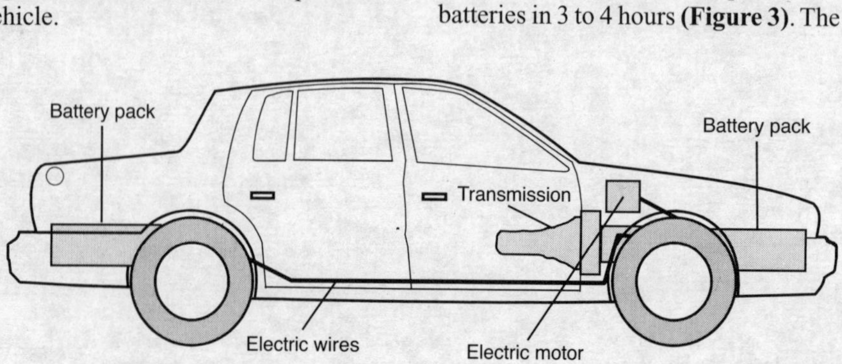


**Figure 2** *The electric vehicle is powered by an electric motor that receives its energy from battery packs.*

**Figure 3** *Recharging the EV's battery.*

plastic paddle is inserted into the charging port located at the front of the vehicle. Power is transferred by magnetic fields.

If the batteries are fully charged, the EV1 accelerates from 0 to 60 mph (97 km/h) in 9 seconds and has a top speed of 80 mph (129 km/h). The EV1 is equipped with a Galileo electronic brake system that employs a computer and sensors at each wheel to direct power-assist braking, regenerative/friction brake blending, four-wheel antilock braking, traction assist, tire pressure monitoring, and system diagnostics. In 1998, GM installed nickel metal hydride batteries in the EV1 vehicles, which extended the driving range between battery charges to 160 miles (257 km).

The driving range of electric vehicles is their biggest disadvantage. Much research is being done to extend the range and to decrease the required recharging times. Currently, the use of nickel metal hydride or lead-acid batteries and permanent magnet motors has extended the operating range. Another disadvantage is that the battery pack adds substantial weight to the vehicle **(Figure 4)**. Other features such as **regenerative braking** and highly efficient accessories

**Figure 4** *Positioning the EV's battery pack for installation.*

(such as a heat pump for passenger heating and cooling) are also being installed on electric vehicles. Regenerative braking means that the braking energy is turned back into electricity instead of heat. Other disadvantages are the cost of replacement batteries and the danger associated with the high voltage and high frequency of the motors.

## Hybrid Vehicles

The first alternative vehicle attempted was the electric vehicle, which had zero emissions and ran primarily on battery power. However, the battery has a limited energy supply and restricts the traveling distance. This limitation was a major stumbling block to most consumers; thus the use of EVs is mainly limited to fleets. One method to improve the electric vehicle was the addition of an onboard power generator that is assisted by an internal combustion engine. The result was the **hybrid electric vehicle (HEV)**. An HEV has two different power sources. In most hybrid vehicles, the power sources consist of a small displacement gasoline or diesel engine and an electric motor. The addition of the internal combustion engine meant the vehicle could not be classified as a ZEV. However, it does reduce emissions levels significantly and increases fuel economy.

Basically, the hybrid electric vehicle relies on power from the electric motor, the engine, or both **(Figure 5)**. When the vehicle moves from a stop and has a light load, the electric motor moves the vehicle. Power for the electric motor comes from stored electricity in the battery pack. During normal driving conditions, the engine is the main power source. Engine power is also used to rotate a generator that recharges the storage batteries **(Figure 6)**. The output

Figure 5 *Hybrid power system.*

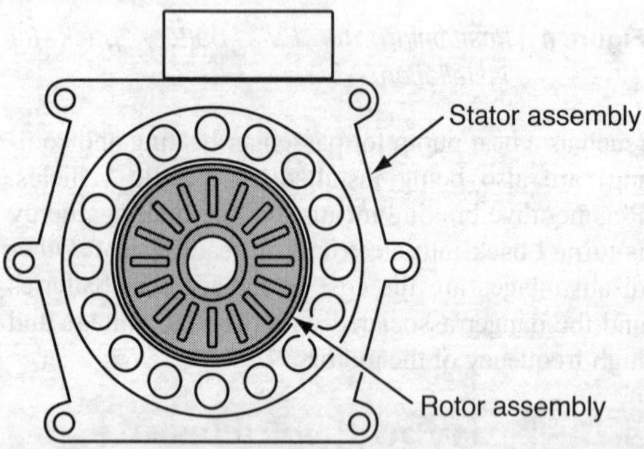


Figure 6 *Engine power is used to rotate a generator that recharges the storage batteries to drive the vehicle. The rotor assembly is a very powerful magnet that induces voltage into the stator windings as it is rotated.*

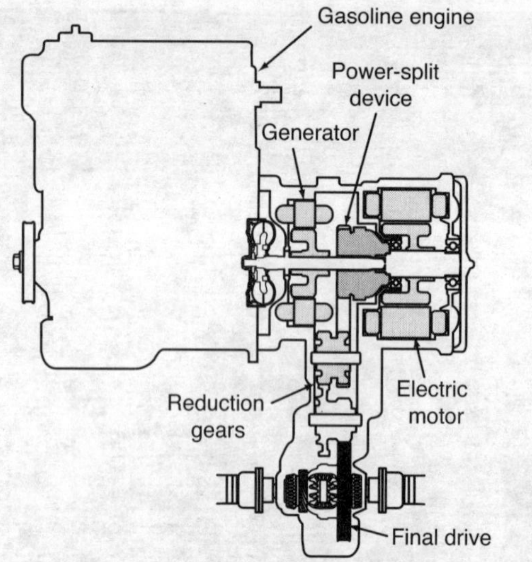


Figure 7 *Hybrid power system with gasoline engine and electric propulsion motor.*

Figure 8 *Toyota Prius was the world's first mass-produced hybrid vehicle.*

from the generator can also be used to power the electric motor, which is run to provide additional power to the power train **(Figure 7)**. A computer controls the operation of the electric motor depending on the power needs of the vehicle. During full-throttle or heavy-load operation, additional electricity from the battery is sent to the motor to increase the output of the power train.

The components of a typical hybrid vehicle include the following:

- Batteries. Some types of battery that are being used or experimented with now are the lead-acid battery, the nickel metal hydride battery, and the lithium ion battery. In the development of the battery, thermal management must be taken into consideration. The temperature can vary from module to module so the performance of the battery is dependent on the temperature. An imbalance in temperature will affect the power and capacity of the battery, the charge acceptance during regenerative braking, and vehicle operation. Also, passenger safety is a major concern. The batteries must be kept in sealed containers in order to ensure complete protection. The Toyota Prius **(Figure 8)** seals its noncaustic, nonflammable nickel metal hydride battery in a carbon composite case positioned in the rear of the vehicle **(Figure 9)**.
- **Electric motors**. One source of power is the electric motor. The motor converts electrical energy to mechanical energy. This mechanical energy is what drives the wheels of the vehicle. This motor is designed to allow for maximum torque at low rpm. This gives the electric motor the advantage of having better acceleration than the conventional motor.

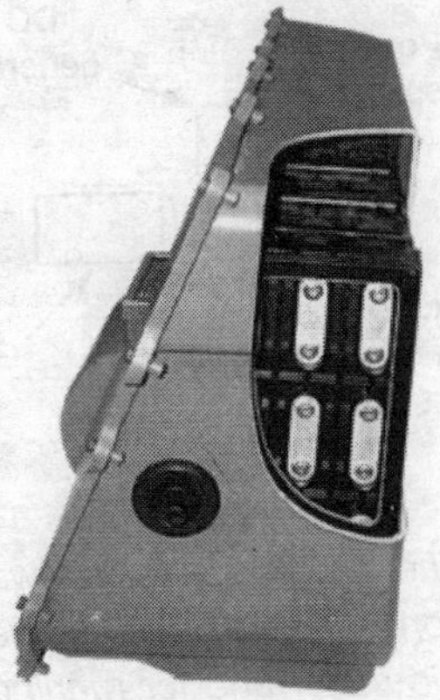

**Figure 9** *Toyota Prius battery pack.*

- **Regenerative braking**. About 30 percent of the kinetic energy lost during braking is in heat. When decreasing acceleration, regenerative braking helps to minimize energy loss by recovering the energy used to brake. It does so by converting rotational energy into electrical energy through a system of electric motors and generators. When the brakes are applied, the motor becomes a generator by using the kinetic energy of the vehicle to store power in the battery for later use.
- **Ultracapacitors**. The **ultracapacitor** is a device that stores energy as electrostatic charge. It is the primary device in the power supply during hill climbing, acceleration, and the recovery of braking energy. To create a larger storage capacity for the ultracapacitors, the surface area must be increased and, in turn, the voltage is increased. However, because the voltage drops as energy is discharged additional electronics are required to maintain a constant voltage.

## Propulsion

There are two typical ways to arrange the flow of power in an HEV. If the combustion engine is capable of turning the drive wheels as well as the generator, then the vehicle is referred to as a **parallel hybrid (Figure 10)**. In a parallel hybrid configuration

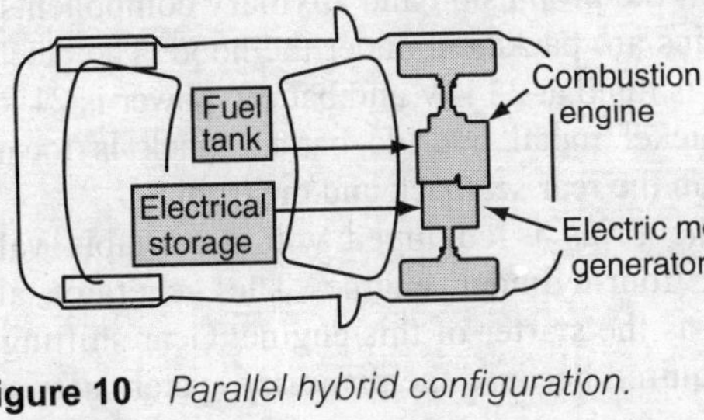


**Figure 10** *Parallel hybrid configuration.*

there is a direct mechanical connection between the engine and the wheels. Both the engine and the electric motor can turn the transmission at the same time.

There is a further distinction between a **mild parallel hybrid** and a **full parallel hybrid** vehicle. A mild parallel hybrid vehicle has an electric motor that is large enough to provide regenerative braking, instant engine start-up, and a boost to the combustion engine. A full parallel hybrid vehicle uses an electric motor that is powerful enough to propel the vehicle on its own.

Other configurations of the parallel hybrid vehicle include the use of an engine to power one axle and an electric motor to power the other **(Figure 11)**. Another concept is to use a combination where the engine, coupled with an electric motor, powers one axle and another electric motor powers the other axle **(Figure 12)**.

Most parallel hybrid vehicles use the electric motor to accompany the engine to help drive the wheels. For example, the engine is used for long driving periods while the electric motor is used for short, low-intensity drives. In other words, the engine is ideal for highway driving and the electric motor is ideal for a trip around town. The electric motor also provides the vehicle with acceleration. This acceleration, however, is only sustained until the vehicle reaches a certain speed. After this speed is reached, the engine is started

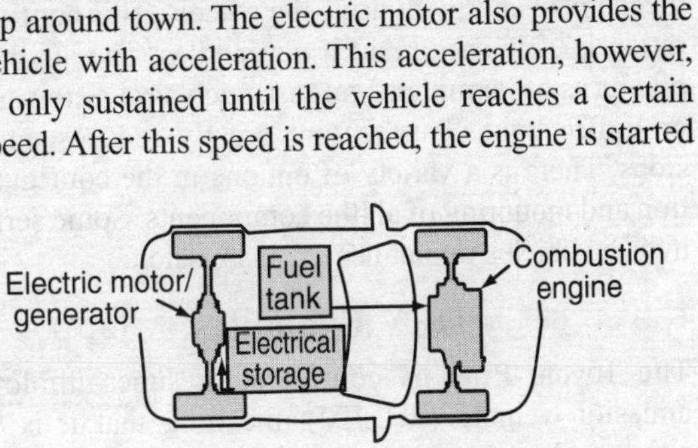


**Figure 11** *Parallel hybrid configuration using an engine to power one axle and an electric motor to power the other.*

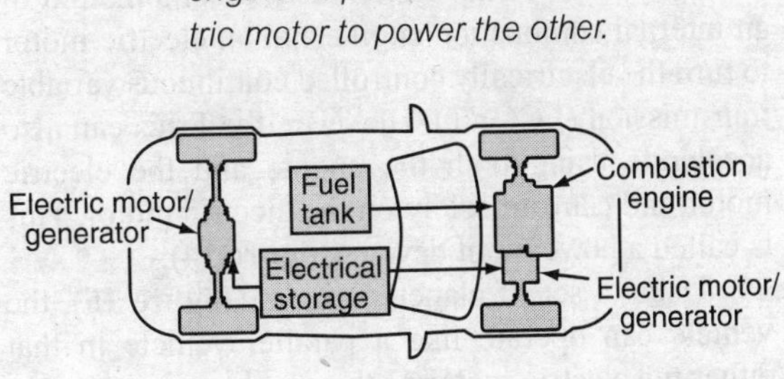


**Figure 12** *Parallel hybrid configuration using a combination where the engine, coupled with an electric motor, powers one axle and another electric motor powers the other axle.*

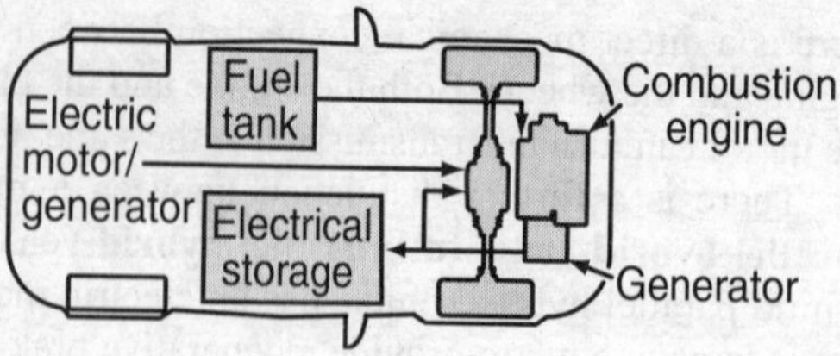


**Figure 13** *Series hybrid configuration.*

and replaces the electric motor. The parallel hybrid combines the alternator, starter, and wheels to create a system that will start the engine, electronically balance it, take power from the engine and turn it into electricity, and provide extra power to the drive line when a power assist is needed for hill climbing or quick acceleration.

In the **series hybrid** vehicle, there is no mechanical connection between the engine and the wheels. The engine turns a generator and the generator will either charge the batteries or power the electric motor, which in turn drives the transmission. Therefore, the engine never directly powers the automobile **(Figure 13)**.

The power used to give the vehicle motion is transformed from chemical energy into mechanical energy, then into electrical energy, and finally back to mechanical energy to drive the wheels. This configuration is efficient in that it never idles. The automobile turns off completely at rest, such as at a stop sign or traffic light. This feature greatly reduces emissions. There is a variety of options in the configuration and mounting of all the components. Some series hybrid vehicles do not use a transmission.

## HEV Examples

The Toyota Prius is considered a super-ultralow-emission vehicle (SULEV), meaning that it is 90 percent cleaner than an ultralow-emission vehicle. The Prius uses a combined hybrid-electric structure to supply power. This system uses a combination of an internal combustion engine and an electric motor to turn the electrically controlled continuous variable transmission (ECCVT). However, the Prius can also accelerate using both the engine and the electric motor and can run solely on the electric motor. This is called a power-split device **(Figure 14)**.

Using a set of planetary gears **(Figure 15)**, the vehicle can operate like a parallel vehicle in that either the electric motor or the gasoline engine powers the vehicle or they both do. However, the vehicle can also operate as a series hybrid where the engine can operate independently of the vehicle speed, either charging the batteries or providing power to the wheels when needed.

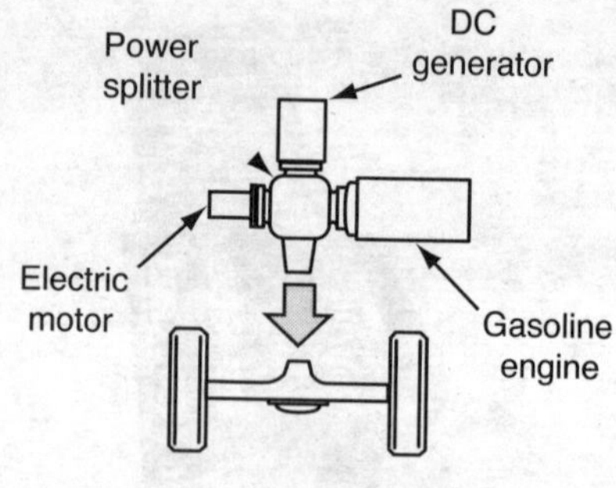


**Figure 14** *The power splitter allows for acceleration using both the engine and the electric motor and for running solely on the electric motor.*

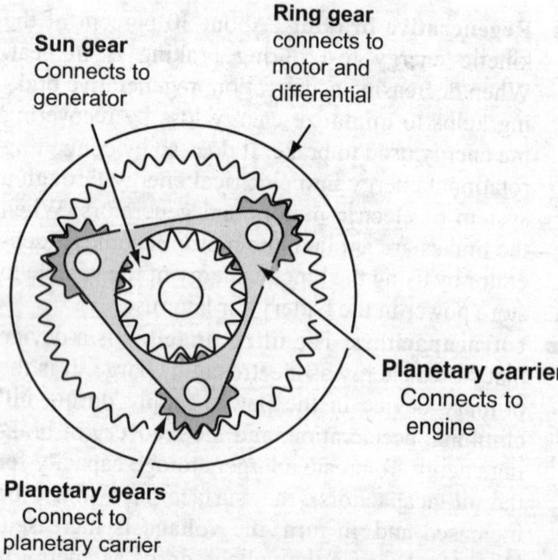


**Figure 15** *Planetary gears are used to transfer power to the drive wheels.*

### You Should Know

*Operating the engine independently of the vehicle speed means that even though the vehicle is traveling at highway speeds, the engine can be close to idle speed since it is only acting as a generator.*

All the propulsion and auxiliary components of the Prius are packaged under the hood. The electric motor is rated at 33 kW and battery power is 21 kW. The nickel metal hydride battery pack is located between the rear seatback and the trunk.

The Prius is equipped with a variable valve timing four-cylinder engine. The generator also works as the starter of this engine. Gear shifting is not required because the planetary system acts as a continuously variable transmission to keep engine rpm in the range of best efficiency. Instead of a normal transmission, the engine drives the planet carrier of a planetary gear set. The sun gear of that set connects to a motor/generator and the ring gear drives both the front wheels and a second motor/generator.

The Prius is also equipped with a drive-by-wire accelerator. The driver inputs to the motor management how much speed is requested. The management system then decides where the necessary power should come from: the engine, the battery, or both. The same accounts for the brake-by-wire system: the driver calls for the appropriate amount of retardation and the motor management coordinates this between the wheel brakes and the regenerative braking system. The computer also makes sure that the generator runs frequently enough to keep the battery charged.

When the battery is fully charged and the engine temperature is acceptable, the engine can be shut off (if the vehicle speed is low enough). If the driver gently presses the accelerator, the vehicle is moved by battery power. If the driver requests a quicker acceleration, the engine is started and powers the vehicle. In normal driving, the Prius maintains the battery state-of-charge within a narrow window. However, driving the vehicle under heavy loads for an extended time may deplete the battery charge.

The Honda Insight is also a parallel hybrid vehicle. In this system, the gasoline engine provides the majority of the power. The electric motor is used to help the gasoline engine provide additional power during acceleration. The Insight uses regenerative braking technology to capture energy lost during braking. The Insight also has a lightweight engine that uses **lean-burn technology** to maximize its efficiency. Lean burn technology, developed in the 1960s, uses high air-fuel ratios to increase fuel efficiency.

The Honda Civic Hybrid uses an integrated motor assist (IMA) system to power the vehicle. The system comprises a gasoline and electric motor combination. The electric motor is a source of additional acceleration and functions as a high-speed starter. The electric motor also acts as a generator for the charging system used during regenerative braking. This way the Civic Hybrid ensures efficiency by capturing lost energy by using regenerative braking, much like the Prius and the Insight.

## Fuel Cells

A **fuel cell** produces current from hydrogen and aerial oxygen. Fuel-cell-powered vehicles have a very good chance of becoming the vehicles of the future. They combine the reach of conventional internal combustion engines with high efficiency, low fuel consumption, and minimal or no pollutant emissions. At the same time, they are extremely quiet. Because they work with regenerative fuel such as hydrogen, they reduce the dependence on crude oil and other fossil fuels.

A fuel-cell-powered vehicle **(Figure 16)** is basically an electric vehicle. Like the electric vehicle, it uses an electric motor to supply torque to the drive wheels. The difference is that the fuel cell produces and supplies electric power to the electric motor instead of the battery. Most vehicle manufacturers and several independent laboratories are involved in fuel-cell research and development programs. Manufacturers have produced a number of prototype fuel-cell vehicles, with many being placed in fleets in North America and Europe.

Fuel cells electrochemically combine oxygen from the air with hydrogen to produce electricity. The oxygen and hydrogen are fed to the fuel cell as "fuel" for the electrochemical reaction. There are different types of fuel cells but the most common type is the **proton exchange membrane (PEM)**. Normally, hydrogen and oxygen bond with a loud bang but in fuel cells a special PEM impedes the oxyhydrogen gas reaction by ensuring that only protons (H+) and not elemental hydrogen molecules ($H_2$) react with the oxygen.

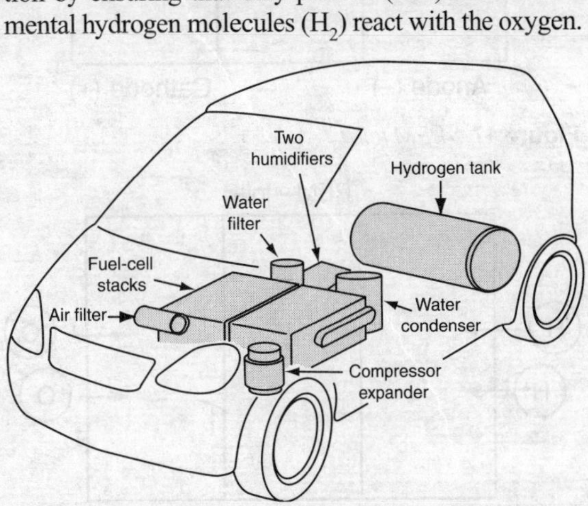


**Figure 16** *Fuel-cell vehicle components. Technology has allowed engineers to design fuel-cell vehicles without the loss of passenger and cargo space.*

## How the Fuel Cell Works

The PEM fuel cell is constructed like a sandwich **(Figure 17)**. The electrolyte is situated between two electrodes of gas-permeable graphite paper. The electrolyte is a polymer membrane. Hydrogen is applied to the anode side of the PEM and ambient oxygen is applied to the cathode side **(Figure 18)**. The membrane keeps the distance between the two gases and provides a controlled chemical reaction.

The anode is the negative post of the fuel cell. It conducts the electrons that are freed from the hydrogen molecules so that they can be used in an external circuit. It has channels etched into it that disperse the hydrogen gas evenly over the surface of the catalyst. The cathode is the positive post of the fuel cell. It also has channels etched into it that distribute the oxygen to the surface of the catalyst. It also conducts the electrons back from the external circuit to the catalyst where they can recombine with the hydrogen ions and oxygen to form water.

A fine coating of platinum is applied to the foil to act as a catalyst. This is used to accelerate the decomposition of the hydrogen atoms into electrons and protons **(Figure 19)**. The catalyst is rough and porous so that the maximum surface area of the platinum can be exposed to the hydrogen or oxygen. The platinum-coated side of the catalyst faces the PEM.

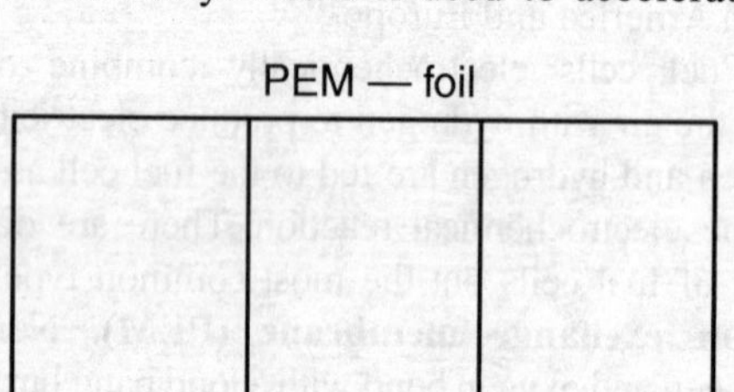


**Figure 17** *PEM foil.*

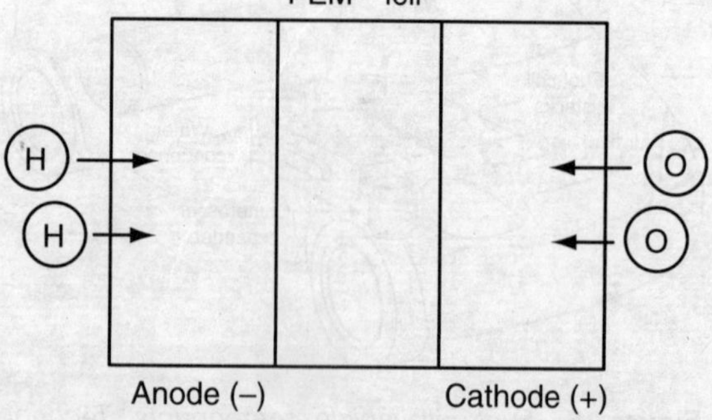


**Figure 18** *Hydrogen is applied to the anode side of the PEM while oxygen is applied to the cathode side.*

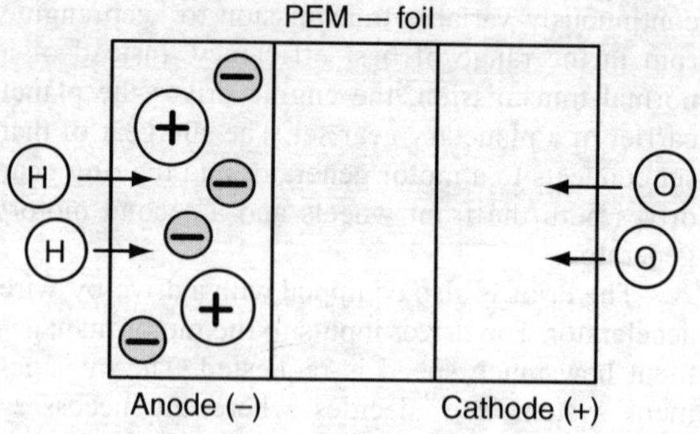


**Figure 19** *The catalyst breaks down $H_2$ into protons and electrons.*

The PEM is the electrolyte. This specially treated material (which looks similar to ordinary kitchen plastic wrap) only allows the protons to move across from the anode to the cathode **(Figure 20)**. As a result, the anode has a surplus of electrons and the cathode has a surplus of protons. If the anode and cathode are connected outside of the cell, current flows through the conductor **(Figure 21)**. The electrons move through the conductor to the cathode. The electrons then recombine with the protons and the oxygen to produce water.

The entire process is illustrated in **Figure 22**. Pressurized hydrogen gas ($H_2$) enters the fuel cell on the anode side. This gas is forced through the catalyst by the pressure. When an $H_2$ molecule comes in contact with the platinum on the catalyst, it splits

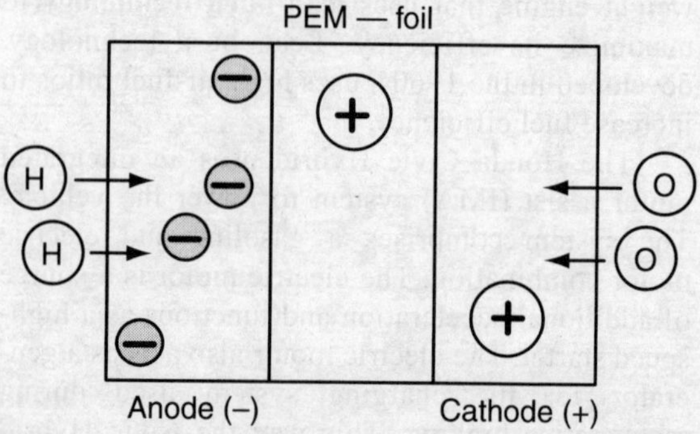


**Figure 20** *The PEM foil only allows the protons to migrate to the cathode, leaving the electrons on the anode.*

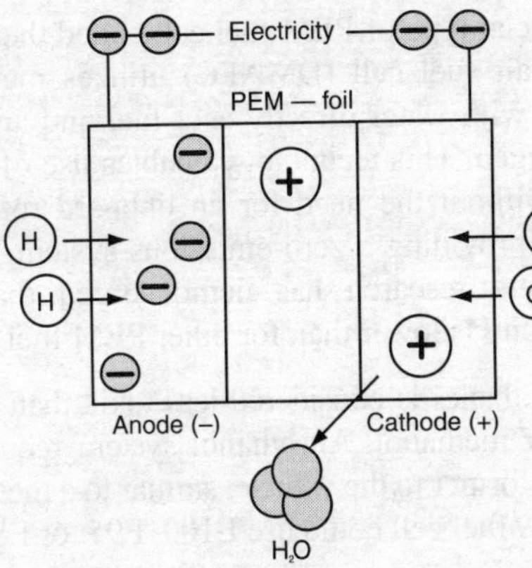


**Figure 21** *With an excess amount of electrons on the anode and an excess amount of protons on the cathode, current flows through an external conductor. The electrons then react with the protons and oxygen to produce water.*

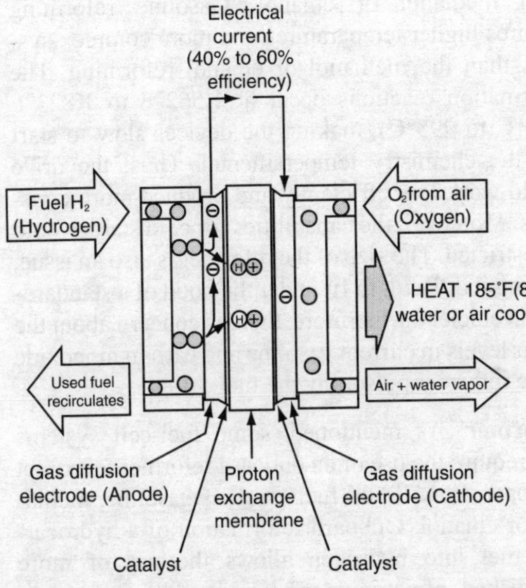


**Figure 22** *PEM fuel cell.*

into two H+ ions and two electrons (e–). The electrons are conducted through the anode where they make their way through the external circuit (doing useful work such as turning a motor) and return to the cathode side of the fuel cell.

Meanwhile, on the cathode side of the fuel cell, oxygen gas ($O_2$) is being forced through the catalyst where it forms two oxygen atoms. Each of these atoms has a strong negative charge. This negative charge attracts the two H+ ions through the membrane, where they combine with an oxygen atom and

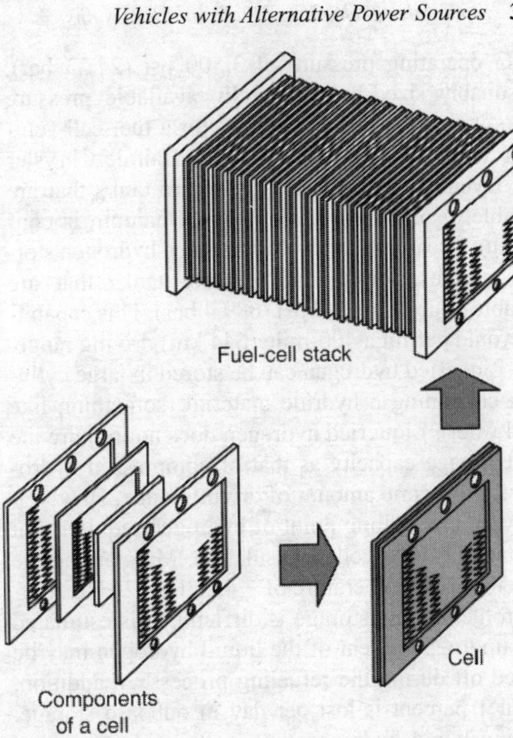


**Figure 23** *PEM fuel-cell stack.*

two of the electrons from the external circuit to form a water molecule ($H_2O$).

This reaction in a single fuel cell produces only about 0.7 volt. For this voltage to become high enough to be used to move the vehicle, many separate fuel cells must be combined in series to form a fuel-cell stack **(Figure 23)**.

## Fuels for the Fuel Cell

A fundamental problem with fuel-cell technology concerns whether to store hydrogen or convert it from other fuels onboard the vehicle. All four principal fuels that automotive manufacturers are considering (hydrogen, methanol, ethanol, and gasoline) pose some challenges.

**Hydrogen Fuel**. One solution is to store hydrogen onboard the vehicle. The ability to use hydrogen directly in a fuel cell provides the highest efficiency and zero tailpipe emissions. However, hydrogen has a low energy density and boiling point, thus onboard storage requires large, heavy tanks. There are three types of hydrogen storage methods under development: compressed hydrogen, liquefied hydrogen, and binding hydrogenate to solids in metal hydrides or carbon compounds.

Compressed hydrogen offers the least expensive method for onboard storage. However, at normal

CNG operating pressure of 3,500 psi (241.3 bar), reasonably sized, commercially available pressure tanks will provide limited range for a fuel-cell vehicle (about 120 miles or 193 km). DaimlerChrysler and Hyundai are now using pressure tanks that are capable of 5,000 psi (348.7 bar). Quantum is conducting research of high-performance hydrogen storage systems, looking at pressure tanks that are capable of up to 10,000 psi (689.4 bar). This capability would permit a 400-mile (644 km) driving range.

Liquefied hydrogen can be stored in large cylinders containing a hydride material (something like steel wool). Liquefied hydrogen does not require the high storage capacity as that of compressed hydrogen for the same amount of driving range. However, the very low boiling point of hydrogen requires that the tanks have excellent insulation. Maintaining the extreme cold temperature of −423°F (−253°C) during refueling and storage is difficult. It is estimated that up to 25 percent of the liquid hydrogen may be boiled off during the refueling process. In addition, about 1 percent is lost per day in onboard storage. Storing liquid hydrogen on a vehicle also involves some safety concerns. As the fuel tank warms, the pressure increases and may activate the pressure-relief valve. This action discharges flammable hydrogen into the atmosphere, creating a source of danger and pollution.

**Methanol**. Several automotive manufacturers are using methanol to power their fuel cells. It is believed that methanol fuel cells could bridge the gap over the next few decades while a hydrogen distribution infrastructure is being built. Using "methanolized" hydrogen as fuel has the advantage that it can be stored on the vehicle similarly to gasoline. For the reaction in the fuel cell, a **reformer** onboard the vehicle produces hydrogen from the methanol fuel. A reformer is a high-temperature device that converts hydrocarbon fuels to CO and hydrogen ($H_2$). To produce the hydrogen, the methanol fuel is mixed with water. When it evaporates, it is decomposed into hydrogen and carbon dioxide. Prior to sending the hydrogen and carbon dioxide to the fuel cell, it is purified in additional steps.

Methanolized hydrogen contains more hydrogen atoms and has an energy density that is greater than that of liquid hydrogen. Like hydrogen, methanolized hydrogen is also independent of mineral oil. Compared with hydrogen, vehicles driven by methanol are not completely emissions-free but they produce very little pollutants and much less carbon dioxide than internal combustion engines.

A special type of PEM fuel cell called the direct methanol air fuel cell (DMAFC) utilizes methanol combined with water directly as a fuel and ambient air for oxygen. This technology enables use of a liquid fuel without the need for an onboard reformer while still providing a zero-emissions system. However, current research has demonstrated that the power density is lower than for other PEM fuel cells.

**Ethanol**. Ethanol is considered less toxic than either gasoline or methanol. An ethanol system requires a adding a reformer to the vehicle, similar to a methanol system. The fuel cell could use E100, E95, or E85.

**Gasoline**. Fuel cells can be driven with a special, more pure gasoline. Using reformers for onboard extraction of hydrogen from gasoline is one approach to commercialization of fuel-cell vehicles since the gasoline infrastructure is already in place. However, producing hydrogen from gasoline in a vehicle system is much more difficult than producing hydrogen from methanol or ethanol. Gasoline reforming requires higher temperatures and more complex systems than the methanol or ethanol reforming. The reformation reactions occur at 1,562°F to 1,823°F (850°C to 995°C), making the devices slow to start and the chemistry temperamental. Thus, the drive would work less efficiently and produce more emissions. Moreover, the capabilities for cold starts would be restricted. The size of the reformer is also an issue, making it difficult to fit under the hood of a standard-sized vehicle. Furthermore, there is concern about the sulfur levels in current gasoline and carbon monoxide in the reformer poisoning the fuel cell.

**Reformer**. As mentioned, some fuel-cell systems may require the use of an onboard reformer to extract hydrogen from liquid fuels such as gasoline, methanol, or ethanol. Onboard reformation of a hydrocarbon fuel into hydrogen allows the use of more established infrastructures but adds additional weight and cost and reduces vehicle efficiency. In addition, the reformer does create some emissions.

PEM fuel-cell reformers combine fuel and water to produce additional $H_2$ and convert the CO to $CO_2$. The $CO_2$ is then released to the atmosphere. Reformer technologies include steam reforming (SR), partial oxidation, and high-temperature electrolytes reforming.

Steam reforming uses a catalyst to convert fuel and steam to $H_2$, CO, and $CO_2$. The CO is further reformed with steam to form more $H_2$ and $CO_2$. A purification step then removes CO, $CO_2$, and any impurities to achieve a high hydrogen purity level

(97 to 99.9 percent). SR of methanol is the most developed and least expensive method to produce hydrogen from a hydrocarbon fuel on a vehicle, resulting in 45 to 70 percent conversion efficiency.

Partial oxidation reforming is similar to stream reforming since both technologies combine fuel and steam but this process adds oxygen in an additional step. The process is less efficient than steam reforming but the heat-releasing nature of the reaction makes it more responsive than steam reforming to variable load. Heavier HC can be used in POX but it has lower carbon-to-hydrogen ratios, which limits hydrogen production.

Hydrogen can also be obtained by **electrolysis** from water. Electrolysis is the splitting of water into hydrogen and oxygen. The drawback to this process is it requires a great deal of electrical energy. Recently, the development of high-temperature electrolytes that can operate at temperatures in excess of 212°F (100°C) has shown some positive results. The benefits of high-temperature electrolytes include:

- **Improved CO tolerance**. This allows the manufacturer to reduce or remove the need for an oxidation reactor and for air bleed. Since these requirements can be reduced, system efficiency is increased by 5 to 10 percent. There is also a considerable reduction in start-up time. The remaining CO will be combusted in a catalytic tail-gas burner to prevent emissions of CO.
- **Facilitated stack cooling**. This reduces the size of the radiator and reduces the fuel-cell stack cooling plate requirements.
- **Humidity-independent operation**. Generally, high-temperature membranes require humidifiers and water recovery, whereas this system does not.

**Solid Oxide Fuel Cells**. Planar solid oxide fuel cells (SOFCs) operate at high temperatures of 932°F to 1,472°F (500°C to 800°C) and can use CO and $H_2$ fuel. SOFCs have a good tolerance to fuel impurities and use ceramic as an electrolyte. BMW is currently developing an auxiliary power unit (APU) with Delphi and Global Thermoelectric using an SOFC. Currently, SOFCs use gasoline fuel and require a reformer.

**Sodium Borohydride**. DaimlerChrysler introduced a new idea in fuel technologies with a concept minivan called the Chrysler Town & Country Natrium (Latin for sodium). This research was driven by the lack of a safe, compact way to contain hydrogen. In addition, most fuel-cell vehicles are not "on demand," meaning that there is a start-up time required before the reformer can begin producing hydrogen.

A fairly simple chemical process mixes sodium borohydride powder with water. The sodium borohydride powder produces free hydrogen for the fuel cell **(Figure 24)**. Unlike the fuel-cell systems that use methanol or gasoline, the Natrium produces no pollution and no carbon dioxide.

### Interesting Fact

*The advantage of sodium borohydride is that it can be easily stored and transported in lightweight plastic tanks at normal temperatures and pressures when dissolved in water. Also, the borate solution is not poisonous, flammable, or explosive.*

The sodium borohydride powder holds more hydrogen than the most densely compressed hydrogen tank. The prototype minivan attains a top speed of 80 mph (129 km/h) and an operating range of

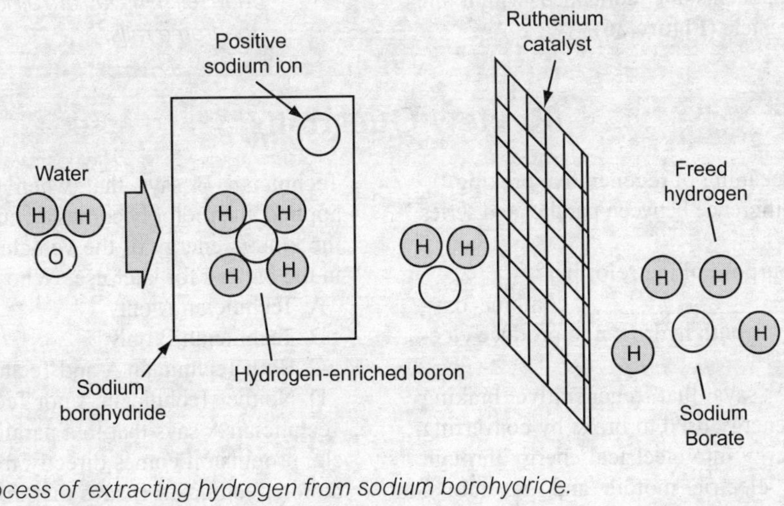


**Figure 24** *Process of extracting hydrogen from sodium borohydride.*

almost 300 miles (483 km). The by-products of this process are water and sodium borate, which is basically laundry detergent. After use, the spent powder goes into a storage tank where it can be pumped out and reclaimed.

As the name implies, sodium borohydride ($NaBH_4$) is a salt. The salt that powers this system is not ordinary table salt (sodium chloride ($NaCl$)). Rather, it is a white salt whose molecules contain a relatively large amount of hydrogen. Through the use of a chemical catalyst, the sodium borohydride reaction with water results in elemental hydrogen. A slurry of sodium perborate ($NaBO_2$) forms during this reaction as well. This compound is chemically related to borax, which is used as a bleaching agent in conventional detergents. The $NaBO_2$ slurry is collected in a special tank and can be effectively recycled in a chemical process. In this reverse reaction, the $NaBO_2$ is reclaimed back to $NaBH_4$, which can then be reused as an energy source for the fuel cells.

The catalyst system is called Hydrogen on Demand™ and was developed by Millennium Cell **(Figure 25)**. When stepping on the accelerator, the minivan driver is actually "throttling" a fuel pump. The $NaBH_4$ is pumped into the catalyst, which immediately generates hydrogen. This dynamic process slows or ceases entirely when the fuel pump is throttled or completely shut off. When the driver accelerates once more, hydrogen is immediately produced again and the fuel cells immediately generate electricity.

An AC motor with an output of 35 kW powers the vehicle. A 55 kW lithium-ion battery serves as a storage unit for electrical energy. The battery is recharged by the fuel-cell unit and by regenerative braking. If the fuel-cell system should fail, the battery would be sufficient to drive the electric motor and move the vehicle. All fuel-cell components are contained within the frame rail of the vehicle **(Figure 26)**.

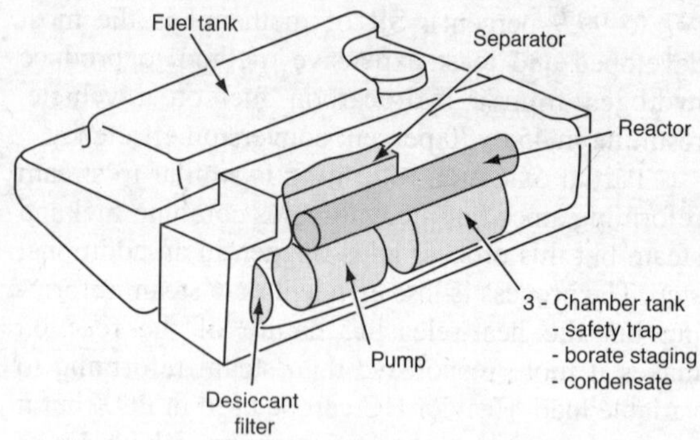


**Figure 25** *The Hydrogen on Demand™ catalyst system is the heart of the drive train. The hydrogen required for the fuel cell to generate power is extracted from sodium borohydride in a reactor.*

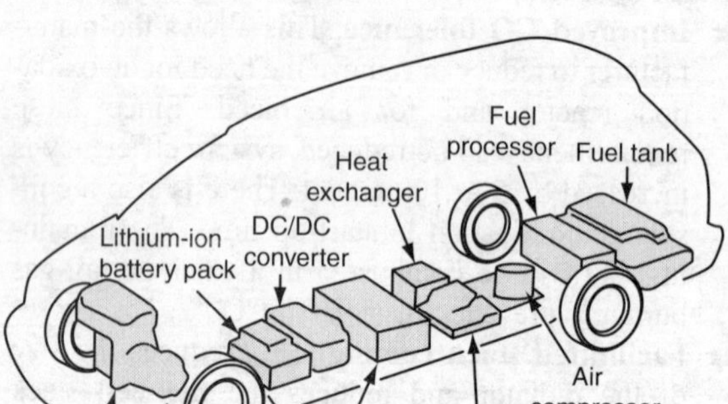


**Figure 26** *Components of the sodium borohydride fuel-cell mini-van are all located between the frame rails.*

## Review Questions

1. Explain the meaning of regenerative braking.
2. Explain the difference between parallel and series hybrids.
3. What is the purpose of the reformer?
4. Fuel cells ________________ combine oxygen from the air with hydrogen to produce electricity.
5. Technician A says that regenerative braking recovers the energy used to brake by converting rotational energy into electrical energy through a system of electric motors and generators. Technician B says that when the brakes are applied, the motor becomes a generator by using the kinetic energy of the vehicle to store power in the battery for later use. Who is correct?
   A. Technician A only
   B. Technician B only
   C. Both Technician A and Technician B
   D. Neither Technician A nor Technician B
6. Technician A says that in a parallel hybrid vehicle, propulsion comes directly from the electric motor. Technician B says that in a series hybrid

vehicle, both the engine and the electric motor can turn the transmission at the same time. Who is correct?

A. Technician A only
B. Technician B only
C. Both Technician A and Technician B
D. Neither Technician A nor Technician B

7. Describe the basic operation of a typical hybrid vehicle.
8. Briefly describe how the proton exchange membrane (PEM) fuel-cell produces electrical energy.
9. __________ __________ recovers the heat energy used to brake by converting rotational energy into ________ energy through a system of electric motors and generators.
10. Most fuel-cell systems require the use of a _____________ to extract hydrogen from liquid fuels such as gasoline, methanol, or ethanol.
11. Electric vehicles power the motor by:
    A. A generator
    B. A battery pack
    C. An engine
    D. None of the above.
12. Technician A says that a fuel cell produces electrical energy by breaking down $H_2$ atoms into electrons and protons. Technician B says that the compressed hydrogen system requires a reformer to cool the fuel cell. Who is correct?
    A. Technician A only
    B. Technician B only
    C. Both Technician A and Technician B
    D. Neither Technician A nor Technician B

# Index

NOTE: Page numbers followed by f refer to figures; page numbers followed by t refer to tables.

**F**